The Behavior and Ecology of Pacific Salmon and Trout

THOMAS P. QUINN

American Fisheries Society
Bethesda, Maryland

in association with

University of Washington Press
Seattle and London

This book is published in memory of Marsha L. Landolt (1948–2004), Dean of the Graduate School and Vice Provost, University of Washington, with the support of the University of Washington Press Endowment.

Publication of this book is supported, in part, by the U.S. Bureau of Reclamation (Upper Columbia Area Office).

Printed in Canada
12 11 10 09 08 07 06 05 5 4 3 2 1

American Fisheries Society
5410 Grosvenor Lane, Suite 110, Bethesda, MD 20814
www.fisheries.org

University of Washington Press
PO Box 50096, Seattle, WA 98145
www.washington.edu/uwpress

Library of Congress Cataloging-in-Publication Data

Quinn, Thomas P.
The behavior and ecology of Pacific salmon and trout / Thomas P. Quinn. — 1st ed.
p. cm.
Includes bibliographical references and index.
ISBN 0-295-98437-6 (hardback : alk. paper)
ISBN 0-295-98457-0 (pbk. : alk. paper)
1. Oncorhynchus—Behavior. 2. Oncorhynchus—Ecology.
I. Title.

QL638.S2Q56 2005
597.5'61568—dc22 2004013612

The paper used in this publication is acid-free and recycled from 20 percent post-consumer and at least 50 percent pre-consumer waste. It meets the minimum requirements of American National Standard for Information Sciences—Permanence of Paper for Printed Library Materials, ANSI Z39.48-1984.

The Behavior and Ecology of Pacific Salmon and Trout

Contents

Preface

Pacific salmon are a remarkable group of animals, and the connections to their ecosystems and to humans may be more complex and profound than any other group of animals, and certainly more than any other group of fishes. First, though perhaps not foremost, they are collectively among the most valuable commercial fishery resources of the United States, with annual landed values that averaged $390 million from 1992 to 2001 according to U.S. Department of Commerce statistical reports. This is matched only by taxa such as crabs and shrimp that are taken from both oceans and include many diverse species. The finfish species that dominate the tonnage landed, walleye pollock and Atlantic menhaden in recent years, are lower in value than salmon despite their volume.

In addition to their commercial value, salmon are the target of recreational fisheries with significant value to local economies. Perhaps more important than the amount of money spent in pursuit of salmon is the psychological uplift (often mixed liberally with frustration) that comes with time spent outdoors fishing alone or in the company of family and friends. Salmon also hold a special place in the culture, nutrition, and economy of peoples native to the coast of the North Pacific Ocean. They were traditionally important for food and for barter, and they continue to be a very important component of the culture and commerce of many groups. The salmon have been adopted as the region's icon by non-native peoples as well. One need only visit the gift shops in San Francisco, Portland, Seattle, Vancouver, Anchorage, and many smaller communities to see that salmon are readily embraced by modern society. Certainly, large trees and snow-capped mountains are also icons of the region, but somehow we do not connect with them as strongly as we do with salmon. The image of the salmon, leaping a waterfall in its heroic but tragic effort to get home, reproduce, and die, is among the most recognizable in the natural world, and it strikes a chord with us.

Salmon are not only important for cultural and consumptive purposes, but their conservation and management presently pervade the regulatory environment of their ecosystem. Past and present human activities, including but not limited to mining, agriculture, hydroelectric production, flood control, forestry, shoreline development, and urbanization, all affect salmon. Increasingly, these activities are regulated because of their effects on salmon. One cannot understand water management in the Columbia River system or forestry on the Oregon coast without understanding salmon. Salmon have also been at the heart of many conceptual and technological advances in fisheries science and management.

Besides the complex roles that salmon play for people, they play equally important and complex roles for other organisms. Most streams they inhabit are nutrient-poor, and the annual return of salmon to spawn and die provides a pulse of food that directly and indirectly enriches the plants and animals in nearby aquatic and terrestrial ecosystems. Finally, the salmon's influence on their ecosystem is not limited to natural processes but they have indirect effects through humans as well. Because salmon are so important, people will modify land-use practices to benefit them when they would have done nothing for amphibians or less charismatic fishes. The northern spotted owl was granted protection under the U.S. Endangered Species Act and was vilified in a way that salmon never will be. Put simply, salmon are special.

The natural history of salmon is important for people seeking to understand these fishes, the North Pacific ecosystems in Asia and North America, and their management by humans. I hope this book will provide insights into the basic biology of salmon to a range of people, including university students and faculty, biologists working in agencies, nongovernmental organizations, and companies devoted to salmon or to some aspect of the natural or human world that interacts with them. In addition to these people with a direct need to know about salmon, I hope the book will also interest members of the public who wish to learn about these fishes or become involved in their conservation. However, this book is not designed for advocacy. My goal is not to sway opinion but to inform and excite the reader. I will have succeeded if I have conveyed some of my enthusiasm for salmon and if I have stimulated readers to question my ideas, formulate and test their own hypotheses, and expand our knowledge of salmon.

The book is entitled *The Behavior and Ecology of Pacific Salmon and Trout*. As will be explained more fully later, the term "Pacific salmon" has traditionally been applied to five species of fishes in the genus *Oncorhynchus* that are native to the North American and Asian coasts of the Pacific Ocean, and to two (or one) species native only to Asia. Trout, notably rainbow (and their sea-run form, known as steelhead) and cutthroat but also lesser-known species such as Apache, golden, and Gila trout, have been included in the genus *Oncorhynchus* since 1989. The fishes of the genus *Oncorhynchus* are the subjects of the book. In addition to this genus, there are two other major genera in the family Salmonidae: *Salmo* (including Atlantic salmon and brown trout, both native only to Atlantic drainages) and *Salvelinus* (the char, including species in all continents around the north temperate and boreal regions). The introductory chapter provides thumbnail sketches of the common fishes in the family found in western North America and Asia.

The rest of the book is focused on the traditional salmon species and steelhead and cutthroat trout, though there are some references to other species. This scope reflects my own knowledge and the richness of the published literature (both of which thin out greatly after the five North American salmon and two trout species). However, I believe that the major points in behavior and ecology of the groups are amply demonstrated in these species and my focus on them is not misleading.

Just the seven principal species of Pacific salmon and trout (often, for convenience, referred to collectively as salmon) are described by a truly vast scientific literature. It is impossible to do justice to the tremendous volume and variety of excellent work that has been done. If I tend to cite my own research it is only because it is familiar to me, not because it is superior to the work of others. It is equally impossible to present all the unusual life-history patterns, habitats, and other ecological circumstances of salmon. I have tried to give both the general patterns and some exceptions that seem instructive, but there will always be some population or site that does not fit the patterns I described. In the interests of a readable book, some compromises were needed.

I have used data from primary, secondary, and unpublished sources, and have made graphs and tables to illustrate important points and highlight selected studies. However, readers are strongly advised to seek out the primary sources when doing their own analyses and should then give credit to those authors rather than to me. In a few cases, such as the compilation of data on fecundity and survival of salmon populations, it is impractical for me to cite every source of information and note every adjustment needed to make my tables comprehensible. I trust that readers will accept my efforts as honest. Finally, I have largely avoided statistical analyses in the text. Unless otherwise stated, patterns that I present as significant meet the general professional standard; there is a less than 5 percent chance that the apparent pattern arose by chance. In a number of the graphs depicting the relationship between two variables, such as body size and number of eggs, I show a value designated r^2. This value indicates the proportion of the variation in the dependent variable (in this case, number of eggs) that is explained in a statistical sense by the independent variable (body size). Thus $r^2 = 0.45$ means that 45 percent of the variation in fecundity can be explained by female size.

Acknowledgments

Many people contributed to this book, in many different ways. Ernest Brannon and Kees Groot supervised my doctoral and postdoctoral work on salmon, respectively, and they educated and inspired me. Since then, my knowledge has been built in large part by the superb graduate students who have worked with me, by my outstanding colleagues at the University of Washington, and numerous fine collaborators at various agencies and companies. Preparation of the book would not have been possible without the help of my wife, Sandie O'Neill, who provided encouragement, data, professional contacts, critical comments on text, photos and illustrations, and assistance with all aspects of the project. Many people provided published or unpublished data for tables or figures, including David Seiler, Thom Johnson, Scott Hinch, Katherine Myers, Bruce Ward, Jeff Cederholm, Joseph Fisher, William Pearcy, Masahide Kaeriyama, Richard Thomson, Richard Brodeur, Robert Francis, Scott Gende, Ted Cooney, Mark Willette, David Lonzarich, Brian Fransen, Daniel Schindler, R. J. Wootton, Sayre Hodgson, John McMillan, James Starr, George Pess, Gregory Ruggerone, Reginald Reisenbichler, John Burke, Carlos Garcia de Leaniz, Ian Fleming, and Sigurd Einum.

I also thank the people who reviewed draft chapters: Andrew Dittman, Mark Scheuerell, Daniel Schindler, Fred Utter, Scott Gende, Robert Lackey, Ted Cooney, Robert Francis, Nancy Davis, Kerry Naish, Scott Hinch, Stephanie Carlson, Walt Dickhoff, Charles Simenstad, Brian Fransen, and Phil Peterson. Jeff Jorgensen, Richie Rich, Bobette Dickerson, Jeramie Peterson, and Peter Westley greatly assisted me by compiling data and finding references. Gregory Ruggerone, Scott Gende, Ernest Keeley, Gregory Buck, Andrew Dittman, Charles Peven, Manu Esteve, Michael Erickson, Trevor Anderson, Katherine Myers, Graeme Ellis, Alan Olson, Carla Stehr, Phil Roni, Susan Johnson, T. Aoyama, Bart Gamett, Martin Unwin, Andrew Hendry, Michael McHenry, John Ford, and Mark Giovannetti generously allowed me to use their photographs. I particularly thank Richard Bell for taking many outstanding photos specifically for the book. I thank Charles Wood and David Ehlert for artwork, Marcus Duke for assistance with production and references, Cathy Schwartz for help with maps and figures, and Michael Duckworth and the University of Washington Press staff for their help and enthusiasm for the project. Preparation of this book was made possible with support from the H. Mason Keeler Endowment. Mr. Keeler's generosity also supported many graduate students in the School of Aquatic and Fishery Sciences at the University of Washington whose research is described in this book, and we are all deeply grateful to him.

I dedicate this book to my parents, Esther and Vincent, who encouraged my interests in fish and in writing; to my brother, Steve, who shares these interests; and to my wife, Sandie, and children, Mackenzie and Brian, whose love and patience made this book possible and worthwhile.

PLATE 1. Conceptual diagram of a northern lake ecosystem, showing *Cyclotella* (a diatom, single-celled alga) at the top left, two types of crustacean zooplankton that consume algae (*Cyclops,* a copepod, below, and *Daphnia,* a cladoceran, to the right). Zooplankton are consumed by juvenile sockeye salmon and their competitor, the threespine stickleback. These small fishes are vulnerable to avian predators such as the Arctic loon on the surface, and larger fishes such as the Arctic char in deep water.

The Behavior and Ecology
of Pacific Salmon and Trout

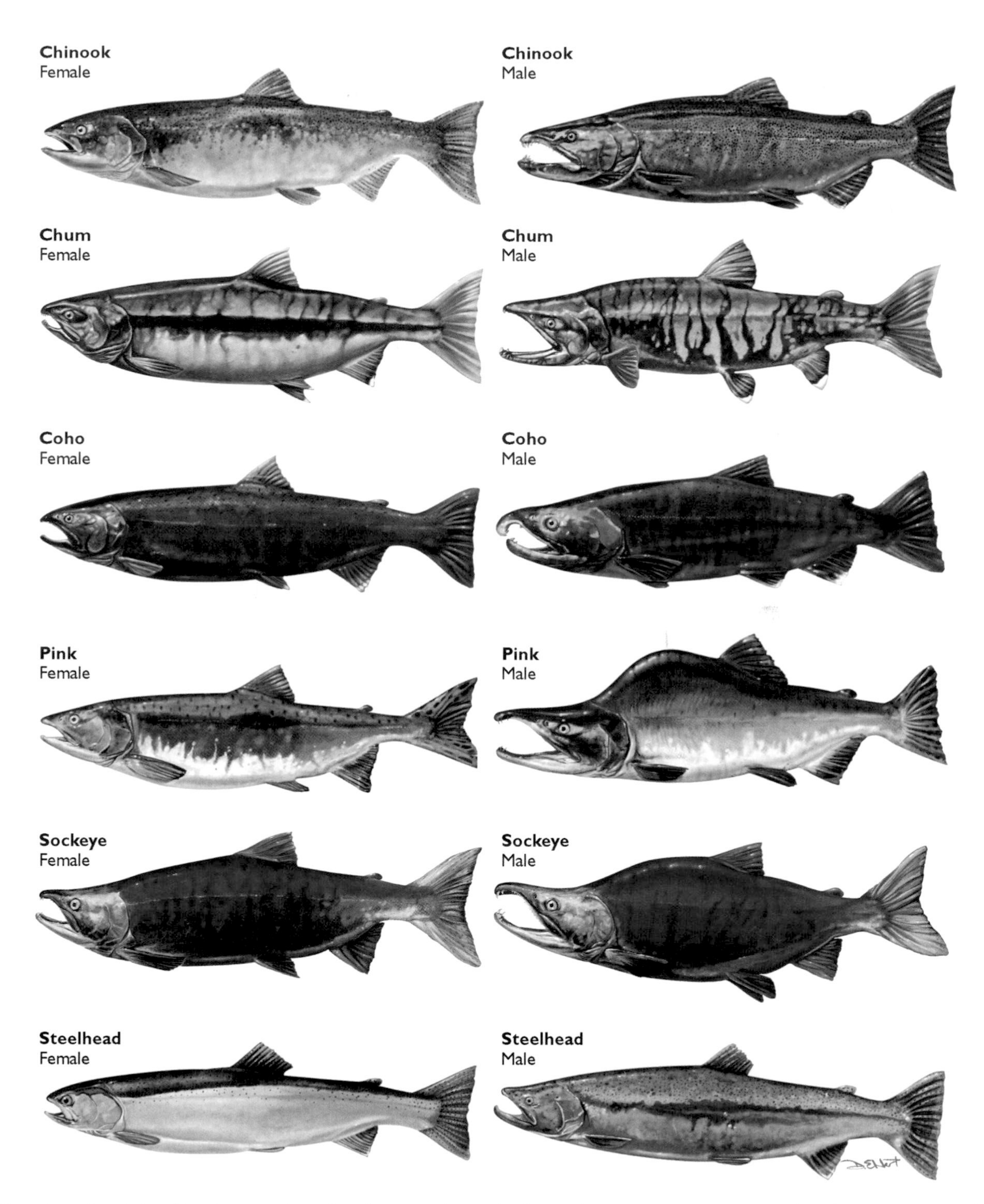

Female and male Pacific salmon and steelhead trout at the peak of sexual maturity (not drawn to scale). Thomas Quinn and David W. Ehlert, M. A. M. S., © University of Washington.

1

Introduction

As with most fields of knowledge, there is a certain amount of terminology associated with the life cycles of salmon, trout, and char (collectively termed *salmonids*). I do not wish to dwell on semantics but some common usage is needed for ease of understanding. The term *egg* refers to the unfertilized ovum, produced by the female. Once fertilized by a sperm cell (mixed with fluids from the male, collectively called *milt*), the egg becomes an embryo, the cell divisions begin, and development proceeds. The embryo is immediately buried by the female in a gravel nest, termed a *redd*, in a stream or lake beach. The redd is composed of several pockets of eggs, deposited and buried by the female in a sequence of spawning events. The embryo develops within the egg membrane for several months and, at an appropriate stage of development, it hatches. The hatchling is termed an *alevin*, with a large, external yolk sac for nourishment. As the alevin grows, the yolk is metabolized until it is fully or largely gone and the young salmon can feed on its own. It then wriggles up through the gravel and emerges into the stream or lake as a fry. Depending on the species, the fry might migrate directly to sea (chum, pink, and some sockeye and chinook), migrate to a lake (sockeye), or remain in the stream (most salmonid species). Those staying in freshwater tend to have vertical brown-green bars on their sides that provide camouflage. These bars are known as *parr* marks and the fish are called parr. There is no set stage when fry become parr (though some people restrict the term fry to fish in their first year of life), and the term *fingerling* also refers to little salmonids.

After some period in freshwater (days, months, or years, depending on species and population), the salmon migrate to sea. To accomplish this transition, they must alter many aspects of their bodies, including color, shape, osmoregulatory (salt balance) physiology, energy storage, patterns of drinking, urination, and behavior. The fish in this transitional stage are termed *smolts*. Smolts can be found in freshwater readying themselves for migration, migrating in freshwater, and in the nearshore marine environment. Because it is a transitional stage, there are no sharp distinctions between smolts and non-smolts. However, the term is not used to describe salmon that have been feeding for long at sea; salmon at sea are generally just termed immature.

There is also a bit of jargon associated with nonanadromous salmonids. Populations are called residents if they spend their entire lives in the stream where they were spawned. In contrast, fluvial populations rear for some time in the natal stream, then migrate to a larger river to grow, and return to the small stream to spawn. Adfluvial populations rear in the natal stream, migrate to a lake for further growth, then return to the stream to spawn. These migrations can be short but in some cases cover several hundred kilometers.

At some point the salmon at sea begin a complex set of physiological processes that will lead them to migrate back to freshwater, spawn, and die. Salmon that have made this "decision" are referred to as maturing fish. They include males and females, of course, and also what are termed *jacks* in many species and populations. Jacks are sexually mature male salmon representing an age group younger than the youngest females in the population. For example, female coho salmon (and most males of the species) spend one full year and a summer at sea, so jacks are males that spend only one summer at sea before maturing. Most sockeye salmon spend two or three years at sea, so those spending only one full year would be jacks. There are only very rare instances of *jills* (females of such young age), for reasons discussed later. Jacks are not to be confused with grilse, a term used by Atlantic salmon biologists to refer to salmon spending one full year at sea. Both males and females can be grilse, and some populations are largely or entirely composed of grilse. In the case of the "traditional" species of salmon (coho, chinook, chum, pink, and sockeye), all individuals die after spawning. However, in rainbow and cutthroat trout, some individuals survive after spawning and are known as *kelts* during their downstream migration.

Several terms associated with fishing so pervade the scientific literature that they too must be explained. The salmon run is the total number of adults surviving the natural mortality agents and heading back to freshwater to spawn. Some are caught (the catch) and the others that evade the fishing gear and spawn are called the escapement. Depending on the dynamics of the population and the management regime, the ratio of catch to escapement can vary greatly. Fishery is a term referring to a type of gear operating on one or several species in a particular area. For example, one might speak of the gillnet fishery for sockeye salmon in Bristol Bay, Alaska, and the troll fishery for coho and chinook salmon off the Oregon coast.

Key themes in the biology of salmon, trout, and char

This book is about one genus of fishes, *Oncorhynchus,* within the family Salmonidae. This family has two other main genera, *Salmo* and *Salvelinus*. Later in this chapter I introduce these genera and their primary species in North America. However, it is possible to get quickly lost in the diversity of life-history patterns among the species within these genera, and then to become even more baffled by the myriad population-specific variants. It is therefore important to understand that there are three key themes in the biology of this family. Each theme is broadly distributed among salmonids but each has interesting and important exceptions.

Anadromy

All salmonids spawn in freshwater and some spend their entire lives there. However, many migrate to sea to grow to their final size and then return to freshwater to spawn. This life-history pattern, known as *anadromy*, leads to rapid growth and high density of salmon relative to nonanadromous salmonids. All Pacific salmon species are anadromous, but some species (notably sockeye and masu salmon, and rainbow and cutthroat trout) have nonanadromous populations and there may be nonanadromous individuals (typically males) in masu and some chinook salmon populations.

In a thoughtful review, Rounsefell (1958) pointed out that anadromy is not an all-or nothing matter; rather, there are degrees of anadromy and closely related life-history traits. He proposed six criteria to assess the extent of anadromy: the spatial extent of migration at sea; the duration of residence at sea in relation to the duration in freshwater; the state of maturity attained at sea; spawning habits and habitats (e.g., use of intertidal areas); postspawning mortality; and the occurrence of nonanadromous forms of the species. Integrating information on these aspects of life history and behavior, he classified pink, chum, and chinook salmon as "obligatory anadromous" species (though chinook less so than the others), and coho and sockeye were termed "adaptively anadromous." He used the term "optionally anadromous" for rainbow and cutthroat trout, Atlantic salmon, brown trout, Dolly Varden, brook trout, and Arctic char.

Hoar (1976) reinforced these themes and showed that seaward migration and the transition from parr to smolt stages is part of a complex set of behavioral and physiological adaptations that vary among species. More recently, Hutchison and Iwata (1997) assessed the extent of aggressive behavior by juvenile salmonids from nine species. They found that the more aggressive species were those that spend more time in freshwater prior to migration, so anadromy is truly connected to the entire life cycle of the species, not just to migration.

Anadromy is a subset of a broader category of migratory life-history patterns termed *diadromy* (McDowall 1988). Diadromous fishes have regular migrations between breeding grounds in freshwater or the ocean and feeding grounds in the other environment. These fishes are scarce in proportion to the fishes that spend their entire lives in freshwater or at sea. They are distinguished from euryhaline fishes that are tolerant of a broad range of salinities and often reside in estuaries but are not specifically migratory. Among the diadromous fishes, about half are anadromous, including not only salmonids but clupeids (including American shad, *Alosa sapidissima*), striped bass (*Morone saxatilis*), lampreys (Petromyzontidae), sturgeons (Acipenseridae), smelt (Osmeridae), and others.

The opposite pattern from anadromy is *catadromy*: species that spawn at sea and rear in freshwater. The most famous catadromous fishes are the anguillid eels but there are other examples as well. Finally, some fishes display *amphidromy*; they spawn in freshwater, migrate to sea as larvae and feed there for a while, and then return to freshwater for further growth before spawning. Examples include the galaxiids of the Southern Hemisphere. There are about 20,000 species of fishes, and according to McDowall (1988), only 160 (0.8%) of them are diadromous. Of the diadromous fishes, about 87 (54%) are anadromous, 41 (25%) are catadromous, and 34 (21%) are amphidromous.

Gross (1987) theorized that animals migrate when the benefits of being in some new habitat exceed the benefits of the present habitat, minus the cost of moving. We may infer that migration between salt- and freshwater, regardless of the direction, is very costly because so few fishes do it. Diadromy requires physiological adaptations for ion regulation in freshwater and at sea. Survival in both environments may also require changes in body plan and behavior, in addition to a shape suited to migratory performance. There are thus many reasons not to migrate. Among diadromous fishes, anadromy is more common at higher latitudes and catadromy more common at lower latitudes (McDowall 1988). Northcote (1978) and Gross et al. (1988) pointed out that at higher latitudes productivity (hence growing opportunities) tend to be greater at sea than in freshwater, and the more rapid growth of salmon at sea than in streams and lakes supports this. At low latitudes, freshwater environments are often more productive than marine ones, and this may explain the tendency for diadromous fishes to spawn at sea and rear in freshwater in these regions.

Homing

Not only do surviving salmon return from ocean feeding areas to streams for spawning but they almost invariably return to the site where they were spawned. This trait, known as *homing*, leads to reproductive isolation of salmon populations (i.e., little interbreeding between salmon from one river and another). These isolated populations, exposed to different physical factors such as temperature, flow, and gravel size, and biotic factors such as predators, prey, competitors, and pathogens, evolve specializations to improve survival in their home river. In addition to such genetic adaptations, the populations may also vary in abundance at carrying capacity and in productivity (number of offspring produced per spawning female at low density). These differences in population dynamics necessitate that they be managed and conserved as discrete populations rather than as species, and this greatly complicates fisheries management. It must be noted, however, that if all salmon homed, new habitat would never be colonized. Much of the present range of salmon was glaciated within the last 10,000 to 15,000 years, so colonization (straying) was also an essential element in the evolution and present distribution of salmon. Such straying continues to provide gene flow among existing populations and allows for colonization after natural disasters extirpate populations.

Semelparity

Not only do salmon migrate to sea and come back to their natal stream to spawn but death inevitably follows reproduction in many Pacific salmon species. This life-history pattern, termed *semelparity*, transfers millions of kilograms of salmon flesh from the ocean to nutrient-poor freshwater ecosystems, reversing the gravity-driven tendency for water and nutrients to flow seaward. However, the trout species often survive spawning, and their life-history pattern is termed *iteroparity* (as in iterative reproduction). Semelparity is not unique to salmon, as some other fishes (e.g., lamprey) and many insects and other invertebrates are semelparous. However, there may be no other group of semelparous animals that are as large as salmon and whose synchronized death contributes as much to the local ecology. Semelparity probably evolved in response to increased adult mortality from the rigors of anadromy and long-distance migration. The adults expend all their energy during migration and reproduction rather than retaining some to assure their own survival. In females, this extra energy for reproduction is put into producing especially large eggs (Crespi and Teo 2002) and guarding the nest after spawning.

Are salmonids typical fishes?

These three key themes provide an important introduction to salmon behavior and ecology. However, a bit more information may be helpful before we consider species-specific details. There are several notable life-history traits common to Pacific salmon in particular and salmonids in general. Taken together, they make salmon quite different from other fishes spawning and rearing with them in freshwater. Some of these differences stem from the anadromous life history or at least are most exaggerated in the anadromous forms, and many of the traits are interrelated. Within salmonids, the species are arrayed along continua related to body size and longevity, dependence on freshwater rearing, extent of anadromy and iteroparity, and breeding season. Wootton (1984) reviewed the life-history patterns of Canadian freshwater fishes (Scott and Crossman 1973) and his summary makes a very useful basis for comparing salmonids to other sympatric species (see also Winemiller and Rose 1992 for further comparisons of fish life-history patterns).

Salmonids grow rapidly but do not live long. The marine waters occupied by salmonids provide superior growing conditions compared to freshwater habitats, and salmonids have high metabolic rates, allowing rapid growth if food is available. Growth facilitates survival and reproductive potential (fecundity and egg size in females, and competitive ability in males). However, salmonids (especially Pacific salmon) have rather short life spans. They are much larger for their age than other freshwater fishes but sacrifice longevity for growth rate (fig. 1-1). Some char and nonanadromous trout, however, can live 10 years or more.

FIGURE 1-1. The relationship between age at first maturity and adult body size of freshwater fish species from Canada (from Wootton 1984; salmonid data updated by Quinn).

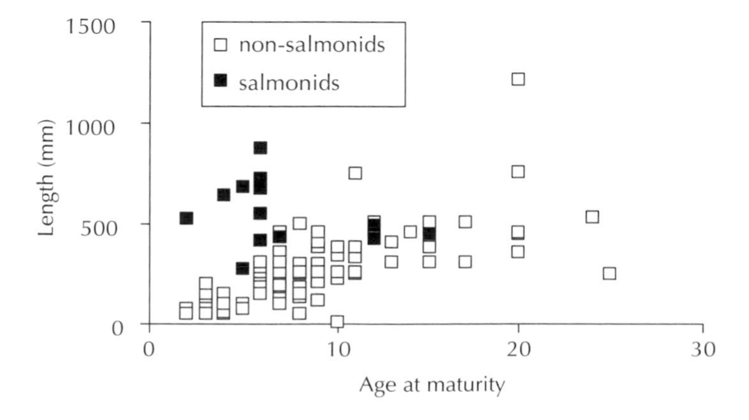

FIGURE 1-2. Modal month of spawning by freshwater fishes in Canada (from Wootton 1984).

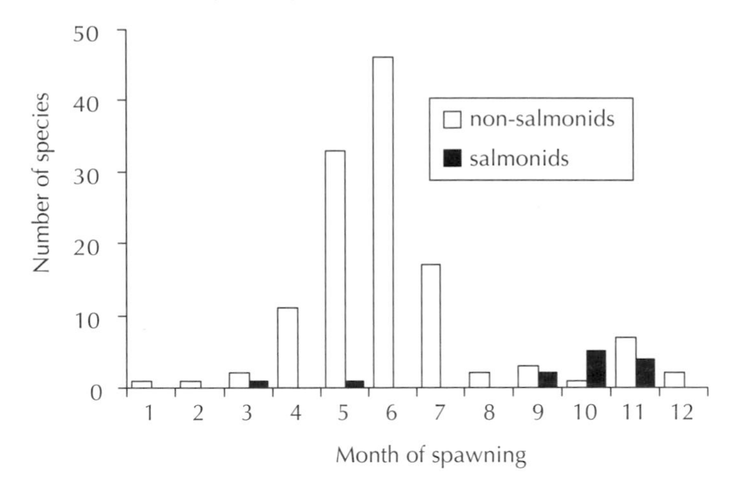

FIGURE 1-3. Number of days from fertilization to reliance on external food sources (hatching, or emergence for salmonids) for Canadian freshwater fish species (from Wootton 1984).

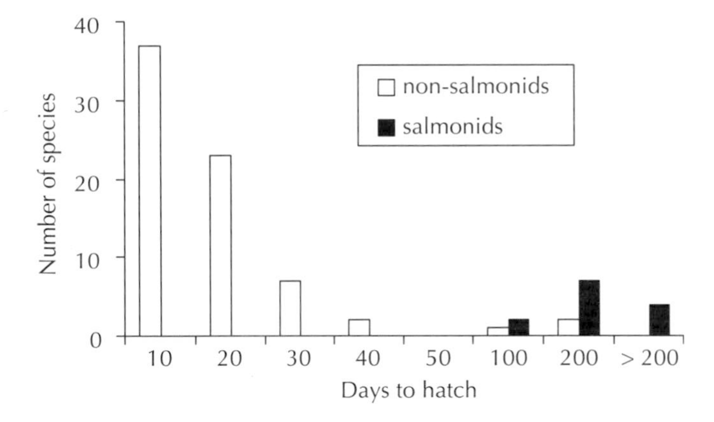

Pacific salmon spawn in the fall. This is quite unusual; most freshwater fishes in the north temperate region spawn in the spring (especially May and June; fig. 1-2). The definition of fall, with respect to spawning season, does vary. Spawning may be as early as July for sockeye in Alaska and as late as February for chum and coho salmon in Washington. Rainbow and cutthroat trout spawn in spring but again, the seasons are defined with respect to local conditions and the actual months can vary. Other than salmonids, the other main group of fall-spawning fishes is the family Coregonidae or whitefish, and they are related to salmonids. Salmonid eggs, spawned in the fall, develop slowly so the juveniles begin feeding in spring when the long days and intense sunlight melt the ice, warm the water, and food (e.g., insects and zooplankton) becomes more abundant. The eggs of spring-spawning fishes develop more rapidly than salmon (fig. 1-3), so the larval fish can also take advantage of warm temperatures and abundant primary and secondary production for rapid growth.

FIGURE 1-4. Distribution of egg sizes among freshwater fishes in Canada (from Wootton 1984).

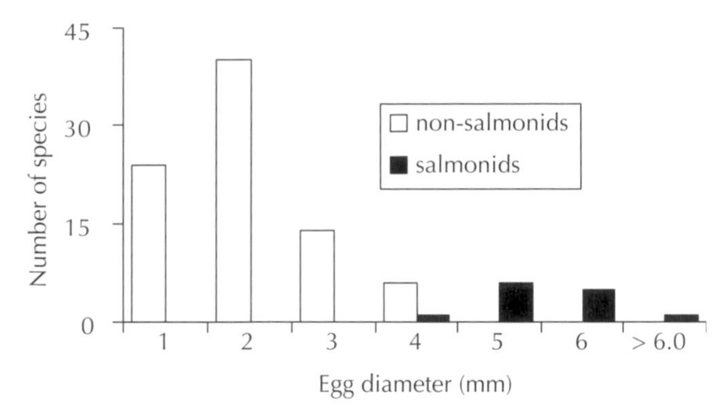

FIGURE 1-5. Relationship between fecundity and mean body length among freshwater fishes in Canada (from Wootton 1984; salmonids updated by Quinn).

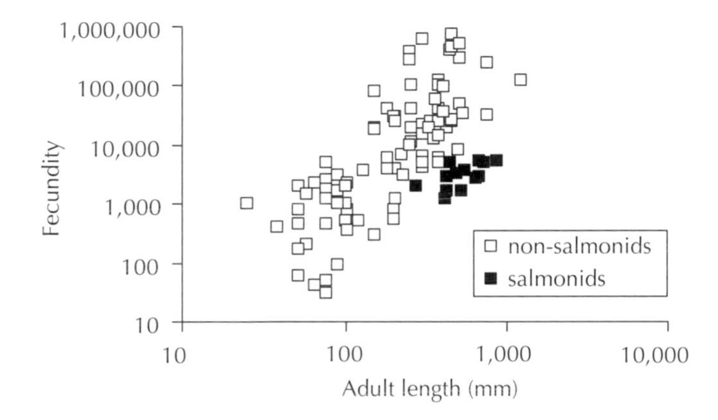

Salmonids have very large eggs compared to other bony fishes (fig. 1-4). The large eggs produce large fry, with higher survival rates than smaller fry. The tradeoff, however, is that the female can only devote about 20% of her weight to gonads. The investment in reproduction must balance number against size of eggs to maximize the number of surviving progeny. Thus salmonids also have fewer eggs, for their size, than other freshwater fishes (fig. 1-5). These eggs provide the embryos and alevins with enough yolk to survive the long winter.

Salmonids display female parental care, in the form of egg burial. Both of these features are atypical in fishes. Parental care is shown in only about 21% of the families of bony fishes (Gross and Sargent 1985), though it is more common among freshwater than marine fishes. Among the fishes with parental care, the provider is more often the male rather than the female (61% vs. 39% of families). The evolution of parental care is a complex matter and beyond my scope, but the benefits and drawbacks are obvious in a general way. Care increases the odds of survival of the offspring in the present generation but is costly to the parent (in depleted energy, risk of predation, lost feeding opportunities, etc.), and so reduces survival and future reproductive opportunities.

Female salmon provide care by preparing a depression in the gravel, winnowing out fine gravel by turning on their side and rapidly sweeping the tail up and down. When the eggs have been fertilized the female buries them and (in the semelparous species) guards them from disturbance by other females until she dies. Egg burial is a very unusual form of parental care. Some fishes such as grunions, *Leuresthes tenuis*, bury their eggs in the upper intertidal region of sandy beaches, and larval emergence is synchronized to the next spring tide series. This seems risky enough, but burying embryos in stream gravel for many months seems like a poor form of care because sediment transport and scour commonly occur. The large size of fall spawning salmonids may enable them to bury their eggs deep enough to avoid the effects of winter floods (Montgomery et al. 1996).

Salmonids are generalists. Juveniles use the whole range of freshwater habitats available to them, and many species commonly occupy both streams and lakes. They are opportunistic feeders; small salmonids eat primarily insects and zooplankton and larger individuals eat fishes and invertebrates. Salmonids also have a generalized body shape suited to mobility, lacking some of the specialized features such as the armor and spines of sticklebacks (Gasterosteidae) and large mouth of sculpins (Cottidae).

Salmonid life at sea is spent in the epipelagic (near-surface) coastal and offshore waters, where they feed on a diverse diet of zooplankton, macro-invertebrates (e.g., krill, squid), and small, schooling fishes such as herring, eulachon, and sand lance. In the region occupied by salmon, the salmon themselves are among the most abundant fishes. They are preyed upon by various fishes when they are small and by sharks and marine mammals when they are adults.

Finally, *salmon populations tend to be very productive.* This term does not mean that they are abundant or that they occur at high density, though they may have these attributes too. Rather, it means that when the population is below its carrying capacity, each salmon produces many surviving offspring. Thus salmon can support higher fishing rates than most species. Consider the fact that most fish and game–management regulations are designed to make sure that all or most individuals of the regulated species have a chance to breed at least once before they are subjected to exploitation (e.g., size limits for recreational fishing, mesh sizes for commercial fishing, hunting regulations, etc.). However, all salmon caught are virgins, and sustainable fisheries often catch at least 50% of the run. Thus in the absence of fishing, competition (for spawning habitat by adults or food and space by juveniles) would exert stronger control over the abundance of salmon than we see today.

The stage: The physical environment occupied by salmon

Before introducing the characters in the play, we might briefly set the stage. Salmon and trout are products of their environment, and they spawn and rear in bodies of water ranging from tiny creeks above waterfalls in the mountains, or streams discharging straight into saltwater, to large rivers; and from small beaver ponds and ephemeral wetlands to the largest lakes of the region. Their native range is from northern Mexico to the Arctic Ocean on one side of the Pacific, and from Taiwan, southern Japan, and Korea to the Arctic Ocean on the other side (fig. 1-6), though salmon are present only as scattered populations in the Arctic. They are found in a number of large rivers (table 1-1) as

FIGURE 1-6. Map of the North Pacific Ocean, showing the coastal extent of spawning by Pacific salmon and trout (shaded) and some of the major rivers and other geographical features.

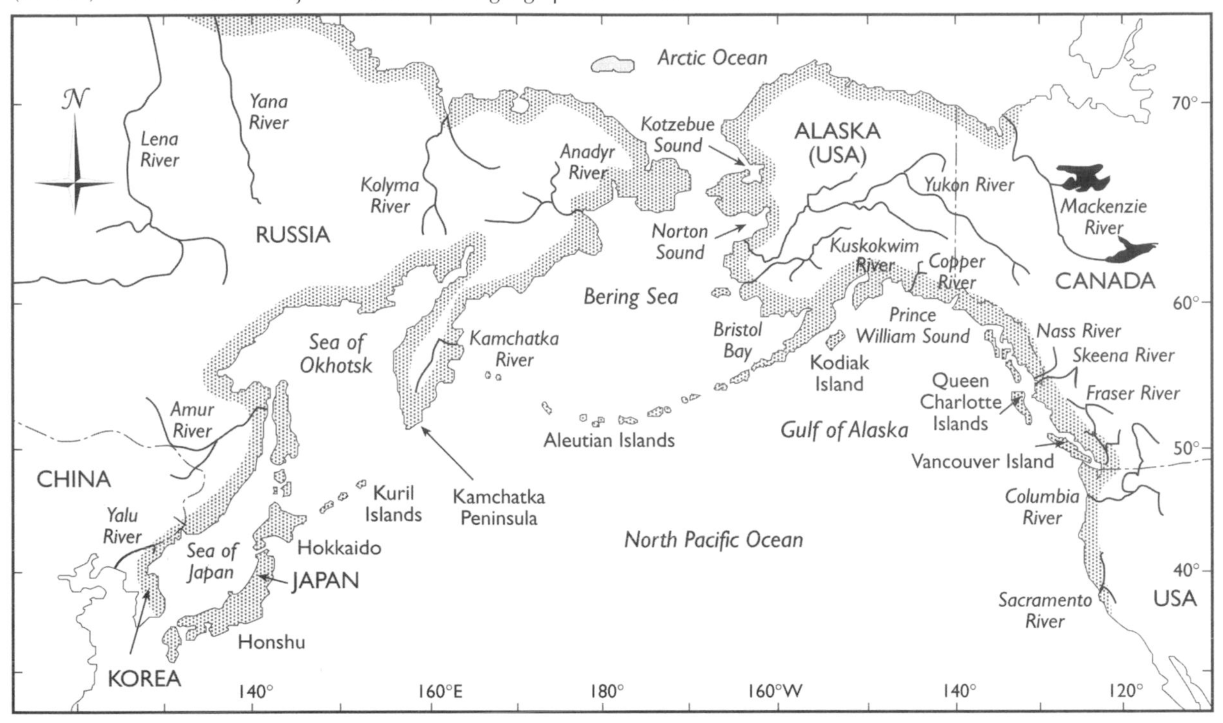

TABLE 1-1. Mean annual discharge (m^3/sec) and length (km) of major North American rivers with Pacific salmon (from Favorite et al. 1976 and unpublished public records). For those marked with an asterisk (*), discharge was not recorded at the mouth.

River	*Mean discharge (m^3/sec)*	*Length (km)*
Mackenzie	7500	4250
Yukon	7000	3185
Columbia	6650	2250
Fraser	2938	1360
Kuskokwim*	1280	1165
Copper*	1040	462
Skeena*	920	570
Nass*	830	380
Sacramento*	650	607

well as in thousands of smaller streams. Of course, there are also great rivers in Asia that support salmonids. The Lena River flows from just north of Lake Baikal 4270 km to the Arctic Ocean, with a discharge of about 16,400 m^3/sec. The Amur River flows 4510 km from Mongolia to the Sea of Okhotsk with a discharge of 12,500 m^3/sec. Thus the Lena has roughly twice the discharge of the Mackenzie, and the Amur has nearly twice that of the Yukon, so these are indeed major rivers of the world. The Anadyr (1117 km) and Yalu

TABLE 1-2. Average daily air temperatures (°C) in January and July, annual total precipitation (cm) and snowfall (cm) for a series of coastal and interior locations representing the North American range of Pacific salmon. Snowfall was adjusted for moisture when included in total precipitation. Latitude (°N) and longitude (°W) are listed, with the period or record. Data for Canadian locations are from Environment Canada; American data are from the Western Regional Climate Center.

City and state	*Jan. temp.*	*July temp.*	*Total Precip-itation*	*Snowfall*	*Latitude*	*Longitude*	*Record*
Coastal							
Barrow, AK	–25.5	4.3	11.6	75.7	71.18	156.47	1949–2001
Nome, AK	–14.6	10.7	41.0	156.0	64.30	165.26	1949–2001
Bethel, AK	–14.6	12.9	42.0	136.7	60.78	161.80	1949–2001
King Salmon, AK	–9.7	12.8	49.2	116.8	58.68	156.65	1955–2001
Cordova, AK	–5.0	12.3	235.9	301.8	60.50	145.50	1949–2001
Juneau, AK	–4.7	13.4	143.4	249.9	58.22	134.35	1949–2001
Prince Rupert, BC	1.3	13.1	246.9	126.3	54.17	130.26	1971–2000
Port Hardy, BC	3.3	13.9	180.8	56.0	50.40	127.21	1971–2000
Seattle, WA	4.3	18.2	97.2	29.7	47.27	122.18	1931–2001
Astoria, OR	5.7	15.6	174.2	10.7	46.90	123.53	1953–2001
Newport, OR	6.8	14.1	173.6	2.8	44.63	124.05	1931–2001
Eureka, CA	8.8	13.8	99.0	0.8	40.59	124.60	1948–2001
San Francisco, CA	9.4	17.0	51.0	0.0	37.37	122.23	1948–2001
Interior							
Fairbanks, AK	–23.6	16.7	26.6	172.2	64.49	147.52	1949–2001
Whitehorse, YT	–17.7	14.1	16.3	145.0	60.42	135.04	1971–2000
Prince George, BC	–9.6	15.5	41.9	216.1	53.53	122.40	1971–2000
Kamloops, BC	–4.2	21.0	21.8	75.5	50.42	120.26	1971–2000
Spokane, WA	–2.7	21.0	40.9	106.2	47.38	117.31	1890–2001
Boise, ID	–1.5	23.4	30.2	52.1	43.34	116.13	1940–2001
Bend, OR	–0.6	17.6	29.8	84.1	44.40	121.19	1928–2001
Redding, CA	7.9	28.5	100.0	12.2	40.31	122.18	1931–1979
Sacramento, CA	7.9	24.0	46.1	0.0	38.31	121.30	1890–2001

rivers (800 km) are also very large, and the Kamchatka River (more than 700 km) has tremendous salmon populations.

In addition to the rivers, we may consider the region's diverse climatic zones. Table 1-2 provides weather data from coastal and interior sites covering the general range of anadromous salmon in North America. Not surprisingly, areas in the south are warmer than those in the north, and coastal areas have much smaller differences in temperature than those inland at the same latitude. There is also an increase in precipitation from south to the north, reaching a maximum in southeast and south-central Alaska, and then precipitation decreases farther north.

The cast of characters: Brief life-history summaries of the major salmonid species

This section briefly describes the general life-history patterns of the major salmonids of the Pacific Rim, including the common names, distinguishing features, and native ranges. Keys to identifying the species can be found in books on fishes of the region such as those by McPhail and Lindsey (1970), Scott and Crossman (1973), and Hart (1973).

Oncorhynchus

O. gorbuscha: Pink salmon, also known as humpback salmon or humpies for the exaggerated dorsal hump in mature males. Their native range in North America is from north-central California (e.g., the Sacramento River; Hallock and Fry 1967) to the Mackenzie River (Heard 1991). The California populations have been at most marginal, and pinks are scarce south of Puget Sound (Hard et al. 1996). Norton Sound is the northern extent of significant populations, though some are found in Kotzebue Sound and the Arctic Ocean. In Asia they extend to Hokkaido, Japan, and to Korea, but are scarce in the Arctic. The heart of their range is the coast of Russia, including Kamchatka Peninsula, and in North America from central Alaska south to the Fraser River. They are the most abundant salmon species and also the smallest at maturity. All pink salmon migrate to sea and all die after spawning. They are 2 years old at maturity in their native range, with only very rare exceptions (e.g., Ivankov et al. 1975). This results in odd-year and even-year populations that do not interbreed. Generally, even years dominate in Queen Charlotte Islands and Alaska whereas odd years dominate in Fraser River, Puget Sound, and southern waters of North America, and in Asia. Some rivers have only odd-year or even-year runs, and this has puzzled scientists for many years. The odd-year runs seem better adapted to warmer water and the differences between lines may have resulted from their colonization from northern (even) and southern (odd) glacial refuges.

Pink salmon produce small eggs, generally in the lower reaches of rivers. The fry emerge from the gravel about 29–33 mm long, without parr marks, and are fully adapted for seawater. They are slim bodied and their color and shape are adaptations for countershading and locomotion in open marine water. They migrate directly to sea and make essentially no use of freshwater for rearing. Pink salmon can therefore spawn at very high densities because food, rather than space for incubating embryos, tends to limit the density of juvenile salmon in streams. At sea, pink salmon migrate to offshore waters, and the Asian populations tend to migrate farther toward North America than vice versa. Males and females are similar in average size (about 2 kg) but males vary much more (about 0.5–6 kg) than females. The species displays extreme sexual dimorphism, notably in the dorsal hump and elongated jaws of the males (photo 1-1). Adults at sea have small scales and large, somewhat indistinct spots on the tail and back.

O. keta: Chum salmon, also known as dog salmon. They range in North America from the Sacramento River (but are scarce in California; Hallock and Fry 1967) into the Arctic Ocean (Mackenzie River) but are scarce north of Norton Sound. In Asia they range west to the Lena River in the Arctic and south in Russia and Japan and the east coast of Korea (Salo 1991; O. W. Johnson et al. 1997). They are the third most abundant species, after pink and sockeye, and typically mature at 3, 4, or 5 years of age (older in the

1-1 Male pink salmon; note the exaggerated dorsal hump and white belly. Photograph by Manu Esteve, University of Barcelona.

north than the south). They generally spawn in the lower reaches of rivers soon after leaving the ocean but exceptional populations migrate more than 2000 km up the Yukon and Amur rivers, and there are small but apparently persistent populations in the upper Mackenzie River system. Like pink salmon, all chum salmon are anadromous and all die after spawning (photo 1-2).

Chum salmon eggs are large, producing relatively large fry (about 32–38 mm). They have small parr marks and elongated bodies, and they may migrate directly downstream to saltwater or stay in their natal stream for a few days or weeks and grow. They then go to sea, use estuaries in many cases for a few weeks, and then migrate to the offshore waters. Adults are quite large at maturity, averaging about 3–5 kg. Adults at sea do not have spots but are distinguished by iridescent pigment in the form of "rays" on the tail, and a relatively long, thin caudal peduncle.

O. nerka: Sockeye salmon, commonly called red salmon in Alaska, and blueback on the Columbia River. They are the second most abundant species, with an historical range in North America from the Sacramento River, California (Hallock and Fry 1967), to Kotzebue Sound, but their primary spawning range is from the Columbia River to the Kuskokwim River. In Asia they range from the Kuril Islands to the area of the Anadyr River, but the heart of their distribution is the Kamchatka Peninsula and tributaries of the Bering Sea (Burgner 1991). They spawn in coastal systems and also ascend as far as 1600 km to Redfish Lake, Idaho.

Sockeye eggs are small, so their fecundity is large for the female's size. Fry are small, usually about 26–29 mm, and almost always go to a lake immediately after emergence and reside there for one or two years. Most populations, therefore, spawn in tributaries or outlets of lakes or on beaches of the lakes themselves (Foerster 1968). However, some "ocean-type" populations go to sea in their first year of life, and "river-type" populations rear in rivers for a year before going to sea (Gustafson et al. 1997). These forms spawn in rivers without lakes, and there is a continuum of rearing patterns rather than

1-2 Two male chum salmon; note the white tips on the ventral and anal fins, and the vertical bars. Photograph by Ernest Keeley, Idaho State University.

two distinct types. I use the term "ocean-type" for consistency with the chinook salmon life-history form, as opposed to the term "sea-type," which other authors have used for sockeye. I use the term "river-type" for sockeye rather than "stream-type," the term traditionally used for chinook salmon, to better represent the larger rivers in which these sockeye tend to spawn.

Typical sockeye smolts make very limited use of estuaries but ocean-type juveniles may use estuaries more extensively (Birtwell et al. 1987). All types have an offshore (as opposed to coastal) marine distribution, usually spending 2 or 3 years at sea, and maturing at about 2–4 kg. Adults at sea do not have spots (but they may in freshwater). Unlike chum and pink salmon, sockeye commonly occur as freshwater resident populations, called "kokanee." These may be physically landlocked but are often sympatric with sea-run sockeye. Both anadromous sockeye and kokanee inevitably die after spawning. Mature adults have bright red bodies and green heads (photo 1-3).

O. kisutch: Coho, commonly called silver salmon, and sometimes blueback in British Columbia. They presently range in North America from Scott Creek, just north of Santa Cruz, California, to Point Hope, Kotzebue Sound, Alaska (Sandercock 1991; Weitkamp et al. 1995), but are scarce north of Norton Sound. In Asia, they seem to be absent, at least as self-sustaining populations in Japan; but they are found in Russia, primarily on the Kamchatka Peninsula, but also on Sakhalin Island, the coast of the Sea of Okhotsk, and occasionally elsewhere. They are more abundant than chinook and the anadromous trout but less so than the other salmon species, and they tend to spawn in small streams of moderate gradient along the coast and the interior.

Fry are about 30 mm when they emerge and are heavier bodied than pink, chum, and sockeye fry. They are also more colorful, with orange dorsal and anal fins with black and white markings on the edges. They typically reside in streams for one or two years but some go to sea as fry in spring, and some populations reside in lakes for all or a portion of their lives in freshwater. Smolts spend little time in estuaries, and subadults feed

1-3 Female sockeye salmon; note the green head and red body. Photograph by Thomas Quinn, University of Washington.

primarily in coastal waters though some are found offshore. Most spend one full year (two summers) at sea but some males return after only a summer at sea and some individuals of both sexes stay out a second year. Adults are generally larger than pink and sockeye but smaller than chum and chinook salmon. Adults at sea have spots only on the upper lobe of the tail and lack the black mouth of chinook. At maturity they are often red along the body (photo 1-4).

O. tshawytscha: Chinook salmon, also known as king salmon, spring salmon (chiefly in British Columbia), and quinnat salmon (a holdover from earlier nomenclature, still used in New Zealand). The term "tyee" (a native word meaning chief) is a sport

1-4 Male coho salmon; note the spots on the back and the red body. Photograph by Ernest Keeley, Idaho State University.

1-5 Male chinook salmon. Photograph by Mark Giovannetti, stammphoto.com.

fisherman's designation for very large chinook, typically more than 30 lbs. Chinook salmon traditionally ranged in North America from the Ventura River on the coast of south-central California and the upper reaches of the San Joaquin River in the south to Kotzebue Sound in the north (Healey 1991b; J. M. Myers et al. 1998). They have been reported in tributaries of the Arctic Ocean but these are apparently not self-sustaining populations. In Asia the dominant populations are in the Kamchatka River but they range from the Anadyr River southward to the Amur River, including the Kamchatka Peninsula and continental side of the Sea of Okhotsk. They have been reported from northern Hokkaido but seem to be rare in Japan. They spawn in large gravel in medium to large rivers in coastal and inland locations and are the least abundant of the five semelparous Pacific salmon species.

Chinook fry are the largest of the salmon (about 33–37 mm) and have two basic patterns of freshwater residence. "Ocean-type" chinook migrate downstream immediately after emergence, or after a few months in the river, and commonly reside in estuaries for a few weeks or more. In contrast, "stream-type" chinook spend a full year in the river before migrating downstream, rapidly exiting the estuary and moving out to sea. In North America, ocean-type chinook are almost exclusively found south of 56° north latitude and tend to be in the lower reaches of coastal rivers. Stream-type chinook are the only form in the northern part of the range. Where they overlap with ocean-type, the stream-type chinook tend to be found farther inland, spawning and rearing at higher elevations.

In the ocean, chinook salmon feed in offshore and coastal waters for 1–5 (typically 2–4) years. Chinook are the largest of the salmon (to 57 kg) but those more than 20 kg are considered large in most populations. At sea they are identified by spots on the upper and lower lobes of the tail, on the back, and by a black gum line inside the mouth (hence the common name "blackmouth"). Mature chinook are typically olive green-brown in color (photo 1-5), though populations (especially in the northern part of the range) can be quite red. Their return migration to freshwater can be long in advance of spawning

1-6 Male anadromous masu salmon, San-nai River, Japan. Photograph by T. Aoyama, Hokkaido Fish Hatchery.

and tends to co-vary with juvenile life-history type. Spring-migrating adults often produce stream-type juveniles whereas fall-running adults tend to be from ocean-type juveniles in the area of overlap. All populations spawn in the fall except so-called winter chinook in the Sacramento River system, which migrate in winter and spawn in spring. Besides this exception to the rule of fall spawning in salmon, chinook display one other unusual trait: some males in stream-type populations mature as parr, having never migrated to sea. Like the other species, chinook all die after spawning, though under controlled conditions the mature parr can survive to spawn in subsequent years.

O. masou: Masu, cherry salmon, yamame (freshwater form); and ***O. rhodurus***: Amago. These two closely related salmon species are native only to Asia, and some authors consider them a single species. Masu are found in rivers leading into the Sea of Japan and Sea of Okhotsk from Korea, China, Japan, and Russia, and their marine distribution is largely limited to these coastal seas (Kato 1991). There are nonanadromous populations as far south as Taiwan. In masu, both males and females often remain in freshwater until maturity at the southern end of the distribution. In the middle of the range, most females go to sea but a large number of males remain in freshwater, so up to 90% of the smolts and upstream migrants are females. At the northern end of the range the sex ratio is more equal, as both males and females migrate to sea. Anadromous masu tend to spend a full year at sea, typically reaching 1–3 kg, and are semelparous, whereas those remaining in freshwater and maturing as parr may survive and spawn more than once and may later migrate to sea (photos 1-6 and 1-7).

Amago are less numerous than masu and more limited in distribution, being found only in central Honshu and Shikoku islands, Japan. The lacustrine form, found in Lake Biwa, is called "biwamasu" whereas the term "amago" refers to the stream-dwelling form. Amago are primarily freshwater residents but some migrate to sea in fall and winter, returning the following spring, having not gone far out to sea.

O. mykiss (formerly ***Salmo gairdneri***): The Asian form (previously known as ***Salmo mykiss***) used to be considered a separate species, and when the fish were reclassified

1-7 Female masu salmon from Sakhalin Island. Photograph by Trevor Anderson, University of Washington.

into the genus *Oncorhynchus* in 1989, both forms were given the Asian species' name because it had been described first. Rainbow trout are the freshwater form, and steelhead are the anadromous form (photo 1-8). This widely distributed, phenotypically variable, iteroparous species has a native range from northwest Mexico (including northern Baja California) to the Kuskokwim River, Alaska, but it is not anadromous in Bristol Bay or northern Alaska or south of Malibu Creek, California (Behnke 1992; Busby et al. 1996). Many southern populations are isolated in higher elevations whereas northern nonanadromous populations often coexist with anadromous salmon. In Asia they occur as both freshwater and anadromous forms, but are native only to Russia, mostly on the Kamchatka Peninsula.

1-8 Steelhead trout; note the numerous, small spots on the body and tail. Photograph by Ernest Keeley, Idaho State University.

Rainbow/steelhead eggs are smaller than those of salmon and are deposited in spring on ascending temperature cycles, unlike those of salmon. Fry emerge in late spring or early summer and are distinguishable from salmon by the spots of the dorsal and adipose fins. Juveniles initially reside in streams, and some remain there but others show fluvial, adfluvial, or anadromous life histories. Steelhead typically spend 1–3 years in freshwater and 1–3 years at sea before returning to spawn. Age and size at maturity vary greatly, depending on habitat and life-history pattern. Males can mature as parr at age 1 and resident females are often small, but adfluvial trout are reported to reach 26 kg in Jewel Lake, British Columbia, and steelhead can exceed 16 kg (Scott and Crossman 1973). The ocean distribution of steelhead is very broad, especially for those going to sea for the first time. Adults in seawater have many small spots on the back and tail and large, square tails.

Rainbow trout exist in isolated areas and divergent forms have evolved, some of which are recognized as different species whereas others are usually classified as subspecies or distinct "races." The taxonomy of these forms is not entirely certain, and they can hybridize, further confusing matters. Behnke (1992) listed Mexican golden trout as *O. chrysogaster*, Gila and Apache trout as *O. gilae gilae* and *O. gilae apache*, respectively, and six subspecies of rainbow trout. The hypothesized relationships among these trout are detailed later in this chapter. According to Behnke, steelhead is the anadromous form of the coastal rainbow trout, *O. mykiss irideus*. They may be divided into ocean maturing and stream maturing (sometimes known as "winter" and "summer" steelhead, respectively). Ocean-maturing (winter) steelhead enter freshwater in winter or spring (e.g., March–April in Washington) in an advanced state of sexual maturity and spawn in spring of that year, perhaps within a month (Busby et al. 1996). In contrast, stream-maturing (summer) steelhead enter freshwater in summer (e.g., August or September) in a much less advanced state of maturity, remain in freshwater through the fall and winter, and spawn the following spring. The ocean-maturing form is typically found in lower elevation, coastal areas (often with fall chinook), whereas the stream-maturing form predominates in the interior (often with spring chinook). Rainbow and steelhead differ from the traditional Pacific salmon by spawning in the spring rather than the fall, and they are also iteroparous. However, the proportion of adults that survive to spawn two or more times varies greatly. Repeat spawning is rare among stream-maturing steelhead, somewhat less so in ocean-maturing steelhead, and common in rainbow trout. Freshwater forms may spend their entire lives in streams or may reside in streams for a few years before migrating to a lake for subsequent rearing.

O. (formerly ***Salmo***) ***clarki***: Cutthroat trout. Like the rainbow, this species was previously classified in the genus *Salmo*, with Atlantic salmon and brown trout. Its native range is restricted to western North America, where they exist as anadromous and freshwater forms along the coast from the Eel River of northern California to the Kenai Peninsula of Alaska. Nonanadromous populations are found in northern Mexico. Unlike the other Pacific species, they are native to both sides of the Rocky Mountains, being found in the upper reaches of the Colorado, Rio Grande, and Missouri (Yellowstone Lake) systems, and other isolated areas. Like rainbow, cutthroat trout are spring spawners. Fry initially reside in streams and can remain as residents or show adfluvial, fluvial, or anadromous migration patterns. Juveniles are very similar to rainbow trout (virtually indistinguishable in their first summer) but have a somewhat longer lower jaw (past

1-9 Coastal cutthroat trout; note the red coloration under the jaws and spots on the sides and tail. Photograph by Michael Erickson, University of Washington.

the middle of the eye), larger spots, and red marks on the throat (hence the name). They are also iteroparous and have a very complex population structure associated with isolated, divergent forms. These isolated populations (Behnke 1992 listed twelve extant subspecies) can differ greatly from each other in appearance. The coastal cutthroat trout, *O. clarki clarki* (photo 1-9), is sympatric with Pacific salmon, and includes resident and anadromous populations (Trotter 1989). Sea-run cutthroat commonly spend about 3 years in a stream before going to sea, typically only for the summer (O. W. Johnson et al. 1999). They do not seem to range far out to sea but their marine distribution is poorly known. Adfluvial cutthroat can be very large (formerly to 19 kg in Pyramid Lake, Nevada) but stream residents are often small. Likewise, sea-run cutthroat can grow to 8 kg but most are much smaller, about 1–2 kg.

Salvelinus

Salvelinus alpinus: Arctic char(r). Salmonids in this genus are distinguished by having light markings on a dark background, whereas *Salmo* and *Oncorhynchus* have dark spots on a light background. Arctic char has the most northerly distribution of any freshwater fish, with records as far north as 82° latitude on Ellesmere Island (L. Johnson 1980). Their distribution is circumpolar but they are sympatric with Pacific salmon across the Arctic, in Alaska to the Kenai Peninsula and Kodiak Island, and in Russian rivers flowing into the Sea of Okhotsk. They are fall (September–October) spawners with small eggs and high fecundity (3000–5000 or more). Embryos are intolerant of temperatures above 8°C. Like cutthroat trout, Arctic char display extremely variable patterns of color, morphology, and life history, including anadromous and nonanadromous forms, lake and river spawners, and up to four highly divergent forms within a single lake (Jonsson and Jonsson 2001).

Growth rates are usually very slow and the fish are often old (10 to more than 20 years) and iteroparous. Anadromous forms tend to mature at an older age and larger

1-10 Mature bull trout. Photograph by Ernest Keeley, Idaho State University.

size than nonanadromous populations (L. Johnson 1980; Tallman et al. 1996). They seem to be more often anadromous in northern areas (e.g., north of the Yukon River). Seaward migration is in spring and upriver migration is in fall, as they seem to seldom overwinter at sea. However, they do not spawn every year, so char returning from the sea may seek suitable overwintering habitat (preferably lakes) that are not their natal site.

S. malma: Dolly Varden (typically capitalized, as the name comes from a flamboyantly dressed character in a book by Charles Dickens). Dolly Varden are iteroparous, fall spawners, distributed in Asia between the Anadyr and Yalu rivers, and in North America from Seward Peninsula south to the coast of Washington. They may have existed in the McCloud River in the Sacramento River system but they are apparently extinct there now. There are two forms of Dolly Varden: the northern form overlaps with and resembles Arctic char in many aspects of appearance and biology, and the southern form resembles bull trout (photo 1-10). Consequently, the scientific literature can be very confusing because it is not always clear which species was actually being studied (see Armstrong and Morrow 1980 for an excellent review of the species). Adults spawn in streams and juveniles rear there initially, though many populations are anadromous. Most sea-run individuals spend only the summer at sea and are not thought to go far (though little is known about their movements at sea). They often skip a year between spawning events. Anadromous fish may overwinter at nonnatal sites when they are not going to spawn, complicating analysis of their population structure and dynamics. Residents typically spawn annually but are smaller than anadromous ones, and there are some residual males (progeny of sea-run parents that mature without going to sea).

S. confluentus: Bull trout. Since 1978, these fish have been recognized as a separate species from Dolly Varden, which they closely resemble. Bull trout are distributed more to the south and interior whereas Dolly Varden are more prevalent in coastal and northerly areas, though their distributions overlap (U.S. Forest Service 1989; Haas and McPhail 1991). Their range once included northern California's McCloud River, western Montana, Nevada, Idaho, British Columbia, and Alberta, but the range seems to be constricting at the southern end. Adults grow to 60 cm or more and spawn in September and October in streams. Bull trout display anadromous, adfluvial, fluvial, and resident life histories, migrating as much as 250 km within large systems such as the Flathead Lake drainage (Fraley and Shepard 1989). They are intolerant of warm water and are seldom found in streams with summer temperatures above 18°C (photo 1-10).

S. namaycush: Lake trout (also known as mackinaw). Exclusively a freshwater species, native only in North America, lake trout are found in U.S. lakes in New England, the Great Lakes region, Montana, Idaho, and Alaska, and in Canada in all provinces and territories except Newfoundland. In the southern part of the range their lakes are usually large and deep but the species may be in shallower lakes and even rivers in the Arctic. Among salmonids they are the least tolerant of saltwater but are occasionally found in water to about 12 ppt. They spawn numerous eggs (up to 18,000) in fall on rocky reefs in lakes (and, rarely, in rivers). Lake trout reach sexual maturity at age 6–7 or greater and live to be more than 20. A 46.4 kg specimen was caught in Lake Athabasca, Saskatchewan (Scott and Crossman 1973), and others weighing more than 35 kg have been caught there.

Taxonomy and evolution

The focus of this book is on the ecology and behavior of salmonids, but a brief perspective on their taxonomy and evolutionary history is helpful in understanding their present conditions. As described by J. S. Nelson (1984), salmon and trout are teleost (bony) fishes in the order Salmoniformes. This order contains four suborders: one of Northern Hemisphere freshwater fishes including the northern pike, pickerel, and muskellunge, and also the mudminnows; two rather obscure groups, one marine and the other including a single Australian freshwater species; and the Salmonoidei. This latter suborder contains 7 families, 35 genera, and about 150 species, many of which are anadromous.

The suborder Salmonoidei contains two families of freshwater and anadromous fishes from the Southern Hemisphere: the Retropinnidae (southern smelts and graylings) and the Galaxiidae (galaxias, inanga, kokopu, etc.). In the Northern Hemisphere, there are three families from Asia: the Plecoglossidae (with only a single anadromous species, the ayu, from Japan, Korea, and China); the Salangidae (icefish or noodlefish, with anadromous and freshwater species in Russia, Japan, Korea, and China); and the Sundasalangidae (Sundaland noodlefish, one genus with two freshwater species in Borneo and Thailand). The final two families are the Osmeridae (Northern Hemisphere smelts, with marine, anadromous, and freshwater species) and the Salmonidae.

J. S. Nelson (1984) divided the Salmonidae into three subfamilies, all Northern Hemisphere freshwater and anadromous fishes. These correspond to full families in other fish-classification schemes. Without passing judgment, I will refer to these as families. Figure 1-7 is an attempt to draw a consensus evolutionary tree showing the ancestry of these families and their major genera and species. The Coregonidae (or whitefishes) includes *Stenodus leucichthys* (the inconnu), the genus *Prosopium* with six species of whitefishes, and the genus *Coregonus* with about twenty-five species of whitefishes and ciscoes. The second family, the Thymallidae (grayling), has a single genus *Thymallus* and four species, one from North America (*T. arcticus*).

FIGURE 1-7. Hypothesized relationships among salmonids and closely related fishes, based on numerous studies of genetics. Species and genera believed to be more closely related are paired together, and longer horizontal lines indicate greater separation. Dashed lines indicate alternative possible patterns of relatedness; the relationships among sockeye (*O. nerka*), chum (*O. keta*), and pink (*O. gorbuscha*) salmon are also ambiguous, as indicated by the two sets of lines. There is some controversy as to the numbers of species in the family Coregonidae and in the *Salmothymus* and *Hucho* genera. In addition, not all *Salvelinus* species are indicated.

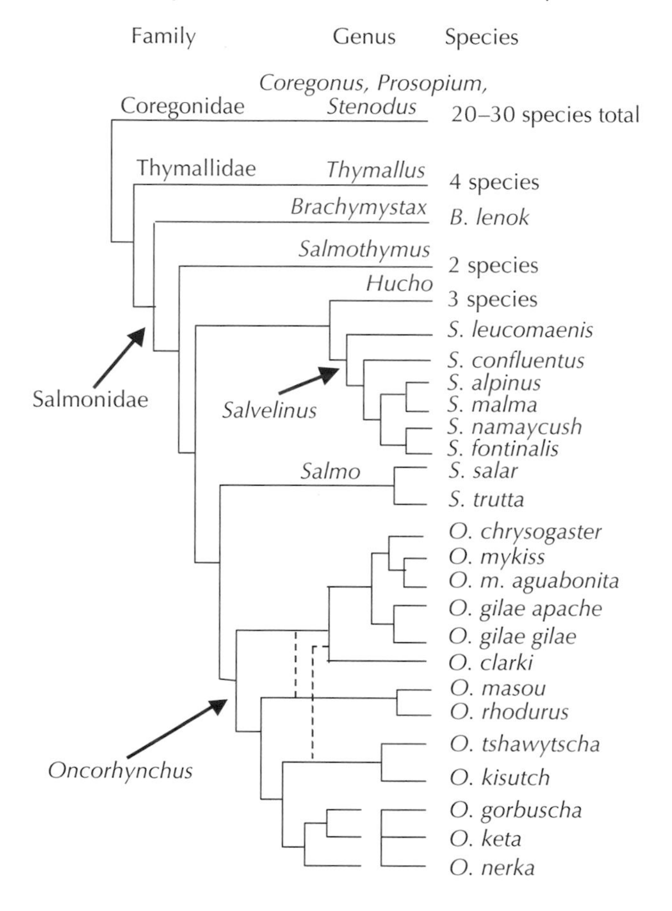

Our focal group, the Salmonidae, contains six genera, including three that are relatively poorly known (at least in North America): *Brachymystax lenok* (an Asian freshwater species), *Salmothymus* (with two European freshwater species), and *Hucho* (three freshwater and anadromous species in Eurasia). The precise numbers of species in these genera, and indeed the status of the genera themselves, are subjects of ongoing debate and research (e.g., Oakley and Phillips 1999). The more widely distributed and more extensively studied genera are *Salvelinus*, *Salmo*, and *Oncorhynchus*. *Salvelinus* includes the chars (or charrs), including brook (also, speckled) trout (*S. fontinalis*) from eastern North America, lake trout, Arctic char, Dolly Varden, and bull trout in North America. Asia has the whitespotted char (*S. leucomaenis*) and other species. The genus *Salmo* includes only two species, both native only to the Atlantic basin. Atlantic salmon, *S. salar*, are found in North America and Europe but brown trout, *S. trutta*, are native only to Europe. Brown trout include a form in northwestern Europe and a very divergent southern form that was once listed as a different species, *S. marmoratus*. The Pacific trout species (cutthroat, rainbow, Mexican golden, Apache, Gila) and subspecies were formerly classified in the genus *Salmo*, but they are now in the genus *Oncorhynchus* (see below). That genus includes the five traditional Pacific salmon species in North America and the two in Asia described above. Based on rather limited fossil evidence, it seems that the species were differentiated about 6 million years ago (McPhail 1996).

All the Salmonidae spawn in freshwater and only some are anadromous, so it is commonly assumed that they have a freshwater rather than marine origin. In general, the breeding areas of fishes seem to be less plastic than rearing areas. For example, eels breed at sea but some species are catadromous and rear in freshwater, and the clupeids (e.g., shad, alewife) breed in freshwater but some feed at sea. It is reasonable to surmise that a freshwater species, rearing in a growth-limited environment, gradually came to exploit estuarine and marine areas for feeding but returned to the natal site for breeding. This may have taken place, though McDowall (1997, 2002) concluded that it is not safe to assume so. McDowall (2001) also reviewed arguments related to the linkage between anadromy and homing and pointed out several alternative hypotheses related to the evolution of these important life-history traits. Unfortunately, our understanding of anadromy and homing is so dominated by work on salmonids that we are in a weak position to generalize about other fishes, and we are in a weak position to determine the true origins of this group.

The active glacial history of the region presently occupied by salmon and trout means that most of the populations found today have relatively recent origin, so one might ask why we need to know about the ancient lineage of the group. However, depending on whether salmon and trout had a marine or freshwater origin, we might view their evolution and population structure differently. If they were fundamentally freshwater fishes that secondarily evolved anadromy, then the ancestral condition would have been many isolated populations. The migration patterns would have been restricted to fluvial or adfluvial movements. These isolated populations would have experienced progressively more gene flow with other populations when feeding migrations took them down to the ocean, enabling them to stray among drainage basins. On the other hand, if they were ancestrally marine, the population structure would have been much more homogeneous (see comparisons of gene frequencies in marine, anadromous, and freshwater fishes by Gyllensten 1985 and Ward et al. 1994). Expansion into freshwater habitat for spawning would have led to increasing levels of population differentiation, as straying gave way to homing. I find myself inclined to the view that salmon and trout are fundamentally freshwater fishes. The geological history of the region, perhaps combined with some propensity for anadromy embedded in the Salmonoidei, has caused them to take advantage of favorable growing conditions at sea but they remain tied to freshwater for reproduction. This anadromy has been essential in the repeated colonization of river systems after glaciation extirpated populations.

For many years, Ferris Neave's 1958 paper, "The origin and speciation of *Oncorhynchus*," was the prevailing hypothesis for the evolution of Pacific salmon. Neave noted that (1) rainbow trout (genus *Salmo* at the time) and *Oncorhynchus* have many morphological and ecological similarities (e.g., freshwater and marine habitats, diet, breeding habitats);

(2) many salmon species are interfertile (that is, can produce hybrid offspring, at least under artificial conditions); (3) salmon are strictly Pacific but trout and chars have much wider ranges; and (4) there are no trout in Japan, though trout exist in Russia and salmon in Japan. From these observations, Neave concluded that salmon evolved since the Pleistocene epoch (about 1–0.6 million years ago). A rainbow trout–like ancestor in the Sea of Japan region, isolated by glacial events and lower sea level, differentiated from trout into a masu-like fish. Further speciation occurred and the more advanced forms radiated back to North America but masu remained in Asia.

Biologists studying the Pacific salmon were quite content to regard them as different from the sympatric trout species because a great deal of importance was attached to two differences: salmon spawn in fall whereas trout spawn in spring, and salmon are semelparous whereas trout are iteroparous. Atlantic salmon (*Salmo salar*) and brown trout (*Salmo trutta*) are fall spawners, masu can be iteroparous, and the rainbow and cutthroat trout share the Pacific Ocean with *Oncorhynchus* rather than with brown trout and Atlantic salmon. However, these facts did little to disturb our sense of the relationships among these groups.

The 1970s saw the rapid development and deployment of biochemical techniques for exploring the evolutionary relationships among organisms. There are many proteins that occur in two or more forms, each of which seems to function equally well. To the extent that these so-called polymorphic proteins or allozymes are selectively neutral, they can provide evidence of reproductive isolation between populations and species. The greater the difference in proportions of various allozymes, the greater the inferred genetic difference. Analysis of such proteins by Utter et al. (1973) was the basis for estimating the present degree of relatedness among species and (by implication) the patterns of evolutionary divergence. These data indicated that sockeye and pink salmon were closely related and that chum salmon were most related to this pair of species. Coho and chinook salmon were also paired together, and they were grouped with sockeye, pink, and chum salmon. Cutthroat and rainbow trout were most closely related to each other. Masu were intermediate between the trout and salmon groups, but more closely related to the trout. The genetic analysis broadly supported Neave's model, though the trout and salmon species were not highly differentiated.

This was the general understanding of the relationships among salmonids until Thomas et al. (1986) examined genetic differences among salmonid species using variation in mitochondrial DNA rather than proteins. They drew the revolutionary conclusion that coho and chinook salmon are more closely related to rainbow trout than they are to sockeye, chum, and pink salmon. This finding called into question the inclusion of rainbow trout in the genus *Salmo* and suggested that the Pacific trouts were actually a form of Pacific salmon. Subsequent research by Ankenbrandt (1988), which included a larger number of salmonid species, supported the major finding: that coho and chinook salmon are more closely related to rainbow trout than they are to pink, chum, and sockeye salmon. A number of papers have been published since that time on salmonid phylogeny (e.g., Phillips et al. 1992; Shedlock et al. 1992). The consensus of these and other studies is that the Pacific trout, formerly *Salmo*, are deeply embedded in the genus *Oncorhynchus*. Smith and Stearley (1989) rearranged the taxonomy, and cutthroat trout became *O. clarki*. Rainbow trout had been recognized as two species, *Salmo gairdneri* in North America and

S. mykiss in Russia; these two species were consolidated into a single species, and because the Russian species had been named first, *mykiss* was retained for rainbow trout in Asia and North America (Stearley and Smith 1993).

Based on many studies of the genetics of the family Salmonidae and the genus *Oncorhynchus* (e.g., Utter and Allendorf 1994), several generalizations and uncertainties are evident, and I have tried to summarize them in figure 1-7. It seems clear that coho and chinook salmon are most closely related to each other. Pink, chum, and sockeye form another group, but it is not clear if all are equally related to each other or if pink and chum are most closely related to each other and a bit more distant from sockeye. These patterns are consistent with life history and morphology. Coho and chinook are stream rearing and are deep bodied and darkly colored as juveniles, whereas the other three species are more pale and slim bodied and primarily feed in the pelagic zone of marine waters or lakes after emerging from the gravel. There is some uncertainty whether the coho-chinook group is more closely related to the pink-chum-sockeye group or to the rainbow-cutthroat group. The position of masu and amago (closely related to each other, or perhaps a single species) is also unclear. Some phylogenies place them near the pink-chum-sockeye group whereas others place them nearer the coho-chinook-trout group. It is likely that further discoveries in genetics will clarify the relationships among these fishes but for the time being this diagram should suffice for our purposes.

There is also a whole constellation of diversity in the rainbow-cutthroat trout group. There are several nonanadromous species closely related to rainbow trout, including *O. chrysogaster* (Mexican golden trout) and *O. gilae gilae* (Gila trout) and *O. gilae apache* (Apache trout). There are also various isolated trout groups, and Behnke (1992) recognized six subspecies of rainbow trout. There is even greater diversity within cutthroat trout, as the populations occupying drainages of the Colorado River, Rio Grande, and Yellowstone (Missouri) River systems, and other areas are completely isolated from natural exchange with each other and with coastal populations. Behnke (1992) listed fourteen subspecies of cutthroat trout, two of which are extinct or probably so. Readers particularly interested in these numerous forms of trout should consult Behnke (1992 and 2002). Excellent illustrations by Joseph Tomelleri in these books reveal the great diversity of color patterns among these trout populations. Unless otherwise stated, the cutthroat trout described in this book are the coastal subspecies, *O. clarki clarki*, but the principles apply to the others as well.

Although the phylogeny of salmon and trout is interesting in its own right, the important evolutionary processes for many populations are more recent. Comprehensive reviews were written on the glacial history and zoogeography of North American regions used by salmon, including western Alaska to the Arctic (Lindsey and McPhail 1986), the region from the Columbia River to the Stikine River ("Cascadia": McPhail and Lindsey 1986), and the Oregon and California coasts (Minckley et al. 1986). The major rivers of the region (Columbia, Fraser, Skeena, Nass, Stikine) are believed to have existed in roughly the same courses since the Pliocene (1.8–5 million years ago), though the Snake River may have flowed southwest and not connected with the Columbia. The Quaternary period had four glaciations in North America, each with a mild, ice-free interglacial period. However, this sequence is not as clear in Cascadia as in other regions. Most recently, the Cordilleran ice sheet reached its maximum extent about 15,000 years

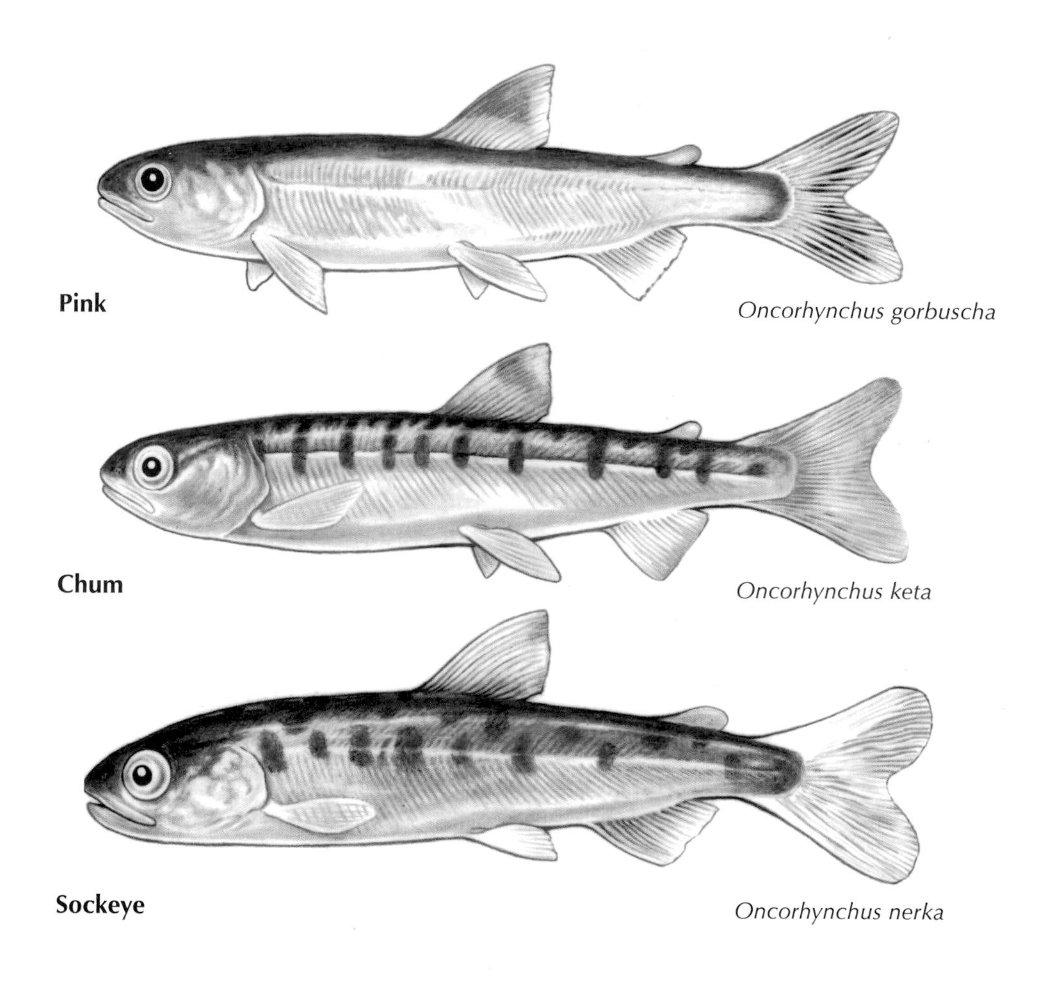

PLATE 2. Illustrations of juvenile pink, chum, and sockeye salmon shortly after emergence from the gravel. The pink salmon are distinguishable by the absence of parr marks. Sockeye and chum fry are similar but the parr marks of chum fry tend to be more regular in size and to be above the lateral line whereas those of sockeye are less regular. Copyright: Charles D. Wood, Ph.D.

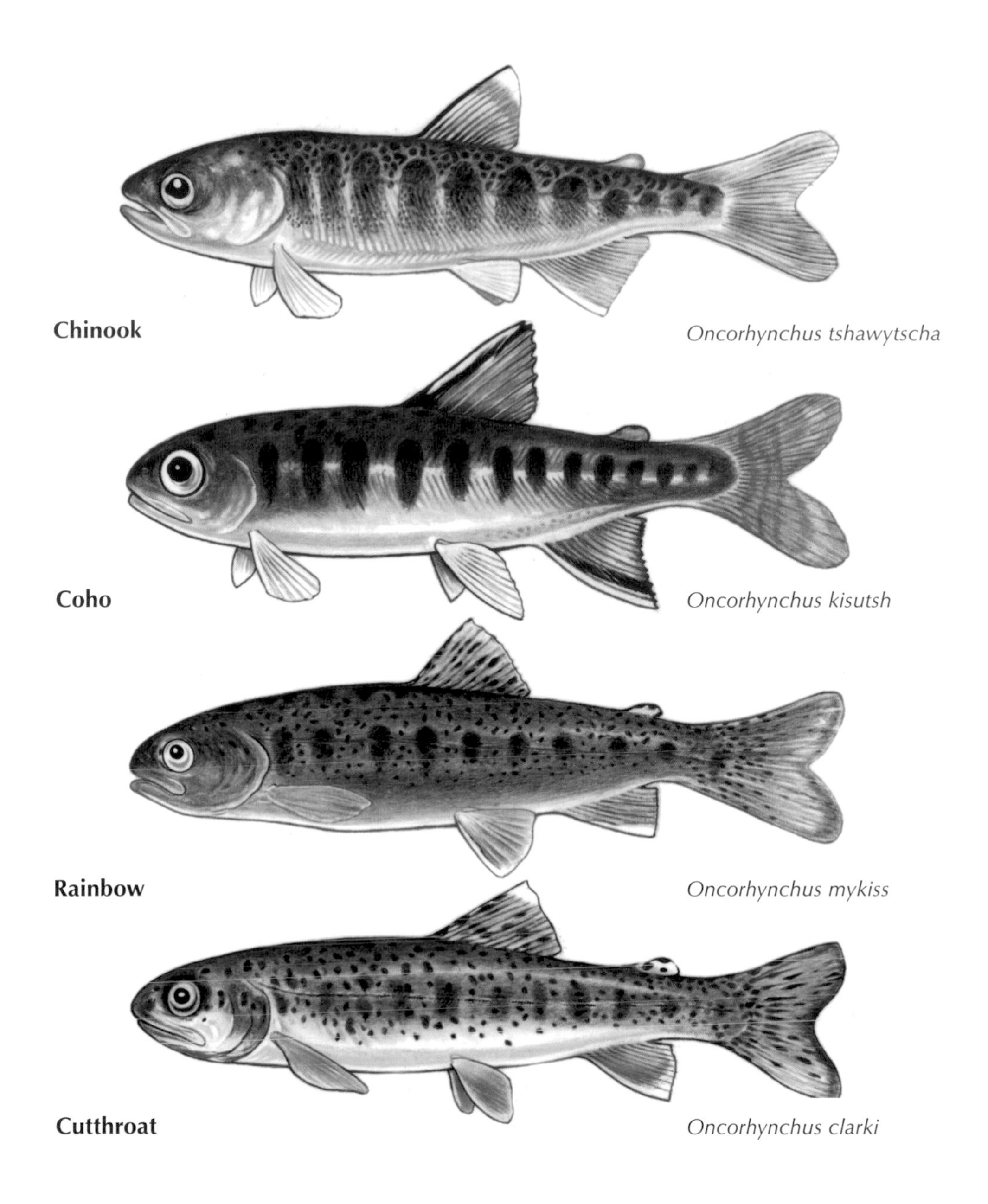

PLATE 3. Illustrations of juvenile coho and chinook salmon, coastal rainbow/steelhead trout, and cutthroat trout. The coho salmon have a hooked anal fin and a white margin with a black inner line on the dorsal and anal fins whereas chinook salmon have a black leading edge on the dorsal fin and the anal fin is not elongated. In chinook salmon, the center of the adipose fin is clear whereas in coho it is opaque. Both trout species have spots on the dorsal and adipose fins, though they may not be apparent early in life, when the species are very similar. Cutthroat trout jaws extend past the eye whereas those of rainbow do not. Both species show great variation in color among populations. Copyright: Charles D. Wood, Ph.D.

PLATE 4. Illustrations of the head and anterior portion of the body of male (below) and female (above) pink salmon at the height of sexual maturity. Note the elongated jaws and exaggerated dorsal hump of the male, and the white belly and large spots in both male and female. Copyright: Charles D. Wood, Ph.D.

PLATE 5. Illustrations of the head and anterior portion of the body of male (below) and female (above) chum salmon at the height of sexual maturity. Note the very large teeth of the male and the mottled, calico colors.

PLATE 6. Illustrations of the head and anterior portion of the body of male (below) and female (above) sockeye salmon at the height of sexual maturity. Note the green heads and red bodies. The males in some populations may have a large hump on the back, reminiscent of pink salmon. However, the dorsal fin of sockeye salmon is at the apex of the hump, whereas in pink salmon the fin is behind the apex of the hump. Copyright: Charles D. Wood, Ph.D.

PLATE 7. Illustrations of the head and anterior portion of the body of male (below) and female (above) coho salmon at the height of sexual maturity. Note the robust jaws of the male, compared to male sockeye and pink salmon, the darker shade of red compared to the sockeye, and the presence of spots on the body. The extent of red color varies among populations. Copyright: Charles D. Wood, Ph.D.

PLATE 8. Illustrations of the head and anterior portion of the body of male (below) and female (above) chinook salmon at the height of sexual maturity. Olive-brown color is typical of southern chinook salmon but northern populations may be more red in color. Copyright: Charles D. Wood, Ph.D.

PLATE 9. Illustration of an adult Arctic char in fresh water. Members of the genus *Salvelinus* have light spots or markings on a green background, unlike the dark spots on a light background seen in adult salmon and trout. Copyright: Charles D. Wood, Ph.D.

ago, covering Cascadia including the upper two-thirds of the Columbia basin. However, most of southern British Columbia was ice-free about 20,000 to 25,000 years ago.

During the most recent glacial period, huge ice dams built giant lakes such as Lake Missoula (some 300 km long), which repeatedly filled and emptied in cataclysmic floods that scoured out the Columbia River scablands. The lower Columbia River was a refuge and the Chehalis River region was also ice-free, as were the rivers to the south. Much of the present range of salmon was glaciated, but isolated areas were ice-free and may have supported salmon. The much lower sea levels exposed part of the Bering Sea and Bristol Bay, which was ice-free. It is obvious that the colonization of newly inhabitable rivers by anadromous salmon could have taken place when maturing adults went up the new river rather than their natal one. However, colonization could have taken place by land as well as by sea. For example, the interior of the Fraser River system flowed into the Columbia River and salmon probably moved between what are now distinct drainage systems.

The postglacial colonization of Cascadia by fishes came primarily from the Columbia, which has 52 native fishes (McPhail and Lindsey 1986). The 39 species in the Fraser River's community are entirely common to the Columbia, indicating colonization from the south. Likewise, the Skeena River, to the north of the Fraser, has 32 species, all apparently derived from the Columbia. The Nass also has a basically Columbia-derived community of 27 species, but the Stikine's 27 species represent both the Columbia and Bering refuges. Even the large islands have lower diversity than the mainland (20 native freshwater fishes on Vancouver Island and 14 on the Queen Charlotte Islands). Up the coast from the Stikine, the diversity is relatively similar, with 20 species in the Taku, 19 in the Alsek, 20 in the Copper, 22 in the Susitna, 27 in Bristol Bay, 24 in the Kuskokwim, and 28 in the Yukon (Lindsey and McPhail 1986). However, the Mackenzie drains a large portion of the center of the continent and has a much more diverse fish fauna, with 48 species. South of the Columbia, the Chehalis River has 34 species, the Klamath has 28, and the Sacramento has 41 (Minckley et al. 1986).

Summary

Salmonid fishes differ from typical bony fishes in many attributes. Their anadromous life cycle allows salmonids to grow very rapidly at sea, and at maturity they are large for their age compared to most freshwater fishes. Their large eggs, tendency to spawn in the fall, female parental care, and egg burial are all unusual traits. After a long period of incubation, the offspring emerge in spring and rear in freshwater or migrate to sea. The high degree of homing to the natal site for spawning isolates populations, despite the opportunities for exchange provided by anadromy. Death inevitably follows reproduction in many species, and this semelparity is linked to the high productivity of many populations.

As a taxonomic group salmon are old, and their relatives include many other migratory and diadromous species. Within the family of salmonids, the genera and species vary in their expression of anadromy, semelparity, and associated traits. Though much can be learned about the ancestral lineages of salmon, the region salmon occupy has had an active glacial history and most present populations were founded by strays within about 10,000 years. Thus the species have existed for several million years but the populations are recent and seem to have evolved quite rapidly.

2

Homeward Migration of Adults on the Open Ocean

As with any description of a life cycle, it is hard to know where to begin the cycle of salmon. I chose to begin with adult salmon on the open ocean, in their final summer at sea, migrating homeward. The details of salmon distribution patterns and migrations at sea were poorly known until the middle of the twentieth century when, after World War II, Japanese fishermen expanded their fishing for salmon on the open ocean. Largely driven by American and Canadian concerns that their salmon were being caught in unregulated international waters, the International North Pacific Fisheries Commission (INPFC) was established in 1953 and included Japan, the United States, and Canada. The INPFC was responsible for a long-term program of research on the distribution and ecology of salmon on the high seas, as well as research on other marine fishes. The bulletins of the INPFC provide great detail on the descriptive oceanography of the North Pacific Ocean and salmon distribution patterns, including a bulletin for each of the major species (coho: Godfrey et al. 1975; sockeye: French et al. 1976; chum: Neave et al. 1976; chinook: Major et al. 1978; pink: Takagi et al. 1981; masu: Machidori and Kato 1984; steelhead: Burgner et al. 1992). In addition, Hartt and Dell (1986) reviewed work on juvenile salmon at sea. The initial focus was on determining general distribution patterns and continent of origin of salmon caught at sea (i.e., were fisheries at given locations catching Asian or North American salmon?). The program provided information on the distribution patterns of such major stock complexes as Bristol Bay (Alaska) and Fraser River sockeye, Japanese chum, and so on, and also a wealth of data on travel rates, size, diet, and other features of the behavior and ecology of salmon in this poorly understood but important phase of their lives.

Techniques

The distribution and migration patterns of salmon at sea were inferred from a variety of techniques. First, vessels were chartered to fish for salmon using various types of gear including purse seines, gillnets, and longlines (photos 2-1, 2-2, 2-3, and 2-4). All types of gear provided indices of abundance (catch per unit of effort), though each type is selective in one way or another (e.g., fish size, vertical distribution). The shifting patterns of relative abundance of immature and maturing salmon in time and space allowed scientists to plot likely migration routes. Estimation of migration routes was further indicated by catches of salmon in purse seines and gillnets designed to detect the direction of movement.

The second technique employed to determine distribution and migration patterns was tagging of salmon caught at sea and recovery of the tags in fisheries in coastal waters, estuaries or rivers, or at spawning grounds and hatcheries. A variety of tags were tried but most of the fish were tagged with individually lettered plastic disk tags, attached below the dorsal fin by a metal pin ("Petersen" disks), or by flexible, lettered strands of plastic attached by a dart-shaped or T-shaped tip inserted into the musculature of the fish below the dorsal fin (reviewed by Davis et al. 1990). These tagging programs had to contend with natural and handling-related mortality, tag loss, nonreporting of tags by fishermen, and the numerous spawning grounds where salmon could spawn and die unnoticed. Nevertheless, they provided convincing evidence of the distribution patterns of salmon from specific geographical areas and also gave information on minimal migration speeds. In addition to the tagging of immature or maturing salmon at sea, since the 1970s, significant numbers of salmon have been marked in freshwater and recovered

2-1 Purse seine fishing for salmon near Adak, Aleutian Islands. Photograph courtesy High Seas Salmon Research Program, University of Washington.

2-2 Gillnetting for salmon on the North Pacific Ocean from a Japanese research vessel. Photograph courtesy High Seas Salmon Research Program, University of Washington.

2-3 Setting a floating longline (baited with salted anchovies) for salmon from Hokkaido University's research vessel, *Oshoro maru*. Photograph courtesy High Seas Salmon Research Program, University of Washington.

2-4 Steelhead caught on the high seas, showing card used for scale samples. Photograph courtesy High Seas Salmon Research Program, University of Washington.

at sea. These salmon were chiefly produced at hatcheries, but to a much lesser extent wild populations have been marked. The capture of these salmon on the high seas has further demonstrated the marine distribution of salmon from specific areas.

The third approach for determining the distribution of salmon at sea was to use natural variation among salmon in traits closely linked to a continent or specific region. Certain parasites have intermediate hosts that are only found in freshwater habitats within part of the range of salmon species, thus infected salmon caught at sea must have originated from that area (Margolis 1963). For example, the life cycle of the cestode (*Triaenophorus crassus*) involves freshwater copepods, usually *Cyclops*, which eat the parasite. Salmon (chiefly sockeye) eat the copepod, get infected, and then pass the parasite to the final host, northern pike (*Esox lucius*). Pike only co-occur with sockeye salmon in western Alaska (mainly Bristol Bay), so the parasite is useful for determining the oceanic distribution of sockeye from this area. A trematode parasite, *Nanophyetus salmincola*, is only found in salmonids such as steelhead from the U.S. Pacific Northwest because of the limited range of its intermediate host, the snail *Juga plicifera* (Dalton 1991). In other cases the parasite may be found over a wide range but the prevalence levels differ dramatically. *Myxobolus arcticus* was found in 57–94% of Asian chinook salmon but was rare in North American populations, allowing Urawa et al. (1998) to determine that most of the chinook in the central Bering Sea were from North America.

In addition to parasites, patterns of variation in polymorphic proteins and DNA markers such as those used to determine the relationships among salmonid species are also useful in distinguishing salmon that originated from different areas. Finally, detailed analysis of salmon scales reveals patterns of variation in the portion of the scale produced during freshwater rearing. Such scale pattern analysis, validated with samples of known origin, has been another very useful tool in identifying salmon at sea (Davis et al. 1990).

FIGURE 2-1. Oceanic distributions of maturing chum salmon, based on 652 salmon tagged at sea and recovered in Japan and 781 recovered in North America (based on INPFC data, provided by Katherine Myers, University of Washington).

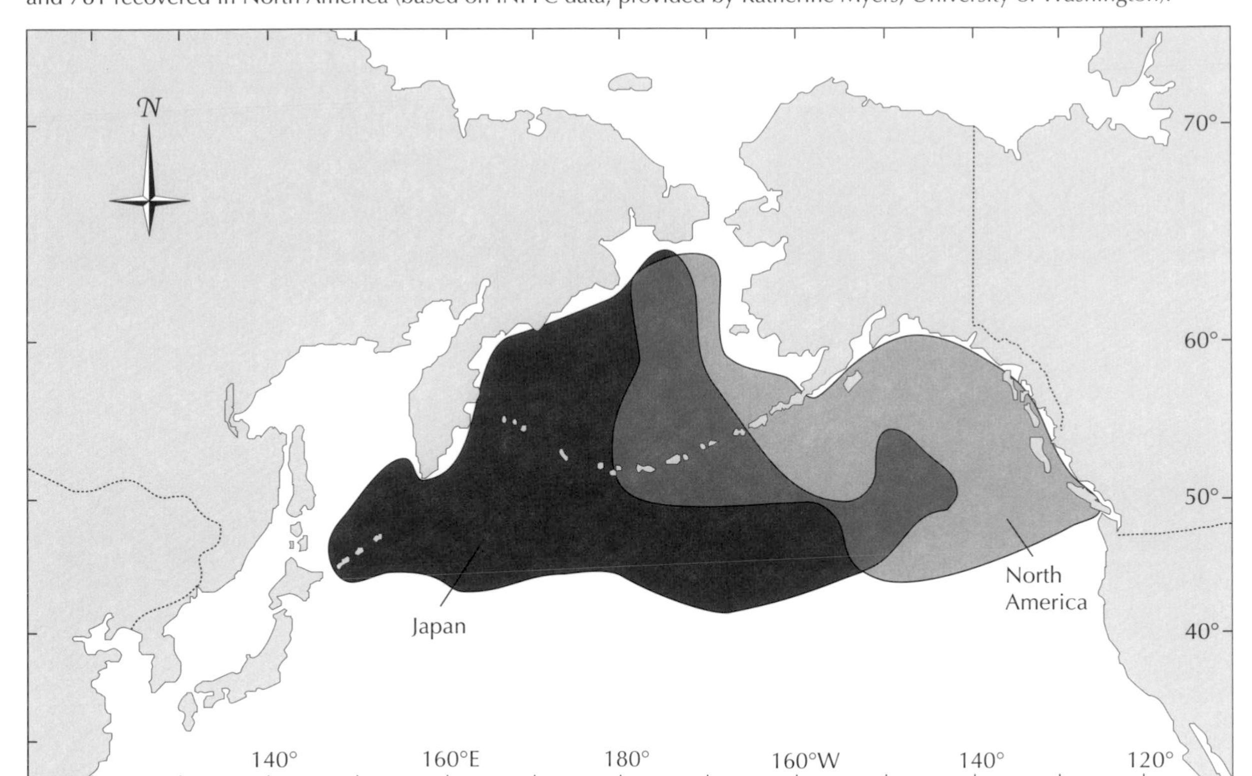

Patterns

Taken together, these types of data suggest four main patterns (with some exceptions) of migratory behavior among the anadromous Pacific salmonids. The first pattern is displayed by sockeye, chum, and pink salmon. These most abundant species tend to move rather rapidly northward and westward through North American coastal waters after entering the ocean and have largely moved off the continental shelf into open waters of the North Pacific Ocean, Gulf of Alaska, and Bering Sea by the end of their first fall at sea. They remain in offshore waters until they mature and then seem to move rather rapidly toward the coastal waters where they may hold for a period of time before moving upstream to spawn. Tagging studies revealed net travel rates (point of release to point of recapture divided by number of days at large) of up to 60 km per day or more for individuals of these species. The details of the routes taken by individuals are poorly known, and we are hampered by very limited information on winter distributions. Nevertheless, the distributions of species and populations from different areas overlap but do not fully coincide, as figure 2-1 on Japanese and North American chum salmon and figure 2-2 on Bristol Bay and British Columbia sockeye salmon show (Neave et al. 1976; French et al. 1976; K. W. Myers et al. 1996). During their period at sea, including the homeward migration, salmon are distributed in the near-surface waters, as revealed by

FIGURE 2-2. Oceanic distributions of maturing North American sockeye salmon, based on 2329 salmon tagged at sea and recovered in Bristol Bay, Alaska, and the Aleutian Islands, and 1241 recovered in British Columbia (based on INPFC data, provided by Katherine Myers, University of Washington).

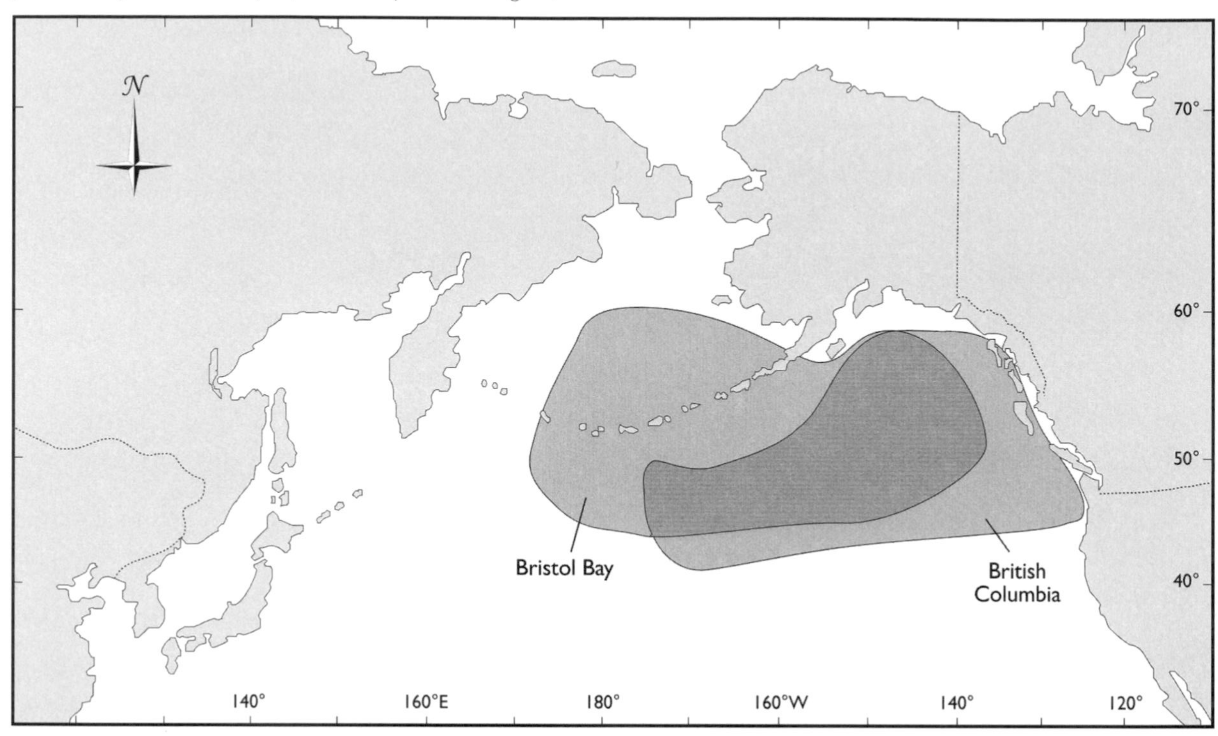

ultrasonic telemetry (e.g., Ogura and Ishida 1995). Salmon also exhibit a general tendency to move toward the surface at night and deeper during the day.

The second pattern of migration is shown by chinook and coho salmon. Many populations of these species remain largely or entirely in coastal waters. In most cases they are generally distributed to the north of their river of origin, but some populations remain relatively close to their natal river and some migrate southward (Nicholas and Hankin 1989; Healey 1991b; Weitkamp et al. 1995; J. M. Myers et al. 1998). In particular, populations from California and Oregon would experience warm, unproductive water if they migrated straight out to sea (fig. 2-3). Thus they remain in the colder and more productive zone of upwelling along the coast and make an appropriate return migration. However, this does not mean that there are no coho or chinook salmon on the high seas; some substantial populations migrate to offshore waters. Fishing on the high seas, though once common, is very limited now and the tendency to migrate offshore is inferred in part from the dearth of catch records in coastal waters and in part from research operations. Chinook and coho salmon seem to move more slowly homeward than pink, sockeye, and chum salmon. They do not necessarily swim more slowly but they probably swim in a less directed manner and feed more extensively while migrating.

The third marine migration pattern, shown by anadromous cutthroat trout, Dolly Varden, and Arctic char, seems to be quite restricted in time and space. These species generally migrate to sea in the spring and return the same fall, overwintering in fresh

FIGURE 2-3. Map of the North Pacific Ocean, showing the major currents and oceanographic domains (based on Favorite et al. 1976; Neave et al. 1976; Pearcy 1991, 1992; Francis and Hare 1994). Numerals indicate approximate current speeds in km/d. Shading indicates the coastal range of salmon.

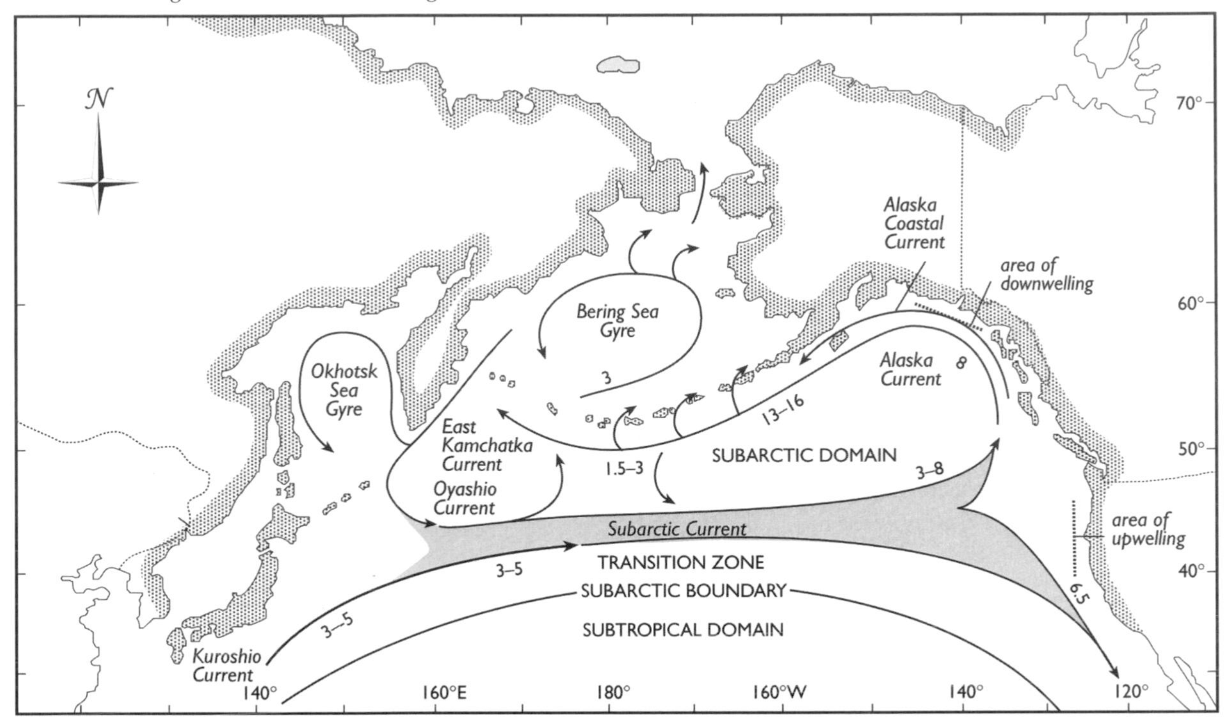

water even if they do not mature and spawn. Information on these species' distributions and movements of individuals at sea is very limited. They are much less abundant than the semelparous salmon and few are caught in commercial fisheries so we know little about their movements. It is generally believed that they remain within estuarine or coastal areas near their natal stream but the evidence for this is largely indirect. From a phylogenetic viewpoint it is tempting to conclude that these species, having many nonanadromous populations, have simply not fully adapted to the full range of opportunities and habitats for growth presented by the ocean. However, this may be misleading, as the fourth pattern, shown by steelhead, is very different. As smolts, steelhead apparently migrate rapidly offshore rather than along the continental shelf, and their return migration is also rapid. At least some populations of steelhead migrate very far across the ocean, and many individuals spent 2–3 years at sea (Burgner et al. 1992).

Orientation

Migration may be defined as the movement of individuals in a species or population, coordinated in space and time. The spatial aspect has two components: orientation and locomotion; and the temporal aspect has two components: initiation and cessation. Orientation is often separated into three processes: piloting, compass orientation, and navigation. Piloting is the ability to find one's way home using information acquired on the outward journey. Orientation, in this strict sense, refers to the ability to move in a fixed (e.g., compass) direction without reference to local landmarks; whereas navigation is the ability to determine one's position, relative to a distant goal, and to move in that direction without use of clues emanating from the source such as odors. Hasler (1966) recognized that the freshwater and marine phases of the homeward migration of salmon are probably guided by different mechanisms. The freshwater phase clearly relies very heavily on olfactory information learned prior to or during seaward migration years earlier (see chapter 5). This piloting, however, is probably of very limited value at sea because salmon do not retrace their outward journey when they return (Royce et al. 1968; Groot and Cooke 1987). The migrations of salmon from distant locations at sea to their natal spawning grounds are among the classic examples of animal migration and orientation. There have been several hypotheses erected to explain what guides the salmon on these migrations. However, before evaluating the hypotheses themselves it is useful to note five general features of salmon migration at sea which any hypothesis must explain.

First, the species and populations of salmon are mixed at sea, and they then *diverge* to their respective home rivers. This divergence of salmon means that individuals experiencing common environmental conditions (current, temperature, salinity, food resources, etc.) must respond in different ways in order to reach their home rivers. At this point I have not presented the evidence that the salmon indeed return to their natal river to breed. I ask you to take my word for it, and the evidence will be presented in chapter 5. Two examples of this divergence phenomenon will suffice to make the point. Neave (1964) reported that a group of sockeye salmon, caught in a single set of a purse seine in the Gulf of Alaska, tagged, and released there, dispersed to populations over most of the North American range of the species, including the Fraser, Skeena, and Nass rivers of British Columbia; Cook Inlet, Kodiak Island, and Chignik Lake in the Gulf of Alaska; and Bristol Bay. In addition, chum salmon taken in a 2° latitude by 5° longitude area during one month diverged to almost the entire range of the species in both continents. Thus, common, species-specific responses to environmental conditions cannot explain the migrations because populations experience the same conditions but migrate in different directions.

Second, not only do comingled populations diverge but representatives of a given population *converge* from a wide area toward their natal location. Thus salmon experiencing very different conditions at sea will migrate in different directions to reach the same coastal location where they will then migrate upriver. Sockeye salmon, for example, converge on the Fraser River and Bristol Bay from a very large area, as inferred from the locations where salmon of those populations are found in their final spring at sea (see fig. 2-2).

Third, the *swimming directions* of salmon at sea are not random. Information on direction of movement comes from the research gillnet and purse seine sampling by the INPFC. Gillnets hang vertically in the water like a sheet on a clothesline and fish are caught as they encounter the net. By keeping track of the direction the net was set (e.g., along a north-south axis) and the direction the fish were caught (e.g., coming from the east or west), one can test the null hypothesis that movement at sea is random. Likewise, a purse seine net is towed between two boats for a period of time in a U-shaped configuration before the fish are encircled. Fish swimming in the direction the boats are moving

FIGURE 2-4. Minimal migration rates (mean + standard deviation) of maturing sockeye salmon tagged at sea and recovered in Bristol Bay, Alaska, or the Fraser River, British Columbia (based on INPFC data, provided by Katherine Myers, University of Washington).

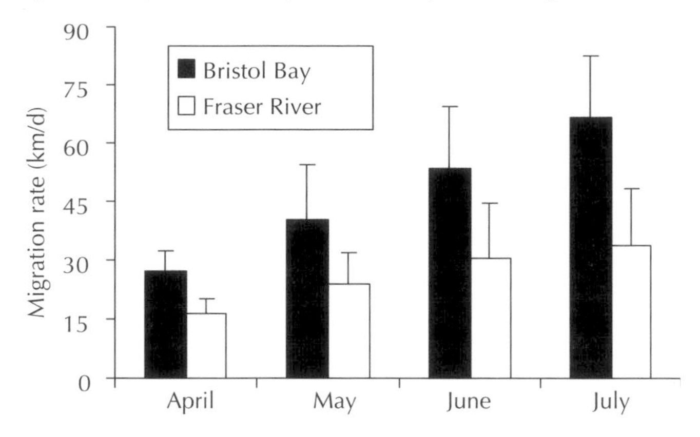

are less likely to be caught than those swimming into the open net. Sets were systematically made in different directions and the relative catches were recorded. We must bear in mind that immature and maturing salmon will not have the same migration patterns, and not all the maturing fish at a given region at sea are going to the same coastal area, so there are good reasons why the patterns of movement at sea might seem to be random. However, the records indicated that salmon at a given location and time were not moving randomly. For example, in gillnet sampling reported by R. C. Johnsen (1964), the ratio of fish caught on the two sides of the net differed significantly from what one would expect by chance in 61 of 97 sets for sockeye, 82 of 103 sets for chum, and 34 of 49 sets for pink salmon. Repeated sets in the same area and month consistently showed directional movement (Dunn 1969).

Fourth, a certain level of orientation can be inferred from the *travel rates* accomplished by salmon at sea. When salmon are tagged at one location and recovered (e.g., in a fishery), it is possible to draw a straight line between the points, divide by the number of days at large, and infer a minimum travel rate. Rates for maturing fish vary among species (see above) and will underestimate actual travel speeds because salmon may be holding in coastal or estuarine waters prior to capture. Nevertheless, travel rates of some stock complexes such as Bristol Bay sockeye salmon are often about 40–60 km/d (French et al. 1976; fig. 2-4). Extensive work by Brett (1983) indicated that the most efficient swimming speed of adult salmon is about 1 body length per second. For a 60 cm salmon this would be 2.16 km/h or 51.8 km/d if the fish swam continuously in a straight line. Thus the observed travel rates are roughly equivalent to energetically efficient and perfectly oriented travel. If the orientation is imperfect, then swimming speed must be increased to compensate. The maximum sustainable speed is about 2.5 lengths/sec or about 5 km/h. Even at this speed the fish would need to have a considerable measure of homeward orientation to accomplish the observed travel rates. In fact, telemetry indicated that salmon swim close to their most efficient speed rather than their maximum speed in marine waters (Quinn 1988), and tracking on the high seas indicated salmon's ability to swim in relatively straight lines for several days (Ogura and Ishida 1992, 1995).

Analysis of temperature records from data storage tags attached to chum salmon indicated that the chum probably made little progress at night (when they were apparently near the surface) but that they moved actively during the day (Friedland et al. 2001).

Finally, the precise *timing of arrival* in coastal locations also indicates a measure of orientation at sea. The most noteworthy example of this is the sockeye salmon returning to Bristol Bay. More than 80% of the fish are caught in about two weeks, having been spread over a large portion of the Gulf of Alaska and North Pacific Ocean only a few months before (Burgner 1980; Quinn 1990a). Fraser River sockeye also show a very orderly sequence of arrival in coastal areas. If the salmon were not well oriented it is difficult to see how they could converge on specific coastal areas from broad oceanic distributions in such a timely manner.

Hypotheses

There is a history of competing hypotheses that have been proposed to explain the migrations of salmon on the high seas. Saila and Shappy (1963) proposed that salmon migrations are largely random. They used a Monte Carlo simulation and concluded that only a very weak homeward bias in the movements of fish was needed to get some salmon from the open ocean to the vicinity of river mouths (where olfaction was presumed to take over as the orientation mechanism). However, many of their assumptions about salmon migrations were in error, and neither this nor a more recent "random" model (Jamon 1990) can plausibly explain the patterns of movement described above (Hiramatsu and Ishida 1989).

If the movements are not random, what might direct them? Leggett (1977, 299) recognized that some fishes are "capable of obtaining directional information from the sun, polarized light, and geomagnetic fields." However, he concluded that other processes play a larger role: "An impressive body of literature supports the hypothesis that fish migrations involve a continuous optimization of physiological and neurological states in response to a multiplicity of environmental stimuli. The nature of this optimum state varies seasonally, and with ontogeny. . . . Racial and species differences in the timing and pattern of these responses are assumed to have developed through natural selection." In the case of salmon, movements at sea might reflect seasonal shifts in preferred temperature, and the homeward migration might involve increasing preference for or tolerance of lower salinity water and other physical features.

There is no doubt that salmon and other fishes move in response to temperature, salinity, dissolved oxygen, pH, and other attributes of their immediate environment. It is also clear that the distribution of salmon at sea is not random with respect to temperature. However, the gradients of physiologically important environmental conditions do not run smoothly in a shoreward direction, and the kind of searching or to-and-fro movements that would be associated with this kind of orientation do not seem consistent with at least some of the tagging data. The divergence and convergence of salmon populations is difficult to reconcile with a physiological optimization hypothesis. How can the myriad populations perceive and react uniquely to the common conditions at sea? Moreover, ultrasonic tracking of salmon on the open ocean revealed that they can swim for several days in relatively straight directions (Ogura and Ishida

FIGURE 2-5. Temperatures experienced by a maturing chum salmon tagged in the Bering Sea and recovered off Hokkaido, Japan, recorded by a data logger over 41 days (Walker et al. 2000).

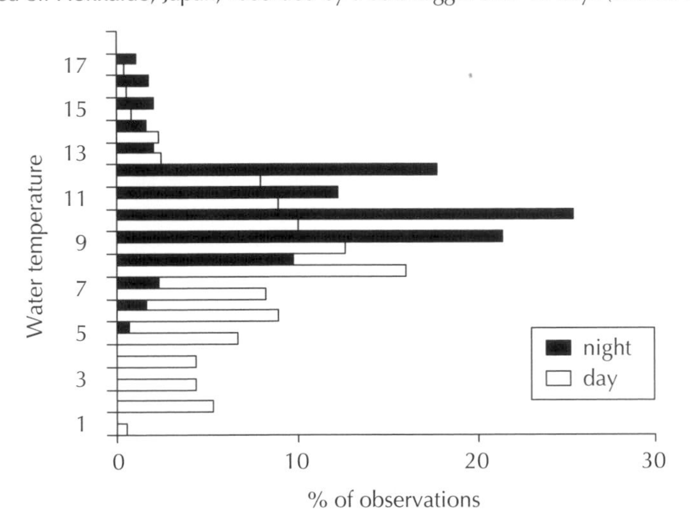

1995), though the origins of the fish were unknown so it was not clear whether they were swimming toward home or not. In addition, data storage tags have revealed that individual salmon routinely experience large, rapid changes in ambient temperature during their vertical movements (e.g., Walker et al. 2000; fig. 2-5). Thus the homeward migrations are probably not guided by local environmental conditions, though the distributions of salmon at sea are influenced by such conditions, and there may thus be correlations between ocean temperature and migration patterns.

In contrast to the models of Saila and Shappy (1963) and Leggett (1977), Larkin (1975, 5) reviewed the information on salmon migrations at sea and concluded, "It thus becomes necessary to postulate that salmon have a bi-coordinate system of navigation that enables them to know where they are and where they are to go (and when to leave in order to get there on time)." The lines of force of the earth's magnetic field could provide some of the "map" information as well as the compass information needed for bi-coordinate navigation (Quinn 1982). Indeed, a sense of location may be needed to help synchronize the migrations. Sockeye salmon migrate toward Bristol Bay in June when day length varies by up to 2.5 hours from 50° to 60° north latitude. If the timing of migration is determined by day length or rate of change of day length, then the salmon at different latitudes would experience very different patterns of day length. Despite this possible source of confusion, Bristol Bay sockeye converge from a very large area with remarkable precision. Perhaps they somehow count the days from the equinox or winter solstice and are genetically programmed to return at some time period after one of these seasonal points.

There are many hypotheses and models but no clear consensus on how salmon accomplish their migrations at sea. Juvenile sockeye salmon orient to the earth's magnetic field in experimental arenas (Quinn 1980; Quinn and Brannon 1982), and presumably all species share this sensory capability and use it to assist their migrations at sea. There has been some exciting recent work on rainbow trout (Walker et al. 1997; Diebel et al.

2000), but our understanding of the sensory systems underlying magnetoreception is incomplete in salmonids, and in animals in general. The earth's magnetic field is used for compass orientation by many animals, including pigeons, salamanders, and sea turtles. However, a compass is not sufficient for navigation on the open ocean. How do the salmon know which direction to swim? Some perception of their location at sea may be needed (a map to go with the compass). There is evidence that some animals, including sea turtles, can use the fine structure of the earth's magnetic field (e.g., its intensity or vertical inclination as opposed to the general horizontal direction of the field) to indicate their location and to orient their oceanic migrations (Lohmann et al. 2001). As of now, the mechanistic basis for a map that salmon might use is unknown, and the linkage between navigation and migratory timing also remains unresolved.

Some insights into salmon navigation at sea are provided by correlations between ocean conditions and the migratory routes and timing of salmon arriving in coastal waters. Burgner (1980) pointed out that the catches of sockeye salmon in Bristol Bay tended to be earlier after a warm spring and later after a cold spring. Subsequent data supported this pattern, though the absolute level of variation was not great (Quinn 1990a; Hodgson 2000). In a seminal paper, Blackbourn (1987) not only confirmed Burgner's results but reported the opposite pattern for sockeye salmon migrating to southern rivers such as the Fraser River (fig. 2-6). He hypothesized that the feeding distribution of salmon at sea is determined by temperature conditions, thus after a warm winter and spring, the salmon are farther to the north than they are after colder seasons. He hypothesized further that the salmon have a sense of date (i.e., time of year) and compass orientation but no sense of their actual location. Therefore, all populations commence homeward migration at their appointed date every year but arrive early to northerly regions like Bristol Bay because they have a shorter distance to go when the ocean is warm (fig. 2-7). Similarly, warmer conditions cause southerly populations like the Fraser River to be distributed farther north, so they arrive late.

FIGURE 2-6. Relationship between the arrival timing (day of the year) in coastal waters of sockeye salmon from the Chilko River (Fraser River system) and from Bristol Bay, Alaska, (weighted average of populations), and the deviation from the average sea surface temperature in the region at sea normally occupied by those salmon (from Hodgson 2000).

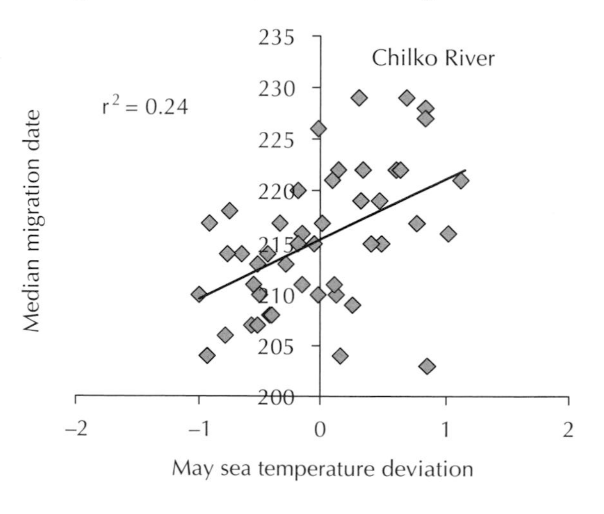

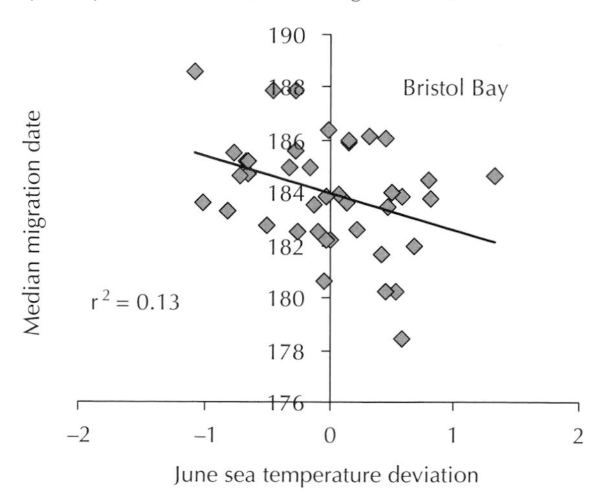

FIGURE 2-7. Hypothesized relationship between ocean temperature, center of distribution of sockeye salmon at sea, and their migration timing at Bristol Bay, Alaska (early in warm years; late in cold years) and the Fraser River (the opposite pattern; modified from Blackbourn 1987).

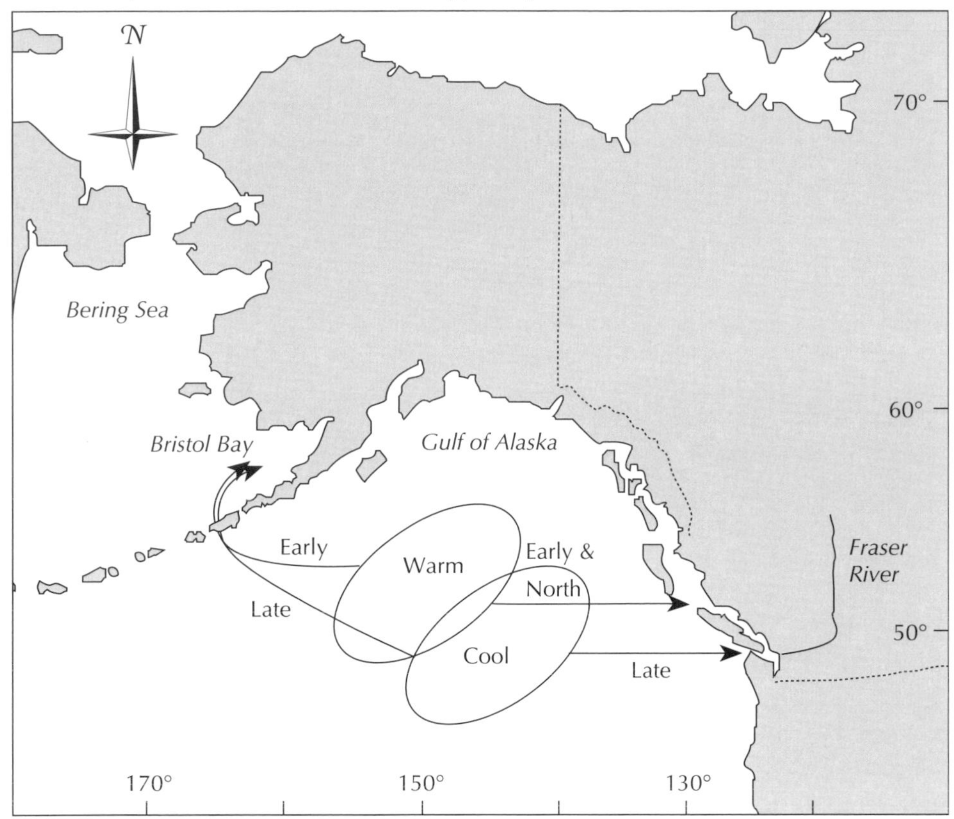

This hypothesis is also consistent with the variation in migratory route of Fraser River sockeye salmon (Groot and Quinn 1987), among the most closely studied salmon population complexes. These abundant and valuable populations can migrate home entirely in Canadian waters if they approach from the north via Queen Charlotte Strait, Johnstone Strait, and then into the Strait of Georgia; or they can pass through American waters if they migrate around Vancouver Island from the south, through the Strait of Juan de Fuca, the San Juan Islands, and into the Strait of Georgia (fig. 2-8). The proportion of sockeye using the northern route was insignificant until 1958, when about 35% were estimated to use that route (fig. 2-9). The variation in route attracted little further attention until the late 1970s when increasing proportions used the northern route, reaching a maximum of about 80% in 1983, which was an El Niño year. The Fraser River itself is in Canada but American fishermen were accustomed to catching these sockeye, and restoration of the runs in the river was accomplished in large measure by the joint efforts of the United States and Canada in the International Pacific Salmon Fisheries Commission (later the Pacific Salmon Commission; Roos 1991). In cool years the sockeye tended to make landfall around the middle of Vancouver Island and migrate south, through the Strait of Juan de Fuca. After a warm spring at sea, the salmon tended to make landfall near the north end of Vancouver Island or the British Columbia mainland

FIGURE 2-8. Map of southern British Columbia and northern Washington.

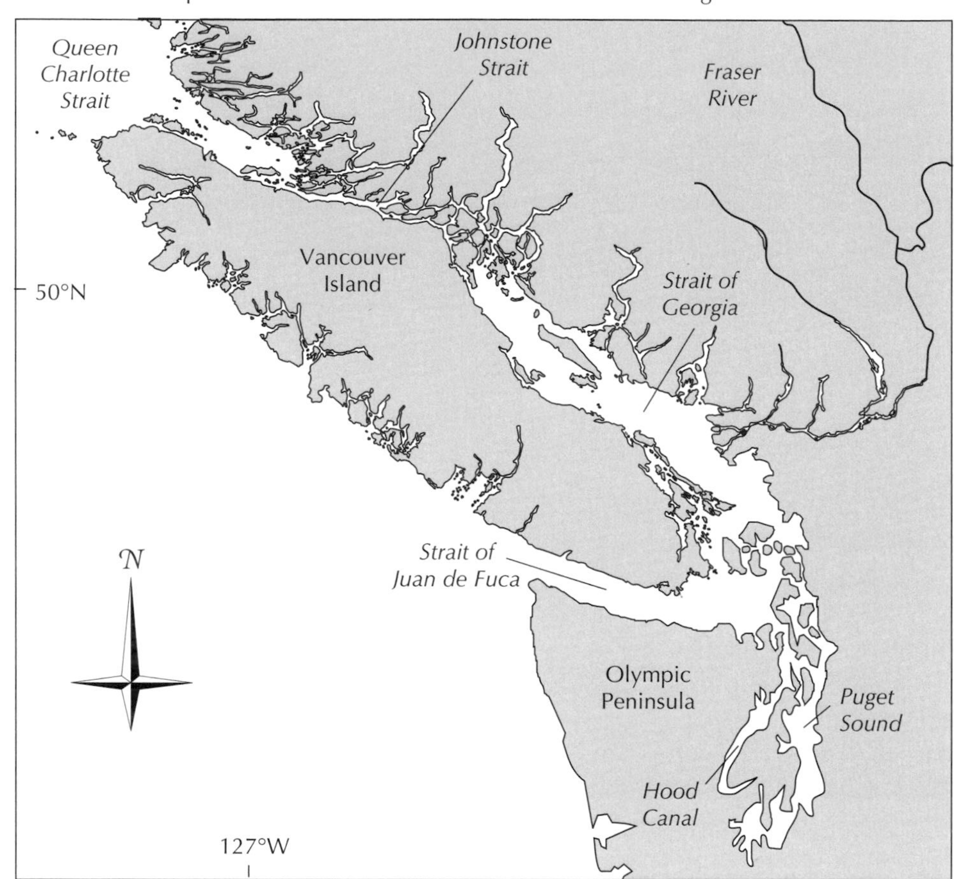

and migrate south through Queen Charlotte Strait and Johnstone Strait (Groot and Quinn 1987). Indeed, there is a correlation between the timing of migration and the route used (fig. 2-10).

The patterns of variation in migration timing and route by Fraser River sockeye salmon support the view that salmon have a sense of direction and tend to swim east-southeast but that they have no sense of their location. Thus they cannot (or at least do not) compensate for location by adjusting either the date when they leave or their route. Recently, simulation models allowed scientists to examine the roles that ocean currents might play in the timing and landfall of Fraser River sockeye salmon. These models indicated that compass orientation was sufficient to explain the migration of these fish (Dat et al. 1995) and that the variation in landfall and travel rate among years might be explained by the patterns of current variation that are associated with the interannual variation in temperature at sea (Thomson et al. 1992, 1994).

The conclusion that salmon have only a compass but no map may be consistent with conceptual (Blackbourn 1987) or simulation models (Dat et al. 1995), but this does not resolve the migration of populations such as Bristol Bay sockeye. Depending on their location prior to initiating homeward migration, they might migrate east (from the

FIGURE 2-9. Proportion of sockeye salmon returning to the Fraser River via Johnstone Strait, as opposed to the Strait of Juan de Fuca (data from International Pacific Salmon Fisheries Commission and Pacific Salmon Commission annual reports).

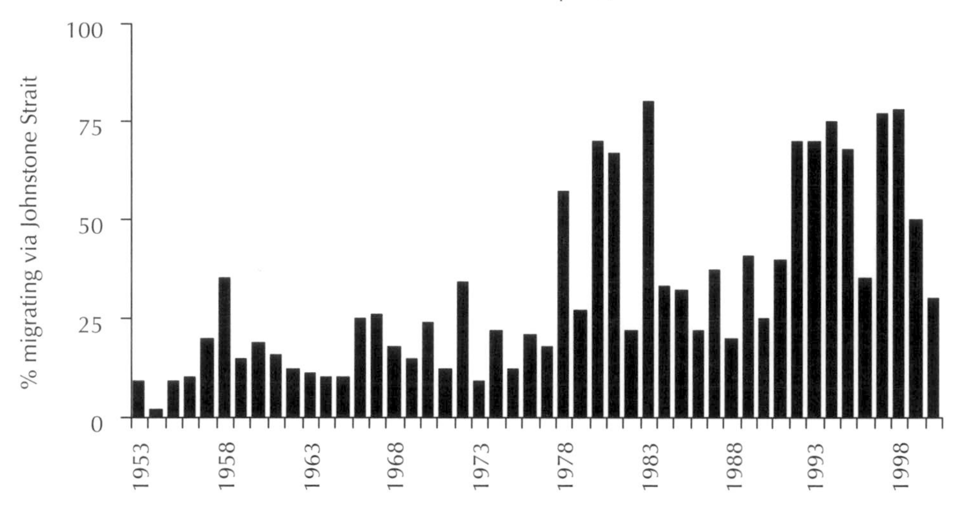

Bering Sea), north through Aleutian Island passes, and then east (from the central and western North Pacific), or west, then north, and then east (from the Gulf of Alaska; see fig. 2-2). It is difficult to explain this complex migration without some sense of location as well as direction. Fundamentally, we still have little direct information on the movement patterns and orientation mechanisms used by salmon on the open ocean. Almost all models and most empirical data pertain to a small number of numerically dominant stock complexes, notably Bristol Bay and Fraser River sockeye. The generality of these models and data is open to question. Are they characteristic or exceptional?

FIGURE 2-10. Relationship between the proportion of adult sockeye salmon migrating to the Fraser River via Johnstone Strait and the median arrival date of one of the main populations, Chilko Lake (data from the International Pacific Salmon Fisheries Commission and Pacific Salmon Commission).

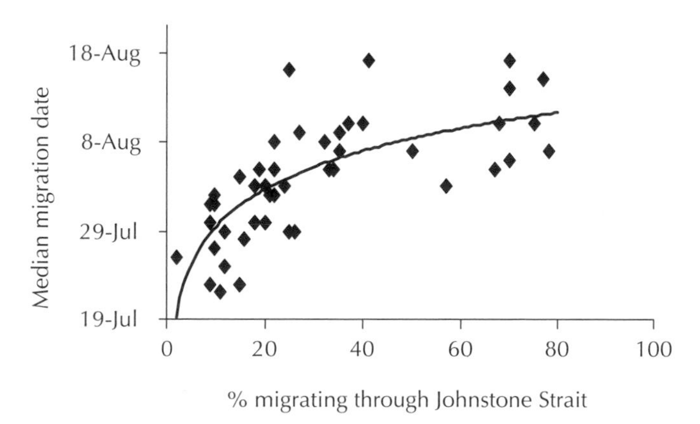

Summary

The ocean is used in common as a feeding ground for salmon from many locations in Asia and North America. The homeward migrations of maturing salmon at sea are characterized by a divergence of salmon experiencing common conditions to their respective home rivers. There is also a convergence on the home river of salmon that had been experiencing very different environmental conditions at sea, because the salmon originating from each freshwater location dispersed while they were at sea. Research fishing indicates nonrandom movement by salmon at sea as they head homeward. Travel rates of at least some salmon on the high seas are so rapid (40–60 km/d) that they seem implausible without invoking relatively accurate orientation because the fish tend to swim about 2–2.5 km/h. The timing of arrival in coastal waters of populations is precise, and this also seems evidence of orientation. However, salmon are affected by the regimes of temperature and currents during their feeding and homing migrations at sea. These physical features are correlated with arrival timing in coastal waters and approach routes, suggesting direct effects on movement (by currents) and indirect effects on the region where the salmon were feeding when they commenced homeward migration. Experiments conducted on juvenile salmon indicated that they orient their movements with respect to the apparent transit of the sun across the sky, the plane of polarization of light, and the magnetic field of the earth. The extent to which these clues are used by salmon to migrate on the high seas has been the subject of considerable speculation and modeling but the actual mechanisms of migration by salmon at sea are still poorly understood.

3

Migrations in Coastal and Estuarine Waters

The migrations of salmon in coastal and estuarine waters have been known to fishermen for centuries based on patterns of catch in space and time. Since the early 1900s the primary scientific tool for determining migration routes was application of tags on salmon in one area and recovery elsewhere, and this is still an important source of information. In the early period the objective of such studies was to determine the distribution patterns of maturing salmon from different regions, their homeward routes, and the patterns of interception. For example, Clemens et al. (1939) summarized 12 years of tagging, chiefly on chinook salmon, and reported a clear tendency for maturing salmon to migrate south along the coast (implying that they had migrated north as juveniles). These migrations included two salmon that returned to the Sacramento River after having been tagged more than 1200 km away. The authors also pointed out that salmon were vulnerable to fisheries for a long time and that 82% of the chinook salmon tagged in British Columbia were recaptured in the United States, indicating that "... migrations are definitely international in aspect" (52). This may seem obvious now but knowledge of salmon migrations was primitive at the time, so these were important discoveries.

Many tagging studies were conducted during the twentieth century, but the review by Anderson and Beacham (1983) of work on chum salmon migrating between Vancouver Island and the British Columbia mainland serves as an example. Most salmon were tagged in Johnstone Strait and recovered in fisheries in the Strait of Georgia and Puget Sound, the mouth of the Fraser River, and spawning grounds there and elsewhere. Such studies tend to underestimate the actual travel rate of salmon if we divide the total distance from release site to recapture site by the number of days at large (see also fig. 2-3) because the fish may have arrived at a given location prior to capture. Nevertheless, the data indicated that chum salmon populations migrated at different time periods and rates along different portions of the route. For example, Fraser River salmon traveled from upper to lower Johnstone Strait (about 110 km) at about 25 km per day but migrated slower (16 km/d) in the northern part of the Strait of Georgia and only 10 km/d in the southern part. Their migration from upper Johnstone Strait to the mouth of the Fraser River averaged 24 days, and they had completed their upriver migration to spawning grounds, spawned, and died in another 21 days. In contrast, chum salmon migrating to Puget Sound traveled through Johnstone Strait much more rapidly than the Fraser River populations but showed a comparable reduction in travel rate once they reached the Strait of Georgia.

A second example of migration patterns is provided by the sockeye salmon populations of Bristol Bay, Alaska, revealed by many years of research fishing and tagging (Straty 1975). Unlike the British Columbia chum salmon that negotiate a maze of inlets and islands for several hundred kilometers before reaching home, these salmon migrate to their home streams from the Bering Sea (swimming northeast, either north or south of the Pribilof Islands) or converge on passes between some of the Aleutian Islands from a broad distribution in the Gulf of Alaska and then migrate northeast. Bristol Bay sockeye do not follow the shoreline but rather are concentrated near the surface several hundred kilometers offshore, coming closer to shore (about 40–80 km) as the bay narrows. Straty (1975) concluded that the sockeye heading for five major fishing districts (centered around one or more lake systems) are initially fully mixed but gradually become more segregated as they leave the main migration corridor for their respective rivers (fig. 3-1). Catch rates in a test fishery operated by the Fisheries Research Institute about 200–300 km from the home rivers showed a distinct migration corridor (fig. 3-1). Recent analysis of age composition data from the FRI test fishery indicated that some segregation of populations takes place at this point along the migration (Flynn et al. 2003). As the salmon get closer to home they are progressively more segregated by river of origin, though some salmon "overshoot" their home river and have to backtrack.

Challenges to migration

The migrations in coastal and estuarine waters involve substantial changes from those on the open ocean in physical environment, physiology of the salmon, and migratory behavior and orientation. The primary physical changes are a transition from slow, gyral, oceanic currents to faster, reversing, tidal currents in coastal waters. Current speeds in the middle of the Gulf of Alaska are on the order of a few kilometers per day (see fig. 2-3), but in coastal waters they commonly approach or even exceed the typical swimming speed of salmon (about 2 km/h), so salmon movements in coastal waters are more affected by currents than they are at sea. There is also a transition from relatively cold, saline, oceanic water to warmer, fresher, and generally more stratified water in coastal regions and estuaries. Eventually, the reversing tidal currents and stratification in the estuary give way to a unidirectional current and fully freshwater as the fish migrate upriver.

Along with the changes in the characteristics and movement of water, coastal and estuarine waters confront salmon with complex shorelines composed of myriad islands, coves, bays, and points of land. For salmon that have been tens, hundreds, or thousands

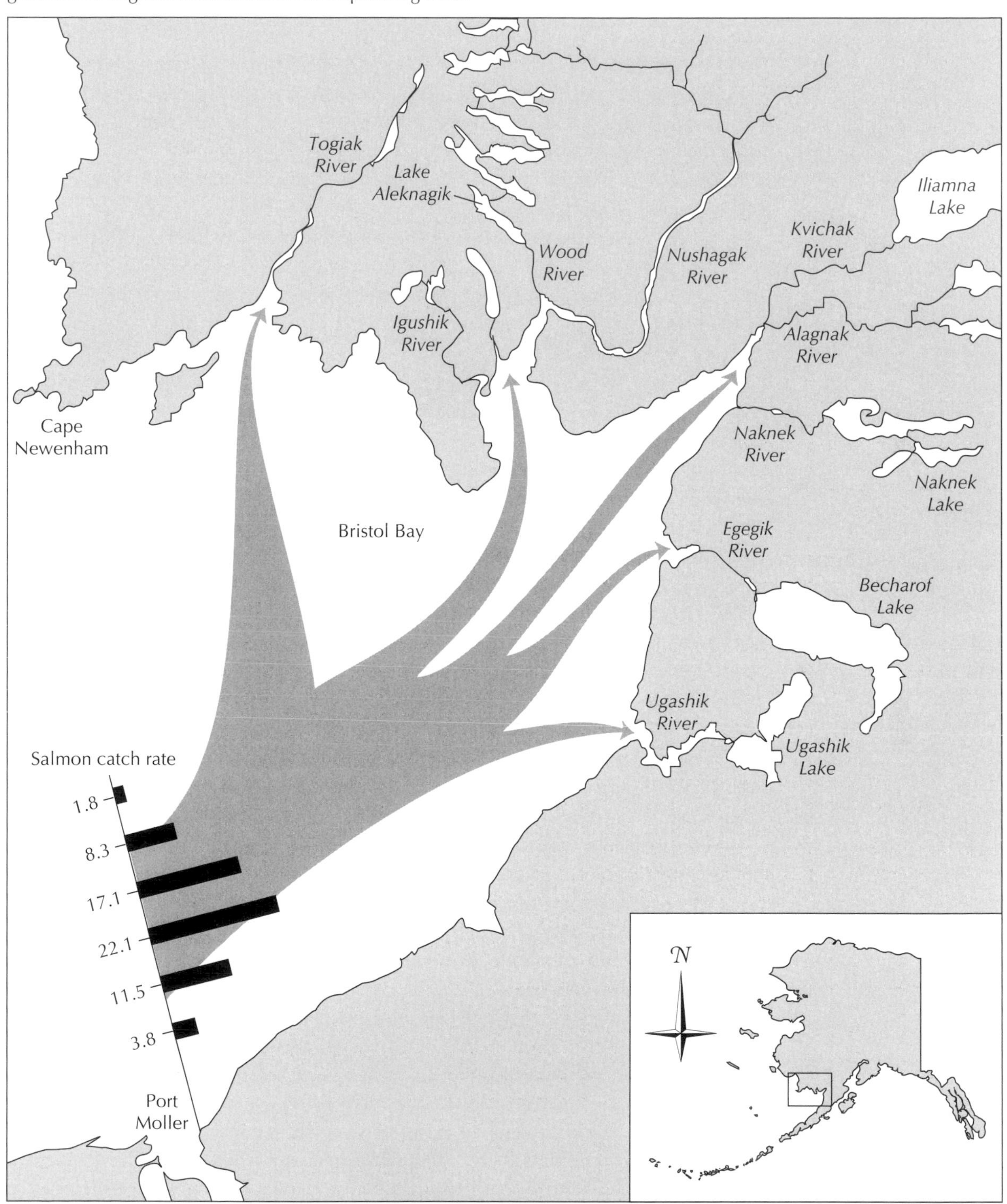

FIGURE 3-1. Map of Bristol Bay, Alaska, showing the major river systems producing sockeye salmon. The bar graph and associated numbers indicate the catch rates of maturing sockeye (fish per hour per 100 fathoms of net) at research stations along a transect offshore from Port Moller (Fisheries Research Institute data, reported in Quinn 1991). The shading indicates generalized migration routes to the main spawning areas.

of kilometers from land and hundreds or thousands of meters from the bottom of the ocean for 1 or more years, this must be an abrupt transition. It would seem impossible for individuals to retrace the route they took on the outward journey, and indeed they do not (see below). When they encounter land, how do they know if it is a small island that they can easily swim around or the tip of a long peninsula? Are they in a small cove or a huge inlet? In addition to the physical structure of land, the water movements are very complex. Tides do not simply rise and fall but rather they pulse with the varying strengths of the lunar cycle, freshwater sources, and wind, causing currents to eddy as they wrap around the land.

The changes in physical environment are accompanied by three important and complex biological changes. One of the most noteworthy is the cessation of feeding. The exact point when salmon stop feeding is not well defined, and it varies greatly among populations, depending on the chronology and route of their homeward migration. Some populations may spend many months in freshwater prior to spawning (e.g., summer steelhead and spring chinook salmon) and they may feed until they leave the ocean and also feed to a limited extent in rivers. At the other extreme, some chum salmon spawn within a few days of entering small coastal streams and they probably stopped feeding weeks earlier. The starvation does not result from an absence of food (though there is certainly far less for salmon to eat in freshwater compared to the ocean) but it is a form of "anorexia" (Kadri et al. 1997). Salmon not only cease feeding but they experience major endocrine changes associated with maturation during the homeward migration (see review by McKeown 1984). Work on Fraser River sockeye by Truscott et al. (1986) and by Ueda et al. (1991) on Japanese chum salmon showed that there is a sequence of hormonal changes that is appropriate for the pattern of holding in coastal waters, duration of migration upriver, and holding prior to spawning. Thus not all salmon leave the ocean in the same internal state, just as the extent of color changes signaling maturation can vary greatly among populations.

The third physiological transition is the osmotic one, from salt- to freshwater. The vast majority of fishes spend their entire lives in either salt- or freshwater. The euryhaline fishes, able to either tolerate a range of salinities, and the diadromous fishes are the exceptions. The large size of adult salmon (i.e., low surface to volume ratio) gives them some protection from rapid osmotic changes, but they must undergo complex changes in drinking, urination, chloride cells in the gills, and kidney function to maintain the appropriate internal osmotic concentration for physiological processes (about 10 parts per thousand salt) in sea water (about 33 ppt) and freshwater (nearly 0 ppt). In freshwater, the fish do not drink and excrete dilute urine, and they also must actively pump ions from the water across their gills.

The complexity of the migration in coastal waters results from not only the intricate geography of islands and points of land, and rapidly changing physiology of the fish, but also from the three roles played by water itself. First, water is the medium surrounding the fish, and it has direct physiological effects on them. Temperature, salinity, and dissolved oxygen are all properties of water that vary in coastal and estuarine systems and fish have responses ranging from preference to tolerance, avoidance, and death. Second, the water is an increasingly important component of locomotion. Do the salmon drift passively, selectively ride favorable tidal currents, or migrate with respect to the land and not modify their behavior when the tides change? Third, the water is the source of critical guidance information that is needed when compass orientation becomes useless in the final stages of migration. Details will be provided in chapter 5 but for now let it suffice to say that juvenile salmon learn odors at one or more locations and stages of their lives in freshwater, and they are attracted to these odors as maturing adults. Given the thousands of rivers along the coasts of North America and Asia that have salmon, this is no small task. There may be complex interactions between physiological responses to water and the challenges of orientation. Temperature and salinity vary horizontally and vertically, and the freshwater, layered on the surface, may contain the odors of the natal stream as well as other properties that may attract or repel salmon. Salmon may show a progressive attraction to freshwater but must also avoid all rivers except their home stream. How can salmon possibly find their way through this maze of conflicting signals and unfamiliar landmasses? I will summarize detailed studies on the migrations of Fraser River sockeye salmon to illustrate some of the hypotheses regarding migration in coastal waters.

Fraser River sockeye salmon

Tagging studies by Verhoeven and Davidoff (1962) indicated that many salmon tagged at Sooke, at the southern end of Vancouver Island, were recaptured at the mouth of the Fraser River (about 200 km away) within a week. However, there was a great deal of variation in timing, and some salmon were not recovered at the mouth until a month later. Interestingly, the early (July) arrivals approached the river much faster than those arriving in August (table 3-1). It subsequently became clear that the earliest migrants would swim about 1000 km upriver to the Stuart Lake and Bowron Lake systems (fig. 3-2) and spawn in late July. They migrate rapidly through marine waters, and also migrate rapidly upriver (see chapter 4). For reasons that are not clear, the populations that spawn later do not remain on the open ocean but rather return to coastal waters and move back and forth in the Strait of Georgia for about a month before migrating upriver.

This mix of populations, with very different destinations, schedules, and patterns, makes descriptive work confusing to interpret. This does not necessarily mean that the

TABLE 3-1. Median travel rates of sockeye salmon from Sooke to the mouth of the Fraser River (about 200 km), based on tagging in July (474 salmon recovered) and August (491 recovered) from 1938 to 1945 (Verhoeven and Davidoff 1962).

	% traveling at that rate	
Travel rate (km/d)	*July*	*August*
57.1	48.7	25.3
19.0	28.7	21.4
11.4	7.8	10.6
8.2	7.0	8.4
6.9 or slower	7.8	34.4

FIGURE 3-2. Map of the Pacific Northwest, showing some of the major lakes producing sockeye salmon in the Fraser River and some of the major tributaries of the Columbia River.

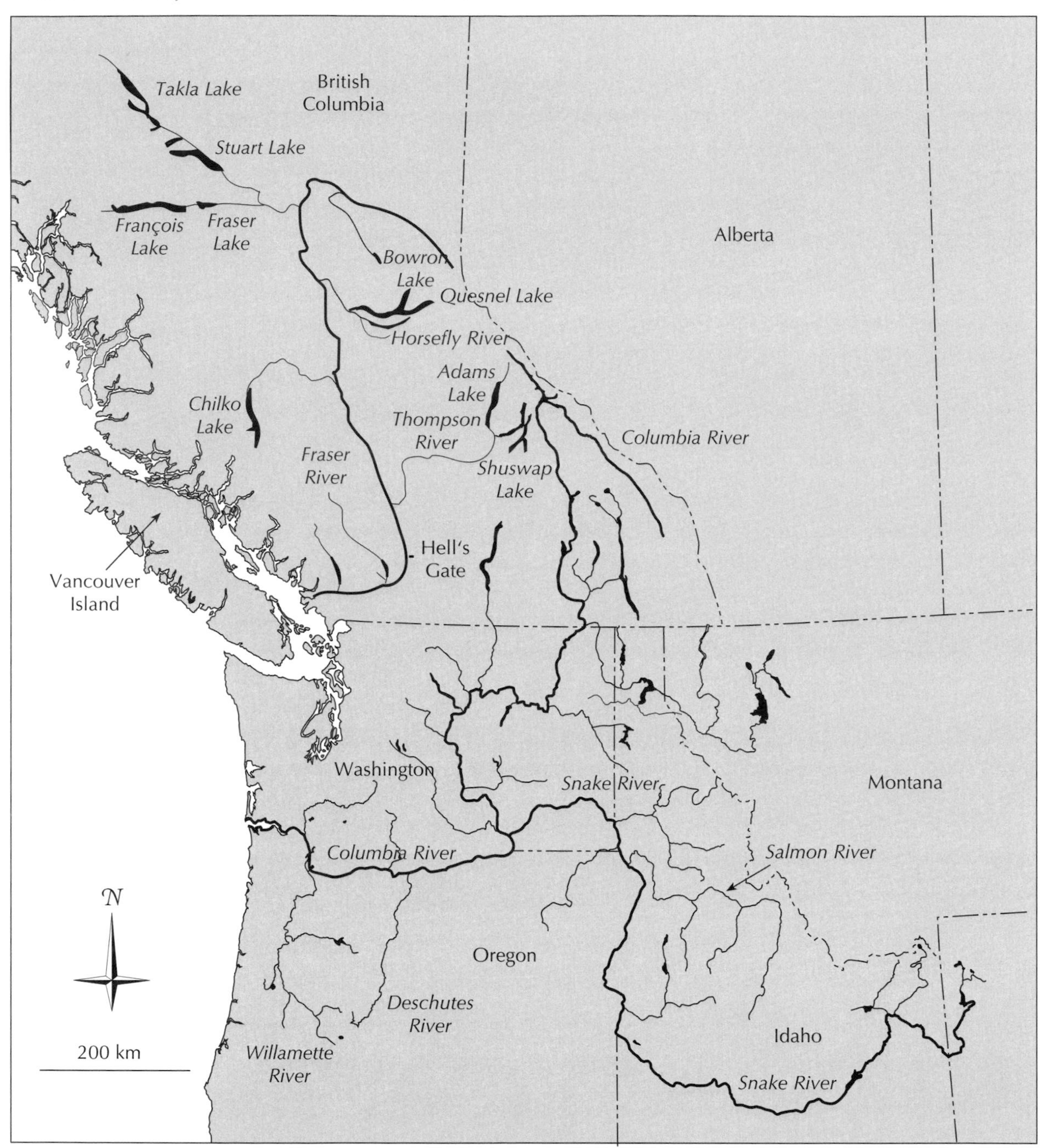

fish are confused. Indeed, there are remarkable patterns in their migratory routes. For example, Verhoeven and Davidoff (1962) tagged sockeye approaching the Fraser River from Johnstone Strait and from the Strait of Juan de Fuca. They recovered sockeye at four areas at the river's mouth. Fully 85% of the fish that had approached via the southern route were caught in the two southerly areas whereas only 53% of those that had approached from the north were taken in the southern areas. Thus despite all the opportunities for mixing during the time spent in the Strait of Georgia, somehow the fish using the two approach routes maintained some degree of separation.

Conventional mark-recapture work provides a modest amount of information on a large number of fish but provides no details on their movements between capture and recapture locations. Tagging can be complemented by telemetry, technology that provides a great deal of data on a few fish. In the 1970s, radio and ultrasonic transmitters came into wide use. Ultrasonic transmitters are battery powered, sending signals at acoustic frequencies (about 60–80 kHz), and operate equally well in fresh- and saltwater but must be detected by a hydrophone in the water. In contrast, the radio transmitters used on wildlife and fish in freshwater operate at much higher frequencies (e.g., 152 mHz). Radio signals travel well in air (and so can be detected from antennae on land, in vehicles, and from the air) but do not travel through saltwater and so cannot be used at sea except for animals that surface, such as marine mammals.

Some of the pioneering ultrasonic telemetry with salmon was conducted on Fraser River sockeye. Salmon were caught, fitted with transmitters, and followed in the waters around the San Juan Islands (Stasko et al. 1976). Some sockeye were active, swimming about 1 body length per second. However, the directions of active movement were in both the homeward and opposite directions, and fish released from the same location did not always move along the same paths. In addition, some "passive" salmon seemed to drift with the prevailing tidal currents. Madison et al. (1972) and Groot et al. (1975) found similar patterns in sockeye salmon off the Skeena River. The authors of both studies seemed predisposed to find evidence of orientation to celestial and geomagnetic clues, but the movements of the salmon were so complex and variable that they could neither conclusively support nor refute hypotheses about guidance mechanisms.

These studies on salmon and similar ones on other fishes led Leggett (1984, 159–160) to criticize the field of fish migration research as having ". . . an excessive fixation on the concept of precision of oriented movement by individual fish [and a] paucity of generalized predictive theory, developed *a priori*, which defines testable hypotheses. . . ." There is much merit in this argument. Maps showing the tracks of salmon (and other fish in estuaries) often resemble a plate of spaghetti rather than the straight lines that might be expected based on the legendary homing ability of these fishes. How can we reconcile the apparently confused movements of salmon in coastal waters with the rapid and apparently directed movement on the ocean and the precision of homing to natal streams?

To shed further light on the movements of Fraser River sockeye and, by extension, the general salmon movements in coastal waters, several colleagues and I conducted a large-scale tracking study along the northern route in 1985 and 1986 (Quinn and terHart 1987; Quinn 1988; Quinn et al. 1989). Like the earlier work, sockeye were caught, transmitters were inserted into their stomachs, and they were released and followed for as long as possible. The boat tracking the fish was followed by a second boat, from which we took periodic profiles of temperature and salinity and also deployed current drogues to determine the speed and direction of the water in which the salmon were swimming. There are virtually no sockeye salmon populations migrating along this route other than Fraser River fish, so we could assume that all our salmon were heading there. This assumption would be invalid for other species of salmon in these waters, and it greatly simplifies interpretation of the results.

The horizontal movements shown by the sockeye can be viewed at several spatial scales. At the simplest scale one might calculate the direction of gross movement, based on where the fish were released and where they were either lost or abandoned. Such analysis indicated that most fish moved in the general direction of the Fraser River (roughly southeast), but some moved in the opposite direction. Very few fish moved perpendicular to the homeward axis, and those that did so were only tracked for brief periods. Many individuals moved in the homeward direction (figure 3-3 shows two of the better examples), but in some cases fish tracked in the same location moved in the opposite direction. Examination of longer tracks revealed some fish that initially moved away from the homeward direction later changed direction and began swimming homeward.

These movements, over the ground, were the resultant of two vectors: the speed and direction of the salmon's swimming and the speed and direction of the water mass in which they were swimming. Using data on water currents from the drogues it was possible to estimate the speed and direction of the fish themselves. The currents, not surprisingly, run roughly northwest-southeast, parallel to shore, and can be quite swift. The average salmon swimming speed, relative to the water, was 67 cm/sec (Quinn 1988), similar to the 1 body length/sec documented in the lab by Brett (1983) as their most efficient swimming speed. However, even after considering the effects of currents it was clear that some salmon swam in the opposite direction from home. Indeed, one individual swam northwest for hours into a flooding tide that would have swept it rapidly homeward.

Our conclusions about the horizontal movements of these salmon came from a combination of our own data, conversations with commercial fishermen, examination of data from test fishing, and a simulation model. Our data showed that most fish swam in the homeward direction after release but some swam in the opposite direction. Those fish tracked long enough tended to turn, usually clockwise, in open water and eventually moved in the homeward direction. The commercial fishermen we spoke with claimed that salmon were not found in open water but migrated exclusively along shorelines. We were skeptical and so we asked (and paid) them to set the purse seine net offshore in open water (photo 3-1). They typically caught only a couple of fish, and their reaction was, "See? No fish!" My response was, "See? Fish!"

From the fisherman's perspective there were no salmon migrating in open waters because it was inefficient for them to set the net and catch at most a few salmon. Indeed, in 30 sets made offshore in Queen Charlotte Strait and upper Johnstone Strait in 1985 and 1986, only 184 salmon were caught. Seven sets caught no sockeye, 17 sets caught 1–9 fish, 6 caught 10–49, and none caught more than 50 (K. Cooke et al. 1987). However, integration of the area encircled by a purse seine (12,820 m^2, or only a little more than a hectare) over the 10 minutes that the net was fishing indicates that in fact many salmon were migrating homeward in open water. The 70 sets made onshore caught 10,066 salmon. However, this total was skewed by 22 sets than caught more than 50 fish each, for 95% of the total catch. There were still 27 sets (39%) that caught fewer than 10 salmon. Thus most of the water is occupied by a scattering of salmon and they are truly concentrated in only a few points of space and time. The places where fishermen tended to fish (and sometimes catch hundreds or thousands of salmon in a set) were islands and coves, especially ones open to the northwest. These seem to be places where salmon encountered land, and became concentrated.

FIGURE 3-3. Horizontal movements of a Fraser River sockeye salmon tracked in Johnstone Strait, (top panel) and one tracked in the Strait of Georgia (bottom panel). The inserts show the proportion of time the fish spent at different depths and the average temperature (T) and salinity (S; in parts per thousand) at those depths (from Quinn and terHart 1987).

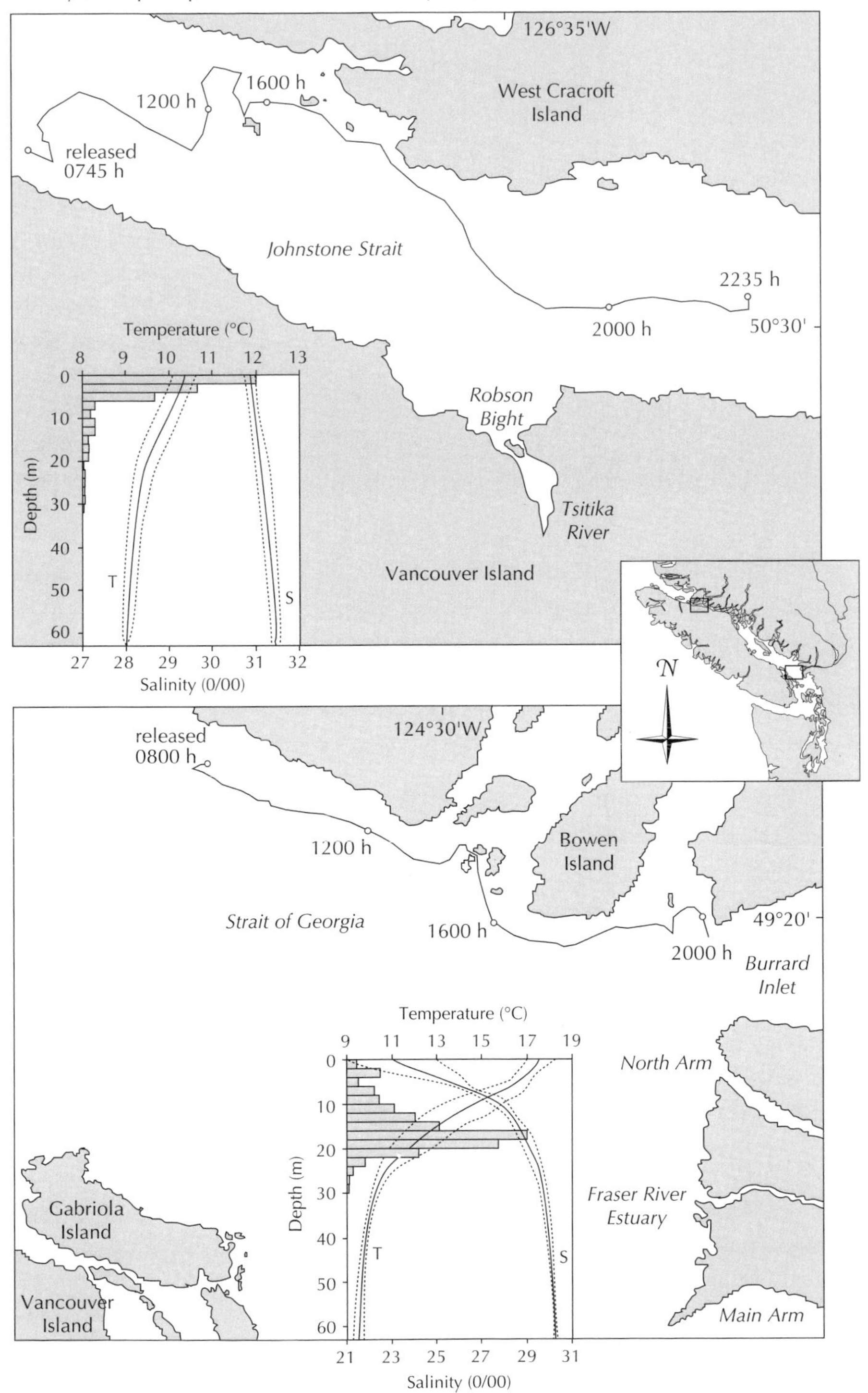

3-1 Purse seine vessel used for research on Fraser River sockeye salmon migrations. Photograph by Graeme Ellis, Department of Fisheries and Oceans, Canada.

When salmon encountered land, their response was not to follow the shoreline. Given the myriad islands and incredibly complex coastline, that strategy would be highly inefficient or ultimately unsuccessful. Rather, they tended to turn around, head in the opposite direction for a few hours, turn back to the southeast and "take another run at it." When caught by the purse seine the salmon sometimes reacted as though they had encountered land; they moved in the opposite direction (northwest) and eventually turned around and resumed homeward migration. These behavior patterns also explain the discrepancy between observed swimming speeds and net travel rates in the area. The salmon actually swam about 2–2.5 km/h. Even accounting for the reduction in swimming at night, they might travel about 40 km/d. Tagging studies indicate that some salmon do so (Verhoeven and Davidoff 1962) but most do not. I believe that all the sockeye salmon have a southeast orientation and some are lucky and miss many of the islands and peninsulas, like a pinball rolling down the slope and missing the bells and bumpers. Other salmon encounter land and their retrograde movement results in slower net travel rates than they accomplished on the open ocean, even though they swim no slower.

Simulation modeling (Pascual and Quinn 1991) supported this "pinball" hypothesis. We started hundreds of computer-generated salmon on their homeward journey to the Fraser River from Queen Charlotte Strait; we programmed individuals to swim in different compass directions and also programmed a range of responses when the fish encountered land. The success of the simulated behavior patterns was evaluated by the proportion of salmon that reached the mouth of the Fraser River and by the length of time it took them to get there. Directional orientation of 140–155° produced the highest proportion of salmon reaching home, and a direction of 120–145° produced the fastest average speed. A strong directional orientation (i.e., steadfast tendency to swim in the prescribed direction) caused many salmon to fail to reach home, though the successful few got home rapidly. The rest became trapped in the many long inlets along the mainland coast and never emerged. The salmon that tended to escape these "traps" in the simulation were those that were programmed to move randomly or in the reverse direction for several hours after they encountered land before resuming homeward direction. However, no combination of behavioral responses by the simulated salmon was adequate to match the travel speed and success of real salmon. The salmon may avoid freshwater other than home; such a response would keep them from becoming trapped in inlets and estuaries and would speed their homeward migration.

One of many open questions about salmon migrations in coastal waters is whether they take advantage of the reversing tidal currents. A number of fish species migrating in shallow coastal areas show such "selective tidal stream transport." They move up in the water column and swim actively when the current is running in the direction favoring their travel, and they move to the bottom where currents are slower when the tides change rather than fight the current. In contrast, salmon seem to swim in their chosen direction without regard to the direction of tidal currents (Atlantic salmon: G. W. Smith et al. 1981; sockeye: Quinn et al. 1989). It may be that the water is too deep for them to efficiently find velocity refuge by moving to the bottom. However, horizontal movement toward shore would achieve the same result yet they did not consistently do so. Alternatively, the salmon may have no way of knowing that the tide has changed. For fish in open water, a change of tide does not mean that they are immediately exposed to different (e.g., less salty) water because the tide affects the whole water mass in which they are moving. It is not obvious how a fish could detect a change in the movement of such a water mass without a fixed visual reference unless the fish had some complex inertial guidance or other navigation system. Our findings contrasted with the "common knowledge" of recreational and commercial fishermen that salmon catches are related to tidal patterns. Salmon have been reported to move up in the water column to take advantage of flooding tides when ascending the estuarine reaches of rivers (e.g., sockeye salmon in the Fraser River: Levy and Cadenhead 1995). The responses of salmon in coastal and estuarine waters to tides are a fascinating subject needing further study.

In addition to the patterns of horizontal movement, the telemetry revealed the average depth of travel and vertical excursions. In Queen Charlotte and Johnstone straits the sockeye spent most of their time in the upper 10 m of the water column (e.g., fig. 3-3 top), though the water was typically several hundred meters deep. These areas are dominated by cold, saline, well-mixed water, as shown in the figure panel. Therefore, the fish would experience similar temperatures and salinities regardless of their depth of travel. However, when they moved into the Strait of Georgia they spent less time near the surface. In this region, the Fraser River's discharge dominates the upper part of the water column (because freshwater is lighter than saltwater) and so the upper 15–20 m are less saline and, in summer, warmer than the deeper waters. The average depth of the sockeye was deeper, not because the maximum depth increased (it did not), but because less time was spent near the surface (e.g., fig. 3-3 bottom).

The stratification of water in coastal areas and estuaries compels salmon to experience some combination of temperature and salinity at any given depth (for example, the chinook salmon tracked in the estuary of the Columbia River; photo 3-2; fig. 3-4).

3-2 Gillnetting chinook salmon in the Columbia River estuary. Photograph by Alan Olson, R2 Resource Consultants.

Did the salmon avoid the surface because it was too warm or not sufficiently saline? The tolerance of the fish for freshwater is presumably increasing during the final phase of the marine migration and the vertical distribution may reflect optimization of the environment for physiological reasons. From the perspective of orientation, the fresh surface water would contain the odors of the home river and would be moving seaward, relative to the colder, saltier layer below it. Westerberg (1982) hypothesized that the vertical movements of salmon in stratified waters allow them to detect fine structure of the current fields at the interface between fresh, odor-rich waters near the surface and colder, saltier, nonhome waters below. In addition, there may be some heightened sensitivity to the odors as the fish move in and out of the odor field, much as one smells the bread in a bakery most strongly upon entry. Døving et al. (1985) presented evidence supporting this hypothesis. Atlantic salmon made numerous small-scale vertical movements near the thermocline but salmon experimentally deprived of their sense of smell moved up and down through the whole water column, as if searching for familiar odors.

Any hypothesis regarding the role of vertical movements in orientation (or physiological regulation) must consider differences in the depth of travel by salmon of a given species over the seasons and in different regions. Early in the season, chum salmon off the coast of Japan avoided warm surface waters and traveled to depths below 100 m but later in the fall they were near the surface (Ueno 1992; Tanaka et al. 2000). In addition to variation in depth distribution within species, there are also differences among species. Fishermen know that coho salmon are caught closer to the surface than chinook, and test trolling in the Strait of Juan de Fuca revealed clear differences in depth of capture among species. Coho were closest to the surface, pinks and sockeye were at intermediate depths, and chinook were the deepest (table 3-2).

These results are consistent with ultrasonic tracking in Johnstone Strait, where chinook swam much deeper than sockeye (Candy and Quinn 1999). The chinook also lacked the

FIGURE 3-4. Vertical distribution of a fall chinook salmon tracked in the Columbia River estuary and the ambient temperature (T) and salinity (S; in parts per thousand) in the water column (modified from Quinn 1990b, data from Olson and Quinn 1993).

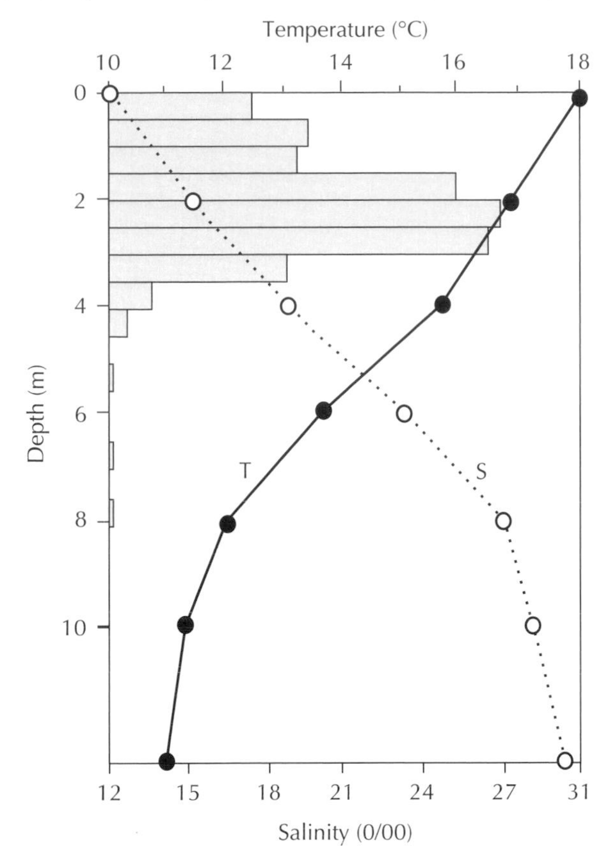

pronounced reduction in activity at night shown by sockeye (Madison et al. 1972; Quinn et al. 1989), suggesting that they may be better adapted for low light conditions. Steelhead stayed extremely close to the surface and greatly reduced their activity at night (Ruggerone et al. 1990). Interpretation of these results is limited, however, by the lack of coordinated studies on all species in the same area at the same time. A comparison between chum salmon in Japan, chinook salmon in Johnstone Strait, and steelhead in Dean Channel, all studied in different years, is obviously flawed.

The final stage of migration before salmon enter their natal river is often in an estuary. In these waters, the salinity is markedly reduced by riverine inputs. In some cases the salmon swim right through the estuary and upriver, whereas other populations in the same or another river may delay for weeks. As with coastal waters, estuaries are often areas of heavy fishing, and so the movements and duration of residence by salmon greatly affect their chances of getting caught. The salmon are likewise presented with complex regimes of salinity, temperature, and odors, and these factors may affect their vertical distribution (e.g., fig. 3-4). In addition to the physiological constraints and need to orient to the odors of the home river, the shallowness of the estuary may constrain the salmon's vertical movements much more than the deeper coastal waters. Some tracking

TABLE 3-2. Catches (percentage of total for each species) of Pacific salmon caught by trolling at different depths in the Strait of Juan de Fuca from June to October of 1967 and 1968. Sample sizes: 1369 coho, 81 sockeye, 561 pink, and 916 chinook salmon (Beacham 1986).

Depth range (m)	*coho*	*sockeye*	*pink*	*chinook*
0 to 9.0	28.1	9.9	7.3	2.2
9.1 to 18.3	26.3	12.3	16.9	6.6
18.3 to 27.4	19.6	32.1	28.3	14.6
27.4 to 36.6	15.4	28.4	26.9	29.1
36.6 to 45.7	6.3	8.6	11.6	13.0
45.7 to 54.8	4.2	8.6	8.9	34.5

studies show considerable to-and-fro movement, as though the fish are strongly affected by tides during a period of either disorientation or lack of motivation or readiness for upriver migration (e.g., Groot et al. 1975; Olson and Quinn 1993).

Summary

Salmon migrations in coastal and estuarine waters are characterized by changes in physical habitat, migratory behavior, and internal processes, relative to migrations in the open ocean. The transition from oceanic to estuarine environments entails (1) changes from slow, gyral, oceanic currents to faster, reversing, tidal currents; (2) changes from colder, oceanic water to warmer, fresher, and more stratified water; and (3) movement from an open environment to one characterized by physical obstacles (islands, bays, reefs, and points of land). The key biological changes are a reduction and eventual cessation of feeding, activation of hormones associated with gonadal development, and preparation of the osmoregulatory system for entry into freshwater. Changes in migratory behavior are required to meet the environmental changes of obstacles, reversing tidal currents, and complex gradients of temperature and salinity. The fish must have a combination of innate and learned responses to guide them through waters that they may or may not have directly experienced, from the open ocean to their home river. Map and compass orientation seem to give way to olfaction-based migration in rivers, but the mechanisms used in this "overlap" zone are unclear. Horizontal movements seem to be guided by compass orientation, coupled with the flexibility to reverse directions when encountering land. The vertical distribution patterns and movements may reflect the efforts of the salmon to optimize physiological processes through behavioral regulation of temperature and salinity. However, the salmon may also move up and down through the thermo/halocline to obtain guidance information from the seaward movement of the surface waters, relative to the deeper waters. The study of salmon in coastal and estuarine waters is of great practical as well as theoretical importance because the vast majority of fisheries take place in these regions. However, research is hampered by the complex factors motivating fish movements, the complex physical oceanography of these waters, and the complex population structure (hence route and schedule of migration and spawning) of salmon sampled in many coastal areas.

4

Upriver Migration and Energetics

Pacific salmon may begin upriver migration at every month of the year if one includes all species and populations, but the patterns for individual populations are much more predictable. There is a chain of events starting with adult salmon leaving the rich food resources of the marine environment and entering freshwater, migrating upriver, and spawning, and the chain culminates the following spring with the emergence of their offspring. The timing of each event is in large part an evolved (i.e., genetically controlled) set of responses in the ocean, estuary, and migratory corridor that will take the salmon to the spawning grounds upriver. Unlike many freshwater fishes, whose migration and reproduction are strongly influenced by environmental conditions such as temperature, flow, or rainfall, in salmon these events are more strongly controlled by genetic factors, as adaptation to the long-term average conditions that prevail in the waters affecting each population (Quinn and Adams 1996). From the ocean, the salmon cannot assess conditions in the river, so selection favors salmon migrating at the right time. The right time, given typical conditions, will allow them to reach the spawning grounds with enough energy to spawn at a date that, given the river's typical temperature and flow regime, will allow their offspring to develop and emerge in time to feed and grow in spring.

Processes leading to complete sexual maturation in salmon begin many months before the fish reproduce. In addition to the long period of adult maturation, salmon have unusually protracted periods of embryonic development compared to most fishes. Finally, the spawning grounds may be very distant from the mouth of the river. Thus the two ends of the chain—initiation of maturation and entry into freshwater and the emergence of offspring—are greatly separated in time and often also in space (Quinn and Adams 1996). As a consequence of this temporal and spatial separation, conditions (e.g., flow and temperature) experienced by adults as they enter a river are at most weakly

TABLE 4-1. Counts of adult salmonids (percentage of the annual total) migrating up the Columbia River at Bonneville Dam, 1939–1948 (U.S. Army Corps of Engineers, Portland, Oregon, annual fish passage reports). These records should not be taken to represent the natural patterns, as significant habitat modification and reduction in population structure had taken place by 1939, and salmon spawning below the dam were not counted.

	chinook	*sockeye*	*coho*	*steelhead*
January	< 0.1	0.0	< 0.1	0.1
February	< 0.1	0.0	< 0.1	0.2
March	0.1	< 0.1	< 0.1	1.2
April	8.4	0.1	0.0	4.4
May	8.6	0.3	0.0	1.9
June	4.1	27.2	0.0	1.5
July	3.9	70.3	< 0.1	21.1
August	10.0	2.1	10.1	30.9
September	63.8	< 0.1	87.9	36.1
October	1.1	< 0.1	1.5	2.0
November	0.1	0.0	0.3	0.4
December	< 0.1	0.0	0.1	0.2
total	3,738,510	784,867	83,307	1,306,966

connected to the conditions that will affect their offspring many months later and perhaps hundreds of kilometers away. Thus they need genetic control over timing to ensure that reproduction occurs at an appropriate date. This control is evident from the ease with which selective breeding can shift the timing of reproduction (e.g., Siitonen and Gall 1989). Transplanted populations tend to retain their ancestral timing rather than adopt the timing of their new site (Ricker 1972), and experimental breeding studies have demonstrated genetic control over the timing of migration and breeding among and within populations (e.g., Smoker et al. 1998; Quinn et al. 2000).

Given these facts, the diversity of migration into freshwater may be considered among species and among populations within species. There are consistent differences in timing among species within particular rivers, as illustrated by counts of salmon migrating up the Columbia River at Bonneville Dam (table 4-1). Early runs of chinook enter in significant numbers in April. Sockeye come next, also before the river temperatures peak, followed by steelhead and then the main runs of chinook and coho. The river also has chums entering in November but they are essentially all below the dam and so not counted here. The comparative patterns of migration and spawning among species have received little attention but there are many fascinating aspects, reflecting in part the constraints of accessing the suitable spawning grounds and the temperatures experienced by embryos during development. For example, chinook often spawn in large rivers where there is always enough water, so they enter before the coho, whose smaller tributaries may have inadequate flow until later in the fall. Coho compensate for this delay with more rapid development rates of their embryos (see chapter 8). There are also many puzzles.

4-1 Female chum salmon struggling up a small stream near Juneau, Alaska, on an ebb tide. Photograph by Thomas Quinn, University of Washington.

In southeast Alaska, chum and pink salmon often spawn together, with chum sometimes coming first, whereas in Puget Sound pink salmon normally precede chum and often spawn in larger rivers. Generally speaking, coho tend to spawn late, chinook, sockeye and pink tend to spawn early, and chum salmon can be quite variable.

All other things being equal, salmon should stay at sea as long as possible before spawning to maximize growth because rivers are poor places for salmon to grow, especially if they arrive at high densities. The growth at sea can be converted into reproductive potential (number and size of eggs by females and competitive ability for both sexes). There is natural mortality at sea but the rate is presumably comparatively low for adult salmon because they are large (see chapter 15). Many pink and chum salmon populations spawn very close to the ocean or even in the intertidal zone. They typically enter freshwater in a fully mature state (photo 4-1), spawn within a few days, and die in another week or two. Upriver migration to distant spawning locations (hundreds or even thousands of kilometers) obviously takes time, but some populations enter freshwater many months prior to spawning and would not need nearly this much time to simply migrate upriver. Notably, summer steelhead often enter rivers in summer and early fall but do not spawn until late the following spring. Table 4-1 shows many steelhead migrating from July through September. These steelhead are also termed "stream maturing" because they enter freshwater in a relatively immature state and complete maturation in freshwater (Busby et al. 1996). The alternate form (winter or "ocean maturing" steelhead) enters rivers in late winter (e.g., March to May) and spawns shortly thereafter.

The ocean-maturing (winter) steelhead of the Columbia River spawn almost exclusively in tributaries below Bonneville Dam, in the coastal region, and they are the prevalent form along the coast from California to southeastern Alaska (Busby et al. 1996).

These rivers have predictably high flows in winter and the steelhead can migrate shortly before they spawn. It seems likely that the stream-maturing steelhead sacrifice growing opportunities at sea and migrate in late summer in order to access suitable spawning grounds that could not be reached shortly before the spawning season. The date of spawning is fixed because prevailing temperatures will control when fry will emerge (see chapter 8) and the emergence date of fry must be optimized. Thus the premature migration is a case of "making the best of a bad situation" rather than the ideal chronology of migration and spawning. There has been no comprehensive assessment of the particular patterns of flow and temperature that might retard or preclude migration in a range of rivers to test this hypothesis, but certainly some rapids and waterfalls are passable only at certain flows, and temperatures can also block migration. Radio tracking indicates that stream-maturing steelhead may migrate hundreds of kilometers inland but spend the winter in larger rivers rather than on the actual spawning grounds (Lough 1983). Small streams in the continental interior may be too cold or have too little water for the adults so the fish cease migration and seek suitable habitat for holding until conditions change in spring and they complete maturation and move to the spawning grounds.

This pattern of premature migration is displayed by other species of salmon, notably chinook, which are often classified as spring, summer, and fall, depending on the timing of entry into freshwater. These populations all spawn at about the same time (fall) but differ in duration of holding in freshwater prior to spawning. As with steelhead, the populations that enter freshwater early tend to be found in the interior (e.g., mid and upper Columbia and Snake rivers) and in the upper reaches of rivers closer to the ocean. Some of these areas where spring chinook spend the summer are quite warm, and there is a metabolic cost to the fish because they are expending energy to stay alive in freshwater and are not feeding, unlike their fall-run counterparts who are still at sea, feeding and growing. Spring chinook minimize their energy expenditures by seeking pockets of cool water and thereby lowering their metabolic rate (Berman and Quinn 1991). Some rivers such as the Snake have spawning populations of both fall and spring chinook. In such cases the fall chinook tend to spawn in the main stem of the river and so can migrate from the ocean directly to the spawning grounds in late summer, whereas the spring chinook spawn in smaller streams that might be inaccessible to them in late summer and so the chinook must enter earlier than might otherwise be needed.

Chinook salmon have an especially great range of migration and spawning dates and are unique among salmon in having populations that spawn in spring as well as in fall. In the Sacramento River system (table 4-2), winter chinook enter in winter and spawn in the upper river from April to July (Slater 1963). This unusual pattern seems to have evolved to accommodate both the very warm temperatures in the lower river that the fish had to migrate through and the cold temperatures where they spawn. The salmon displaying this pattern were apparently confined to the region cut off by Shasta Dam in 1945; the population has persisted but is currently listed as endangered under the U.S. Endangered Species Act.

Sockeye salmon tend to spawn in tributaries of lakes, and in the southern end of the range these lakes (e.g., Lakes Quinault and Ozette on the Washington coast, Baker and Washington in Puget Sound, and several lakes on the west coast of Vancouver Island) can get quite warm. Sockeye migrating to the spawning grounds via the lake in late summer

TABLE 4-2. Characteristic peak timing of life-history events for the seasonal runs of chinook salmon in the Sacramento River system, California (Fisher 1994). Spring chinook (*) apparently enter the ocean in two periods: March–June and November–March.

	Seasonal run			
Characteristic	*Late fall*	*Winter*	*Spring*	*Fall*
Adult migration	December	March	May–June	Sept–Oct
Spawning	early Feb	early June	mid-Sept	late Oct
Fry emergence	April–June	July–Oct	Nov–March	Dec–March
Ocean entry	Oct–May	Nov–May	Nov–June*	March–July
Spawning habitat	upper river, mainstem	spring-fed streams	headwaters	lower river and tributaries

might experience stressful or lethal temperatures. These populations often migrate into the lake before the temperatures peak (e.g., May or June; Hodgson and Quinn 2002) and spend the summer below the thermocline in comparatively cool water prior to spawning in September or October (Jennifer Newell, University of Washington, unpublished data). However, some populations that spawn late (November or December) are able to enter freshwater after the lake and its outlet have cooled down. In more northerly (cooler) lakes, the temperatures do not routinely get warm enough to pose a problem for the sockeye, so they enter freshwater about a month before they begin to spawn (photo 4-2).

Thus the variation in timing of entry into freshwater and upriver migration among populations within species seems to be related to the fundamental constraint of spawning date and the problem of access to the spawning grounds. For example, in south-central California some coastal streams are entirely blocked by sand bars until fall rains arrive, and the timing of entry by coho salmon is constrained by access (Shapovalov and Taft 1954). Having said this, there are many populations whose timing patterns are hard to explain by such general hypotheses. Intraspecific and interspecific variation in the timing of migration and spawning are almost entirely unstudied and they seem to be a ripe area for connection between flow and thermal regimes, species-specific habitat preferences and tolerances, and competition.

The genetic control over initiation of maturation and migration is not complete, of course, and environmental stimuli do influence migration. Unfortunately, there has been no comprehensive review of this large and inconsistent literature. For example, Hunter (1959) reported an extremely strong connection between stream discharge and pink salmon migration up Hooknose Creek, British Columbia. At the beginning of the season there was a period of high flow but no salmon arrived, presumably because they had not yet arrived near the mouth of the creek or were not ready to ascend. Thereafter, three high-flow events were perfectly matched with pulses of salmon. Subsequent increases in flow brought no migrants, presumably because there were no more fish waiting to ascend the creek. On the other hand, Neave (1943) reported that the migration of coho and chinook salmon up Skutz Falls, British Columbia, was independent of flow. Davidson et al. (1943) concluded that the different relationships between flow and upstream migration by pink salmon in three creeks were best explained by a combination

4-2 School of maturing sockeye salmon in Iliamna Lake, Alaska, prior to spawning; note that the fish have not attained the deep red color or exaggerated morphology of fully mature salmon. Photograph by Thomas Quinn, University of Washington.

of innate (population-specific) behavior patterns and differences in flow. In some very shallow creeks, the fish may hold off the mouth and only enter when fully mature and when flows permit, whereas in deeper rivers the fish may ascend without regard to flow and undergo the final stages of sexual maturation in pools within the stream. Allen (1959) reported that entry into the University of Washington hatchery, where flows were constant, was correlated with nighttime rain for coho and falling barometric pressure for chinook. I do not know whether this fascinating pattern has been documented elsewhere. On the other hand, high flows may delay migration, depending on hydraulic conditions. These site-specific features further complicate matters for researchers seeking general patterns. We should also bear in mind the migrations of fluvial and adfluvial populations of nonanadromous salmonids. Their migrations can be upstream or downstream (e.g., Brown and Mackay 1995a; Meka et al. 2003) and can cover 100 km or more.

The relationship between time of day and migration is also complex and awaits comprehensive assessment. Neave (1943) reported that migration occurred almost exclusively in midday at Skutz Falls, and the fish ladders on the Columbia River see very few salmon at night compared to the daytime passage rates. However, migration rates up the Klamath River, California, were higher at night (Johnston and Hopelain 1990). Gaudet (1990) used hydroacoustic equipment to assess migration up a series of turbid rivers in Alaska and found diurnal migration in some rivers, nocturnal migration in others, and irregular patterns in still other rivers. To make matters still more complicated, Daum and Osborne (1998) reported a clear diel migratory pattern on one side of the Chandalar River, Alaska, but uniform migration throughout the day and night on the other side.

In addition to the species-specific and population-specific patterns of arrival and upriver migration described above, two other fascinating patterns may be observed. First,

FIGURE 4-1. Percentage of 5-year-olds among the chum salmon counted during migration up the Tuluksak River, part of the Kuskokwim River system, Alaska (from Molyneaux and DuBois 1998).

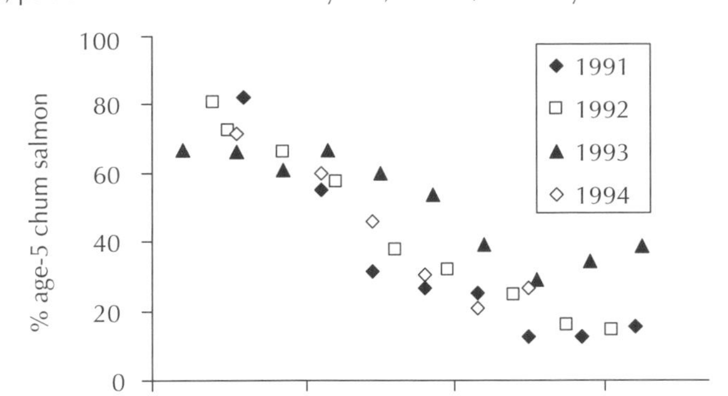

older or larger fish often precede smaller fish on the migration. For example, chum salmon sampled at a weir on the Tuluksak River, part of the Kuskokwim River system in Alaska, included ages 3, 4, 5, and 6 but the great majority were ages 4 and 5 (Molyneaux and DuBois 1998). Over the course of the run, the proportion of 5-year-olds dropped dramatically (fig. 4-1). In addition, the mean length of fish of a given age also decreased over the migration period, magnifying the decrease in size (fig. 4-2). The phenomenon of older and larger fish migrating into freshwater before smaller ones is even more dramatic in Atlantic salmon (e.g., Shearer 1990), where older "multi–sea winter" fish arrive months before those salmon that had spent only one winter at sea (called "grilse"). One might hypothesize that because larger fish swim faster, they would get home sooner if all ages and sizes of fish left the oceanic feeding grounds at the same time of the year. This explanation seems inadequate, however. If the optimal arrival date did not vary with size, the smaller fish could leave the feeding grounds earlier and arrive at the same time as the older fish.

There are two more plausible explanations that assume size-dependent tradeoffs, one based on conditions at sea and the other based on conditions on the spawning grounds. First, the optimal return date may vary with size, reflecting the relative benefits of prolonged feeding at sea. Larger (and older) fish will grow less, proportionally, after an extra week at sea than smaller fish. So, large fish may return at the optimal date for migration and reproduction and the smaller fish may stay at sea a bit longer, grow a bit more, and accept some loss of fitness associated with migration or spawning after the ideal date. In this case, differences in migration and spawning timing reflect the delay of small fish at sea. Alternatively, conditions on the spawning grounds may drive the pattern. Under this scenario, competition with large fish causes small fish to have a later optimal spawning date than larger fish. For example, larger females dig deeper nests than smaller females, and so a small female spawning at the beginning of the season would be more vulnerable to having her nest disturbed than one spawning at the end of the season, when such disturbance would be very unlikely (see chapter 6).

Besides the shift in age and size over the course of the run, the chum salmon population also exemplified another pattern: the tendency for males to precede females (fig. 4-3).

FIGURE 4-2. Length at a given age for two groups of chum salmon counted during migration up the Tuluksak River, part of the Kuskokwim River system, Alaska (from Molyneaux and DuBois 1998).

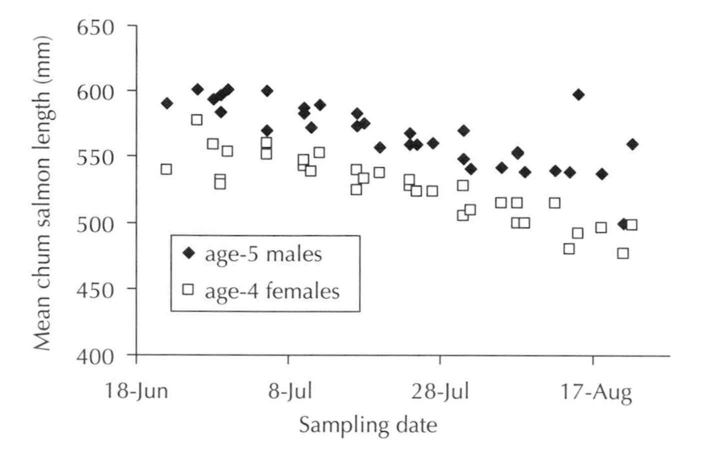

This pattern is sometimes seen in catches of salmon in coastal waters (McKinstry 1993) and is quite commonly documented along migratory routes and onto spawning grounds. Males do not select or prepare nest sites, so one might have expected females to arrive before males rather than the reverse. I think the explanation for this pattern lies in the reproductive biology of salmon, and the migration pattern merely reflects the paradoxical selection for males to arrive before females (see chapter 6).

Changes in migration timing

The timing of upriver migration is not fixed, but varies among years. As indicated by the relationship between ocean temperatures and the migration route and timing of sockeye salmon in coastal waters (Blackbourn 1987), there are oceanic influences on arrival

FIGURE 4-3. Percentage of males among the chum salmon counted during migration up the Tuluksak River, part of the Kuskokwim River system, Alaska (from Molyneaux and DuBois 1998).

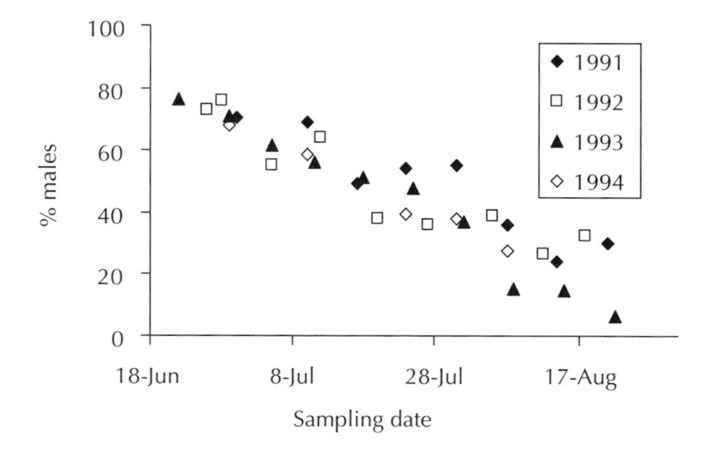

FIGURE 4-4. Trends in the median passage date of adult sockeye salmon at Bonneville Dam, Columbia River, and the first day of the year that reached 15.5°C (60°F; Quinn and Adams 1996 and U.S. Army Corps of Engineers, Portland, Oregon, annual fish passage reports).

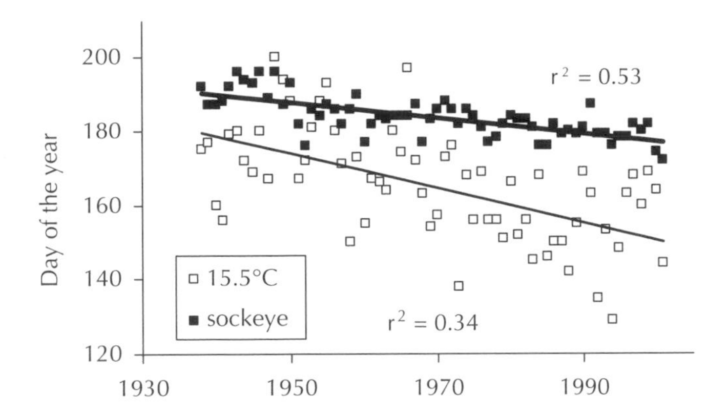

timing at the mouths of rivers. In addition, river conditions can affect travel rates. Probably the best time-series of data on this subject is available for Columbia River salmonids. Bonneville Dam (234 km upriver from the mouth) was completed in 1938 and there are daily counts of adult salmon since then. Data on sockeye salmon are particularly instructive because there are relatively few populations (the basin had more but they were largely eliminated prior to construction of Bonneville Dam), they all migrate to spawning grounds far above Bonneville Dam, and they are wild rather than hatchery produced.

The Columbia River's physical environment has been greatly modified by construction of storage dams over the years. These dams have been designed to dampen the high spring flows (caused by melting snow in the upper part of the basin) by storing the water and then releasing it in winter when demand for electricity in the region is greater and flows would normally be low. Consistent with this plan, the June flows (about the time of sockeye migration) are now only about 40% of their former levels. This reduction in flow has been associated with an unintended increase in temperature, and the two features of the river are closely correlated (high discharges are associated with low temperatures). Over the years, the timing of sockeye salmon migration (indicated by the day when 50% of the annual total passed the dam) has become about 2 weeks earlier (fig. 4-4).

One might hypothesize that the sockeye move upriver once a threshold temperature is reached, and that the change in migration timing over the years merely reflects a tendency for the river to warm earlier. This is only partially correct. The first day of the year when the river warmed to 60°F (or 15.5°C) has been getting earlier, but the shift in fish migration has lagged behind the environmental change (Quinn and Adams 1996). Thus the salmon seem to be adapting to the warmer springs by migrating earlier but are still experiencing warmer temperatures than in the past. This lag suggests that the adaptation by salmon is a genetic one rather than a proximate response to some environmental cue. Similar analyses showed shifts in migration timing of steelhead over the past decades (Robards and Quinn 2002).

FIGURE 4-5. Travel rates of thirty tagged adult sockeye salmon from Sooke (Vancouver Island) and ninety from Hell's Gate (Fraser River canyon) to the spawning grounds of the Bowron Lake system (data from Killick 1955).

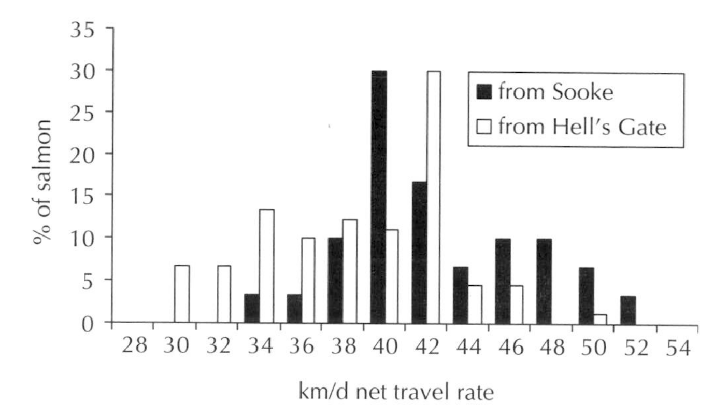

Rate of migration and energetics

Gilbert (1922) was among the first scientists to document the rate of upriver travel by Pacific salmon and his words convey a sense of respect for the salmon. "As regards the rate at which [chinook salmon] ascend the river, we have more reliable and complete data for the Yukon than have been secured in any other stream. Records were obtained of their first appearance at a large number of localities. . . . The first king salmon to reach Dawson in the middle of July, 1920, had been traveling against a consistently rapid current for 29 days, at the rate of 52 miles per day, and during this period, as always within the river, had taken no food" (319). Gilbert stated futher that chum salmon "entered the river about a week later than the kings, at Tanana they were not more than 10 days behind the latter, and at Dawson they were some 14 days behind the kings. The lower 800 miles of the river, as far as Tanana, were traversed at the rate of 50 miles per day and the next 700 miles, between Tanana and Dawson, were covered at the rate of 35 miles per day" (326). Such observations of the first arrivals tend to represent the fastest fish rather than the average, but more recent tagging and radio-tracking studies corroborated the impressive migration rates of Yukon chinook salmon. Travel rates averaged 30–40 km day and some fish traveled much faster (Milligan et al. 1985).

A great deal of information has also been gathered on upstream migration of Fraser River sockeye populations, and they too show impressive rates of travel. Tagging studies documented that the early run to the Stuart Lake system migrated from Lummi Island (near the estuary of the Fraser River) to the spawning grounds at Forfar Creek (1074 km) in 27 days or 43 km/d (Killick 1955). Sockeye showed similar rates of travel to the spawning grounds in the Bowron Lake system, 1058 km upriver and at an elevation of 945 m. They were tagged at Sooke, on the southern end of Vancouver Island, about 200 km from the mouth of the Fraser River, and also at Hell's Gate, about 208 km upriver from the mouth. The fish were recaptured at a weir near the spawning grounds, 1256 km from Sooke and 850 km from Hell's Gate. The average travel rates of tagged fish were 43 km/d from Sooke and 39.2 km/d from Hell's Gate (fig. 4-5). Based on tagging and the sequential arrival of salmon at different areas, Killick (1955) estimated that the early

TABLE 4-3. Estimated rates of upriver travel by sockeye salmon to selected spawning grounds in the Fraser River system (from Killick 1955).

Migration period	*Population*	*km/d*	*km upriver*	*Elevation (m)*
July	Early Stuart	45.8	1074	678
July	Horsefly	51.2	720	725
July	Bowron	51.5	1058	945
August	Chilko	34.4	640	1172
August	Stellako	36.0	976	640
September	Lower Adams	27.2	480	347

runs (primarily migrating in July) to Stuart Lake, Bowron Lake, and Quesnel Lake (Horsefly River) traveled the fastest, followed by August runs to the Chilko and Stellako rivers; and the late run to the lower Adams River was the slowest (table 4-3).

Considerable energy is required for migration and reproductive behavior such as courtship and nest defense after the migration has ended. Taken together, migration and reproduction deplete the salmon of almost all their fat and about half their protein (table 4-4). These data show the expenditure of energy in populations with very arduous migrations. The three populations entered freshwater with different levels of fat but all had about the same amount (3–4% of body weight) when they reached their respective spawning grounds. In contrast to these very long migrations, Hendry and Berg (1999) studied the energetics of Bristol Bay, Alaska, sockeye salmon sampled at four locations and times: off Port Moller, in the outer part of the bay; shortly after the fish entered freshwater at the mouth of Lake Aleknagik; at Pick Creek (98 km upriver and 22 m elevation, as the fish entered the spawning grounds); and after the fish had died (table 4-5). There was an almost total loss of fat and substantial loss of protein, with accompanying increase in water. It should be noted, however, that the samples at Port Moller included populations other than those in the Wood River system, and the sockeye sampled at Lake Aleknagik included Wood River populations other than Pick Creek. Nevertheless, when the Pick Creek data are combined with data from the Fraser River populations, and data on other sockeye, pink, and chum salmon, we see that the level of fat in

TABLE 4-4. The percentage of fat and protein remaining in sockeye salmon at death on the spawning grounds, relative to the salmons' condition at Albion, in the lower Fraser River (Idler and Clemens 1959; Gilhousen 1980).

Population	*Sex*	*% fat*	*% protein*	*Distance (km)*	*Elevation (m)*
Early Stuart	males	9	64	1152	693
	females	5	45		
Chilko	males	17	61	725	1170
	females	7	41		
Lower Adams	males	15	70	483	347
	females	13	61		

TABLE 4-5. Body composition of adult Bristol Bay, Alaska, sockeye salmon sampled in Port Moller, the entrance to Lake Aleknagik, the entrance to the spawning grounds at Pick Creek, and after death (Hendry and Berg 1999).

Sex		*Port Moller*	*Lake Aleknagik*	*Pick Creek*	*Death*
Males	% water	68.7	70.7	78.4	83.2
	% fat	7.9	6.6	1.6	0.1
	% protein	21.3	20.6	17.8	14.5
Females	% water	68.2	70.6	78.0	83.4
	% fat	8.8	6.8	1.5	0.2
	% protein	21.1	20.7	18.2	14.3

salmon when they enter freshwater is related to the distance and elevation that they must migrate (fig. 4-6). In addition, females in populations with more arduous migrations allocate less energy to gonads (Linley 1993; Kinnison et al. 2001).

Interestingly, although salmon populations seem to have evolved patterns of fat storage that prepare them for their migration, and swimming speed is a function of body size, the populations with lengthy migrations are not especially large. Roni (1992) compiled data on the average lengths of chinook salmon populations (all ages combined) and the distances upriver that they migrate. Of the eight-five populations, most (seventy-seven) migrate less than 775 km and these actually showed a significant negative correlation between length and migration distance (fig. 4-7). The largest bodied populations (averaging more than a meter in length!) actually migrate very short distances. The remaining eight populations migrate very far (about 1500 to almost 3000 km) and their average length was similar to that of fish migrating less than 200 km (females: 94.7 vs. 93.1 cm). Similarly, Linley (1993) reported that the populations of sockeye salmon with the most lengthy migrations in the Fraser River are smaller than those spawning downriver, and the long-distance migrants in the Columbia River are also rather small bodied (Gustafson et al. 1997).

FIGURE 4-6. Average percentage of body fat in eighteen salmon populations at the time of entry into freshwater, as a function of the distance and elevation they will have to swim to reach their spawning grounds (data complied from sockeye, chum, coho, and pink salmon studies in Washington, British Columbia, Alaska, and Russia).

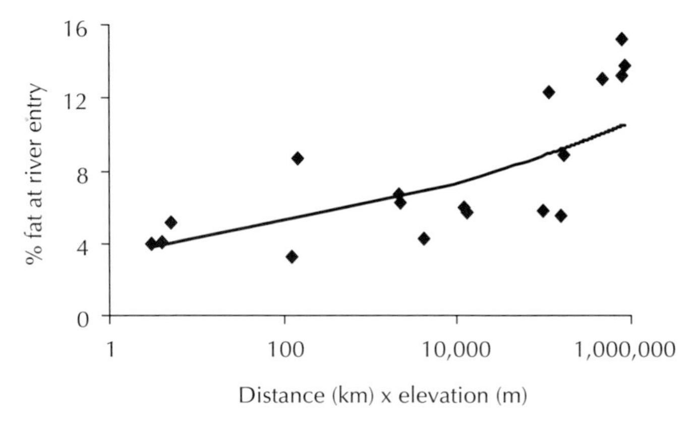

FIGURE 4-7. Average population-specific fork length of female chinook salmon as a function of the distance they migrate to their spawning grounds (from Roni 1992; eight populations migrating more than 1400 km and averaging 94.7 cm were omitted).

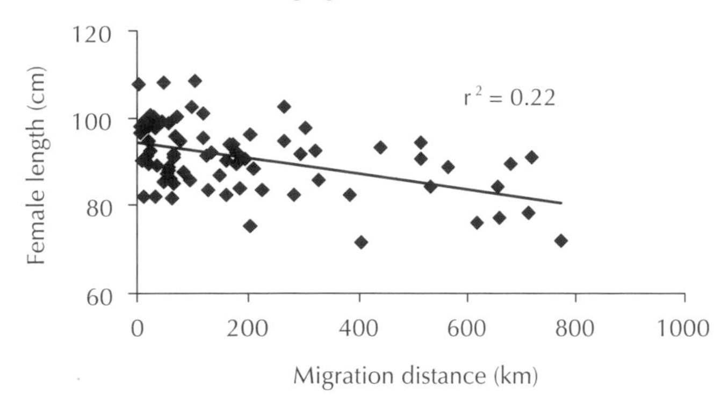

Insights into the upriver migration of salmon in general and sockeye in particular are provided by the remarkably thorough research by John Roland Brett on sockeye salmon energetics. He conducted a series of laboratory studies documenting the resting and active metabolic demand (oxygen and caloric consumption) of salmon as a function of body size, temperature, and swimming speed (reviewed by Brett 1983, 1995). This work revealed that the metabolic demand increases with swimming speed until the fish reaches a maximum sustainable speed. Fish can burst faster than this maximum speed but they cannot sustain such speeds. Thus the least amount of energy is used per unit of time at the slowest speed (analogous to a car using the least amount of gasoline while parked with the engine running). However, this is not efficient from the perspective of energy expended per unit of distance traveled. For a standard adult sockeye salmon (2.27 kg, 61 cm long), Brett estimated that the "least cost" or most efficient speed was 1.8 km per hour or just under 1 body length per second, and the maximum sustainable speed was about 4.8 km/h (depending on temperature; fig. 4-8). The estimated swimming speed of sockeye salmon in coastal waters, after factoring out the effect of currents, was only slightly faster than their optimal (least-cost) speed (Quinn 1988). However, a salmon swimming 1.8 km/h relative to the water would make little progress against a rapid river like the Fraser. The upriver migration would take so long that the total energy consumed would be greater than if it swam faster, using more energy per unit of time but getting there sooner.

With his information on the rate of energy expenditure at different swimming speeds, and the information on loss of fat and protein provided by Gilhousen (1980), Brett could estimate the equivalent speed that sockeye salmon were swimming to get to upriver spawning grounds such as the Stuart Lake system. For the 61 cm sockeye salmon, the estimated velocity against the water was 4.3 km/h, integrated over the entire migration (Brett 1983). Thus the populations are swimming much faster, against the water, than their most efficient speed and are close to their maximum sustainable speed. There are places where they swim more efficiently but there are also places where they have to sprint. Information on these areas of particularly difficult passage was provided by a novel technique: electromyogram telemetry. Hinch et al. (1996) and Hinch and Rand (1998) placed radio transmitters in the sockeye salmon with electrodes connected to the muscles powering swimming. The transmitters provided information on muscular

FIGURE 4-8. Energetics of migration by sockeye salmon. The curve indicates the energy required by a typical adult sockeye salmon to swim a fixed distance at a given speed, relative to the water, including speeds that can only be accomplished as bursts (i.e., not sustained; from Brett 1983). The arrow at the low point in the curve indicates the optimum (least-cost speed, about 1.8 km/h), and a second arrow indicates the average speed of sockeye estimated in coastal waters (Quinn 1988). The third arrow indicates the speed against the water, inferred from electromyogram telemetry, of sockeye salmon migrating up the canyon of the Fraser River (from Hinch et al. 1996 and Hinch and Rand 1998). The circle on the curve shows the estimated swimming speed, based on observed energy loss, during migration up the Fraser River (modified from Brett 1983).

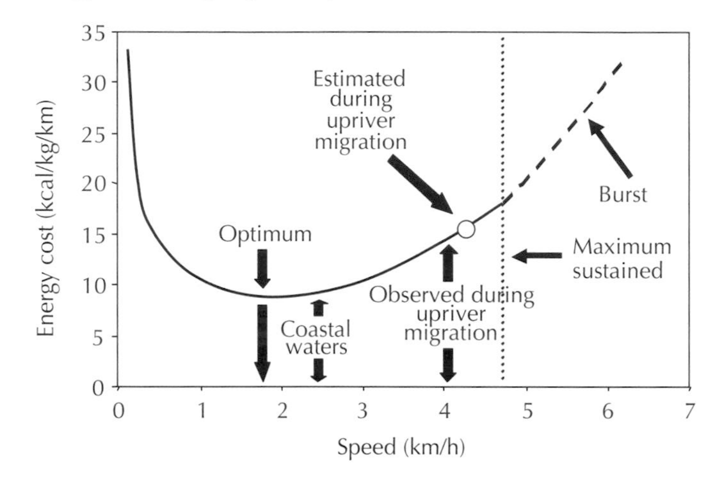

contraction that was calibrated to oxygen consumption and tail-beat frequency in laboratory tests. These transmitters indicated that at places the salmon were swimming about 2.5 body lengths/sec (150 cm/sec or 5.4 km/h for a 60 cm fish; fig. 4-8). The median swimming speed in areas where the water's flow was constricted (e.g., by islands, rock outcroppings, etc.) were faster than those in unconstricted areas (about 90 vs. 80 cm/sec), and both were much faster than Fraser River sockeye swam in coastal waters (60 cm/sec: Quinn 1988; fig. 4-8). However, the passage of sockeye at constricted sites was marked more by delay than higher speeds (Hinch and Rand 1998), and this reduced efficiency at such sites, as measured by energetic cost per distance traveled.

Not surprisingly, the salmon adapt their migration patterns to minimize energy expenditure. They do so by avoiding the fastest water and so generally swim near shore and near the bottom. These behavior patterns facilitate visual counting and, in clear rivers, salmon can be easily counted from towers on shore as they swim along the bank (e.g., Becker 1962). Smaller salmon (both among and within species) tend to swim closer to shore than larger fish. The large fish swim near the bottom in deeper water, and Hughes (2004) hypothesized that the reduced wave-drag experienced by large fish in deeper water makes this efficient, despite the higher current velocities offshore. The telemetry work provided further insights into the differences in migration between sockeye and the smaller pink salmon (Hinch et al. 2002). Average speeds of these species were similar but they seemed to use different tactics. Pink salmon tended to swim steadily whereas sockeye would often burst at up to 6–9 body lengths/sec but then swim very slowly. At constricted sites, pinks were more efficient at energy use than sockeye, perhaps compensating for their smaller size by their behavior, reminiscent of the tortoise and the hare. More generally, the differences in swimming performance, jumping ability, and other forms of migration among species need further study. Chum salmon, for example, are commonly considered to be poor jumpers. Does this result from their physiology, morphology, motivation, energy reserves, or other factors?

Hell's Gate on the Fraser River

Studies on the upriver migrations of salmon took on special meaning in British Columbia in the early 1900s. The runs of salmon, and especially sockeye, to the Fraser River were very abundant. The river has a low elevation, low gradient, coastal region flowing westward, and a much steeper region upriver as it cuts through the coast range of mountains. Above the canyon, the Fraser is joined by the Thompson River. The great majority of the system's sockeye were produced above the canyon, and they had to pass an especially narrow stretch of water called Hell's Gate (208 km upriver, 68 m above sea level). At this point the river is only 33.5 m wide.

Hell's Gate was probably always a hard stretch of river for fish to ascend, depending on when they arrived and the river's level and discharge. Sockeye migration took place from July to October, and the river routinely varies by 15 m of vertical elevation at Hell's Gate during this period. Railroad tracks were built by the Canadian Pacific Railroad on the right bank in the 1880s. Rock dumped from these operations may have constricted the flow and impeded passage but there is no direct evidence to this effect. In 1911–1912 the Canadian Northern Railroad began laying tracks on the other side of the river and rock was again dumped in. The water was noticeably faster in the summer of 1913 and most salmon were unable to negotiate Hell's Gate. Roos (1991) reviewed the historical records, including the definitive work of Thompson (1945) and concluded that "large numbers of sockeye were observed in creeks and rivers [where they did not normally spawn] for many miles downstream from Hell's Gate in 1913. Included among the sockeye were large numbers of pink salmon. While some sockeye were able to pass upstream in 1913, the vast majority failed in their struggle. The stage was set for a staggering decline in future production" (24).

The water velocities confronting the upstream migrants in 1913 exceeded their swimming capacity, and most died unspawned trying to migrate upriver or produced few offspring when they spawned at high densities in unsuitable habitats below Hell's Gate (Roos 1991). This was particularly tragic because the run in 1913 was the largest on record, with a catch of more than 30 million and an estimated escapement of 6.61 million as well (Ricker 1987). The Fraser River's upriver sockeye salmon populations almost all mature at age 4, and the major populations were on synchronous 4-year cycles with peaks in 1901, 1905, 1909, and 1913. Ricker (1987) estimated that from 1896 to 1915 the average catch on the 1901 cycle was 22.5 million, with catches on the 1902 to 1904 cycles averaging 6.1, 4.7, and 3.9 million sockeye. The Fraser River populations were extraordinarily productive, with 4.38 returning adults per spawner (i.e., able to sustain exploitation rates more than 75%). In stark contrast, the 1913 escapement of 6.61 million barely replaced itself with only 7.67 million returning adults (1.16 per spawner) 4 years later. Considering that the production below Hell's Gate was unaffected by the slide, the effects upriver must have been devastating.

4-3 Vertical-slot fishways at Hell's Gate, on the Fraser River, British Columbia; note the water's velocity and vertical drop at this site. Photograph by Thomas Quinn, University of Washington.

Greatly compounding the problem, a huge rockslide on the left bank was created by tunneling operations on February 23, 1914. The constriction of the channel caused a vertical drop in the river of almost 5 m at the choke point (despite the fact that the river was more than 30 m deep!) and exceptionally high water velocities. Even after removal of material from the channel after the slide in 1914, the vertical drop was about 3 m and velocities at the surface in midchannel ranged from 5.0 to 6.75 m/sec. The populations of sockeye salmon were greatly reduced by a combination of difficult passage upriver and overfishing until fishways initiated in 1944 along both banks of the river at Hell's Gate were completed in time for the 1946 runs (photo 4-3). These vertical-slot fishways were designed to accommodate not only the great densities of salmon but also the great range of water levels over the period of migration. Combined with other fishways, such as those at Yale Rapids and Bridge River Rapids on the Fraser River and Farwell Canyon on the Chilcotin River, the Hell's Gate fishways greatly contributed to the rebuilding of the Fraser River sockeye salmon runs (Roos 1991).

A rockslide, less famous than the ones at Hell's Gate but nevertheless devastating, occurred in a very remote reach of the Skeena River in 1951. Godfrey et al. (1954) estimated that only 31% of the 461,000 sockeye that escaped the fishery successfully negotiated the slide and reached Babine Lake, 60 km upriver of the slide, and even heavier losses of pink salmon were estimated. In 1952 the sockeye salmon and pinks also experienced mortality and delay at the site of the slide (despite efforts to clear it). In addition, only 30–42% of the sockeye that reached a counting fence at the outlet of Babine Lake migrated to their streams and completed spawning (Godfrey et al. 1954). This slide was apparently caused by natural geological processes rather than human activities. Indeed, a 100 km road had to be built to reach the site so that the slide could be cleared to aid the migrating salmon. The slide was cleared in time for the 1953 run (Godfrey et al. 1956), and the salmon populations eventually recovered. Such slides (and lesser ones) are part of the variable landscape that the salmon must contend with on their upriver migration, and they serve as a useful reminder that natural events are not always benign.

Mortality during upriver migration

The upriver migration to distant, high-elevation spawning grounds is not only energetically demanding but it can be associated with significant levels of mortality. The term "pre-spawning mortality" usually applies to fish that have successfully completed the migration but died prematurely on the spawning grounds. As Gilhousen's (1990) review of Fraser River sockeye indicated, such mortality is often associated with high temperatures (fig. 4-9), which both weaken the fish and accelerate proliferation of pathogens. Fish may arrive on the spawning grounds with too little energy or die of infection prior to spawning. Heard (1991) also reviewed several instances of pre-spawning mortality of pink salmon that were caused by intense crowding, high temperatures, and insufficient dissolved oxygen. In addition, bears can kill many salmon on the spawning grounds (Quinn et al. 2001a), but predation along the migration route is generally negligible. However, there are some notable (and photogenic) cases in which bears take advantage of waterfalls that give them access to migrating salmon, including chum salmon on the McNeil River and sockeye salmon at Brooks Falls, Alaska. This might more properly be called "en route" rather than pre-spawning mortality.

En route mortality is difficult to study and contentious because it is often inferred from differences in numbers of salmon counted at some downriver site and on the spawning grounds. Possible miscalibration of counters and poaching complicate estimates of mortality unless large numbers of carcasses are found. However, an extreme event occurred in September 2002, when some 20,000–30,000 adult salmon (mostly chinook) died in the lower 50 km of the Klamath River, California. The mortality seemed to have resulted when a combination of low flows and high temperatures stressed fish and concentrated them in a few holding areas where disease proliferated.

The late 1990s also saw the emergence of a very puzzling and serious syndrome of mortality in some Fraser River sockeye populations. It has long been known that many of the late-spawning populations, including those spawning below Hell's Gate (such as Weaver Creek) and at some upriver sites (notably the lower Adams River), tended to arrive in the Strait of Georgia a month or more before entering the river (see table 3-1). They would delay in saltwater and then migrate upriver to spawn. However, since 1996 these salmon have been entering freshwater much earlier in the year, with little or none of the characteristic delay in saltwater off the mouth of the river. It is not obvious that the change in behavior was caused by anomalous conditions in either the saltwater (that might have driven them from the ocean prematurely) or in the river (that would have attracted them upriver too soon). The change would be little more than a curiosity except that very heavy en route mortality was inferred, and considerable pre-spawning mortality was observed in these years (Macdonald 2000; Macdonald et al. 2000; S. J. Cooke et al., 2004). Evidence indicated that the causal agent was a myxosporean parasite, *Parvicapsula minibicornis.* The parasite's life cycle is not well known (it was only

FIGURE 4-9. Percentage of female sockeye salmon found dead on spawning grounds of the Fraser River system without having completed spawning, as a function of the water temperatures experienced along the migration route at Hell's Gate (for the Early Stuart populations) and temperatures on the spawning grounds (for the Raft River population; all data from Gilhousen 1990).

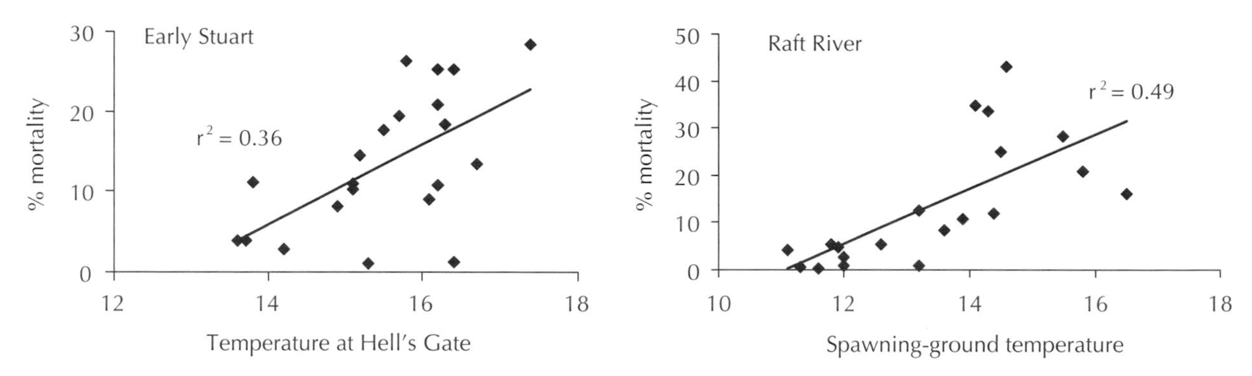

described in 1996), but it appears to infect the fish in the estuary or lower reaches of the river (St-Hillaire et al. 2002). It may have nothing to do with the early migration, but the increased period of time spent in the river prior to spawning now gives the parasite time to proliferate, and the fish die of kidney failure. The fish that enter early are exposed to warm water and high flows in the river, compared to the cooler water they can occupy below the thermocline in the Strait of Georgia (see fig. 3.3), and their energy stores get depleted. Indications are that the large run of Adams River sockeye in 2002 was not devastated by this en route and pre-spawning mortality syndrome but it remains to be seen whether the traditional pattern of delayed migration and low mortality will reassert itself or if early migration and high mortality will prevail for some time (S. J. Cooke et al., 2004).

Summary

The patterns of entry into freshwater seem to reflect the balancing demands of feeding at sea and the need to reach the spawning grounds at the optimum date, and this may require some sacrifice by the adults in the form of lost feeding opportunities at sea. Thus some populations enter freshwater only a few days before spawning whereas others spend many months in rivers or lakes prior to spawning. However, for any given population the timing of entry and migration is quite predictable among years. Salmon populations also vary in the amount of energy in their bodies when they leave the ocean and in their travel rate, as needed for the rigors of their particular migration route and schedule. Some populations migrate a thousand kilometers or more, but most salmon spawn near the coast and some are within tidal influence. In addition to patterns of variation among species and populations, there is a tendency for older and larger salmon to enter first and also for males to enter before females. This may reflect factors related to both reproduction and migration. Long migrations to high elevations are very demanding and some salmon die along the route or on the spawning grounds before they can reproduce. Migration patterns of nonanadromous salmonids to spawning grounds have been less thoroughly investigated and can be complex, depending on where the fish reside and spawn in the river network.

5

Homing and Patterns of Straying

Much of the discussion on migration so far has been predicated on the assumption that salmon are migrating back, not merely to freshwater, but to the site where they were spawned years earlier. At this point it is useful to review some of the history and evidence in support of this assumption. There is evidence that centuries ago a few astute naturalists realized that adult salmon were returning to their natal river. The writings of a Norwegian priest, Peder Claussön Friis, in 1599 indicated an awareness that Atlantic salmon homed, as did those of a Scottish priest, Hector Boece, in 1527 (Nordeng 1989), and Izaak Walton's *The Compleat Angler* (first published in 1653) described salmon that had been marked with thread as juveniles, returning to the same river as adults.

By the mid- to late-nineteenth century, biologists were gaining knowledge of salmon and other anadromous fishes, and some recognized their homing tendency. For example, the 1876 report of the U.S. Commission on Fish and Fisheries stated, "This stream [near Elko, Nevada] is one of the many that form the headwaters of the Columbia River, and to this point, eighteen hundred miles from its mouth, the salt-water salmon come in myriads to spawn. . . . From these facts we may infer that the instinct of location is probably sufficient to attract a colony of fishes as far inland as the headwaters of the longest river, whenever their home has been once established there" (part 3, xxviii–xxix).

Homing was suggested, not just by the persistence of salmon populations in some rivers but by consistent differences in appearance. J. W. Milner (1876, 323) noted "the generally accepted fact in the habits of anadromous fishes that they are disposed to return to almost the exact locality where they passed their embryonic and earlier stages of growth. . . . Observations of the shad brought to the large markets shows considerable difference in the physiognomy and general contour of those from different rivers. The suggestion is natural that they are distinct and separate colonies of the same species, and thus slight characteristics are perpetuated because they breed in-and-in and do not mix with those of other rivers."

Perhaps the earliest large-scale marking study was conducted on sockeye salmon in Cultus Lake, a small lake supporting a modest population in the lower Fraser River system (Foerster 1936, 1968). The idea was to mark all smolts by clipping one of the ventral fins. Thus, unmarked adults entering the lake in the future would be strays and fish with clipped fins would be homing to the lake. It is important to note that the vast majority of Fraser River sockeye salmon spawn upstream of Cultus Lake, so there were many salmon that might stray into Cultus Lake. If the entire Fraser River population returned at random, few marked fish would enter Cultus Lake.

The experiment was first carried out in 1931, when all 365,265 smolts were marked (a very substantial undertaking). Like most of the Fraser River populations, Cultus Lake sockeye predominately return after 2 years at sea, so the marked fish were expected in 1933. In that year, 2856 large marked fish returned, along with 409 small, unmarked fish. These small fish were assumed to have spent only 1 year at sea and so were of uncertain origin. There were also 17 large unmarked fish, presumably having spent 2 years at sea. It is possible that these fish regenerated their fins or that they somehow escaped marking. However, assuming that they had originated from other parts of the system, they constituted only 0.6% of the population. Not satisfied with this apparently conclusive result, the researchers repeated the experiment again with an even larger sample size of 497,598 smolts in 1936. Similar to the previous results, in 1938 they recovered 8980 large marked fish, 4226 small, unmarked fish, and 136 large unmarked fish for an estimated 1.5% strays into the population. There was no effort to see if any Cultus Lake fish strayed to other populations, only if other populations strayed into Cultus Lake, so we do not know if any Cultus Lake fish spawned elsewhere.

By the 1930s there was growing acceptance that salmon migrated significant distances and home back (Moulton 1939). However, there were some skeptics, especially Archibald G. Huntsman. He doubted that many salmon migrated far from the mouth of their home river, and he believed that those salmon found wandering in the open ocean would be unable to return home and so were essentially lost. Before he would be convinced, he wanted to see an example of a juvenile salmon tagged in a river, caught at sea, and then recovered in the natal river as an adult. Given the limited research being conducted on salmon during those years, this was indeed a very high burden of proof. However, Huntsman (1942) was forced to accept the evidence when an Atlantic salmon caught west of Newfoundland on June 17, 1940, was one of 31,359 smolts that had been marked about 900 km away in the Margaree River, Cape Breton in 1938, and the same fish was recovered in the home river on September 21, 1940.

From this point on, the central question was no longer, "Do salmon home?" but "What proportion of salmon home?" and "How is the homing accomplished?" Important evidence pertaining to the first question was provided by an extremely important tagging study in coastal California, south of the Sacramento River (Shapovalov and Taft 1954). Their study was noteworthy for two reasons. First, it involved two species (coho and steelhead); and second, it considered straying between two adjacent streams, Waddell and Scott creeks, less than 8 km apart. The streams had small populations but the study was conducted over 9 years for the steelhead and 7 years for the coho salmon. The key

TABLE 5-1. Homing (indicated by bold numerals) and straying by coho salmon and steelhead between Waddell and Scott creeks, two small coastal California streams (Shapovalov and Taft 1954).

		% recovered	
Species	*Marking site (sample size)*	*Waddell*	*Scott*
Coho	Waddell (369)	**85.1**	14.9
	Scott (56)	26.8	**73.2**
Steelhead	Waddell (485)	**98.1**	1.9
	Scott (960)	2.9	**97.1**

results (table 5-1) show two important points. First, the proportion of salmon and steelhead that homed to their natal river was far more than would have been expected by chance, but was not 100% either. Second, the two species differed in the proportion of fish that "strayed," with more straying by coho salmon than steelhead. Surprising as it may seem, these data—on wild populations of more than one species—are virtually unique.

Tagging studies such as these are very laborious, but there are sometimes natural "tags" on salmon that can serve the same purpose. A number of parasites are found in some regions but not others and these have proved very useful in determining the origin of salmon caught on the high seas. However, some parasites have sufficiently fine-scaled distributions that they can be used to estimate patterns of homing and straying between nearby watersheds. *Myxobolous neurobius* and *Henneguya salminicola*, two protozoans, provide an example of this (Quinn et al. 1987). The life cycles of these parasites are not known in detail but they infect juvenile sockeye salmon, probably during lake residence, remain in them throughout the duration of marine residence, and complete their life cycle when the salmon return and die. The parasites do not seem to do the fish much if any harm, but they serve as a biological tag if they are present in one lake system but absent in another. Long Lake and Owikeno Lake, along the central coast of British Columbia are a case in point. Infection was not detected in any smolts or adults from Long Lake, but in Owikeno Lake (in the adjacent inlet along the coast) fully 98.5% of the adults were infected. Thus, straying of Owikeno Lake (infected) fish to Long Lake must have been exceedingly rare, and at most a small fraction of the Long Lake fish strayed to Owikeno Lake. Barkley Sound, on Vancouver Island, has three major lakes with populations of sockeye salmon: Great Central, Sproat, and Henderson. *Myxobolous* had a very high prevalence in Sproat and Henderson but a very low prevalence in Great Central, indicating little exchange between these populations (table 5-2). *Henneguya* was absent or very rare in Great Central and Sproat but common in Henderson. The combination of the two parasites was sufficient to demonstrate the homing of adults to their lake of origin in this system. Interestingly, these patterns of parasite prevalence are not fixed. In the years since this study was completed, the prevalence of the parasites in Sproat and Henderson lakes has been largely unchanged but both parasites are now much more prevalent in Great Central, reducing their value for identifying the populations (Margolis 1998).

The early fin-clipping experiments and data on relative prevalence of parasites clearly demonstrated that the great majority of salmon that survive the marine period return

TABLE 5-2. Percentage of adult sockeye salmon infected with two kinds of parasites, as evidence of homing (Quinn et al. 1987).

Region	*Lake*	*Myxobolous*	*Henneguya*
BC coast	Long	0	
	Owikeno	98.5	
Vancouver Is.	Great Central	3.9	0.6
	Sproat	99.8	0.0
	Henderson	98.8	66.0

to spawn at their natal site. However, not all salmon returned home, and we term those that do not "strays." The prevalence of straying is difficult to assess, and there are a number of methodological issues to bear in mind. First, there is "straying in" and "straying out." Foerster searched for strays entering Cultus Lake but he did not look for Cultus Lake strays elsewhere. When one considers both forms of straying, it is necessary to decide if we are interested in the number of fish straying or the proportion. If two populations are of very unequal sizes, then a small proportion of strays may still be a large number of fish. For example, assume Population A with 10,000 adults and Population B with only 100. If 100 fish from Population A strayed into Population B, we might say that this was only 1%, or we might say that 50% of the now 200 fish in Population B were strays.

Another problem in studying homing is that the fish are not always given a chance to make a second choice. Often they are trapped at a fence on a stream or at a hatchery and they cannot move downstream. Observations of wild salmon and studies of the mechanisms of orientation in streams (P. B. Johnsen 1982) suggest that a certain amount of "testing the waters" is part of the homing process and so straying may be overestimated in some situations. The other fundamental problem in the early studies of homing and straying was the difficulty in conducting a broad, unbiased search for marked adults. Scientists attempted to recover marked adults in the site where the fish were released and perhaps in a few nearby streams but it is impossible to survey all possible streams, and it is difficult to thoroughly survey more than a few. In addition to these problems, there are only so many combinations of fins one can clip from fish or unique parasites carried by fish, so these techniques are not suitable for large-scale studies.

However, the invention of the coded wire microtag (Jefferts et al. 1963) revolutionized salmon management and also provided an unparalleled opportunity for studies on homing. The invention involves a coil of fine stainless steel wire that has a binary code etched onto the wire in a continuous fashion. The tag is cut into 1 mm lengths and injected into the nasal cartilage of juvenile salmon. The tag is retained within the growing skull of the salmon throughout its life, and its presence is traditionally indicated by removal of the adipose fin. The tags are magnetized and so can be found and removed from adult salmon. The code, read under a microscope, reveals the agency and hatchery where the fish were marked, the year, and the experimental group if there was one.

Coded wire tags are used to estimate the marine distribution (i.e., catch) patterns of salmon from specific regions, the fishing and mortality rates, and other information critical to management. However, the tags quickly became so ubiquitous that virtually

FIGURE 5-1. Map showing the major dams along the Columbia and Snake rivers and the major tributaries (but not the dams on the tributaries). The Snake River dams are passable by salmon up to Hell's Canyon Dam; the Columbia River was passable to Grand Coulee Dam, but Chief Joseph Dam presently limits migration. Modified from an original map by Blake Feist, National Marine Fisheries Service.

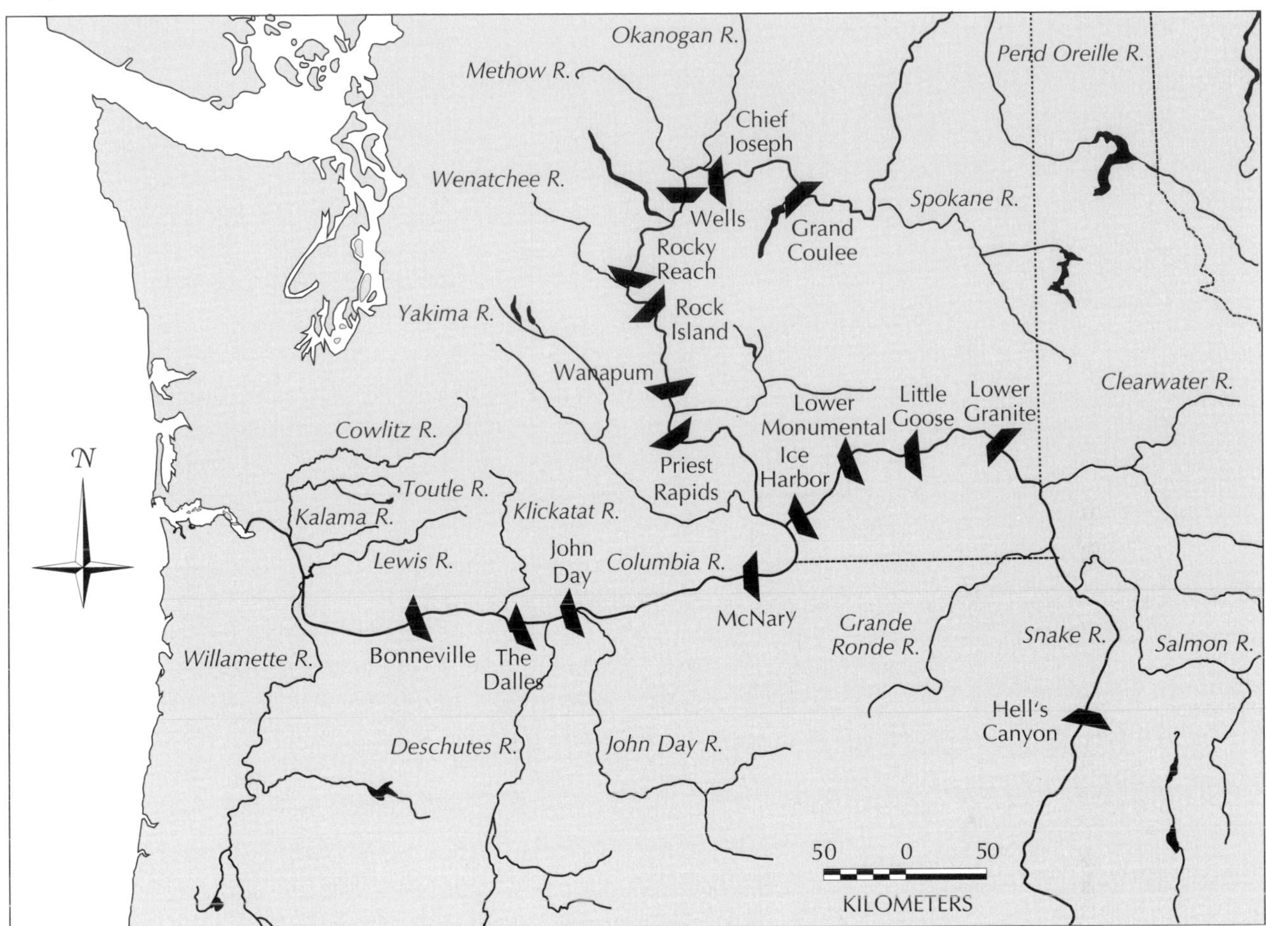

all hatcheries examine adult salmon for tags, and they are sampled at many spawning grounds as well. Thus these tags provide an extraordinary basis for studying the proportion of salmon that home and the patterns of straying. Among the first analyses of straying using coded wire tags was that conducted by Quinn and Fresh (1984), using data from Cowlitz River Hatchery spring chinook salmon from brood years 1974–1977. The Cowlitz River is a tributary of the lower Columbia River, entering below Bonneville Dam on the Washington side (fig. 5-1). Over these years, 1.2 million tagged smolts were released, and 24,139 tags were recovered in hatcheries and spawning grounds. After accounting for fish not sampled, an estimated 41,085 tagged salmon returned. Overall, the vast majority of salmon homed; 98.6% were recovered in the Cowlitz River and 1.4% strayed. However, there were some intriguing patterns in the strays. First, only ten (1.7%) of the strays went outside Columbia River system, and almost all the strays were in the vicinity of the Cowlitz River. However, 76.6% of the strays went to the Lewis River and 18.6% to Kalama River, even though the Kalama is closer to the Cowlitz than the Lewis.

TABLE 5-3. Homing (bold numerals) and straying by ocean-type chinook salmon among five hatcheries in the Columbia River below Bonneville Dam. The "% native" column indicates the percent of fish returning to each hatchery that were from one of these five sites and does not include strays from elsewhere (Quinn et al. 1991; Pascual and Quinn 1994).

	Release sites						
Recovery sites	*Abernathy*	*Cowlitz*	*Kalama*	*Lewis*	*Washougal*	*Total*	*% native*
Abernathy	**618**	1	0	0	0	619	99.8
Cowlitz	1	**4022**	32	48	4	4107	97.9
Kalama	60	46	**477**	54	366	1003	47.6
Lewis	0	137	47	**1388**	212	1784	77.8
Washougal	1	0	3	0	**2794**	2798	99.8
Total strays	97	198	94	104	640	1133	
Total	715	4220	571	1472	3434	10412	
% strays	13.6	4.7	16.5	7.1	18.6	10.9	

Moreover, there was a tendency for older salmon to stray more than younger salmon (age 2: 0.3%, age 3: 1.6%, age 4: 2.0%, age 5: 3.6%).

Further work on the Columbia River hatchery populations revealed some striking patterns in the "ebb and flow" of strays among sites. A comparison of straying among five hatcheries below Bonneville Dam, combining salmon from brood years 1977 through 1984, is illustrative (table 5-3). The average level of straying from the hatchery of origin was 11% (the range was 5–19%), and strays tended to enter a hatchery near their natal site (though not always the closest one). Abernathy and Washougal hatcheries received very few strays from the other hatcheries, though some of their salmon strayed. In contrast, some rivers, notably the Lewis River, were magnets for strays, and these patterns were quite consistent over years. It is unclear why some rivers are more attractive to strays than others, and this is a fruitful area for more research.

The issue of straying among streams raises the fundamental question, "What is the spatial unit to which salmon return?" The answer is linked to the mechanistic aspects of homing, which will be discussed later in this chapter (see "Olfactory imprinting"), but some insights may be given here. Do salmon return only to a major river system like the Columbia or Fraser and then disperse at random? Do they home to basins within such large rivers (e.g., the Snake or Thompson rivers, respectively), to tributaries within those tributaries, and so on? At some point, homing presumably gives way to selection of a spawning site by females and search for mates by males, or do salmon return to the very redd site where they emerged? Wagner (1969) showed that steelhead released in different reaches of a river as smolts tended to be caught there as adults more often than would occur at random (table 5-4). Fish released in the lower reaches were seldom caught in the upper reaches, and some of the fish released in the upper reaches that were caught in the lower reaches as adults may have been migrating, and so were not necessarily going to spawn where they were caught. This suggests that homing can be very precise, and still greater precision may yet be demonstrated.

TABLE 5-4. Patterns of capture of adult steelhead in the Alsea and Wilson rivers, Oregon, that had been released into lower, middle, and upper reaches as smolts. The results are presented as the percentage of fish caught in the different regions of the rivers, calculated by averaging the data from 1964 and 1965 (from Wagner 1969).

		% in each recovery area			
River	*Release site*	*lower*	*middle*	*upper*	*total*
Alsea	lower	52.6	27.6	19.8	3408
	upper	27.7	20.7	51.6	2456
Wilson	lower	34.4	33.0	32.6	1878
	middle	6.5	39.0	54.4	2789
	upper	3.4	12.4	84.3	3316

Our understanding of salmon homing has been largely limited to data on hatchery populations because the vast majority of tagged salmon are produced in hatcheries. McIsaac (1990) presented evidence that hatchery-produced chinook salmon strayed more than did naturally produced chinook in the same river, though Labelle (1992) did not find consistent differences in straying by wild and hatchery coho salmon from Vancouver Island rivers. Nevertheless, information on the precision of homing by wild salmon, incubating and emerging under natural conditions, is needed. These kinds of studies are challenging, but if the spawning/incubation sites differ in thermal regimes the embryos may be distinguishable by banding patterns on the otoliths (Quinn et al. 1999). Natural differences in water chemistry are also a promising way to identify salmon and trout from different parts of a river system (e.g., Ingram and Weber 1999; Wells et al. 2003).

Evolutionary aspects of homing and straying

The vast majority of salmon home but some do not, and these strays are not merely an aberration to be ignored. Much of the present range of Pacific salmon was glaciated some 10,000–15,000 years ago, so most current populations were founded by strays since then. Thus, straying is just as fundamental an attribute of salmon as homing. One might either ask why salmon home or why they stray. From an evolutionary perspective, these behavior patterns must be viewed as adaptations to increase the reproductive success of individuals. That is, fish behaving in a manner that tends to result in many surviving offspring will predominate in a population at the expense of fish behaving in a manner that tends to produce fewer offspring. Initially, individuals that return to their natal site for breeding have their own survival as evidence that this location was suitable for reproduction, and they are likely to encounter other conspecifics there with whom they can breed. An unfamiliar location might be unsuitable or devoid of conspecifics. Moreover, the individuals that return to their natal site (i.e., the survivors) may possess attributes that are compatible with that particular habitat such as the timing of breeding, temperature tolerance, disease resistance, body size, and so on. If these attributes are heritable, then they will be reinforced at the natal site. Individuals that home will be better adapted to the natal site than they would be to some other location where they might stray to, and strays will be less successful than the local fish.

There are countless phenotypic and genotypic differences among populations, and transplant experiments such as those described by Reisenbichler (1988) reveal the superior performance of local populations over outsiders (see chapter 18 for more details). So, how can we explain straying? When new territory is available, the strays will enjoy low levels of competition for breeding and rearing habitat. Indeed, the recession of glaciers in Glacier Bay, Alaska, has been followed by comparatively rapid colonization by salmonids and other organisms (A. M. Milner et al. 2000). Similarly, the rapid spread of pink salmon above Hell's Gate on the Fraser River after the fishways were built to ease upriver migration (Roos 1991) and the spread of pink salmon introduced to the Great Lakes (Kwain 1987) demonstrate how important straying can be.

The most dramatic recent, natural example of the importance of straying was the eruption of Mount St. Helens in Washington on May 18, 1980. The volcanic eruption blew off the top of the mountain and devastated more than 500 km^2 of forest. The associated heat, ash, and mudflows boiled and buried all fishes in the north fork of the Toutle River. The progeny of any Toutle River coho, chinook, or steelhead that spawned there in the fall of 1979 were killed, so only strays of that generation enjoyed any reproductive success. The natural reestablishment of populations in such situations is also by strays.

Presumably, strays are testing the suitability of stream habitats every year, and the stability of species distribution patterns reflects fundamental ecological constraints. Still, the strays persist. I have hypothesized that straying and homing are under genetic control and are maintained in dynamic equilibrium within species and populations (Quinn 1984). Three factors may affect the relative frequency of strays. First, the stability of the river is likely to be important. In this context, "stability" does not necessarily mean low variation in physical factors such as flow per se, but it rather means low variation in recruitment. If there is a high probability that the home site will be suitable, then this is the best place to spawn. However, if it is a good place in some years but bad in others it may be safest for a female to have some of her offspring home and have others stray. This will allow her to maximize the probability of having not merely children but grandchildren (and so on). Thus salmon populations using less stable rivers (e.g., small tributaries) should have higher straying rates than populations in more stable habitats such as the outlets of lakes. To the extent that this factor is important, straying might be most common in pink and chum salmon in "flashy" coastal streams or among coho in steep, unstable streams, and least common in sockeye and chinook salmon spawning in the outlets of lakes and main stems of rivers. Given the variation in habitat use, there may be more variation among populations than species in this factor.

In addition to stability, the extent of specialization for freshwater habitat should influence straying. Species with more protracted freshwater rearing, such as sockeye salmon, show more complex adaptations for migration and disease resistance than species like pink and chum salmon that migrate to sea at emergence. The coast of southeast Alaska has countless small streams with pink and chum salmon spawning in the lower reaches and intertidal zones. One might surmise that the cost of straying from one such stream to another would be slight because the salmon are not highly specialized. To the extent that this factor is important, the order of straying from most to least might be pink = chum > chinook > coho > sockeye, depending on local conditions and life history.

The hypothesized influences of stability and specialization on straying rates are based on the idea that salmon should reduce the risk of having no children. A given female spawns at one place and time so she has put all her eggs in one basket (more or less). However, a mixture of homing and straying by the offspring is a way for the parents to spread the risk because the offspring will not spawn in the same place. Logically, a third factor that might influence straying is the population's variation in age at maturity and extent of iteroparity. For example, pink salmon all mature at age 2, so if every pink salmon homed, a given pair would have all their descendants spawning in the same stream every other year, and there would be a great risk that in some year the entire line would be extirpated. On the other hand, chinook salmon mature at a number of ages, and so even if there was a total loss in one year, the family line would continue if representatives successfully reproduced the previous or following year. Thus variation in age at maturity is a form of straying in time and may influence the evolutionary need to stray in space. To the extent that this factor is important, the order of straying from most to least might be pink > coho > chum = sockeye > chinook, depending on populations. Finally, the extent of iteroparity may be important because such individuals could not only spread out their risk by having children mature at different years but could themselves mature and spawn in more than one year, providing greater protection against a catastrophic loss of offspring from harsh conditions of one sort or another.

Three things must be emphasized at this point. First, there is no suggestion that salmon are "thinking this out" in any cognitive way. Rather, we hypothesize a genetic control over the tendency to home/stray, and the proportion of the population showing these behavior patterns or possessing those genes would vary dynamically among years and among populations. As conditions are stable, the tendency to home might predominate. In the event of a catastrophic loss, the straying tendency might experience a transient rise in abundance, much like fireweed or other colonizing plants after removal of later successional vegetation. Second, the actual process of straying is entirely mysterious. Are some fish "programmed" to stray or do they fail to home? That is, do strays identify the home stream but then go elsewhere, or do they have poorer memory or sensory capabilities and so stray out of ignorance? In either case there might be a genetic association with the trait, though the underlying basis would be totally different. Finally, there is undoubtedly a proximate (environmental) aspect to straying. If conditions in the natal river are sufficiently degraded, salmon may move elsewhere to spawn. However, the homing tendency is remarkably strong, as evidenced by the huge numbers of sockeye salmon that died struggling to ascend Hell's Gate on the Fraser River after the rockslide.

Are any of these hypotheses correct? Unfortunately, there are so few comparative studies of homing and straying, especially on wild populations, that it is difficult to definitively test the hypothesized patterns. The hypothesis that pink salmon stray more than the other species is consistent with the generally low levels of genetic differentiation of their populations (Hendry et al. 2003), but the straying rate of 5.1% reported by Thedinga et al. (2000) in southeast Alaska is not exceptionally high.

Olfactory imprinting

The work of Foerster and others convinced scientists that the majority of salmon surviving to maturity return to spawn in their natal river, but there was no consensus as to the mechanisms and sensory systems by which this was accomplished. The breakthrough came when Hasler and Wisby (1951) proposed the "olfactory hypothesis" to explain salmon homing. They hypothesized that (1) rivers differ in chemical composition because of natural combinations of rocks, soil, and plants; (2) these differences are constant over the seasons and among years; (3) fish can learn the chemical characteristics of their natal river; and (4) fish are attracted to learned odors.

The idea came to Hasler when he was walking in an alpine meadow and smelled mosses and columbine from the vicinity of a waterfall he had not visited since his youth. The odors evoked such strong memories that he hypothesized that salmon might form similarly powerful memories of their natal stream (Hasler and Scholz 1983). Hasler was familiar with the emerging field of ethology, led by Konrad Lorenz, Niko Tinbergen and Karl von Frisch, and with Lorenz's concept of imprinting. Imprinting is a special form of learning that takes place only at a sensitive stage in development, involves no reward or punishment, is retained over long periods of time without reinforcement, and elicits a response only in a specific context. Hasler hypothesized that salmon olfactory learning is an example of imprinting: the smolt stage is the sensitive period when salmon learn the odors of the home river, salmon retain the memory during their years at sea, and they respond to the odors only at sexual maturity.

The first experiments (Hasler and Wisby 1951) were conducted, not with salmon but with bluntnose minnows, a small, nonmigratory fish. These fish did not imprint, but were conditioned (the more common form of "reward-punishment" learning) to discriminate between water taken from one of two rivers. This simple experiment verified the first two components of the hypothesis: that streams differed in chemical composition and that fishes were capable of discriminating such differences based on odors. But do these processes function in salmon? Wisby and Hasler (1954) then conducted a simple experiment at Issaquah Creek, east of Seattle, Washington. They caught adult coho salmon that had ascended the east or main fork of the creek, tagged them, and released them below the confluence of the streams. Some fish (controls) were just tagged, whereas the experimental fish had been deprived of their sense of smell (made anosmic) by having their external nares plugged with cotton and Vaseline (table 5-5). Most of the controls repeated their previous choice (though some from the smaller east fork went up the main fork on the second trial). Some of the anosmic fish from the main fork went up the east fork, and most from the east fork ascended the main fork. The behavior of the anosmic fish clearly differed from that of the controls, though it was not random (they tended to follow the larger current of the main branch). One might criticize the small sample sizes and question the effects of handling and the operation that rendered the fish anosmic, but the simple experiment demonstrated the involvement of the olfactory system in homing by salmon.

Evidence that learning occured during the smolt period came from the transplant of coho salmon that established the population at the University of Washington (UW) hatchery in Seattle (Donaldson and Allen 1957). On January 19, 1952, juvenile coho salmon

TABLE 5-5. Recoveries of adult coho salmon caught in the main fork or east fork of Issaquah Creek, handled as controls or made anosmic, and released below the confluence (Wisby and Hasler 1954). The numbers of fish repeating their first choice are shown in bold.

		Recapture Site	
Treatment group	*Capture site*	*Issaquah*	*East fork*
Controls	Issaquah	**46 (100%)**	0
	East fork	8 (30%)	**19 (70%)**
Anosmic	Issaquah	**39 (76%)**	12 (24%)
	East fork	16 (84%)	**3 (16%)**

from Soos Creek, a tributary of the Green River, which flows into Puget Sound (fig. 5-2), were moved to the University of Washington and the Issaquah Creek hatcheries. The coho were differentially marked by removal of the left or right ventral fin and then released from their respective hatcheries March 18–19, 1952. No marked fish returned to Soos Creek, even though all of them had been incubated and reared there for most of their first year of life. Seventy of the fish reared at Issaquah Creek returned there, joined by only 1 from the UW release group. The UW hatchery got back 124 of its own fish, and not a single fish from the Issaquah Creek release entered the UW site, despite the fact that they had to swim directly past the UW hatchery to get to Issaquah Creek.

Further evidence for the role of olfaction in discriminating home-stream odors was provided by electrophysiological experiments (e.g., Hara et al. 1965; Ueda et al. 1967). In these studies, the olfactory rosettes of adult salmon that had returned to a hatchery were exposed to either the water from that hatchery or from other, more distant sources. Recording of electrical activity from the olfactory bulb of the brain indicated higher levels of activity in response to water from the home site than to water from other sites. These experiments showed that the olfactory system responded selectively to different odors, and it appeared that the authors had found a laboratory assay corresponding to the behavioral responses of migrating salmon to odors. However, this conclusion was criticized because the fish might have responded most strongly to the odors of home merely because this was the water in which they were held prior to testing. Bodznick (1975, 487) concluded that "no relationship was found between the relative [electro-encephalographic] response of the fish to the waters and the actual behavioral responses of the fish in the same waters during migration."

Streams presumably differ in a variety of odors (chemicals) and it is difficult to know which ones might be essential for homing. Hasler and his students had the brilliant idea of finding artificial odors that salmon could detect at low concentrations and using them to experimentally demonstrate imprinting. After trying many different odors they settled on morpholine and phenyl ethyl alcohol (PEA) as suitable. Scholz et al. (1976) did the really conclusive experiment by exposing smolts to an artificial odor and showing that they were attracted to this odor at maturity. Coho salmon were reared in hatcheries in Wisconsin that did not flow into Lake Michigan, and they were exposed to PEA at 10^{-3} mg/l or to morpholine at 5×10^{-5} mg/l in April and May as pre-smolts and smolts. Both groups and a set of controls (exposed to neither chemical) were then trucked to

FIGURE 5-2. Map of eastern Puget Sound showing the locations of the University of Washington (UW), Issaquah Creek, and Soos Creek hatcheries.

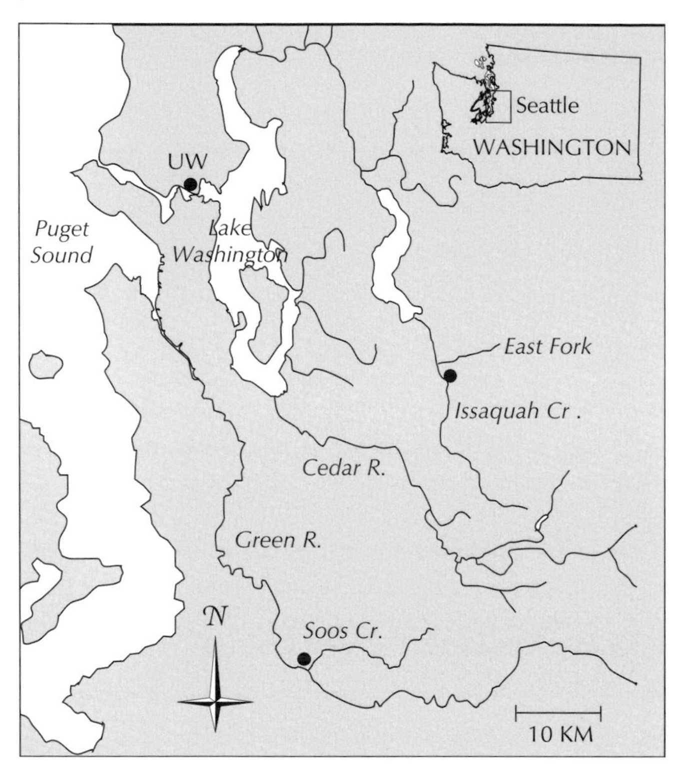

Lake Michigan and released. A year and a half later, when the fish were expected to mature, morpholine was metered into one stream (the Little Manitowoc River) and PEA into another site, the Two Rivers breakwater, at approximately the concentrations to which smolts had been exposed. These two sites, located equidistant from the release locations, were monitored for returning coho salmon, along with seventeen other locations in the area. More than 90% of the salmon treated with morpholine or PEA were caught at the site scented with the odor that they had experienced about 18 months earlier, and the controls were broadly distributed among the recovery sites (table 5-6). This elegant experiment clearly showed that the salmon had learned the odors at the smolt period and had used them to locate a stream that was otherwise entirely unfamiliar to them. Similar experiments on other salmonid species (e.g., brown and rainbow trout: Scholz, Cooper et al. 1978; Scholz, Gosse et al. 1978) revealed that olfactory imprinting is a general phenomenon in salmonids and not unique to coho salmon.

Further insights into the process of imprinting and homing came from experiments by Nevitt et al. (1994) and Dittman et al. (1997). It had been assumed that salmon olfactory epithelial tissue in the convoluted folds of the rosette inside the nares detected the chemicals during the imprinting period and the information was conveyed to the olfactory bulb of the brain for storage. Nevitt et al. (1994) showed that the sensory cells themselves have a form of "memory." They exposed juvenile coho salmon to PEA during the smolt stage and subsequently tested individual olfactory receptor neurons (photo 5-1)

TABLE 5-6. Homing patterns of coho salmon exposed to morpholine, PEA, or no odor (controls) as smolts, before being released into Lake Michigan. Bold numerals indicate salmon that homed; no river contained odors familiar to the controls so there was no "home" site for them (Scholz et al. 1976).

	% from each treatment group		
Recovery site	*Morpholine*	*PEA*	*Control*
Little Manitowoc River (morpholine)	**96.6**	5.5	26.7
Two Rivers (PEA)	1.3	**92.0**	19.3
Release site	0.1	0.0	2.8
Other sites	1.9	2.5	51.2
Number recovered	682	362	285

electrophysiologically for evidence of response to low concentrations of PEA. The cells from fish that had been exposed to PEA responded at lower concentrations than "naïve" cells. This was not because the fish had better olfactory systems in general, as their responses to a novel odor were similar. This phenomenon of "sensitization" allows the organism to learn a range of odors that it might experience, rather than being genetically predisposed to detect a certain odor. It is a very flexible system by which the sensory capacity augments the memory of imprinted odors. This is especially amazing when we consider that olfactory receptor cells are thought to turn over every few weeks or months. Thus the actual cells that were exposed to the odors in the juvenile fish may not have survived to adulthood. Sensitization to imprinted odors must be passed to newly differentiated replacement cells by a mechanism not yet understood.

Using this same cohort of PEA-imprinted fish, Dittman et al. (1997) demonstrated that this cellular sensitization may be due in part to sensitization of an intracellular signaling enzyme, guanylyl cyclase, thought to be involved in odor recognition. PEA stimulation of guanylyl cyclase in olfactory neurons was heightened in adults that had previously been exposed to this odorant as juveniles relative to PEA-naïve fish. Interestingly, this heightened PEA sensitivity was only observed in maturing fish at the time of year when they would have been completing the final phase of their homing migration. These results parallel earlier behavioral and electrophysiological results (reviewed by Hasler and Scholz 1983) that also indicated heightened sensitivity to imprinted odors at the time of homing, not during earlier periods.

Salmon use odors, learned as juveniles, to locate their natal river, but we still need to know how the fish use the odors. Do they seek the highest concentration of the long-remembered odors? P. B. Johnsen (1982) presented a simple "control model" of the process, based on experiments conducted with imprinted coho salmon. He concluded that salmon react to home stream odors as a "sign-stimulus." That is, detection of odors from the natal site stimulates salmon to swim upstream; in the absence of such odors they tend to swim downstream. So, a salmon migrating along the coast would detect freshwater from many rivers and might initially "investigate" them but would not ascend far up the rivers because the stimulus would be absent. Upon reaching the mouth of the natal river system, the fish would detect imprinted odors and migrate upstream. If it came to the

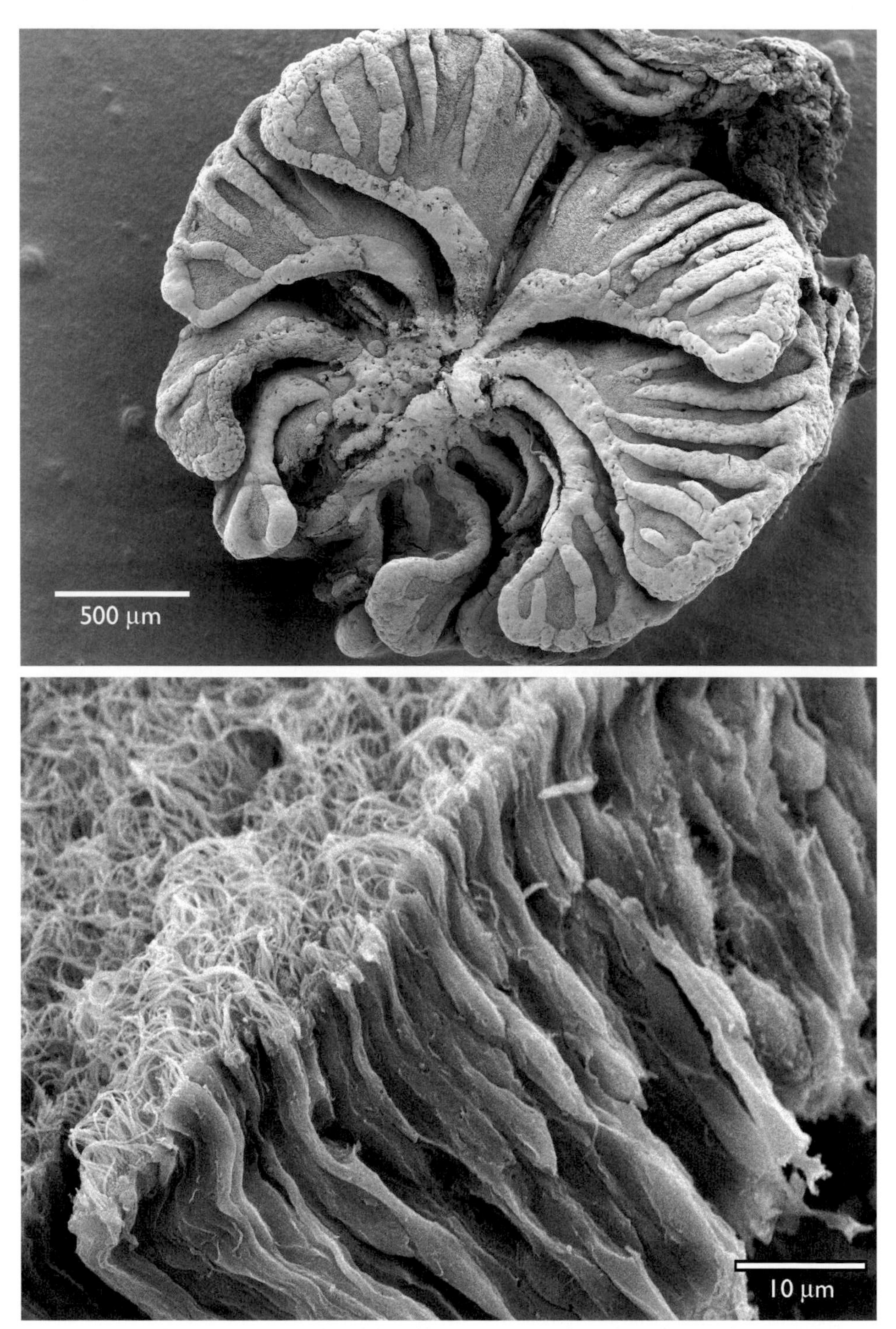

5-1 Scanning electron micrographs of the olfactory rosette of a coho salmon. The top panel shows the whole rosette with its lamellae. Lamellae are covered with olfactory sensory neurons, shown at higher magnification in the lower panel. Note the cell bodies within the epithelium and their cilia extending into the water where odorants are detected. Photographs by Carla Stehr, National Marine Fisheries Service.

FIGURE 5-3. Control model showing hypothesized responses of salmon to the odors of their natal stream that would enable them to migrate up a complex river system (based on Johnsen's 1982 model, with significant modifications).

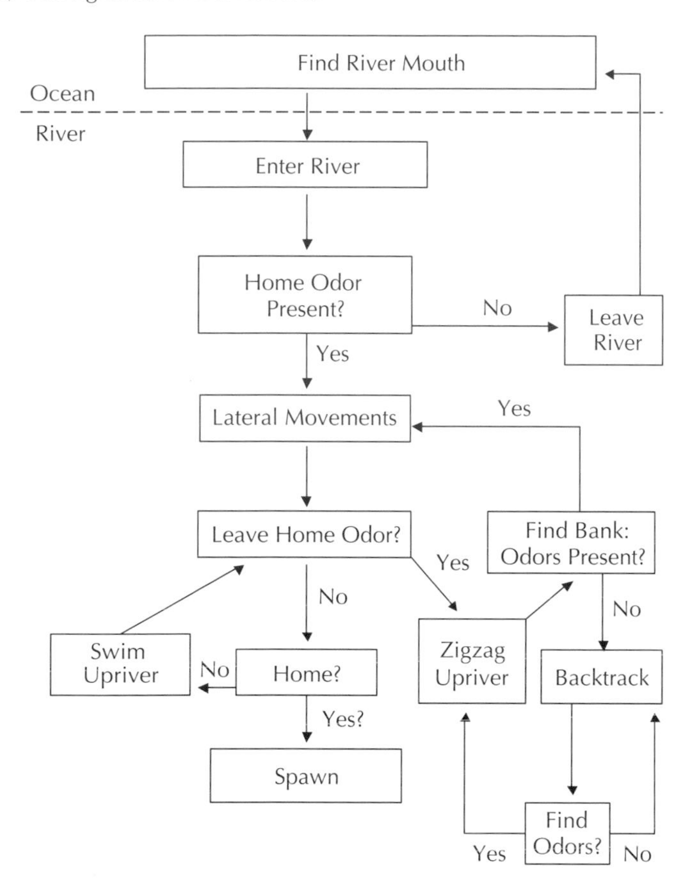

confluence of two branches, it might move up the wrong one for a short distance because eddying currents might carry the odors back and forth. However, the fish would eventually come to water devoid of the natal odors, and it would start to move downstream and zigzag back and forth until it again encountered familiar odors. Thus the fish could work its way up a complex network of streams to the natal tributary (fig. 5-3).

The experiments by Hasler and his students (reviewed in Hasler and Scholz 1983) provided convincing evidence (and a mechanistic explanation) for imprinting at the smolt stage and homing responses by maturing adults. However, some scientists concluded that salmon may (or must) imprint sequentially during their freshwater life-history stages and during seaward migration (e.g., Harden Jones 1968; Wagner 1969). In part this hypothesis sprang from skepticism that a salmon at the mouth of a huge river such as the Columbia or Yukon can really detect the odors of some trickle of a tributary more than 1000 km upriver. The dilution of the natal stream odors in the volume of the whole river seems too great. In addition, consideration of the natural history of salmon indicates that things cannot be as simple as imprinting solely at the smolt period. Sockeye salmon typically spawn in tributaries of lakes but leave these streams as soon as they emerge from the gravel redds. As smolts they migrate from the outlet of the lake and this may be distant from the spawning grounds. The complex population structure of sockeye salmon can only be established and maintained by homing to natal streams rather than to the lake as a whole. In addition, a number of studies of coho (e.g., Peterson 1982a) and chinook salmon (Murray and Rosenau 1989) revealed movement within river systems prior to smolt migration. The persistence of populations seems to require imprinting during the alevin stage or during emergence.

There are surges in thyroxine during the hatching-emergence period, and these may coincide with periods of olfactory learning. Perhaps more fundamentally, there seems to be an integral connection between migration itself and the imprinting process (Dittman and Quinn 1996). It appears that internal rhythms, synchronized by photoperiod, stimulate the smolt transformation, including many aspects of behavior, morphology, and physiology (see chapter 12), including imprinting and downstream migration. Migration brings the fish into contact with new odors, and these odors seem to stimulate surges in thyroxine (Dickhoff et al. 1982), as though the fish are logging a series of "olfactory waypoints" at the confluence of each new river they encounter on their way to sea. Salmon that do not migrate do not seem to learn the odors of their natal stream. Dramatic evidence of this came from an experiment in which coho salmon were raised for their whole lives in a hatchery (thus getting every possible opportunity to learn its odors). As adults they were taken about 1 km away and released but they failed to return, even though members of their cohort that had been released as smolts had no difficulty returning (Dittman et al. 1996). Further evidence that mere exposure to odors at the smolt stage is insufficient for imprinting comes from experimental releases of salmon at different seasons of the year (Unwin and Quinn 1993). The highest proportion of salmon homed if they were released from the hatchery as smolts at about the time of year when they would normally migrate to sea. Not only did more of the fish released earlier stray, but more of the fish held longer (and thus exposed to home odors for a longer time period) strayed as well.

If salmon learn a sequence of odors, why didn't the coho salmon spawned and reared for almost a year at Soos Creek but released by Donaldson and Allen (1957) at the University of Washington and at Issaquah Creek return to Soos Creek? Perhaps the fish return first to what they detected last and then search for the next odor in the sequence. At the UW and Issaquah Creek hatcheries, the salmon would have been unable to detect Soos Creek water, and unable to leave the hatcheries in any case, so they did/could not return to Soos Creek. Consistent with this hypothesis, when salmon are taken from their hatchery and released within the same basin, they are more likely to return to the natal (i.e., incubation) site than if they are released outside the basin.

Transplants of salmon among rearing sites are thus consistent with imprinting at more than one stage, but what happens if the sequence of learning during downstream migration is disrupted? Three kinds of studies, conducted in the Columbia River system, are informative. First, salmon smolts taken from a hatchery, driven by truck to a site at the lower river, and released there (to reduce the mortality associated with downstream migration), returned as adults to the site in the lower river. Specifically, coho salmon trucked from the Little White Salmon hatchery to Youngs Bay returned to Youngs Bay, not to the hatchery (Vreeland et al. 1975). From this we might infer that no learning took place at the hatchery, but is this correct?

Second, transportation of smolts from a hatchery to a range of freshwater and marine locations dramatically demonstrated the effects of a disruption in the learning process. Solazzi et al. (1991) reared coho salmon at Cascade Hatchery, near Bonneville Dam. Coho smolts were released at six locations: below Bonneville Dam (river km 234), at Tongue Point (rkm 29), the bar of the river (rkm 2), 19 km offshore in the river's plume, 19 km offshore outside the river's plume, and 38 km offshore in nonplume water. These six locations, progressively farther from the rearing site in both distance and odor connection, produced the following proportions of salmon that returned to rivers outside the Columbia River system: less than 0.1%, 3.4%, 4.1%, 6.1%, 21.0%, and 37.5%. The fact that more than one-third of the salmon reared entirely in a hatchery on the Columbia River system failed to return to this huge watershed after release outside the river plume was particularly striking. Why had the fish not imprinted properly?

In contrast to the impairment of homing indicated when salmon are transported from their "point of origin," there is evidence that fish captured during their seaward migration can home successfully, despite missing a large segment of the migration route. Chinook and steelhead captured at Ice Harbor Dam on the Snake River and trucked to a release site below Bonneville Dam (see fig. 5-1) did not show significant impairment in homing to upriver sites (Ebel et al. 1973; Slatick et al. 1975). Even transportation 400 km from Little Goose Dam to sites below Bonneville Dam did not seem to prevent the fish from homing back upriver (Ebel 1980). These studies seem consistent with the hypothesis that imprinting is linked to migration.

Pheromones

There is clearly a great deal more to be learned about the mechanisms of imprinting and homing by salmon. The many insights provided by the use of artificial odors distracted many scientists from the question, "What are the natural odors that salmon use for homing?" Most scientists assumed that they were combinations of organic and inorganic molecules (both presence/absence and concentration), though attempts to determine the critical odors were largely unsuccessful. However, Nordeng (1971, 1977) hypothesized that it is not odors from rocks, plants, and soil that salmon imprint on, as Hasler and his students hypothesized, but rather that salmon imprint on population-specific odors from juvenile salmon residing in the stream and migrating to sea. Nordeng hypothesized that each salmonid population has odors that are unique to it and that there is innate recognition of these odors by returning adults. Thus chemicals (termed "pheromones") released by juveniles residing in the stream and migrating to sea may provide the chemical basis for homing.

In some ways the "pheromone hypothesis" was parallel to Hasler's imprinting hypothesis, and both were well defined and readily testable. Nordeng hypothesized that (1) fish release population-specific odors; (2) the chemical differences among populations are detectable by conspecifics; and (3) the odors guide the returning adults. This hypothesis seemed plausible to scientists working on anadromous Arctic char and Atlantic salmon because juveniles of these species commonly rear in streams for at least 2 years before migrating to sea, so there would always be a pool of resident "odor producers." In contrast, sockeye salmon typically leave their natal streams as fry and the streams may be devoid of juveniles when adults return. In many pink salmon populations, the stream would have been devoid of juveniles for about 1.5 years if the population only occurred on one cycle.

It seems hard to believe that the odors of populations could differ, but one will seldom win money betting against the sensory systems of animals. Several studies have shown that juvenile and adult salmon can distinguish the odors of their own population from other populations. For example, Courtenay et al. (1997) used two-choice experimental tanks to show that coho salmon from different Vancouver Island populations, spawned on the same day, incubated and reared in the same water, and fed the same diet, could distinguish members of their own population from the other populations, even if they had not been reared with those individuals. These and related experiments showed that the salmon can distinguish populations based on odors, and therefore chemical differences must exist among populations. The key question, however, is, "Are the odors actually used by salmon for homing?" The answer, in my mind, seems to be "no," for several reasons. First, in pink, chum, and sockeye salmon, there are normally no juveniles residing in the spawning stream when adults return, so the odors would have to be retained in the gravel from eggs and alevins. This is possible, however, as lake trout in experimental arenas tended to construct nests at sites with the odors of egg membranes and feces from fry (Foster 1985). Second, there are many instances of transplants that clearly show the salmon returning to the release site, not to the site where their populations abounded (e.g., the transplant that established the University of Washington hatchery population by Donaldson and Allen 1957; Candy and Beacham 2000). Finally, in a specific experiment designed to test the hypothesis, adult coho salmon swam past the University of Washington hatchery, containing juveniles of their population, and returned to their release site on Lake Washington where there were no conspecifics at all (Brannon and Quinn 1990).

Adult salmon are attracted to the odors of conspecifics, and salmon can distinguish species, sex, reproductive state, population, and even their own family within a population based on odors (Brown and Brown 1996). I am not convinced that these odors are responsible for homing, though they may play some role. Pheromones may be much more important in reproductive behavior, perhaps mediating inbreeding and outbreeding. Individuals might tend to discriminate against their siblings but also discriminate against strays from other populations when choosing mates. This hypothesis, however, remains to be tested.

Genetic aspects of homing

The foregoing review of the mechanisms of homing has been based on the assumption that the odors used for return migration were learned prior to or during the seaward journey. Some interpretations of the pheromone hypothesis call for genetically based recognition of population-specific odors but learning is clearly central to the imprinting hypothesis. Is there also some genetic propensity to return to the ancestral site? The vast majority of transplants show that the salmon return to the site of their release, and this is inconsistent with a genetic basis for homing. However, a pink salmon transplant experiment conducted by Bams (1976) provided the first evidence for the role of genetics. Gametes from male and female pink salmon were collected from the Kakweiken

TABLE 5-7. Patterns of recovery (percentage of total from these locations) of three groups of adult chinook salmon: Bonneville Hatchery salmon released from Bonneville Hatchery (BH controls), Priest Rapids salmon that had been reared and released from Bonneville Hatchery (PR transplants), and Priest Rapids salmon that had been reared and released from Priest Rapids Hatchery (PR controls). Data from Pascual and Quinn (1994).

Recovery locations	*BH controls*	*PR transplants*	*PR controls*
Hatcheries below Bonneville Dam	0.2	0.0	0.0
Bonneville Hatchery	98.5	84.1	0.0
Hatcheries just above Bonneville Dam	1.1	9.2	0.0
Hanford Reach and Priest Rapids Hatchery	0.0	6.4	98.0
Upper Columbia and Snake rivers	0.1	0.4	2.0
Total	3370	3640	4668

River, on the mainland coast of British Columbia, and brought more than 200 km to Headquarters Creek, a tributary of the Tsolum River on the east side of Vancouver Island. Pure Kakweiken stock were incubated, marked, and released from Headquarters Creek, along with the marked progeny of Kakweiken females and Tsolum males. The pure Kakweiken (i.e., nonlocal) fish were about half of the recoveries at canneries (45.4%, indicating comparable survival rates), but they were only 22.0% of the fish recovered in the Tsolum River, and only 8.8% of those actually returning to Headquarters Creek. Thus the nonlocal salmon survived to adulthood but for some reason they were more "reluctant" to enter the site where they had imprinted than fish with local genes. Bams concluded that "addition of the local male genetic complement improved return to the natal river system [but] the male complement alone was not sufficient to achieve normal accuracy of return to the natal tributary within the system" (2716).

This fascinating study was apparently the only experimental evidence for a genetic component to homing until efforts to resuscitate a run of Columbia River chinook salmon provided additional evidence and perspective. Most populations of ocean-type chinook salmon in the lower Columbia return in fall to spawn and are in a comparatively advanced state of maturity. There are other ocean-type chinook known as "upriver brights" that are more silvery when they arrive in the river and migrate farther upriver to spawn, primarily in the main stem of the Columbia River at a region known as the Hanford Reach, below Priest Rapids Dam. Adults from this population were intercepted at Bonneville Dam on their homeward migration, held below the dam at Bonneville Hatchery, and spawned there (see fig. 5-1). The progeny of these fish, of upriver ancestry, were tagged and released from Bonneville Hatchery, along with the normal chinook from this hatchery. In addition, upriver brights produced in the Priest Rapids Hatchery were also marked. The project was not designed to examine the genetic basis for homing, but from that perspective there were Bonneville Hatchery controls, upriver controls, and fish of upriver ancestry reared and released from the lower site (McIsaac and Quinn 1988; Pascual and Quinn 1994).

The results revealed an intriguing mix of learned and innate homing behavior (table 5-7). All the upriver controls returned to the natal area (Priest Rapids Hatchery or the spawning grounds at the Hanford Reach) except a few that strayed to other upriver sites. Virtually all the Bonneville Hatchery (BH) controls returned to Bonneville Hatchery, with a few straying to other hatcheries. Most of the salmon in the experimental group returned to Bonneville Hatchery, as imprinting would dictate. However, a significantly larger number (9.2 vs. 1.1% for the BH controls) ascended Bonneville Dam and entered hatcheries above the dam. Moreover, many more of the experimental fish were caught in gillnet fisheries above Bonneville Dam than the BH controls. Both these lines of evidence indicated a greater tendency to migrate upriver in the Priest Rapids fish than the controls reared at Bonneville Hatchery. Most amazing were the estimated 232 fish that actually migrated to the Priest Rapids Hatchery or the Hanford Reach spawning grounds, where they had never been.

What could possibly have brought some of the upriver brights to their ancestral spawning grounds, despite the imprinting at the Bonneville Hatchery? Population-specific odors (pheromones) from the upriver site might have played a role, though there would have been such odors from members of their population in Bonneville Hatchery as well. It may be that these salmon are "programmed" in some genetic way to swim a long distance upriver and then begin searching for home, causing some to pass their rearing and release site. Alternatively, the population may have a genetic disposition to spawn in the main stem of the river and they search for appropriate substrate, depth, velocity, temperature, and other habitat features. Faced with a conflict between the imprinted odors and habitat features at the entrance to Bonneville Hatchery, most entered but many did not. Given the wide variety of habitats where salmon spawn and the adaptations that they show in traits like adult body size and egg size, it would not be surprising if there were innate, population-specific preferences for the appropriate habitats. Further evidence for a genetic component comes from the report by Hard and Heard (1999) of differences in homing by two populations of chinook salmon reared and released from the same hatchery. The researchers also found that the progeny of parents whose gametes were brought to the hatchery were more likely to stray than the progeny of parents who had homed there.

Summary

Homing to the natal site is the characteristic behavior pattern in all salmonids; about 95–99% of the fish that survive to adulthood do so. It is not clear how much variation there is among species and what factors influence straying. Indeed, it is unclear if straying reflects a failure to home or a decision to spawn elsewhere. In either case, straying has been essential for the persistence and distribution of salmon during the glacial advances and retreats and during lesser disturbances that have taken place within the present range of salmon. From a mechanistic perspective, homing is accomplished by (1) unconditioned, irreversible learning of odors from the natal stream at one or more stages prior to and during seaward migration; (2) retention of these odor memories at sensory and central levels; and (3) stimulation of upstream swimming in response to the odors at maturity. However, the nature of the odors that are learned, the role of pheromones (species-specific or population-specific odors), the genetic aspects, and the interface between homing and habitat selection are still not well understood.

6

Mating System and Reproductive Success

Salmon spend very little of their time in reproductive activities but these few days are inordinately important for them, as they are the culmination of years of growth and predator avoidance. Before describing the salmon mating systems and the factors that determine reproductive success (i.e., number of surviving progeny), some general concepts should be introduced. In most animals, females devote more energy to gametes than males do. Females usually produce a smaller number of large gametes compared to males. As a consequence of this "anisogamy," the maximum number of offspring a female can produce is smaller than the maximum for a male. A female fish can have no more offspring than she has eggs (about 2000–4000 in salmon), but the number of offspring that a male can sire is limited only by his ability to deliver sperm cells to ripe eggs. Females, and their eggs, thus limit the productivity of the population, and males tend to compete for them. With the exception of clonal species, animals have a father and a mother. Therefore, assuming an equal sex ratio among adults, the average reproductive success of males and females is the same. However, there is usually less variation in reproductive success among females than males. The variation in male reproductive success drives competition for access to females and provides females with the opportunity to choose which male(s) to mate with. These processes of sexual selection generally fall into two categories: intrasexual competition (direct competition between members of the same sex) and intersexual selection (mate choice).

Female salmon select, prepare, and guard their redds, and they compete vigorously among themselves (including females of other species) for the best sites. Males do not participate in redd construction or guarding but instead compete for access to ripe females. Females tend to prepare their redds and spawn soon after arriving on the spawning grounds, so they can maximize their time guarding the redd from disturbance by

PLATES 10 and 11. Illustrations of a mature male (above) and female (below) sockeye salmon at the height of sexual maturity. The size of the dorsal hump in the male and the shade of red vary among populations. The red pigment in the muscle of salmon at sea is transferred to the skin at maturity, leaving the muscle white. Females are not as red as males because some of the

red pigment from the muscle is shunted to the eggs rather than the skin. Small spots may be seen on the tail, and the caudal peduncle (base of the tail) is often a lighter shade of red than the body. Copyright: Charles D. Wood, Ph.D.

other females. Males, on the other hand, are reproductively active (or at least potentially so) during essentially their entire lives in the stream. Consequently, there are more males than ripe females during most of the breeding season, and thus competition for females is keen. This fundamental difference in sex roles and reproductive patterns drives much of the behavior seen on spawning grounds.

This chapter deals with the factors determining reproductive success in salmon: the number of viable offspring produced by adults that have survived to breeding age. Needless to say, the success of the male depends in part on attributes of the female, and vice versa. However, the sex roles of salmon are sufficiently discrete that it is helpful to describe the patterns of behavior and selection of each sex separately and then to consider interactions. The general reproductive system is very similar among all salmonids, with only relatively minor variation (mostly between semelparous and iteroparous species). I will first describe the process of spawning-site selection and preparation by the female, then describe courtship and reproductive behavior per se, and end by discussing the factors influencing the number of viable offspring produced by females and males.

Redd site selection and preparation

Each female selects and prepares a redd, and the semelparous species also guard it after spawning. This is a critical aspect of the salmon life cycle because the great majority of lifetime mortality generally takes place during the period of incubation in the gravel, and much of that is related to features of the site (see chapter 8). Females must accomplish something entirely novel to them: choosing a site that will be suitable for precious embryos to incubate for many months after she has died, and preparing the gravel to receive those embryos. There is a large literature describing the physical criteria that salmon seem to use in selecting redd sites. The criteria are typically defined by comparing the range of conditions available to fish and the conditions where redds are found. The most common physical variables measured are stream depth, velocity, and substrate size (sand, gravel, etc.). These features of streams covary (e.g., deep, slow-moving water tends to have fine substrate), so it is sometimes difficult to determine what feature is most important to the fish.

Preferred habitat is determined by the incubation needs of embryos: high flow of oxygenated water through the interstitial spaces in the streambed. The larger the salmon, the larger the gravel that they can move, and larger species of salmon tend to use larger gravels for redd sites (fig. 6-1). Larger fish can also hold their position in the faster water associated with such large gravel. In addition, large females produce larger eggs than smaller females, and such eggs may require more space. Salmon use a limited subset of the available physical habitat, usually avoiding the slowest water and the fine sand and silt found there, and also avoiding the fastest water. Water about 30–60 cm deep, flowing about 30–100 cm per second over coarse sand and small-to-medium gravel (2–10 mm diameter) is generally suitable. For example, measurements in Kennedy Creek, Washington, revealed that the average depths at chum salmon redds were somewhat shallower than at nearby sites without redds (29 vs. 48 cm), but the redd sites were much less variable in depth (standard deviations: 11 vs. 31 cm), reflecting avoidance of very shallow and deep sites. The average velocity at redd sites was similar to that at sites not used

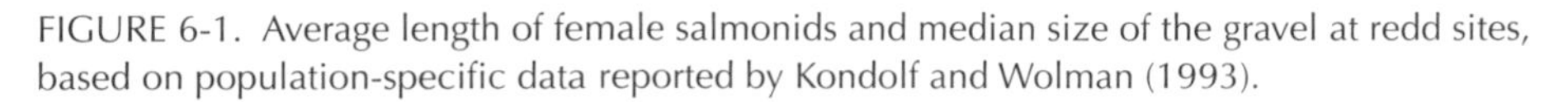

FIGURE 6-1. Average length of female salmonids and median size of the gravel at redd sites, based on population-specific data reported by Kondolf and Wolman (1993).

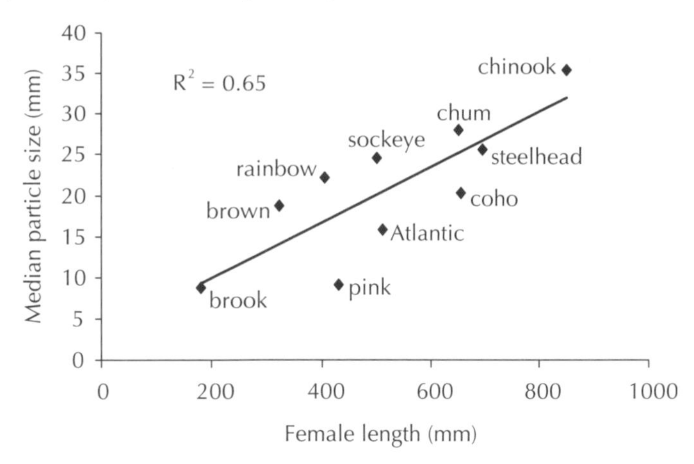

(32 vs. 30 cm/sec), but the sites used represented a narrower range (standard deviation: 22 vs. 27 cm/sec; Quinn, unpublished data, fig. 6-2). However, redd site selection must be considered in the context of available habitat and the sources of mortality that prevail at that stream. In areas where winter freezing can destroy embryos, salmon seem to favor sites with upwelling groundwater, which will be warmer in winter than river water, though it may also have lower levels of dissolved oxygen (e.g., Leman 1993). In addition, females choose sites based on the ambient conditions when their eggs are ripe. The river may later rise or fall, making the chosen site more or less favorable. In semelparous species females rarely make a second redd, though iteroparous females may do so (Barlaup et al. 1994).

We have developed an image of the habitats used by "typical" salmon in "typical" streams, but it is also important to recognize the range of conditions used. In the main stem of the Columbia River, for example, chinook salmon spawn in water as deep as 6.5 m, on substrate 2.5–15 cm in diameter, and at bottom velocities up to 2 m/sec (Groves and Chandler 1999). At the other end of the continuum, small salmonids such as resident

FIGURE 6-2. Water depths and velocities at the locations of chum salmon redds in Kennedy Creek, Washington, and the measurements at regularly spaced points in the same area without redds (Quinn, unpublished data).

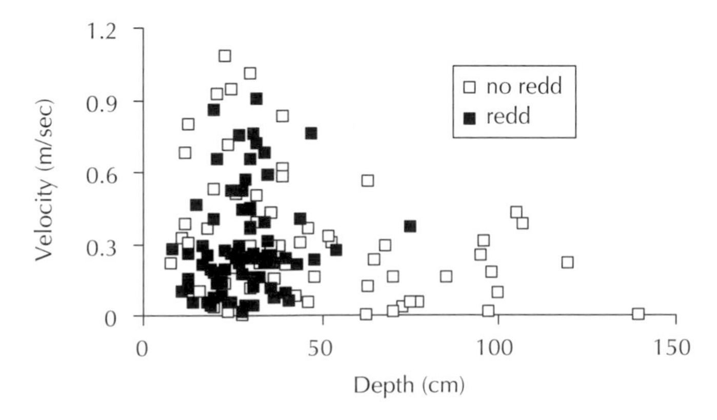

6-1 Female chum salmon digging her redd while the male waits behind her. Photograph by Manu Esteve, University of Barcelona.

golden trout select sites 5–20 cm deep, substrate 0.4–1.2 cm in diameter, and velocities of 30–70 cm/sec (Knapp and Vredenberg 1996). In addition to the variation in stream habitats that may be used, some salmon (notably sockeye) spawn on beaches of lakes. Some of these habitats are in alluvial deposits at the edges of the lake, where subsurface water from a river irrigates embryos. In other cases there is no surface water nearby but groundwater flowing down a hillside upwells on a beach with suitable sand and gravel for spawning. Salmon may dig redds in surprisingly fine sand at these beaches and in ponds fed by springwater (Quinn et al. 1995). On the other hand, Iliamna Lake, Alaska, is known for large numbers of sockeye salmon spawning on beaches of low-lying islands with no groundwater. These females use very large substrates (median diameters of about 6–7 cm) because wind-driven surface currents irrigate the embryos rather than groundwater (Kerns and Donaldson 1968; Leonetti 1997; photo 1-3).

Putting aside the variation in habitats used among species and populations, we can return to the process of redd site acquisition. The first female that arrives in the stream can select whichever site she chooses. Females arriving later have three choices: select another site, drive off the territory holder, or wait until she dies. Territory holders resist eviction (Foote 1990), especially after their eggs have been deposited; and females may not be able to delay egg deposition long, so increasing density results in use of progressively less suitable habitat. Once a site has been chosen, the female digs a redd by facing upstream, turning on her side, arching her back, and sweeping her tail downward several times (Burner 1951). Keenleyside and Dupuis (1988) reported that

6-2 Female chum salmon digging her redd while the male waits behind her. Photograph by Thomas Quinn, University of Washington.

pink salmon averaged 0.75 digs per minute, decreasing to 0.51 per minute immediately before spawning. The arched back seems to help the female stay in place rather than swim upstream, while the tail sweeps do the digging (photos 6-1, 6-2). Water is incompressible; the downstroke loosens sand and silt and they are wafted up into the water column, perhaps lifted by the upstroke, and they drift downstream. Thus the female digs a crater in the streambed gravel and also winnows the material, leaving larger rocks and gravel while displacing the finer particles (Kondolf et al. 1994). These fine materials build up on the downstream side of the crater and are termed the "tailspill." Although this sorting of the gravel may simply be a by-product of the digging process, it increases the flow of oxygenated water through the interstitial spaces in the gravel, facilitating respiration by the embryos.

While the female is preparing her redd she may be attacked by females holding adjacent territories. For this reason, females not ready to spawn generally remain in pools in streams or in larger rivers and lakes, moving onto the spawning grounds only when they are fully ripe. Attacks often are directed at females in the very act of digging, presumably because they are vulnerable and cannot easily retaliate. Males have larger jaws and teeth than females, deceiving some people into thinking that females are less aggressive than males, but this is not really the case. Observations of hundreds of sockeye, pink, and chum salmon at the Weaver Creek spawning channel, British Columbia, showed that the average attack frequencies by males and females were similar, with only somewhat more attacks by males than females (table 6-1). However, male attacks were directed almost exclusively towards other males whereas females attacked males almost as often as they attacked females, including males of other species (photo 6-3).

TABLE 6-1. Number of attacks by male and female salmon against their own and the other sex (all species combined) per 15-minute observation period in the Weaver Creek spawning channel, British Columbia (from Quinn 1999 and additional unpublished data on a total of 144 pairs of chum, 133 pink, and 459 sockeye salmon). "Male-female" refers to attacks by males on females; "female-male" refers to attacks by females on males.

Species	*male-male*	*male-female*	*female-male*	*female-female*
chum	8.2	1.1	2.7	3.3
pink	7.0	1.2	4.7	3.6
sockeye	5.7	0.6	2.2	3.9

Courtship and mating behavior

While the female is preparing her redd she is courted by one or more males (Mathisen 1962). The sole or the dominant male positions himself just downstream of her. This male will frequently "cross over," swimming from one side of the female to the other, remaining just behind her. This position may enable the male to detect odors indicating the female's species, sex, and reproductive state (Emanuel and Dodson 1979; Honda 1982). The crossing over also enables the male to position himself for fertilization and to drive competitors from the female. The male displays to the female, often moving up along her side and rapidly quivering his flanks. He also interacts with other males, both in a side-to-side "parallel" display and also by moving upstream of the competitor, turning perpendicular to the other male, and drifting down against him in a "T-display" (Schroder 1981). If these displays are not sufficient to establish dominance, the males engage in direct combat, chasing each other, ramming with closed jaws, and biting. The attacking male may try to get his jaws around the caudal peduncle of the other male, or even across the middle of his body from back to belly if the victim is small enough (photo 6-4). These forms of attack can result in serious wounds, and the males quickly become very scarred.

If ripe females are scarce (either because more males have arrived than females or because most of the females have already spawned their eggs and are guarding their redds), then hierarchies of males are established. The dominant male, often referred to as the alpha male, positions himself closest to the female. By a series of displays and attacks, he establishes his dominance over other males, and they position themselves at increasing distances from the female, inversely related to their status. These males are collectively termed "satellites" because they tend to orbit to some extent around the female and the dominant male. However, there may be other males, typically small ones such as jacks or sexually mature parr, that do not participate in fights and so are not part of the dominance hierarchy. They tend to remain at the periphery of the courting group, under a bank or other structure, and slip in to fertilize eggs at the moment when they are released. The success of this tactic will be reviewed in the next section.

Only the female is involved in active construction of the redd but males also dig, though much less frequently than females (Quinn 1999). These digs are more perfunctory and do not seem to move gravel or create a pit in any systematic way. Dominant males dig more often than satellites, and males competing with satellites are more likely

6-3 Female pink salmon attacking a male sockeye salmon. Photograph by Manu Esteve, University of Barcelona.

6-4 Male sockeye salmon attacking another male sockeye. Photograph by Gregory Ruggerone, Natural Resources Consultants, Inc.

6-5 Male (left) and female (right) chum salmon spawning. Photograph by Manu Esteve, University of Barcelona.

to dig than those without competitors, suggesting that male digging is a competitive display. However, males courting females sometimes dig in the absence of competitors, so digging may also be a courtship display. There may be more than one form of digging, or more than one function, and this behavior pattern has yet to be fully explained.

When the depth of the redd and the coarseness of the gravel meet the female's criteria, and she is courted by an acceptable male, she will be ready to spawn her eggs. This often occurs within a few days after she has arrived on the spawning grounds. The male signals his readiness by repeatedly crossing over and by quivering alongside her body. These vibrations may be an important signal coordinating spawning (Satou et al. 1991). The eggs, previously connected to a skein of connective tissue in the ovary, are now loose in the female's body cavity and are squeezed out her vent by muscular contractions of the body walls. One or more males will rush in along her flanks and release milt, including sperm cells, as close to the female's vent as possible (photos 6-5, 6-6). Fertilization occurs quickly, and the penetration of the egg membrane by a sperm cell (photo 6-7) is facilitated by the uptake of water by the egg. This "water hardening" renders the eggs incapable of being fertilized after about 20 seconds exposure to water, and the milt is also much less successful in fertilizing eggs after 10–20 seconds in the water (Liley et al. 2002).

The female soon begins a series of gentle "covering" digs, starting just upstream from the pocket containing the eggs, to bury the eggs and begin digging a new pocket. Keenleyside and Dupuis (1988) reported a rate of 5.05 digs/min by female pink salmon in the 5 minutes after spawning. Later digging completely buries the first egg pocket while constructing another pocket. The eggs of a female are not deposited all at once but

6-6 Male (right) and female (left) sockeye salmon spawning. Photograph by Manu Esteve, University of Barcelona.

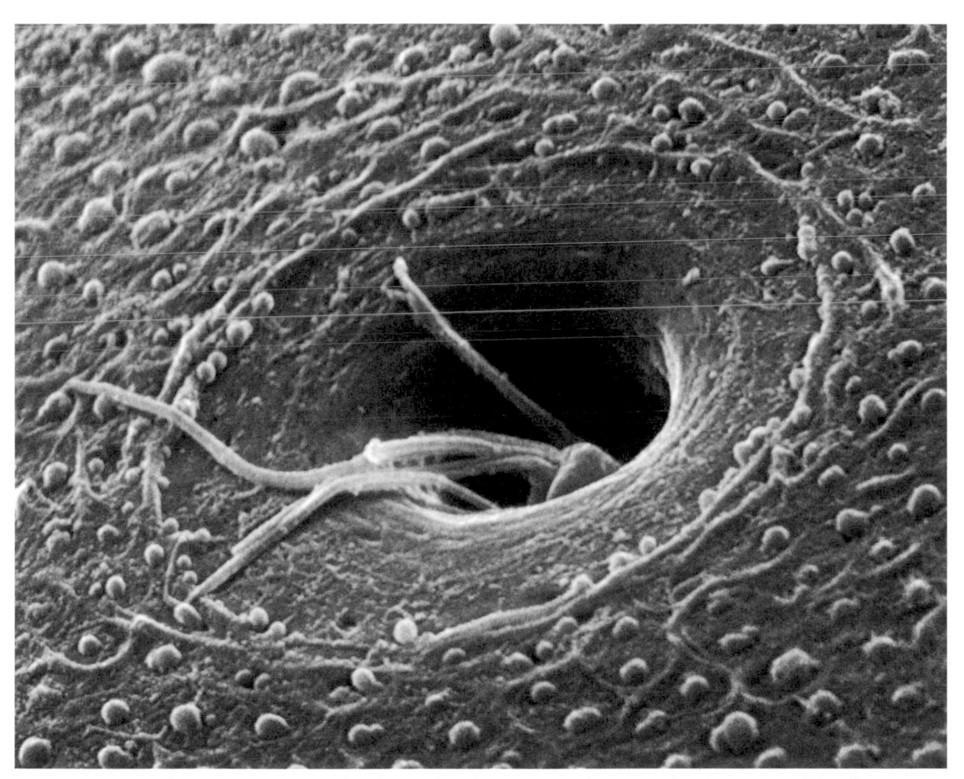

6-7 Scanning electron micrograph of a pink salmon egg, showing the moment of fertilization as sperm cells disappear down the micropyle. Photograph by Carla Stehr, National Marine Fisheries Service.

in a series of pockets. If all eggs were literally in one basket, the ones in the middle might not have access to enough oxygen to develop properly. If they died, fungus might kill the other developing embryos. Thus the multiple egg pockets probably reduce the female's risk of losing all her offspring. In addition, there may be some benefit in mating with several males, in case the sperm from one are of poor quality, or in maximizing the genetic diversity of the female's offspring. Hawke (1978) found seven chinook salmon redds that had been stranded by changes in river level and he meticulously uncovered four to six egg pockets in each. Female Pacific salmon almost never complete more than one redd, though iteroparous species such as steelhead and Atlantic salmon may do so within one season (Fleming 1998). Females release progressively fewer eggs in the second and subsequent spawning events. When all eggs have been released (usually in a few days), the semelparous females guard the redd from encroachment by other females and continue digging to keep it in good shape. When they eventually become too weak to hold position in the stream, they drift downstream and die.

The eggs in separate pockets may be fertilized by the same male(s) or he (they) may court other females. Healey and Prince (1998) found that dominant male coho salmon and jacks tended to move within a 200–300 m reach, averaging less than 100 m between locations, whereas satellites moved much more widely. When densities are low, movement may be the only way for a male to obtain other matings. Our observations of sockeye salmon at high densities in Bristol Bay, Alaska, indicate that males do not move widely in the streams. Rather, when the female that they are courting has spawned all of her eggs, the males seem to wait for another female to arrive. This may minimize the fighting needed to established hierarchies in other locations and is consistent with Foote's (1990) observations on the importance of prior residence in male dominance.

Female reproductive success

A female salmon can have no more offspring than the number of eggs she produces, though many factors affect survival during incubation and subsequent life-history stages. Fecundity increases with body size but the relationship is often quite variable, as illustrated by data on coho salmon from Forks Creek, Washington (Quinn, unpublished data, fig. 6-3). In this case, only 22% of the variation in fecundity was explained by length. Some females had at least twice as many eggs as others of similar size. The relationship with female weight was no better. Larger females tended to have larger eggs as well as more numerous eggs, but individuals have a finite amount of energy that they can devote to gonads (averaging 21.5% among these coho). Individuals can make a large number of small eggs or a small number of large eggs but they cannot make many large eggs. Consequently, females with many eggs, for their length, have to sacrifice egg size, and females with large eggs have to sacrifice fecundity. Large eggs have more yolk than smaller eggs and produce larger fry. These fry may be more resistant to starvation if food is scarce and may have higher survival rates (see evidence in chapters 10, 11, and 15), thus providing advantages to large females in both the number of offspring and their odds of survival.

Fecundity increases with length, but the average length of females varies among species, as does the relationship. Based on data I obtained from many sources, anadromous

FIGURE 6-3. Relationship between body length, fecundity, and egg size of individual coho salmon from Forks Creek, Washington (Quinn, unpublished data).

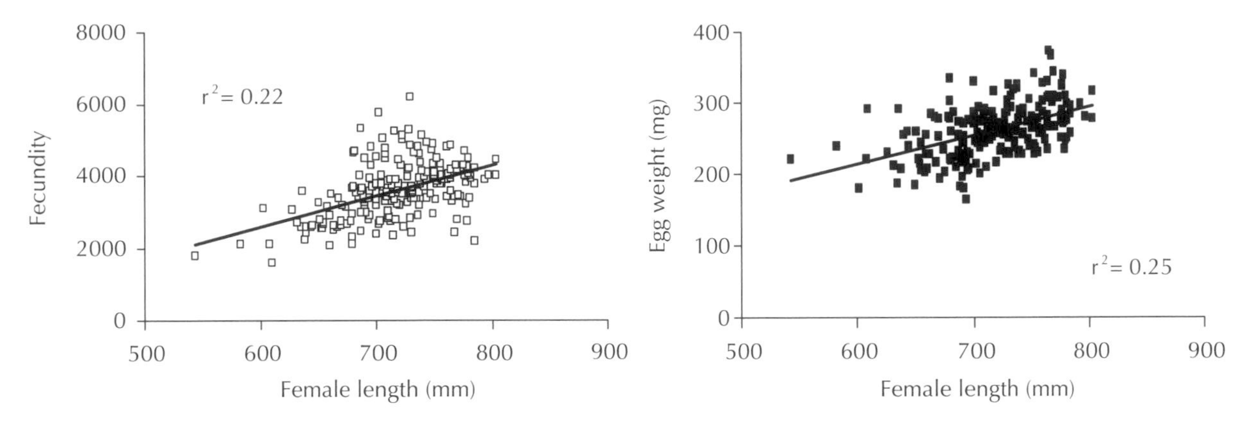

salmonids range from chinook salmon and Atlantic salmon, averaging more than 5000 eggs, to cutthroat and brown trout, with only slightly more than 1000 eggs (fig. 6-4). This plot also reveals that some species fall above the line, indicating higher than expected fecundity for their length (especially sockeye and char), and that some fall below the line (especially chum). These deviations from the overall pattern reflect differences among species in the trade-off between egg size and egg number. Female sockeye are smaller than chum but have more eggs. However, sockeye eggs and fry are smaller than those of chum. The production of eggs requires a great deal of energy, as they may comprise about 20% of a female's weight and are rich in nutrients. Populations with very arduous migrations (in terms of distance and elevation) tend to produce smaller eggs, rather than fewer eggs, compared to populations with easier migrations, and individuals can allocate their energy to egg production or retain more energy in their bodies for other activities (Kinnison et al. 2001).

In addition to having advantages related to the number and size of offspring, large females are more successful in competition for redd sites. However, the "right of prior claim" is strong in salmon, and a small female possessing a territory can resist eviction

FIGURE 6-4. Species-specific average length and fecundity of female salmonids, based on population-specific averages (468 in total) drawn from numerous sources. All the populations are anadromous except kokanee.

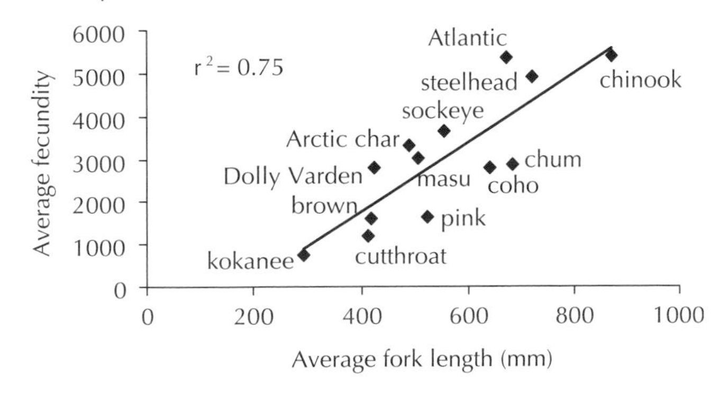

FIGURE 6-5. Relationship between female length and the depth of egg burial. Each point represents a reported average length and average depth of egg burial or pit of the redd for a salmonid population; lines represent studies that reported relationships within a population (from Steen and Quinn 1999, updated with additional references). Species include brook and bull char, cutthroat trout, coho, chum, chinook and Atlantic salmon (one study each); brown trout, golden trout, and *Hucho perryi* (two studies each); sockeye and kokanee (four studies); and Dolly Varden (five studies).

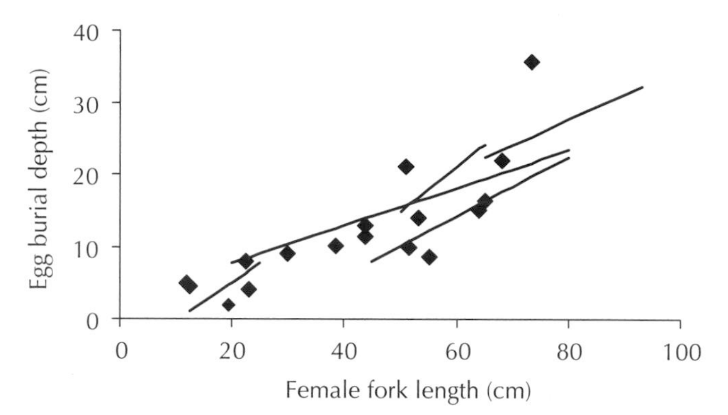

by a larger intruder (Foote 1990). Larger females also tend to construct deeper redds than smaller females (fig. 6-5). These deeper redds are presumably more resistant to disturbance by other females and also less susceptible to scour and intrusion of fine sediment associated with floods (see chapter 8). From the combination of these effects, large females were estimated to have as much as a twenty-three-fold advantage in reproductive success over small females (van den Berghe and Gross 1989; see also Fleming and Gross 1994). Support for this hypothesis was provided by Helle's report (1989) that the production of Olsen Creek, Alaska, chum salmon was positively related to the size of the females in the parental generation, even after accounting for the higher fecundity of such large fish.

However, Holtby and Healey (1986) questioned the hypothesis that larger females are more successful than smaller females. They pointed out that the annual average survival of coho salmon embryos in Carnation Creek, British Columbia, was strongly correlated with physical factors (flow and gravel quality; see chapter 8) but unrelated to density or average female size. They hypothesized that small and large females occupy different habitats and thus their offspring experience different forms of mortality. Large females, they hypothesized, tend to spawn at sites that give their large embryos a high flow of oxygenated water but are more susceptible to scour, whereas small females spawn in areas of lower flow and finer sediment, traditionally considered poorer quality. Such sites may experience less scour, and the water flow may suffice for the smaller eggs of smaller females. Some of these assumptions, regarding habitat use by females of different sizes and size-specific rates of embryo mortality, are open to question and the whole issue of female size and reproductive success needs much more research. However, Holtby and Healey are to be commended for challenging the conventional wisdom about the importance of female size in determining reproductive success.

Another aspect of female reproductive success that needs much more research is "redd superimposition." This term comes from the observation that one redd may be

FIGURE 6-6. Length of life in the stream of pink salmon in Pleasant Bay Creek, southeast Alaska, from date of entry to death (from Dangel and Jones 1988).

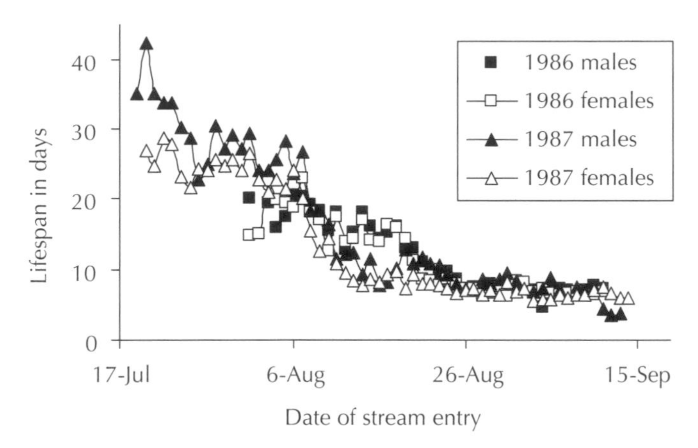

superimposed over another, and this is widely considered to be a major source of density-dependent mortality in salmon populations. The survival from egg to fry stages is density related for many populations, especially those spawning at high densities such as sockeye, chum, and pink salmon. Females expend a lot of energy defending the redd from other females, implying that the behavior affects embryo survival. The embryos are especially sensitive to physical disturbance during about the first 2 weeks after fertilization (see chapter 8), so it is important that early-arriving females live as long as possible to prevent disturbance by other females. Late-arriving females have less risk of redd superimposition, and this may explain why they tend to live only about half as long in the creek as earlier arrivals (fig. 6-6), tend to be smaller, and allocate proportionally more of their energy to gonads rather than body stores (Hendry et al. 1999). There is a strong genetic control over migration and spawning date, and so these interrelated differences have apparently evolved to accommodate the different selection regimes on early and late fish.

Despite the recognition of redd superimposition and density-dependent mortality, the extent to which the embryos of a specific female are lost is hard to demonstrate. It is often assumed that if a large female constructs a redd over one made by a smaller female, the embryos of the first female are destroyed. This may be so, but the variation in burial depth and dynamics of mortality suggest that things may not be so simple. Detailed studies of parentage from DNA or other techniques should reveal the success of individual females in such cases. One final point may be made regarding reuse of redd sites. It is obvious that if one redd is situated in an ideal place, then other females may choose to use that location for the same reason. However, Essington et al. (1998) reported that the probability of redd site reuse between brown and brook trout was greater than could be explained by site selection based on physical features. It is possible that there were important features that the scientists did not measure, but a more interesting explanation may be that the second female uses the site to take advantage of the labors of the first female. Perhaps there is also some element of competition. Oust the progeny of the other female to reduce local competition for one's own offspring?

FIGURE 6-7. Factors known or hypothesized to affect female reproductive success.

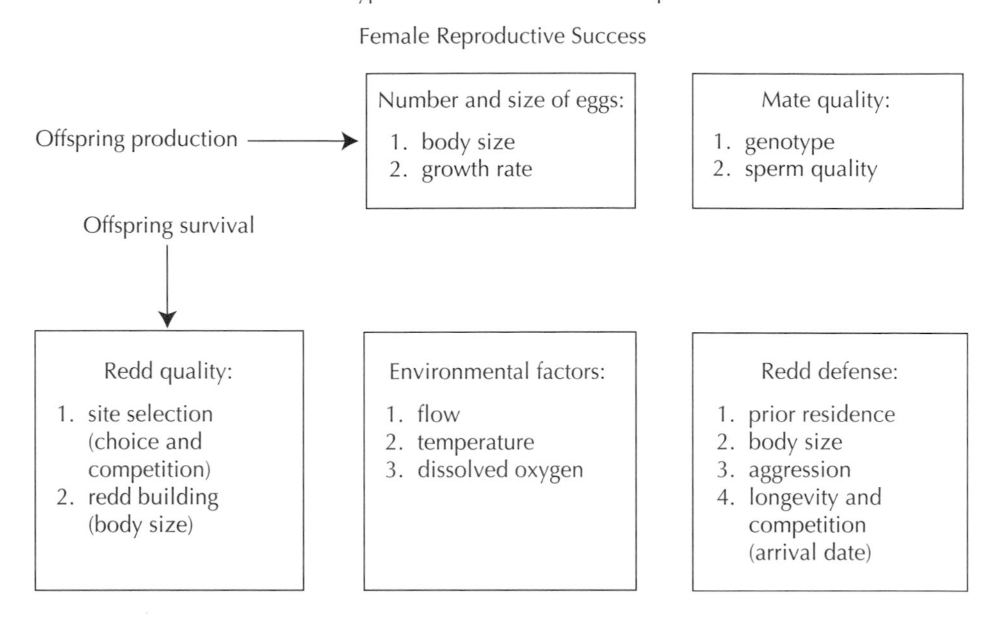

In addition to the complex factors directly affecting reproductive success of females (fig. 6-7), their success is also related to the quality of the males who sire their offspring, and they have some control over their mates. Unlike some animals, in which males gather together for communal displays (leks), enabling females to choose mates from the assembled suitors, the redd construction duties limit a female salmon's opportunities for mate choice. Females cannot circulate through the spawning grounds in search of males but rather must wait for males to come to them. Accordingly, the primary tactic of a female is to delay spawning if she is courted by a male not to her liking. Thus females complete spawning more rapidly if they are courted by large males than by small ones (Foote 1989). For example, Berejikian et al. (2000) selected females that had spawned only once (indicating that they were still ripe) and placed them with smaller males (averaging 46% of the female's weight) or larger males (112% of the female's weight). On average, females spawned within 9.6 h with the larger males but delayed for 16.2 h with the smaller males.

Male reproductive success

Unlike females, the reproductive success of male salmon is not limited by the number of gametes (i.e., sperm cells) but by the ability to fertilize eggs. Because males are able to reproduce over essentially their entire lives in the stream, a few males can fertilize the eggs of many females (Mathisen 1962). This ability is influenced by a combination of their ability to compete with other males and the choice of the female. Male competition is, not surprisingly, affected by body size. All other things being equal, large males tend to dominate smaller males for access to females, as we observed in tagged sockeye salmon of known length, classified on the basis of dominance in competition for females (fig. 6-8). Although there was a significant relationship between male length and

FIGURE 6-8. Relationship between male sockeye salmon length and their dominance status on spawning grounds of two sites in western Alaska. Dominance was estimated by giving the male a score of 1 if he was dominant, 2 if he was competing for access to a female, 3 if he was a satellite, and 4 if he was not involved in courtship, and averaging the scores over all observations (from Quinn and Foote 1994 and Quinn, Hendry, and Buck 2001).

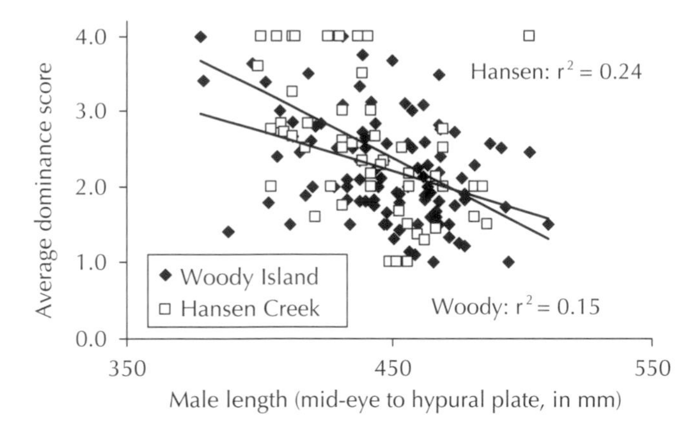

our index of social status, there was also considerable variation. Part of this may be measurement error (that is, the observation did not represent the fish's actual status), but factors besides length also affect dominance. Males presumably vary in strength and vigor, and physiological features that are not obvious to us. However, males clearly vary in sexually dimorphic features such as the length of the jaws and extent of dorso-ventral compression that produces the large "hump" in males. The Woody Island sockeye population spawns on beaches of Iliamna Lake and shows extreme development of the dorsal hump. For this population, males with especially large humps were more often dominant than less well-developed males of the same length (Quinn and Foote 1994). Thus, both shape and overall length were associated with reproductive opportunity. It is not clear if this resulted from female choice or male competition, and much remains to be determined regarding the relationship between size and shape (and other features such as color) and breeding opportunities.

If large and deep-bodied males have greater access to females, one might ask why these traits vary within and among populations (e.g., Beacham and Murray 1985; Quinn, Wetzel et al. 2001). Achieving large size may require spending more years at sea or foraging more aggressively there, and so incurring risk of mortality prior to reproduction. Moreover, males need surplus energy to develop large secondary sexual characteristics, and males that are deep bodied and have large jaws for their length are also heavier than average for their length (Quinn and Foote 1994). Some individuals may have been less successful than others in foraging at sea and may have little surplus energy to devote to these metamorphoses. The restructuring of the body is in conflict with other energetic demands such as migration and competition. Males show the same kinds of trade-offs in energy allocation that females face, and an arduous migration leaves less energy to be devoted to sexual dimorphism (Kinnison et al. 2003).

In addition to the environmental factors affecting the size and shape of male (and also female) salmon, body shape is also a genetic adaptation to the population's regimes of natural and sexual selection. Among sockeye salmon populations in the Wood River and Iliamna Lake systems of Bristol Bay, Alaska, body depth is closely related to the depth of water at the spawning site, apparently reflecting both the difficulty that large, deep-bodied males experience in ascending shallow creeks and the size-selective predation by brown bears (Quinn, Wetzel et al. 2001; Quinn, Hendry, and Buck 2001). Thus, there is genetic control over body shape among individuals within and among populations (Kinnison et al. 2003).

The intrinsic attributes of a male are only one of many factors contributing to his access to females. The passage of time may affect his vigor, as his stores of fat and protein are rapidly depleted during the 1–3 weeks he may live on the spawning grounds. More importantly, however, the levels of competition change rapidly over the course of the spawning season. The level of competition in breeding systems is often expressed as the ratio of sexually active males to sexually active females, or operational sex ratio (OSR). Male salmon are capable of reproducing during virtually their entire lives on the spawning grounds. Females, however, spawn within a few days and then either guard the redd or, in the case of iteroparous species, leave the spawning grounds. In either case, they are not available for reproduction, so there are usually far more males than ripe females late in the breeding season.

The high levels of competition (i.e., OSR) generated by the much shorter reproductive life of females compared to males, provides an explanation for a common, paradoxical phenomenon. No breeding can occur until females arrive and complete their redds; so one might expect that females would arrive before males, but the opposite usually occurs. It is possible that males establish and defend territories after they arrive, in the absence of females, but Foote (1990) reported that male territoriality depended on the presence of females. I think that the phenomenon of early male arrival (termed "protandry" in a review by Morbey 2000) is best explained by the much more severe costs paid by males arriving late compared to those arriving early.

To illustrate this, I made a simple population model with the following plausible assumptions. First, the numbers of males and females arriving each day are normally distributed and identical (i.e., no protandry). Second, 10% of the females complete spawning after 1 day, 40% after 2 days, 70% after 3 days, and 100% after 4 days. Third, the lifespan of males and females is similar, and declines linearly from 14 to 6 days as a function of arrival date. As figure 6-9 shows, the peak number of ripe females occurs only 1 day after the peak of arrival, and during much of the simulated spawning season there are no ripe females at all. Consequently, the OSR quickly goes from about one ripe female per male to a great excess of males and then becomes infinite when no more ripe females are present.

I then created a simple index of male breeding opportunities by taking each day that the male was alive and discounting his likelihood of breeding by the number of ripe females in the population per male on that day, and summing this over his lifespan. This index was highest for the males arriving on the first day and quickly declined for later arrivals. In fact, a male that arrived 1 or even 2 days before the first female would have nearly the same breeding opportunities as one arriving on the first day when females arrived (7.00 and 6.87 vs. 7.07), and they would have considerable advantages over males

FIGURE 6-9. Model of the patterns of arrival and spawning by females and the consequences for male reproductive opportunity. In the left panel, males and females arrive synchronously, resulting in a preponderance of spawned-out females during much of the breeding season. The right panel shows the average male breeding opportunity and operational sex ratio that would result from such arrival and spawning patterns (see text for details).

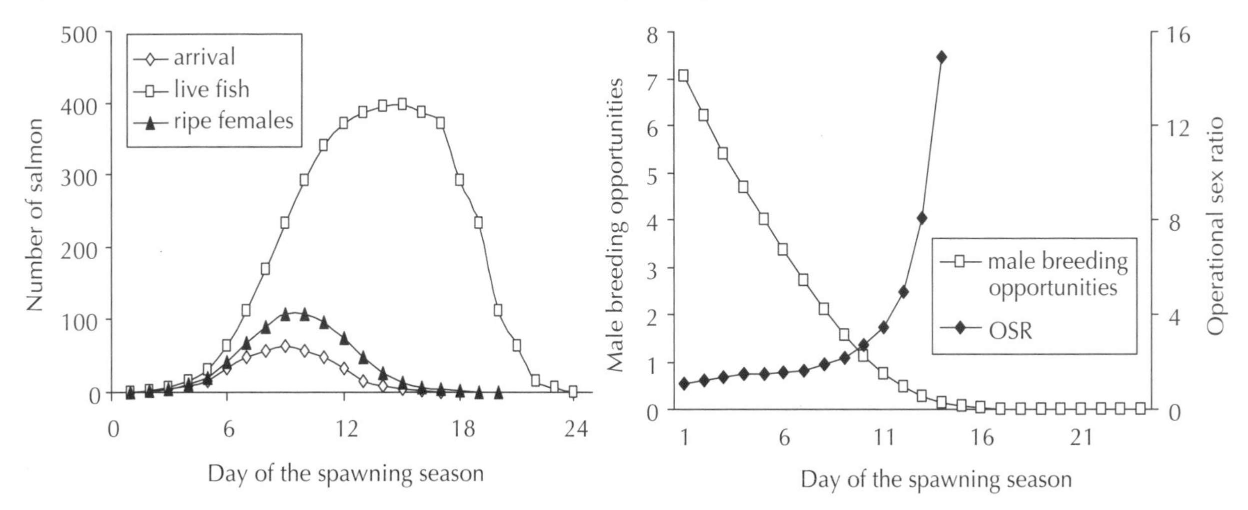

arriving as little as 1 and 2 days after the first females (index values of 6.2 and 5.4, respectively). A male arriving too early wastes only a fraction of his breeding life, whereas a male that arrives too late loses far more opportunities because the females that arrive spawn so soon. One might quibble over the details of this model but the general conclusion seems clear enough. Given the uncertainty as to exactly when females will arrive, males arriving early will fare better than those arriving late, and so males should arrive earlier than females, on average. One other implication of this model is that late-arriving males have negligible prospects for reproduction at the end of the season. Therefore, they might be expected to have reduced longevity, and this is commonly observed (see fig. 6-6).

All other things being equal, there is a tendency for males to court large females, (Foote 1988), and the combination of choice and competition can lead to size-assortative mating. This tendency for large males to court large females and small males to be with small females was observed in a natural population of sockeye salmon by Hanson and Smith (1967). However, competition is usually quite keen, and males seem to court females that are ready to spawn regardless of their size. Accordingly, several studies in natural populations have failed to detect assortative mating (e.g., Quinn and Foote 1994; Dickerson et al., forthcoming), so the generality of this finding is in question.

Given the unbalanced OSRs that generally prevail, male competition and female choice can operate, and they both tend to favor large males. However, the presence of jacks has long been recognized in Pacific salmon. They occur routinely in coho, chinook, and sockeye salmon. For reasons that are not obvious, they seem to be very unusual in chum salmon (Salo 1991) and absent in pink salmon within their native range. In addition, there are some nonanadromous male salmon in some populations, especially masu (Kato 1991; Tsiger et al. 1994) but also chinook salmon (e.g., Mullan et al. 1992). Such sexually mature parr are common in Atlantic salmon (Fleming 1998). There had been a tendency to view jacks as aberrant, and they were often systematically excluded from breeding at hatcheries. However, since the mid-1980s we have learned that they represent alternative life-history patterns under some measure of genetic control. The success of jacks and parr on the spawning grounds is linked to their alternative behavioral tactics. Rather than fight with large males for proximity to females, they position themselves farther from the pair. At the moment when eggs are released they quickly swim next to (or even under) the female and fertilize as many eggs as possible.

Gross (1985) conducted a particularly influential study of jacks in coho salmon. "Typical" male coho salmon spend two summers at sea and jacks spend only one, so their sizes at maturity differ greatly. Gross termed the larger males "hooknoses"; they tend to fight and the jacks tend to sneak. Gross posed the question, can these two life history patterns be evolutionarily stable strategies? That is, can the average reproductive success of males adopting each life-history pattern be similar, even though intuition might suggest that jacks are always dominated by hooknoses? Gross observed that large, "fighter" males were closest to females when eggs were released but that jacks were also quite close. However, there was an intermediate size at which neither sneaking nor fighting seemed successful (judged by proximity to female upon spawning): too small to win fights but too big to successfully sneak. Thus there seemed to be disruptive selection on body size (lower reproductive success at intermediate male sizes). Gross estimated the average reproductive success of jacks and hooknoses to assess whether these forms might be evolutionarily stable strategies.

Gross assumed that survival of juveniles in freshwater is independent of whether they will become a jack or not and that the survival of jacks at sea was 13% versus 6% for hooknoses (based on Washington Department of Fisheries data). Jacks were alive on the spawning grounds for an average of 8.4 days compared to 12.7 for hooknoses, and reproductive opportunity was assumed to be proportional to number of days alive. The distance to the spawning female at egg release averaged 124.6 cm for jacks compared to 93.0 cm for hooknoses, and Gross assumed that proximity determined fertilization success, so he estimated that jacks' success was 66% of hooknoses' per spawning. Combining these data, he modeled the lifetime reproductive success (W = fitness) as W_j (fitness of jacks) and W_h (fitness of hooknoses):

$$\frac{W_j}{W_h} = \frac{0.13}{0.06} \text{ probability of surviving to maturity}$$

$$\times \frac{8.4}{12.7} \text{ breeding lifespan in days}$$

$$\times \frac{0.66}{1} \text{ mating success per spawning (based on distance from the female)}$$

$$= \frac{0.72}{0.76} = 0.95 \text{ or approximately } 1$$

Thus the average jack's lifetime reproductive success was estimated to be approximately the same as the average hooknose's success. These success rates, however, may be

TABLE 6-2. Average length (cm), total body weight (g), and gonadosomatic index (GSI: gonad weight as a percentage of total weight) of male chinook salmon from the University of Washington population (Gunstrom 1968; Quinn, unpublished data).

Age	*Fork length*	*Body weight*	*Gonad weight*	*GSI*
1	23.9	185	26	14.1
2	47.8	1355	130	9.6
3	68.5	3810	284	7.5
4	86.4	7517	425	5.7

frequency dependent. If most of the males were fighters, there might be a greater advantage for the few jacks, and if most males tried to sneak then the large fighters might have a greater advantage. Rivers vary considerably in the proportion of jacks in the population (Young 1999; Healey et al. 2000; Quinn, Wetzel et al. 2001), but this variation has not been adequately explained.

One might wonder how a very small male could fertilize many eggs, given his small gonads. Part of the explanation is that smaller males have proportionally larger gonads (though larger males have larger gonads on an absolute basis; for example, table 6-2). Thus the small males seem to allocate their energy differently than larger males: more to gonads and presumably less for fighting.

As with other aspects of age at maturity, the prevalence of jacks reflects a mix of genetic and environmental influences. Iwamoto et al. (1984) showed that families sired by jack coho salmon had 10.55% jacks whereas families sired by nonjack males produced only 2.15% jacks. However, the fastest growing males from a given cohort tend to return as jacks, and rapid growth and large size of smolts is strongly associated with higher proportions of jacks. The interaction between genetic and environmental factors is illustrated by an experiment by Appleby et al. (2003). Juvenile coho salmon were produced from groups of older males (only 2% jacks) and exclusively from jacks, and they were released at large (about 38 g) and normal (about 24 g) sizes. The percent of the smolts released that survived and returned as jacks ranged from 1.37% for the large smolts sired by jacks, to 0.61% for the large smolts sired by older males, to 0.37% from the normal-size smolts sired by jacks, and 0.16% for the normal-size smolts sired by older males. Control over precocious maturity is part of an overall pattern of genetic and environmental control over age at maturity, discussed in more detail in chapter 17.

The reproductive success of jacks and satellite males can only be guessed from observational studies like those conducted by Gross (1985). However, Schroder (1981) had the brilliant idea of using natural variation in gene frequencies among males to determine which individual sired most of the progeny. Using polymorphic proteins, he was able to distinguish the offspring of two or more males placed in an arena to spawn with a female. His results showed that the dominant male (and the male closest to the female when she released eggs) tended to fertilize most of the eggs (53–100%, among six trials), but that satellite males could also fertilize a significant proportion of the eggs. Subsequently, this genetic approach to determining paternity (using protein or, more recently, DNA markers) has been used in many situations to determine the relative contribution

FIGURE 6-10. Factors known or hypothesized to affect male reproductive success.

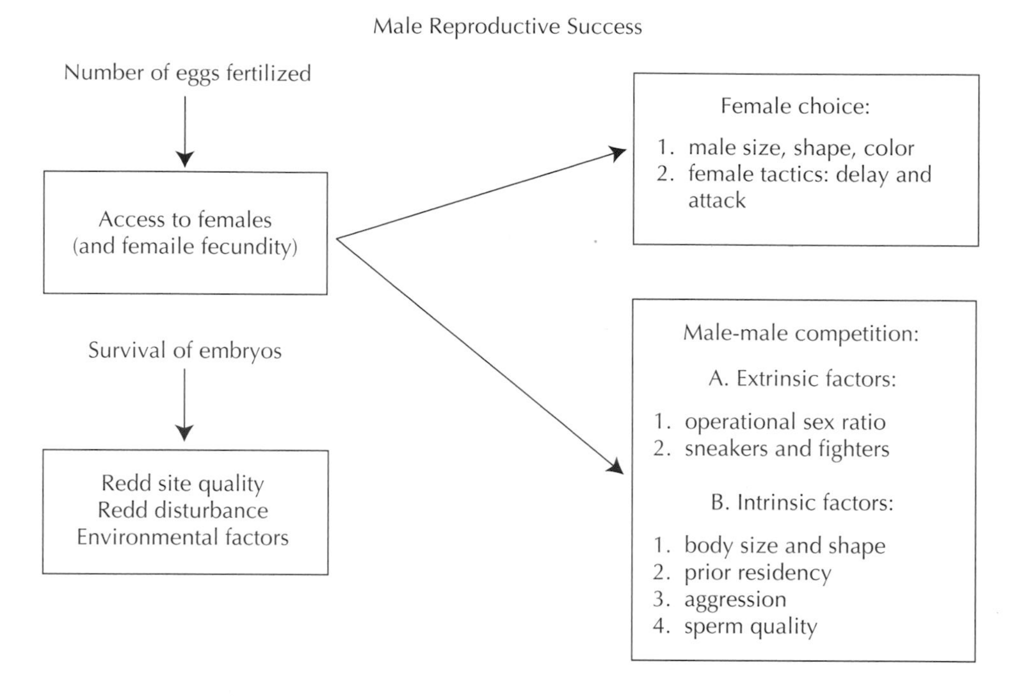

of adults (Seamons et al. 2004), jacks (Foote et al. 1997), and mature parr (Morán et al. 1996) of various salmonid species.

These studies show that, on average, the reproductive success of the dominant male exceeds that of any other male (summarized in fig. 6-10). However, there is considerable variation, and not all of this may be attributable to sampling error or variation in behavior. Variation in sperm quantity, motility, and performance after exposure to water (Liley et al. 2002) may affect fertilization success. Gharrett and Shirley (1985) showed that equal volumes of milt from two males often produced unequal numbers of offspring sired by each one; some males consistently had more "potent" milt. This is not just because some males have higher densities of sperm cells in their milt; some aspect of sperm quality or mobility is involved.

For the semelparous species, the lifetime reproductive success is determined in the single season of breeding. However, in the iteroparous species, an individual's lifetime success is the sum of its success in all breeding seasons. Thus salmonids are confronted with a dilemma. They can put a great deal of energy into the breeding season at hand and save little energy for themselves, which maximizes present breeding opportunities but reduces the likelihood of surviving to breed again. Semelparous species are the extreme example of this. On the other hand, the parent could expend less energy in producing or guarding the present generation of offspring, and save the energy for itself to increase the chances of surviving to breed again, and this is the iteroparous strategy. One might think that because females' gonads are so much larger and more energy-rich than those of males, females would be more likely to die after spawning than males in the iteroparous species. However, the pattern is the opposite; repeat spawning is more

TABLE 6-3. Percentages of the males and females among the steelhead ascending Petersburg Creek, Alaska, to spawn for the first through fifth time (Jones 1972, 1973).

	Number of breeding events					
%	*1*	*2*	*3*	*4*	*5*	*Total*
Males	47.2	46.5	29.9	13.6	0.0	294
Females	52.8	53.5	70.1	86.4	100.0	375
Total	379	198	67	22	3	669

common in females (e.g., table 6-3). Iteroparous females reduce or eliminate the period of redd guarding and go to sea after they breed, whereas males compete actively among each other and stay in freshwater longer, so fewer of them survive to breed again. It should be emphasized that the fish are not making a conscious choice to devote energy towards themselves or their offspring. Rather, these life-history and behavior patterns are under genetic control; individuals displaying patterns that are appropriate for a given environment will have more offspring than others, and so the traits will tend to proliferate in the population.

Summary

The breeding behavior of salmonids reflects the discrete sex roles: females select, prepare, and guard nest sites, and compete among each other. Males provide no parental care and compete among each other for access to ripe females. The reproductive success of a female cannot exceed the number of eggs she produces, but the survival of embryos is affected by many factors, including the physical features at the spawning site (water flow rate, dissolved oxygen, gravel movement, etc.) and disturbance by other females. In addition, egg size and the timing of breeding affect the survival of the offspring after emergence, through connections to territorial behavior and predator avoidance. Male reproductive success is determined by the ability to fertilize eggs, and this results from active competition with other males, female choice, and ability to circumvent these processes and sneak fertilizations. In addition, male success is determined by the fecundity of the female with whom he mates and all the factors affecting her success. The tendency for females to deposit their eggs soon after arrival on the spawning grounds and then guard them (in semelparous species) or leave to seek feeding opportunities elsewhere (in iteroparous species) leads to intense competition among males for breeding opportunities. Recent research suggests that the date of spawning is very important and that body size is less well correlated with overall success than one might think.

PLATE 12. Illustration of a mature male chinook salmon at the height of sexual maturity. Olive brown color is typical of those from southern areas but chinook salmon from northern populations may be more red in color. Copyright: Charles D. Wood, Ph.D.

7

The Ecology of Dead Salmon

After reproducing, female Pacific salmon guard their nests until they become too weak to hold position and eventually drift away and die. Males, too, become increasingly listless, scarred, and emaciated, and they also die. Streams and lakes in temperate and northern latitudes are generally unproductive, limited by phosphorous (P) or nitrogen (N). This is a result of the geology that dominates much of the region, and the inevitable downstream flow of nutrients to the sea. The upstream migration of salmon, followed by their death, is a very important source of nutrients. The quantity of rotting fish flesh (photo 7-1) could not have escaped notice by humans, nor did they fail to see the congregations of animals feeding on the live and dead salmon. Still, research in this subject was quite limited until the past 2 decades, when interest in this subject increased greatly. This research has revealed that the entire ecosystem—from insects to bears and trees, including the salmon themselves—benefits in complex direct and indirect ways from decomposing salmon.

Salmon as fertilizer in aquatic and terrestrial food webs

Early in the twentieth century, scientists studying Karluk Lake, Alaska (known for its especially dense populations of sockeye salmon), pointed out the importance of P and N from decomposing carcasses in otherwise nutrient-poor freshwater systems (Juday et al. 1932). Appreciation of the importance of these nutrients (especially P) in lake and stream productivity grew during the middle of the century, and Donaldson (1967) estimated the P budget of Iliamna Lake. This is the largest lake in Alaska, with an area of 2622 km^2 and a volume of 115.3 km^3. The lake and its associated tributaries had a record escapement of 24.3 million sockeye in 1965, weighing about 2–3 kg each. Donaldson

7-1 Decomposing male chinook salmon, exposed by lowering water levels, Adams River, British Columbia. Photograph by Andrew Dittman, National Marine Fisheries Service.

estimated that they delivered 169.3 metric tons of P to the ecosystem, the primary source in such years of great salmon abundance. The fishery took another 17.8 million salmon that year, so the total run would have brought 42.1 million carcasses to the region, with a concomitant increase in nutrient contribution. Donaldson (1967) pointed out that the smolts migrating to sea constitute a loss to the system, as they weigh about 6–10 g and can number in the hundreds of millions. On balance, sockeye salmon were a major source of nutrients in this system, at least in some years.

Many lakes inhabited by sockeye salmon are nutrient limited. The fertilization of some of these ultra-oligotrophic lakes in coastal Alaska and British Columbia (Koenings and Burkett 1987; Stockner and MacIsaac 1996) shows that addition of nitrates and phosphates increases primary production, zooplankton biomass, and sockeye salmon growth (see chapter 10), though there is increasing variation with each trophic step, so the clearest results are for increases in primary production. However, are the inorganic nutrients from salmon carcasses actually incorporated into the various trophic levels?

Scientists can take advantage of a characteristic of N to trace this element from adult salmon through the aquatic and terrestrial ecosystems. Nitrogen normally has a molecular weight of 14 but some atoms contain an extra neutron, increasing the molecular weight to 15. The proportion of the heavier isotope is greater in marine ecosystems than in freshwater and terrestrial ecosystems. Most salmon put on at least 99% of their total weight at sea, so the bodies of adults reflect the ratio of isotopes in marine waters. By examining the ratio of heavy "marine-derived" N isotopes to lighter atmospheric isotopes in various organisms—including producers (phytoplankton in the water column, rooted aquatic plants, and periphyton on rocks), primary consumers (insects, zooplankton), and higher trophic levels such as planktivorous and piscivorous fishes—one can determine the proportion of N that is marine derived. Similar analyses can be conducted on trees along the riparian corridor, as well as the birds, mammals, and insects that might feed on salmon, indicating the extent to which salmon contribute to the terrestrial ecosystem (Kline et al. 1990, 1993). There are some complexities in the interpretation of such data, owing to the presence of nitrogen-fixing plants such as alder, fractionation of isotopes, seasonal changes in baseline values, and so on (Kline et al. 1990; Bilby et al. 1996). For carbon, marine ecosystems have a higher proportion of ^{13}C relative to ^{12}C (the most common isotope) than do freshwater ecosystems. Unfortunately, there is no marine isotope of P, and it is often the limiting nutrient.

Kline et al. (1990) collected samples from periphyton, insects, and fishes from sections of Sashin Creek, in southeast Alaska, that were within and beyond the range of pink salmon. By comparing the isotope ratios, they estimated that substantial fractions of the N and C in all trophic levels were derived from salmon carcasses. Similar work at Iliamna Lake (Kline et al. 1993) indicated uptake of marine-derived nutrients in juvenile sockeye salmon, especially in years following large escapements of adults to the lake. This supported Donaldson's (1967) hypothesis that the productivity of the lake for sockeye salmon depends, in part, on the carcasses of large numbers of adults. Most of the nutrients from salmon were found in the limnetic zone, but coast range sculpins (*Cottus aleuticus*) apparently derived a significant portion of their diet from salmon eggs and fry. Later work by Foote and Brown (1998) showed that both coast range and slimy sculpins (*C. cognatus*) eat sockeye salmon eggs, and that a single large sculpin can eat almost 50 eggs at one meal and 130 eggs in a week.

Subsequent observation and experimental research has shown that salmon carcasses enhance the abundance of algae in streams and the density of insects. Wipfli et al. (1998) reported that reaches of a creek in southeast Alaska accessible to pink salmon had fifteen times more "biofilm" (microbes covering rocks) and twenty-five times higher densities of macroinvertebrates than reaches of the creek not accessible to salmon. Working in western Washington, Bilby et al. (1996) used stable isotope ratios to estimate that marine-derived N made up about 20% of the N in biofilm and about 15–30% of the N in stream insects. Juvenile salmon eat primarily insects during much of their lives in streams, and it is reasonable that some of the nutrients from carcasses might reach the fish via insects. Bilby et al. (1996) showed that carcasses contributed about 20–30% or more of the N and C in juvenile stream-rearing salmonids (cutthroat and steelhead trout and coho salmon). This uptake might be a case of substituting one source of the element for the other with no net benefit to the fish if the elements are not in short supply. However, juvenile steelhead and coho salmon congregated at sites where carcasses were deposited, and they showed dramatic increases in condition factor (weight for a given length) relative to sites without carcasses (Bilby et al. 1998). Stomach content analysis revealed that the young salmon not only ate insects (hence an indirect link to carcasses) but salmon eggs and even the flesh from dead salmon. The size of juvenile salmon in streams is positively correlated with survival over the winter (e.g., Quinn and Peterson 1996), so a plausible case can be made that large runs of salmon enrich otherwise unproductive streams, to the benefit of the salmon.

There is thus a growing body of evidence that salmon carcasses fertilize aquatic (lake and stream) systems, with direct and indirect feedbacks to salmon populations via various trophic pathways. However, there is also evidence for complex pathways of nutrient transfer to terrestrial ecosystems as well. Bilby et al. (1996) showed that about 20% of the N in foliage from terrestrial vegetation along salmon streams in western Washington had been derived from carcasses, and Helfield (2001) corroborated these findings in southeast and southwest Alaska. Thus trees take up marine-derived nutrients, but do such nutrients contribute to tree growth? Helfield and Naiman (2001) cored and measured trees in areas of the Kadashan River basin in southeast Alaska, above and below a barrier to salmon migration. There are many factors that affect the growth rates of trees, but Helfield and Naiman concluded that the presence of salmon carcasses had a stimulating effect on Sitka spruce growth. Trees within 25 m of the stream in areas with salmon grew 22.9 cm^2 per year versus 6.4 cm^2 at sites without salmon, and the researchers estimated that trees in areas with salmon would grow to a diameter of 50 cm in 86 years compared to 307 years at areas devoid of salmon. They pointed out that trees play several important roles in salmon ecology, such as maintaining stream habitat complexity, retaining gravel, providing structural cover, and trapping finer organic material. Thus there may be a feedback between salmon densities and the habitat conditions that maintain such densities.

How might the nutrients from salmon carcasses be transferred to trees? One pathway is the hyporheic water that flows beneath and alongside the visible stream. Water not only flows down the conventional streambed but also percolates through the ground between bends in the stream and along the edges of the stream, and O'Keefe and Edwards (2002) demonstrated the importance of this pathway in transporting and storing salmon-derived nutrients. In addition to hyporheic flows providing subsurface transport of nutrients from salmon to trees, floods can transports carcasses from the stream and deposit them along the banks. Ben-David et al. (1998) used stable isotopes of N and C to show both the dispersal of carcasses adjacent to streams from floods and uptake of nutrients by vegetation. However, not all streams that support salmon are flood-prone, and the distribution of carcasses from floods is unlikely to extend very far from the stream, so this is probably not a major pathway. Sites with predators (chiefly bears) showed a much wider distribution of carcass-derived nutrients in terrestrial vegetation than sites where they were absent (Ben-David et al. 1998). Eagles and otters consume salmon but are too small to transport large quantities of salmon from streams. However, brown (*Ursus arctos*) and black bears (*U. americanus*) have the means and the motive to move large quantities of carcasses from streams (Reimchen 2000; Gende and Quinn 2004), and there seems to be a very special linkage between bears and salmon that affects both the ecology of these organisms (predator and prey, respectively), but also other components of the ecosystem (photo 7-2).

Bears and salmon

W. K. Clark (1959, 337) noted, "One can walk along almost any Alaska salmon stream in bear country during the summer spawning season and see jaws, heads, and other parts of salmon left by bears." It is common knowledge that bears kill and eat salmon, but what factors determine the magnitude and consequences of the predation for the bears and for the salmon? Records of the numbers of live and dead sockeye salmon, and

7-2 Brown bear with a ripe female pink salmon in Pack Creek, southeast Alaska; note the area on the bank that has become worn by the bear's activity, and the small size of the stream. Photograph by Scott Gende, U.S. National Park Service.

categorization of the dead as bear-killed or senescent in twenty-three different streams within the Wood River lake system, Bristol Bay, Alaska, over more than 10 years (Quinn, Wetzel et al. 2001; Quinn et al. 2003), revealed that the percent of the sockeye killed by bears was strongly related to the size (especially width) of the creek: the smaller the creek, the higher the predation (fig. 7-1). Some rivers were too big for accurate surveys but all indications were that predation there was negligible. Variation in predation among years within streams was related to density, though the strength and shape of these relationships varied among streams. In general, as the number of salmon in the creek increased, the number killed rose toward an asymptote, approximated in some cases by a log relationship (fig. 7-2).

Beyond our interest in how many salmon are killed, there is a growing appreciation of the complexities of bear predation. First, all the salmon will die at the end of a few weeks in the spawning stream, regardless of the presence of bears. Therefore, predation (i.e., salmon killed by bears) does not constitute evidence of an effect on the dynamics of the salmon populations. As indicated in the previous chapter, females tend to spawn within a few days of entering the stream and thereafter defend redds. Pre-spawning predation on females would therefore have much greater consequences for the population that postspawning predation. Predation on males would have little or no effect on the overall production of juveniles because males are essentially always surplus (in a numerical sense) to the needs of the females (Mathisen 1962). However, the reproductive success of an individual male

FIGURE 7-1. Relationship between the width of the stream and the percentage of adult sockeye salmon killed by bears in twenty-three streams in the Wood River system, Alaska (Quinn, Wetzel et al. 2001, Quinn et al. 2003, and unpublished data). Three larger rivers, indicated by open squares, have so little predation that it cannot be quantified and they were not used in the relationship.

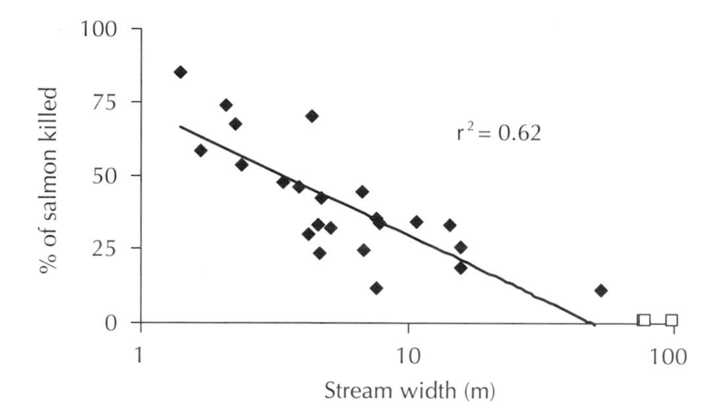

might be reduced by predation, depending on whether the salmon was killed early or late in his natural life in the stream, and his breeding opportunities. Whether the salmon are killed pre- or postspawning depends on several factors.

Bears are omnivorous, eating a wide variety of plants and animals. However, Hilderbrand, Schwartz et al. (1999) showed that the density, size and productivity of bears were correlated with the availability of meat (especially salmon). There is no source of nutritious food that is as easily acquired and predictably available (in space and time) for bears as salmon. During the late summer and fall, bears need to deposit the fat that will sustain them through the winter period of fasting and parturition (Hilderbrand, Jenkins et al. 1999; Hilderbrand, Schwartz et al. 2000). Accordingly, they congregate along streams with salmon at the appropriate time of the year. Bears feed very selectively on salmon, tending to eat the body parts (chiefly eggs from females and brains) that provide the

FIGURE 7-2. Relationship between the number of sockeye salmon spawning during a season and the number killed there by bears, in two small creeks in the Wood River system, Alaska (from Quinn et al. 2003 and additional unpublished data).

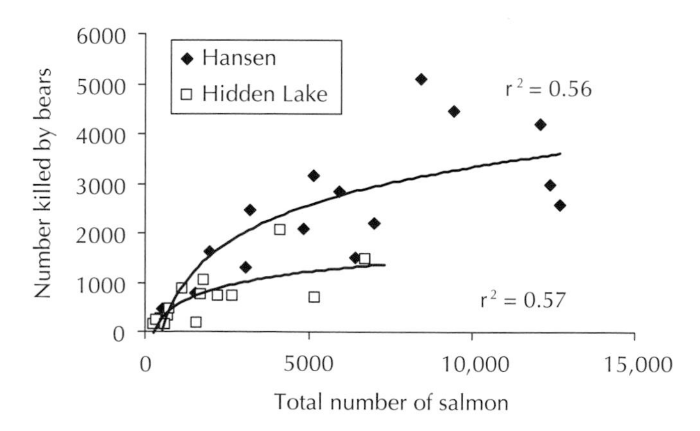

7-3 Male sockeye salmon attacked by a brown bear; note the bite on the salmon's dorsal hump. Photograph by Susan Johnson, University of Washington.

most concentrated amount of fat (Gende et al. 2001; Gende et al. 2004). Salmon lose fat and protein very rapidly after they enter streams (see chapter 4), especially females after egg deposition. Thus bears should prefer to eat newly arrived salmon. In shallow streams, they indeed selectively kill these "fresh" fish rather than older ones, despite the fact that the older, senescent fish are presumably easier to catch as they become weak (Gende et al. 2004). However, in larger streams or ones with more complex habitat, the bears must "take what they can get" and they forage at random or tend to kill older fish. Only in very small streams, where fishing is easy, and when salmon are scarce, do bears seem to kill a large fraction of the ripe females.

Besides tending to kill newly arrived salmon, bears also tend to kill large salmon (e.g., Ruggerone et al. 2000). This may be because larger salmon provide more food for similar capture effort, because they are less maneuverable in small streams and so are easier to catch, or because they are simply more noticeable. Regardless of the mechanism, this selective predation on large salmon from populations in small streams where predation is intense can lead to the evolution of salmon that are younger and smaller than larger streams nearby with lower predation rates (Quinn, Wetzel et al. 2001).

Putting aside this evolutionary force exerted by bears on salmon life history, we return to the carcasses. Given that bears tend to feed selectively, how much salmon is left over? Intensive examination of carcasses over 5 years on daily surveys of Hansen Creek, a small stream with dense populations of sockeye salmon in Bristol Bay, Alaska, revealed an average of 4829 salmon that died of senescence and 3609 that were apparently killed by bears each year (Gende et al. 2001; Quinn, unpublished data). Of those killed, there was only minor consumption (bitten and dropped with no tissue eaten, or consumption of only the brain or skin) in 30%, and another 31% of the fish were partially consumed (typically the belly in females or dorsal musculature in males; photos 7-3 and

7-4 Male sockeye salmon carcass with only a bite of dorsal hump consumed by a bear. Photograph by Thomas Quinn, University of Washington.

7-4). Only 39% of the fish captured by bears had major tissue loss (e.g., all or most of the body eaten). In 1997, detailed measurements and weights were taken on fresh, whole salmon and on fish after a variety of body parts had been consumed. These data allowed us to estimate that, on average, less than 25% of the biomass had been consumed from the 4218 salmon killed by bears. Indeed, salmon sometimes survive after a bear attacks and then releases them (photo 7-5).

Thus at streams with abundant salmon, the bears may kill a large number of salmon but actually eat a small fraction of what is available to them. The uneaten portions of the carcasses may be deposited in the stream, on gravel bars at the edge of the stream, or in the nearby forest, where they are available for terrestrial scavengers and decomposers (Reimchen 2000; Gende et al. 2001). Hilderbrand, Hanley et al. (1999) found that the concentration of marine-derived N was inversely related to the distance from the salmon stream (on Alaska's Kenai Peninsula), and that the distribution of bears tracked with radio transmitters mirrored the distribution of ^{15}N enrichment. The nutrients from salmon are transferred to the forest via excretion (especially urination) by bears (Hilderbrand, Hanley et al. 1999) and via deposition of uneaten body parts. Helfield (2001) also found that trees growing near salmon streams were enriched in ^{15}N compared to more distant sites and sites near streams without salmon. Sites where bears fed and deposited uneaten parts of salmon were especially enriched in ^{15}N.

Bears transport nutrients to the forest for decomposition by microbes, but they also make nutrients available to a wide variety of animals that might be unable to access carcasses in the water, such as other mammals and birds. Cederholm et al. (1989) tagged salmon carcasses in seven western Washington streams and estimated that 51% of the biomass was consumed, 28% remained in the stream, 5% was on the banks, and 16%

7-5 Male sockeye salmon that survived an attack by a bear; note the holes made by the bear's canine teeth on the fish's back and the removal of the upper part of the head, down to the eye. This fish survived for a week with these injuries. Photograph by Thomas Quinn, University of Washington.

could not be accounted for. Forty-three taxa of birds and mammals were seen on the stream, and 51% of these (ranging in size from shrews and winter wrens to bears and bald eagles) were believed to consume salmon. Bears, raccoons, or otters seemed to have retrieved the carcasses from the streams (especially bears in deeper water) and to have consumed selected parts. Secondary scavenging took place over some period of time by small mammals and birds.

Our observations and those of others (e.g., Reimchen et al. 2003) have indicated that scavenging by insects, especially blowflies (Calliphoridae) can be a very important pathway for nutrients (though less charismatic than bears) once the carcasses are removed from the water. Flies deposit eggs in the eyes, gills, and other accessible areas of salmon. When the tiny maggots hatch they feed on the carcass, grow very rapidly, and can reduce an entire salmon to a skeleton in a few days (photo 7-6). The flies depend on bears or changes in stream flow to expose carcasses to colonization, and temperature and moisture seem to affect oviposition and larval development. In three streams in the Wood River system, more than 95% of the carcasses available to flies (i.e., out of water) were colonized, and a 3 kg sockeye salmon carcass supported up to 50,000 maggots (Meehan et al., forthcoming).

The invertebrates that feed on salmon carcasses may provide an important food resource to insectivores. Gende and Willson (2001) found that the densities of passerine birds were higher along stream reaches with spawning salmon than along other streams without spawning salmon, and they hypothesized that elevated densities of insects, particularly chironomids, may support higher densities of riparian birds. The ability of

7-6 Male sockeye salmon carcass being consumed by fly maggots. Photograph by Thomas Quinn, University of Washington.

flies to rapidly locate carcasses, the patterns of maggot growth and development, and the effects of elevated invertebrate densities as a result of the presence of salmon carcasses are fruitful, if unappetizing, areas for further research.

Thus bears seem to play a special role in the ecology of dead salmon. They can kill far more salmon that any other terrestrial predator, and their patterns of partial consumption, carcass deposition, and excretion of wastes transfer nutrients from the stream to the nearby forest or grassland. People living in urban areas may doubt the ecological and evolutionary importance of bear predation. However, the range of black and brown bears was extensive until hunting and human development extirpated them from much of their range, and bears consumed salmon in these regions (Hilderbrand et al. 1996). The vast majority of salmon are (and always were) produced in small streams where bears would have had easy access to them. The densities of bears in coastal Alaska and British Columbia, even in the face of some hunting, and their levels of predation on salmon suggest a strong and long-standing linkage between these organisms.

Salmon as keystone species and the management of salmon abundance

Recent reviews (e.g., Willson et al. 1998; Cederholm et al. 1999; Gende et al. 2002; Naiman et al. 2002; Reimchen et al. 2003) have pointed out the importance of salmon in nutrient cycling, for both aquatic and terrestrial ecosystems, and the feedbacks to salmons' own abundance though algae, insects, and trees. Willson and Halupka (1995) termed salmon

FIGURE 7-3. Relationship between the number of spawning parents and the production of offspring, as modeled by Ricker (1954) and Beverton and Holt (1957).

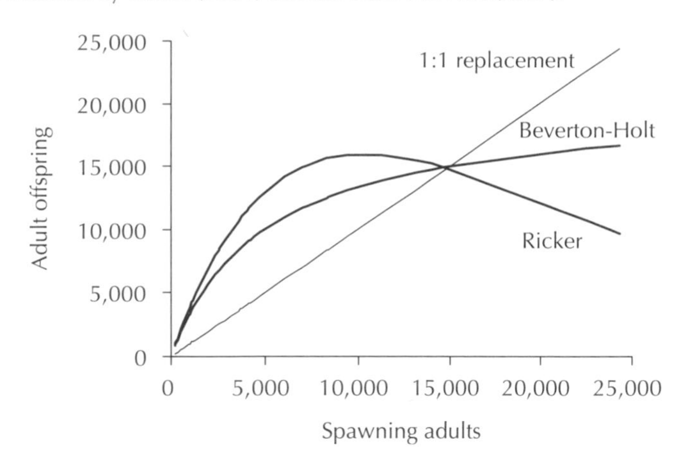

"keystone species" in recognition of salmon's special role enriching otherwise nutrient-poor systems. This new view of salmon as not merely part of a region's biodiversity but as a critical influence on many other species as well has stimulated concern for the adequacy of salmon runs from an ecological perspective. The natural dynamics of salmon populations return more adults to freshwater than the spawning grounds and rearing habitats can support. Competition among females for good nest sites, nest disturbance by other females, and competition among juveniles for food and space all limit the capacity of streams and lakes to produce salmon.

As salmon abundance approaches the carrying capacity specific to each site, the population is increasingly less productive (i.e., fewer surplus salmon are produced per spawner). The primary models for the relationship between numbers of spawning adults and their offspring are those of Ricker (1954) and Beverton and Holt (1957). As illustrated in figure 7-3, the Ricker curve predicts lower production of offspring at very high densities than at intermediate densities, whereas the Beverton-Holt curve predicts that the production of offspring will approach an asymptote. In either case, at low densities, the population produces several times more adults than the number of spawners. From the perspective of fisheries, these are "harvestable surplus" because catching them will maintain the population's productivity rather than diminish it. The general goal of salmon management is to allow the number of salmon to spawn in each river that will produce the maximum number of surplus offspring for fisheries to catch. If this escapement goal is achieved the population should not only sustain itself but support fisheries forever, a kind of biological perpetual motion machine. In this view, any salmon in excess of the number needed to maximize surplus production should be caught because they are otherwise wasted. However, the view of salmon as critical sources of nutrients for freshwater and terrestrial ecosystems makes us wonder whether long-term productivity is reduced by inadequate density of carcasses in streams.

Bilby et al. (2001) explored the idea of using nutrient levels to assess the adequacy of escapement goals. They surveyed a number of streams in western Washington that

FIGURE 7-4. Relationship between the density of adult salmon carcasses and an index of enrichment of juvenile salmonids with nitrogen derived from the carcasses (from Bilby et al. 2001).

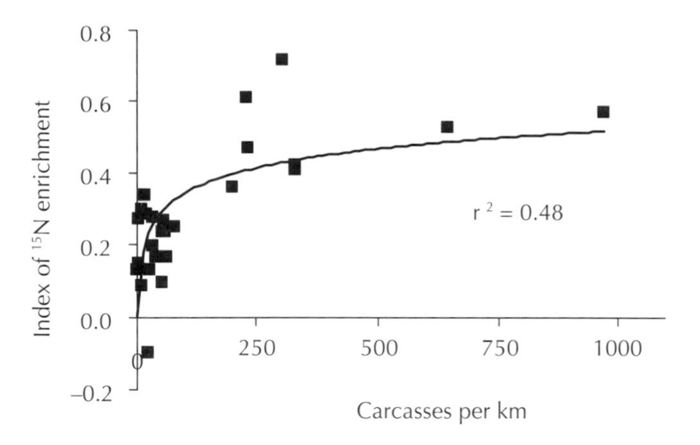

varied in salmon-carcass density and then determined the levels of marine-derived N in juvenile coho salmon. By comparing the isotopic ratio of N in cutthroat trout in the same watersheds, but in areas without salmon, and the ratio in adult salmon, they produced an index of nutrient enrichment that was related to carcass density (fig. 7-4). Bilby et al. (2001) modeled the growth of coho salmon as a function of N and concluded that in these streams the benefits of carcasses would plateau at escapements over about 120 salmon per kilometer. However, many western Washington streams fail to reach this level.

There are many reasons for low escapements, and some of the variation is natural. However, many rivers, especially in the southern part of the range of salmon, have seen great reduction in salmon abundance over at least the past century. First, the fisheries all but guarantee that fewer salmon return to spawn than would have otherwise returned. Many fisheries are managed at about 50% exploitation rate and in some cases at least 75% of the adults are caught. Beyond the effect of fishing, impassable dams have eliminated runs of salmon in parts of many rivers. In much of the remaining habitat, the numbers of salmon being produced have been diminished by factors degrading the streams and lowering their carrying capacity (e.g., logging, mining, agriculture, water diversions, dams, and urbanization: NRC 1996). Thus the streams are less productive than in the past and we catch many of the salmon. The total numbers of salmon in recent years include a large number (sometimes the majority) produced in hatcheries, and until very recently, these carcasses were not returned to the streams.

Gresh et al. (2000) estimated the reduction in carcass deposition over the twentieth century in North America. They took records of the number of salmon caught at the earliest period of efficient Euro-American fisheries as estimates of the "historical" levels of salmon abundance and then obtained records of salmon escapements to rivers during the present period, excluding hatchery populations to the extent possible. Their summary revealed striking variation among areas in the historical and present numbers of carcasses. In the past, the majority of salmon were produced in Alaska (60.5% by number) and British Columbia (23.7%), with smaller numbers farther south. This imbalance

TABLE 7-1. Estimates of the current numbers and biomass of salmon escaping fisheries to spawn in rivers, and the percentage of historic levels that they represent. Data are from Gresh et al. (2000), using the average values given or the average between high and low estimates. Adjustments were made for differences in species composition among regions and differences in weight among species.

	Number of salmon (thousands)		*Biomass (thousands of kgs)*	
Area	*current*	*% of historic*	*current*	*% of historic*
Alaska	187,466	107.0	388,554	93.4
British Columbia	24,800	36.2	59,312	30.7
Puget Sound	1,600	8.0	4,123	7.1
Washington coast	72	1.8	197	1.2
Columbia River	221	1.7	1,201.5	1.3
Oregon coast	213	6.9	662.5	4.3
California	278	4.7	1258	4.7

has greatly increased recently. The researchers estimated that Alaska has roughly the same total number and biomass as before (within 10% of historical levels) but this now constitutes about 87.3% of the total escapement in North America. All other regions have seen dramatic reductions (table 7-1). British Columbia's escapement is on the order of 30–35% of the past, and Washington, Oregon, and California all have less than 10% of their former levels. The researchers did not report data for Idaho but the same would likely be true there as well

The study of nutrient cycling from salmon and the ramifications of carcass density are likely to be fascinating and controversial areas of research and policy for some time to come. To what extent have the carrying capacities of streams been reduced by decades of low escapements, and how reversible might these effects be? Do feedbacks tend to magnify or reduce effects of carcass limitation? Should we incorporate carcasses and nutrients into the calculation of escapement goals, and how might the calculations differ between stream-rearing species like coho and those migrating to sea such as chum and pink? More broadly, what are the ramifications of salmon density for the health of terrestrial and aquatic communities?

Summary

Salmon achieve at least 99% of their final body size at sea and so, despite the mortality that takes place there, there is a tremendous net influx of biomass from the ocean to relatively unproductive stream and lake ecosystems. The millions of adult salmon are preyed upon, scavenged after death, and decomposed by a wide variety of organisms from bears to gulls, fly maggots, and bacteria. Recent research using stable isotope ratios has demonstrated that these "marine-derived" nutrients in the salmon carcasses are an important contribution to the aquatic and terrestrial ecosystems, affecting the growth and density of bears, growth of juvenile salmonids, productivity of lakes, biofilm and insects in streams, and even the growth of trees in the riparian zone. Bears seem to have a particularly strong ecological connection with salmon, as they can not only kill a significant number of salmon but also transfer them from the stream to the riparian zone for scavenging and decomposition by other organisms. Capture of salmon in fisheries, even when well managed, reduces the number of salmon carcasses in streams, and the effect of this reduction on the ecosystem and the long-term productivity of salmon is an area of active research and controversy.

8

Incubation Rate and Mortality of Embryos

The months that young salmon spend below the surface of the stream are an important period for them. During this time they develop from unicellular fertilized eggs to complex organisms ready to emerge into open water and make their way in the world (photos 8-1 through 8-6 show chinook from the eyed egg stage through yolk absorption). Over their entire lives, from fertilization to maturation and spawning, the majority of mortality takes place during this period in the gravel. This critical phase is difficult to study because everything takes place out of our sight and often at times of the year that hinder research. Field observations are difficult at best and laboratory experiments are artificial. However, there have been many important discoveries about the rate of embryo development and the factors causing their mortality.

Spawning date, temperature, and development

Embryonic development begins when the egg is fertilized, and the timing of reproduction is among the most critical adaptations of salmon populations to their environment. Developmental rate and metabolism of salmon increase with temperature, as in all ectothermic organisms. Within the tolerable range, warmer water leads to faster development. The working hypothesis is that adults spawn at the time of year which, given the long-term average thermal regime, results in emergence of fry at a date that optimizes their opportunities for growth and survival. There may be other constraints on spawning date, including physical factors like flooding or freezing in the river, lake level in populations spawning on beaches, and biological factors such as predation on adults (e.g., by bears) or eggs (by sculpins). The relationship between temperature and development is so strong that most biologists assume that this is the dominant factor.

8-1 Chinook salmon embryos at the “eyed” stage. Photograph by Richard Bell, University of Washington.

8-2 Chinook salmon alevin hatching. Photograph by Richard Bell, University of Washington.

8-3 Chinook salmon alevins early in their development. Photograph by Richard Bell, University of Washington.

8-4 Chinook salmon alevin early in its development; note the large yolk sac and lack of pigment. Photograph by Richard Bell, University of Washington.

8-5 Chinook salmon alevin late in its development; note the distinct pigmentation and small yolk sac. Photograph by Richard Bell, University of Washington.

8-6 Chinook salmon fry, after emergence from the gravel, with some yolk remaining. Photograph by Richard Bell, University of Washington.

FIGURE 8-1. Relationship between the median spawning date of chinook salmon populations from California to Alaska (Myers et al. 1998) and estimated temperatures experienced by embryos during incubation (from Brannon et al. 2002).

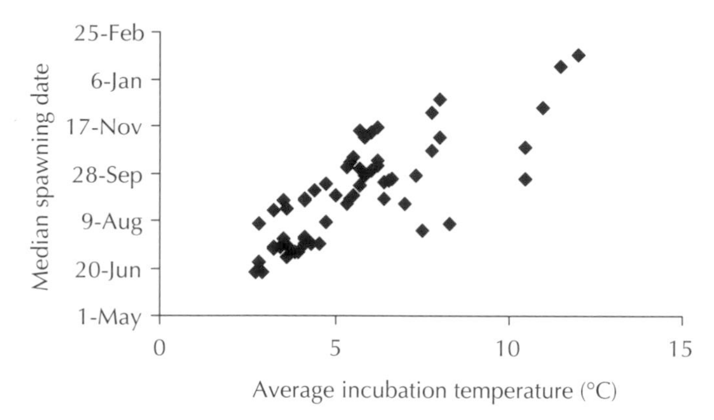

However, there is surprisingly little research on the variation in adult reproductive success as a function of spawning date, and the relationship between fry emergence and survival has only recently received attention. Many researchers have pointed out that populations spawning in cold rivers do so earlier in the year than those using rivers with milder temperatures (fig. 8-1). Salmon also tend to spawn earlier at higher latitudes, and this seems to be an adaptation of populations to their local environment (see also fig. 18-2). For example, the outlets of lakes typically have milder winter temperatures than their inlets, and sockeye tend to spawn later in the outlets (Brannon 1987). This presumably results in synchrony of emergence because all populations need to strike the same balance between food availability, temperature, and predator avoidance in the lake, though this hypothesis is seldom really tested.

Sheridan (1962) provided an example of the importance of matching spawning date to the thermal regime of the site. Pink salmon spawn in the Kadashan and Klawock rivers in southeast Alaska. The Kadashan is colder and the salmon spawn primarily in mid-August, whereas they spawn at the end of September in the warmer Klawock; but in both sites the juveniles emerge and migrate to sea in April and May. Using the observed temperature regimes and the temperature-specific rate of development, Sheridan estimated that if Kadashan (early) fish were to spawn in the warmer Klawock, their progeny would emerge in early November. If (late) Klawock fish spawned in the cold Kadashan, their progeny would emerge at the end of June. Marine survival tends to depend heavily on date of entry into the ocean (see chapter 15), so we may infer that natural selection has adapted these populations to spawn at the appropriate dates, given the thermal regimes in their respective rivers.

The relationship between spawning date and emergence stems from the rough equivalence between time and temperature in controlling embryonic development, and a combined unit known as "degree days" or "temperature units" (TUs) reflects this fact. The developmental rate of embryos can be estimated from the product of the number of degrees Celsius above 0 times the number of days. For example, coho salmon hatch about 500 TUs after fertilization, and this could be 50 days at 10°C or 100 days at 5°C. As

TABLE 8-1. Number of days from fertilization to emergence and from fertilization to hatching for Pacific salmon at constant temperatures, from Murray and McPhail (1988). Data on other species are from Gunnes (1979), Humpesch (1985), Velsen (1987), Crisp (1988), Johnston (2002) and unpublished sources. Methods varied among the studies, and there is also variation among populations, so the values are only generalizations.

	Temperature (°C)				
	2°	*5°*	*8°*	*11°*	*14°*
Days to emerge					
sockeye salmon	282	173	121	90	72
chinook salmon	316	191	115	84	63
chum salmon		161	124	98	86
pink salmon		173	120	91	72
coho salmon	228	139	109	74	61
Days to hatch					
sockeye salmon	206	120	77	52	47
chinook salmon	202	102	67	47	38
chum salmon		97	67	52	46
pink salmon		99	72	47	40
coho salmon	115	87	63	42	32
lake trout	155	100	65		
Atlantic salmon		100	63	43	
brook trout	140	84	60	40	31
Arctic char	139	88	56	40	31
brown trout	148	82	57	35	
rainbow trout	115	68	42	28	22
cutthroat trout		61	45	25	
Hucho hucho		*55*			

table 8-1 shows, the development rate (in days to hatch or complete yolk absorption) is faster in warmer water for all species.

However, development takes only about four times as long at 2°C as it does at 14°C, though this is a sevenfold difference in temperature. This pattern is commonly observed; incubation requires fewer TUs at very low temperatures (and sometimes also at high temperatures, though these are less likely to be experienced in nature) than at moderate temperatures. The result is that progeny spawned late in the season, when the water is cold, do not fall as far behind their counterparts spawned earlier, as would be the case if a fixed number of TUs were required for development. Despite this fact, salmon embryos spawned earlier will still hatch and emerge earlier than those spawned later on an absolute basis because they started sooner. Moreover, in the case of species spawning in the fall on a descending temperature regime, the progeny of females that spawned earlier will have accumulated more TUs in their first days (because the water was warmer) than those of later-spawning females.

Table 8-1 reveals not only variation in developmental rate as a function of temperature (the most dramatic pattern), but also variation among species. Consider the data for 5°C, a fairly representative average temperature experienced by wild salmon embryos. Sockeye, pink, and chinook develop slower than the other salmon species, and coho are the fastest. This is consistent with the general chronology of spawning where these species are sympatric; chinook and sockeye spawn before coho. Chinook, in particular, tend to spawn in large rivers and the outlets of lakes, which have more moderate temperatures than the smaller tributaries where coho spawn. The timing of chum and pink spawning varies considerably. In Puget Sound, pink salmon tend to spawn in September and October and overlap in space and time with chinook salmon in some moderately large rivers, whereas most chum salmon spawn in November and December in smaller, low-gradient streams near saltwater. In southeast Alaska, however, pink and chum salmon are commonly sympatric and chum salmon may spawn earlier than pinks. The patterns of spawning date, particularly among sympatric species, are worthy of more investigation, as they may reflect species-specific patterns, effects of interspecific competition, and seasonal patterns of secondary production in the habitats where the species will feed as juveniles.

More fundamentally, the embryos of fall-spawning species take longer to develop, at a given temperature, than those of spring spawners (notably rainbow and cutthroat trout but also *Hucho hucho*, a large, nonanadromous European salmonid). This allows the trout to emerge in spring, shortly after the emergence of salmon spawned many months earlier. In addition to variation in developmental rates among species and among rivers as a function of different thermal regimes, there is also variation among populations exposed to the same temperatures (e.g., Beacham and Murray 1989). For example, Bush Creek, on Vancouver Island, has distinct populations of early- and late-spawning chum salmon. Despite the separation of spawning dates, all the fry migrate to sea synchronously because the early population develops slower, for a given thermal regime, than the later one (Tallman 1986; Tallman and Healey 1991).

Although temperature exerts the strongest influence on developmental rate, the concentration of dissolved oxygen (DO) also plays a role. The embryo starts out as a single cell with a very large volume of yolk. As the cells divide and differentiate into tissues and organs, the embryo gets larger and the yolk gets smaller but the egg membrane that encloses it stays the same size. Oxygen uptake and the release of metabolic wastes must occur by diffusion across the egg membrane until hatching takes place. The concentration of DO in water at saturation decreases with increasing temperature, from 14.23 parts per million (or mg/L) at 1°C, to 12.8, 11.33, 10.15, 9.17, and 8.38 mg/L at 5°, 10°, 15°, 20°, and 25°C (Davis 1975). As temperatures increase, metabolic demand for oxygen increases but the capacity of the water to hold it decreases.

The DO requirements of embryos increase as they grow, reaching a peak just before hatching. After hatching, the alevins can deal with low DO by pumping water with their gills, moving to areas of higher DO, and also by circulating water around themselves with their fins. Alderdice et al. (1958) estimated that the critical levels of DO (defined as

FIGURE 8-2. Minimal level of dissolved oxygen (DO) needed for normal development of chum salmon embryos as development proceeds (from Alderdice et al. 1958).

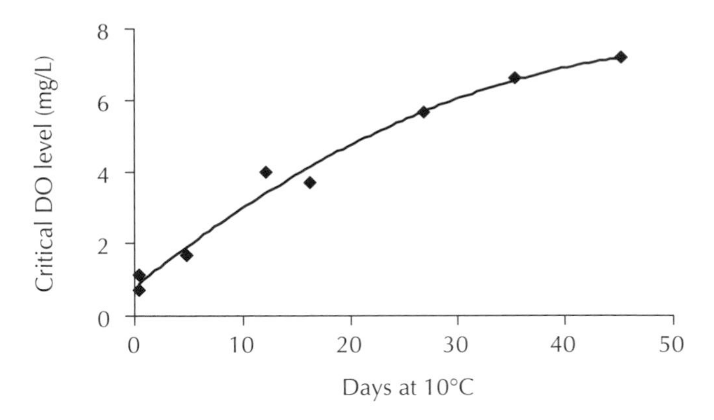

"those at which respiratory demand is just satisfied," p. 248) went from about 1 mg/L just after fertilization to 7 mg/L prior to hatching (fig. 8-2). Between these "critical levels" and the DO levels at which 50% of the embryos died (about 0.5 to 1.5 mg/L, depending on the stage at which the reduction in DO took place), there were levels at which embryos survived but did not thrive and were judged to have been unviable under natural conditions. Shumway et al. (1964) revealed that decreased oxygen levels delayed hatching (at 10°C) from about 35 days to 40 days in steelhead and from 43 to 50 days in coho salmon, or about a 15% delay. Hatching was also delayed by reductions in the rate of water flow around the embryos. The embryos are living, breathing organisms and they require flowing water to deliver the oxygen they need and also to remove metabolic wastes. Thus water that is initially high in DO will be depleted unless there is a sufficient flow. More dramatic than the reductions in developmental rate, however, were the effects of reduced DO on the size of the alevin (not including the yolk) at hatching. Embryos incubated at low DO (2.9 mg/L) were about 50% lighter than those incubated at saturation (fig. 8-3). This may be related to the fact that low DO not only delays development but it also can stimulate premature hatching, as the alevin's best chance to survive. It is unclear whether these alevins eventually thrive or if their chances of survival are compromised. In summary, development rate is primarily determined by temperature, with an important secondary effect of DO.

Factors affecting the survival of embryos

In addition to the rate of development and hatching, survival (or mortality, if you are pessimistic) is the other key issue during the intragravel period. Very low and high temperatures can be lethal but are probably not a major cause of embryo mortality in most situations. Murray and McPhail (1988) incubated embryos of the five salmon species at 14°, 11°, 8°, 5°, 2°C. Each species had a slightly different pattern (and there is also variation among populations), but survival was generally poor at 2° and 14° and good from 5–11°. Salmon generally do not spawn in water near the upper lethal temperatures, and lethal low temperatures are not reached in most coastal systems. However, in the interior and

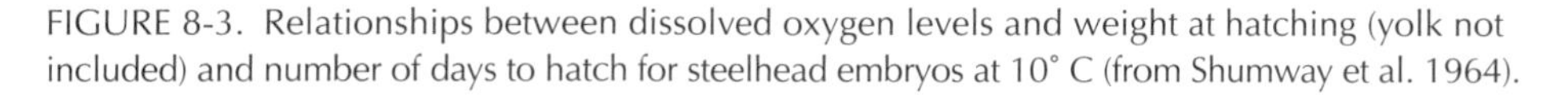

FIGURE 8-3. Relationships between dissolved oxygen levels and weight at hatching (yolk not included) and number of days to hatch for steelhead embryos at 10° C (from Shumway et al. 1964).

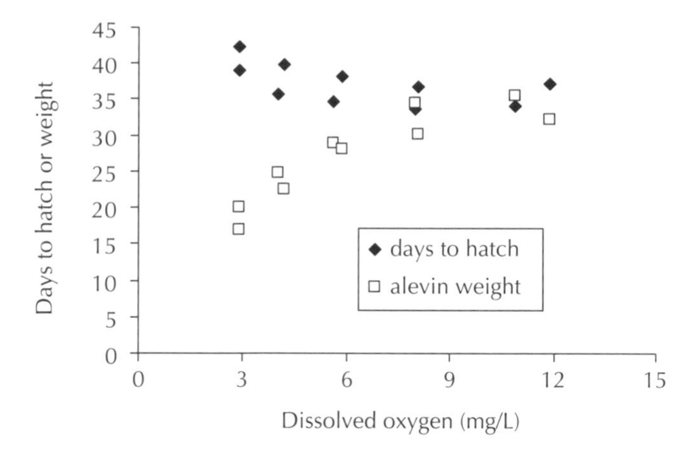

northern rivers, low temperatures and freezing are threats, and the salmon seem to seek areas of upwelling groundwater (generally warmer in winter than river water) for spawning (e.g., Kamchatka chum salmon: Leman 1993). One drawback of groundwater is that it is often lower in DO than surface water, and there is considerable variation in DO in the water irrigating the egg pockets (e.g., Peterson and Quinn 1996). Low DO, either in the water or low delivery rate of oxygenated water, is an important source of mortality, and rainbow trout embryos in water below about 5 mg/L failed to survive at all in the field (Sowden and Power 1985; fig. 8-4).

Many studies have been conducted on the relationship between embryo survival and the size of particles (silt, sand, and gravel) in the stream as a whole or in the egg pocket. Particular attention has been focused on the effects of fine material (usually defined as particles less than 1 mm or less than 0.85 mm diameter). Such fine particles can fill the interstitial spaces of the egg pocket and reduce the flow of oxygenated water, though the proportion of fine particles may not be correlated with DO levels per se (Peterson and

FIGURE 8-4. Survival of rainbow trout (Sowden and Power 1985) and steelhead (Coble 1961) embryos in redds with different dissolved oxygen (DO) levels.

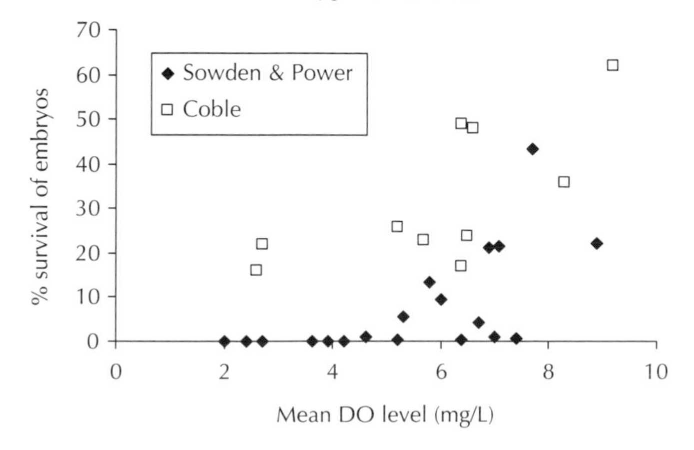

FIGURE 8-5. Annual average egg-to-fry survival of coho salmon in Carnation Creek, British Columbia, as a function of gravel quality (from Holtby and Healey 1986).

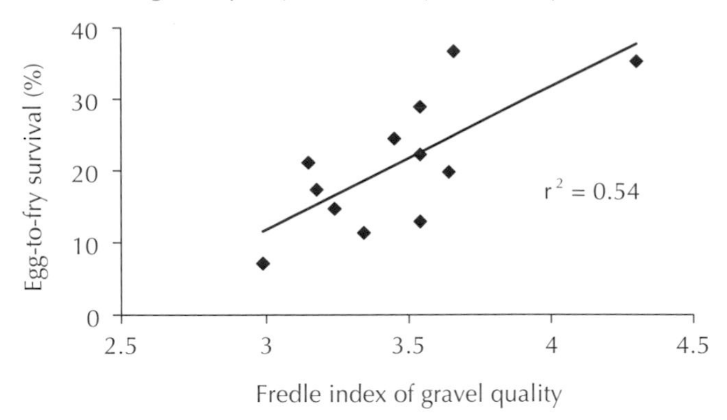

Quinn 1996). Relatively high embryo survival rates are observed in field and lab situations as long as the fine material does not constitute more than about 10% by weight of the gravel mixture or some related index of gravel size composition (Chapman 1988). However, at about 15–20% fine material we see dramatic reductions in survival. For example, Holtby and Healey (1986) reported that the annual average egg-to-fry survival of coho salmon was related to a measure of gravel quality known as the "Fredle index" (fig. 8-5).

Much of the work on fine sediment and embryo survival was motivated by the observation that land-use activities and especially logging increased the proportion of fine sediment in the gravel. Some fine material is washed from exposed land surfaces after removal of trees, but much of it comes from the bare surfaces of unpaved roads and from slope failure associated with logging roads. Cederholm and Reid (1987) related the proportion of fine sediment in a stream to the percent of the drainage basin's area that is road surfaces. A long-term study of salmonid responses to logging conducted at Carnation Creek, a small stream on the west coast of Vancouver Island, showed a dramatic decrease in survival to emergence of embryos from prelogging to postlogging periods (coho salmon: 29.1 to 16.4%; chum salmon: 22.2 to 11.5%; Scrivener and Brownlee 1989), and this was attributed to fine sediments. Thus there is natural variation in levels of intragravel flow rates and levels of DO, both of which affect development and survival of embryos, and fine sediment decreases flow rate and survival.

In addition to fine sediment and low DO, the other main physical factor affecting survival of embryos is the movement of streambed gravel, often referred to as "scour." At high water velocities, the gravel on the surface of the stream starts to move. Not surprisingly, the finer particles are moved most easily and only high velocities move large rocks. The mobilization of particles depends on several physical factors, but scour can be deep enough to reach buried embryos (Lisle and Lewis 1992; Montgomery et al. 1996). The depth of the egg pocket is largely a function of the size of the female (see fig. 6-5), and the timing of floods depends on the hydrology of the region (e.g., winter rains in coastal areas or late-spring snowmelt in the interior). Indeed, gravel movement may be a fundamental constraint on the timing of reproduction and body size in salmonids (Montgomery et al. 1999). Only large-bodied salmonids may be able to dig their eggs deep enough to avoid loss from the common winter floods in lowland streams along the

FIGURE 8-6. Annual average egg-to-fry survival of sockeye salmon in the Cedar River, Washington, as a function of the river's peak flow during the incubation period (brood years 1975, 1976, 1978, and 1979 from Stober and Hamalainen 1980; brood years 1991–1999 from Seiler, Volkhardt, Fleischer et al. 2002).

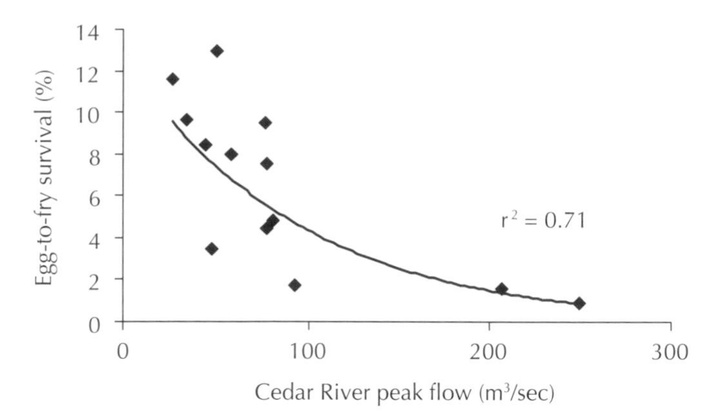

coastal rainforests; and in higher elevations there may be scour to bedrock every year, so the only alternative is to spawn in spring (e.g., cutthroat trout). The effects of high flows on survival are demonstrated by sockeye salmon spawning in the Cedar River, the main tributary of Lake Washington, near Seattle. The river is prone to flood in the winter and high flows clearly increase mortality (fig. 8-6).

The effect of high flows is not a uniform mobility of the streambed, however, because velocities vary with topographical features of the stream. The tail-outs of pools are typically choice locations for redds because the rate of water flow through the gravel is high, resulting from the increasing velocity as the pool shallows up and becomes a riffle, and the water is forced down into the gravel. These features also make such tail-outs vulnerable to scour during high flows, whereas riffles may experience less subsurface flow under ordinary conditions (and so be less desirable) but are less vulnerable to scour under high flows. Female salmon are unable to know whether the upcoming season will bring floods or not, so they must choose redd sites based on what they experience when they are ready to settle. Depending on the level of flooding, production may be primarily from one habitat or another. This subject has been examined (e.g., Holtby and Healey 1986; Rennie and Millar 2000; Schuett-Hames et al. 2000), but much more work on it is needed. It is relevant to the conservation of salmonids because the tendency of streams to flood, at a given level of rainfall, is related to the delivery rate of water from the land. Land-use practices such as logging (Jones and Grant 1996) and urbanization (Moscrip and Montgomery 1997) can greatly increase the peak flows for the same rainfall.

The actual agent of mortality related to flooding may be displacement of the embryos from beneath the gravel to the flowing water of the stream, where they are easily eaten by fishes and birds. There is also mortality from physical shock during sensitive developmental stages. Embryos are quite resistant to shock for the first 2 days after fertilization. They are then vulnerable until the blastopore closes, about 11 days later (depending on temperature), but are later quite tolerant until hatching. Scientists have devised a diverse and sometimes bizarre set of ways to test the tolerance of embryos,

FIGURE 8-7. Survival of coho salmon embryos after experiencing a standardized mechanical shock at different stages of development, as indicated by the number of temperature units after fertilization when the shock took place. Note that the effect occurred at the same stage, regardless of temperature (from Johnson et al. 1989).

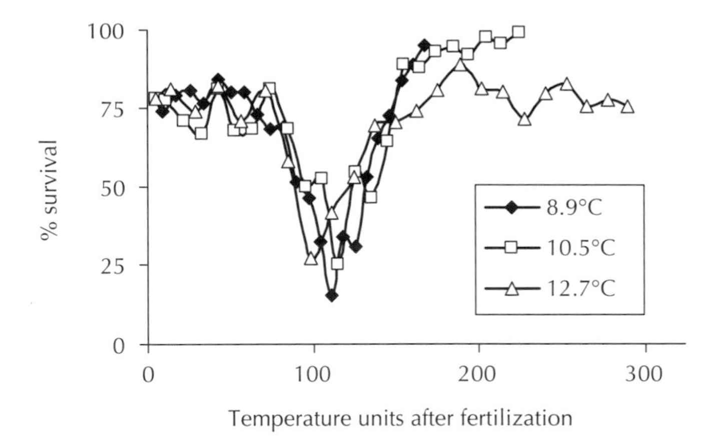

including dropping them from measured heights, squeezing them, or shaking them vigorously. These techniques all indicate that embryos are most vulnerable around 80–100 TUs. For example, S. C. Johnson et al. (1989) incubated coho salmon at three different temperatures, and the mortality after exposure to mechanical shock always occurred at about 80 to 120 TUs (fig. 8-7). However, the effects of scour are not simply (or perhaps even primarily) from crushing embryos or wafting them into the water column to be eaten. Data from Carnation Creek revealed that flooding affects gravel quality, and so both are correlated with survival of embryos (Holtby and Healey 1986). In years with high flows, gravel quality was lower, apparently because the fine sediment transported during high flows was then deposited and intruded into redds as the flows subsided after the flood (fig. 8-8).

FIGURE 8-8. Relationship between peak winter discharge at Carnation Creek, British Columbia (measured as height on a staff gauge) and an index of gravel quality related to embryo survival (from Holtby and Healey 1986).

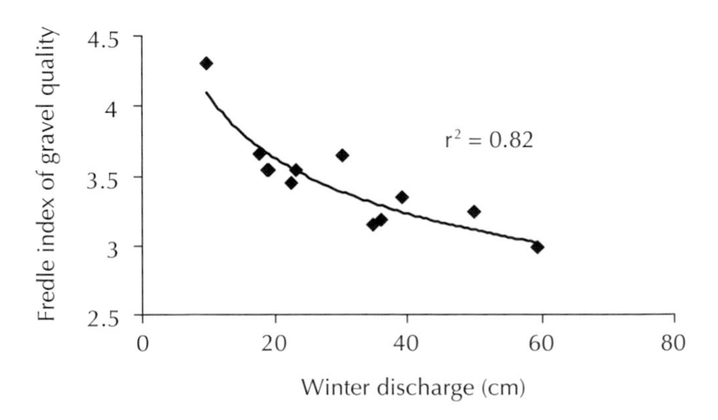

FIGURE 8-9. Annual average egg-to-fry survival of sockeye salmon in the Weaver Creek spawning channel, British Columbia, as a function of the total number of female salmon (mostly sockeye, but some chum and pink; from Essington et al. 2000, updated with unpublished data from the Pacific Salmon Commission).

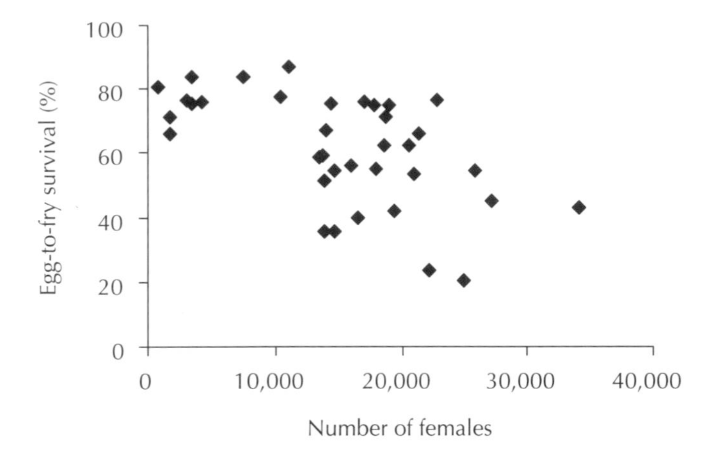

If scour does not occur and the size of the gravel is ideal, up to 80% of the eggs may survive to produce free-swimming fry. This typically only takes place in artificial spawning channels, where presorted gravel and regulated flows provide nearly ideal conditions. In these situations it is possible to clearly identify the other major cause of embryo mortality: physical disturbance from the nest digging activities of other females. Data for sockeye, chum, and pink salmon from the Weaver Creek spawning channel, British Columbia, showed that at low densities, survival was high but the absolute number of fry produced was low. As more and more females spawned, the survival rate went down but the total number of fry increased, though more slowly. Eventually, the carrying capacity was reached and further increases in the number of spawning females produced no additional fry or might even produce fewer fry if some disease or water quality problem affected them all (Essington et al. 2000; fig. 8-9).

This density-dependent fry production is generally viewed as the dominant feature of the population dynamics of pink and chum salmon, whose migration to sea frees them from the constraints of stream productivity that affect such species as coho and steelhead. Sockeye may be constrained by the capacity of the spawning environment or the rearing capacity of the lake, depending on the system. Redd superimposition can also occur in the stream-rearing species even though absolute densities of females are typically much lower than the other species of salmon. As indicated in chapter 6, females tend to use redd sites previously prepared by other females more often than would occur strictly on the basis of physical suitability (Essington et al. 1998). It is commonly assumed that the eggs of the first female are destroyed by the digging action of the second, especially if the second female is larger than the first (and so presumably could dig deeper). However, I am not aware of good evidence of this destruction, at the level of individual redds, and the consequences of redd superimposition for embryo survival need further investigation.

FIGURE 8-10. Annual production of sockeye salmon fry as a function of the number of parents in the Chilko River population (Fraser River system) from 1949 through 1981, excepting 1968 (from Roos 1991). The line is based on a Beverton-Holt spawner-recruit model (Beaverton and Holt 1957).

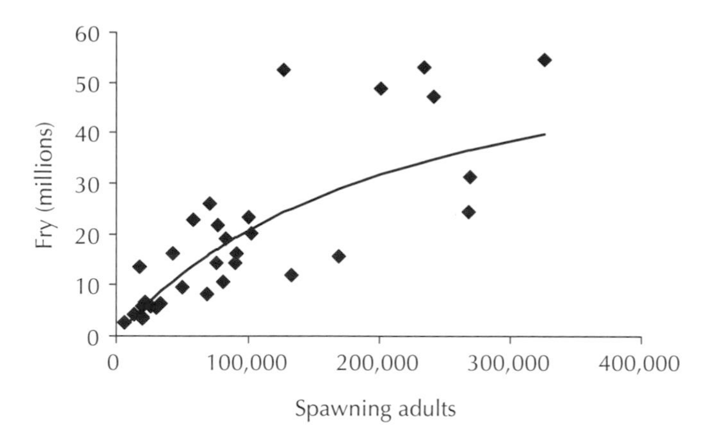

Density-dependent processes are not independent of physical factors such as scour and low DO affecting survival. As densities of adults increase, some redd superimposition will take place but females are also forced to use increasingly inferior sites, where low DO may decrease offspring survival. All these density-related effects vary over the course of the run. The first females get their pick of sites but are vulnerable to superimposition by later females, and the later females have fewer sites to choose from but have less risk of superimposition. Moreover, density-dependent mortality is greatly reduced by fisheries that have operated for about a century (and much longer, if one includes preindustrial fishing). Often about 50% and sometimes 70–80% of the adults surviving to maturity are caught before they reach the spawning grounds. Thus competition was probably a much greater factor in the evolution and dynamics of salmon populations than would be inferred from the densities commonly observed today. Data on the production of sockeye salmon fry in the Chilko River population (in the Fraser River system) reveal the general density-dependent pattern (fig. 8-10). At low densities, there is an apparently linear increase in fry with increasing numbers of parents; the slope of this line is the population's "productivity." At higher densities, the production tends to flatten as survival of embryos declines, indicating that the carrying capacity is being reached. However, variation in production is typical of such relationships for natural populations, as density-independent factors also affect survival.

A final consideration in egg-to-fry mortality and density dependence is disease, and a noteworthy example is the infectious hematopoietic necrosis (IHN) virus. The pathways of transmission are incompletely known, but this virus afflicts salmon, and especially sockeye, from the time of hatching to emergence from the gravel, or shortly thereafter. It is a great concern in hatcheries, where the circulation of water among embryos at high density can rapidly spread the virulent disease. It is unclear how prevalent it is in nature, but major outbreaks affecting alevins and emerging sockeye fry have occurred in the Weaver Creek spawning channel (Traxler and Rankin 1989) and in the

FIGURE 8-11. Factors causing mortality of salmon embryos and alevins. Factors within the circle are proximate causes of mortality; factors outside the circle affect those inside the circle.

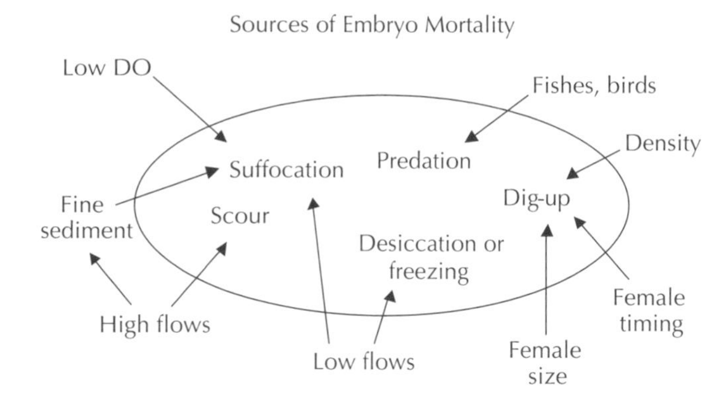

wild Chilko River population (Williams and Amend 1976), suggesting that it may be important. The virus might occasionally cause high mortalities but then subside as densities drop. However, it would remain in the system, ready for another opportunity to spread.

Summary

Incubation rate is primarily a function of temperature (warmer water leads to faster development) and to a lesser extent DO (low levels retard development). Much of the mortality is caused by physical factors, notably the restriction in flow rate of oxygenated water by fine sediment, the physical displacement or damage from scour, or the intrusion of fine sediment in the aftermath of a flood. These physical factors are indirectly affected by density because competition may force some salmon to use poor-quality sites. Nest disturbance by females of the same or other species also causes direct mortality, though information on this process is surprisingly poor. Such "redd superimposition" may be an important factor in the patterns of age and size distribution of spawning within species (small females tend to spawn later than large females) and perhaps among species as well. Figure 8-11 summarizes factors affecting embryo survival.

Given all these sources of variation among years and among sites, what are the overall rates of mortality from egg to emergent fry stages? I compiled a table of average stage-specific survival rates of salmon species, based on values reported in studies cited by Bradford (1995), Groot and Margolis (1991), and a variety of other sources reporting data on wild populations (excluding values from spawning channels or other artificial environments; see table 15-1). Some care must be used when interpreting these averages because some studies had much longer periods of record than others, and the methods varied greatly among studies. Typically, the "potential egg deposition" was estimated for a given female or for a population from average or length-specific fecundity estimates. The number of surviving progeny was estimated from a trap over the individual redd or from the number of fry caught in the stream or migrating from it. Mortality after emergence can be difficult to separate from mortality in the gravel, especially if the fry are

8-7 A sculpin (genus *Cottus*) in the rocky substrate of Woody Island, Iliamna Lake, Alaska, where they prey on sockeye salmon eggs. Photograph by Thomas Quinn, University of Washington.

counted after some period or migration or residence. Notwithstanding these considerable methodological issues, what are the patterns? The species that spawn at high densities show lower average survival rates (pink: 11.5%; sockeye: 12.7%; and chum: 12.9%) than the species spawning at lower densities (coho: 25.3%; steelhead: 29.3%; and chinook: 38.0%). The especially high survival rate for chinook may also result from their large size (hence ability to dig deep redds) and their tendency to spawn in larger, deeper, and perhaps more stable rivers than the other species.

9

Alevin Movements, Emergence, and Fry Migrations

As we have seen from the previous chapter, the timing of hatching depends primarily on temperature but also on dissolved oxygen. Once the embryos have hatched, the alevins can breathe with their gills and can also move. This stage of the life cycle takes place below the surface of the stream, so field observations are very difficult and laboratory studies are rather artificial. However, there are many interesting behavior patterns during this period, and important factors affecting the utilization of yolk and alevin growth.

Alevin movements

In general, alevin movements have three phases. They first move down deeper in gravel after hatching, then move laterally through the gravel, and finally move upward and emerge when their yolk has been largely or entirely absorbed and they are fully formed fry. The orientation responses of alevins are described as tactic and kinetic movements. Tactic responses are directed movements with respect to some specific stimulus such as light or the flow of water. A positive phototaxis means an attraction to light, and negative phototaxis is an avoidance of light, for example. These behavior patterns are innate, not learned, but they change over the course of development. Kinetic responses are changes in activity in response to physical stimuli and will tend to concentrate the organisms in areas with desirable conditions, even though they did not have a directional response to the stimulus. For example, if alevins tended to remain still in dark areas and moved in an undirected manner in illuminated areas, they would become concentrated in the darker areas of their environment.

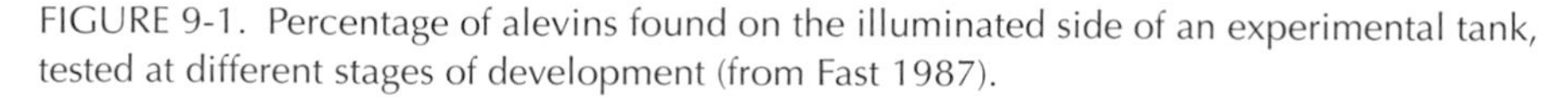

FIGURE 9-1. Percentage of alevins found on the illuminated side of an experimental tank, tested at different stages of development (from Fast 1987).

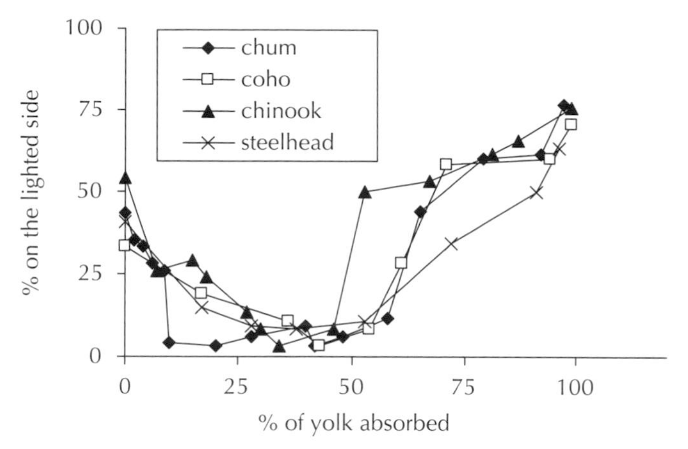

Alevins respond strongly to light but the nature of the reaction depends on the stage of development. Fast (1987) examined this by placing groups of chum, coho, chinook, or steelhead alevins in an aquarium with dark and lighted sides and counting the number on each side. Immediately after hatching the distribution of alevins was nearly random but there was a progressively stronger avoidance of light, reaching a maximum when the alevins had absorbed about 25–50% of their yolk, after which they became more tolerant of and even attracted to light (fig. 9-1). Under ordinary conditions, these responses to light (shown by all species tested) will tend to keep the alevins deep in the gravel where they are safe, rather than near the surface where they would be vulnerable. When their yolk is almost completely absorbed, they must become progressively more tolerant of light to function in the open water.

In laboratory studies, alevins move downward about 10–20 cm after hatching (e.g., coho salmon: Dill and Northcote 1970; rainbow trout: Carey and Noakes 1981). The downward movement is inhibited by small particles, as the newly hatched alevin's maneuverability is restricted by the large yolk sac attached to its belly. Later in development, alevins may move laterally and will eventually move upward to emerge from the gravel as fully formed fry. Interestingly, the initial downward movement and subsequent upward movement do not result merely from responses to light, though avoidance and later attraction to light would cause such movements. Nunan and Noakes (1987) found that newly hatched rainbow trout alevins moved down, then later moved up, regardless of light regime. The alevins not only showed the full suite of responses when held in complete darkness, but they even did so when the tanks were illuminated from the bottom, placing the downward orientation in conflict with the tendency to avoid light. Nunan and Noakes (1987) concluded that an innate sense of down and up, termed "geotaxis," is the dominant process in orienting the vertical movements of alevins. Nunan and Noakes did not indicate how the alevin would know up from down, except that it was not merely a response to the direction of water flow, though alevins indeed show a positive rheotaxis (meaning that they tend to swim into the flow of water). The sense of gravity or pressure that apparently orients them has not been investigated in detail.

FIGURE 9-2. Percentage of chum salmon alevins moving into the arm of a two-choice apparatus with the higher concentration of dissolved oxygen (DO), tested in the middle and late stages of alevin development (from Fast 1987).

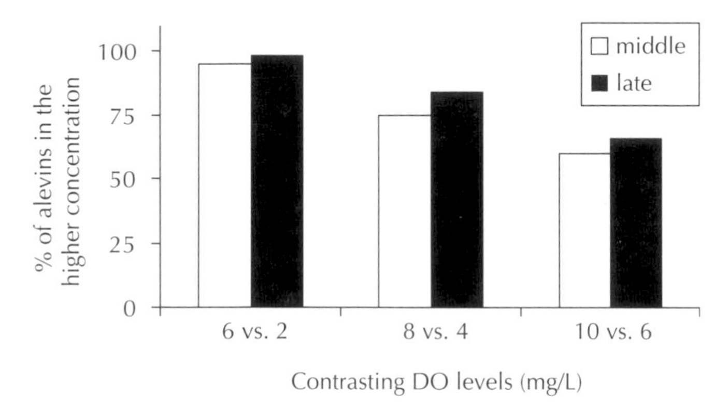

About a month after hatching, alevins are more active, and their mostly lateral movements are facilitated by larger gravel. At this stage, movement is generally positively rheotactic, though immediately after hatching the alevins move downstream in high velocities (Fast 1987). Alevins are also sensitive to dissolved oxygen (DO) and carbon dioxide concentrations. Low DO levels stimulate early hatching to increase the scope for breathing, and alevins move to high DO concentrations. Fast (1987) built an experimental arena with two inlets of water (differing in DO) flowing into a gravel-filled trough. He placed chum, chinook, and coho salmon and steelhead alevins in the lower section and recorded whether they moved toward one of two DO concentrations, differing by 4 mg per liter (6 vs. 2, 8 vs. 4, and 10 vs. 6). Significant preference was shown for the higher concentration in all tests (e.g., chum salmon: fig. 9-2), but the preference was strongest at the lowest concentration (that is, 6 mg/L was preferred over 2 to a greater extent than 8 over 4 and 10 over 6). In addition to these directed movements, a buildup of carbon dioxide around the alevin induces what Bams (1969) termed "ventilation-swimming"—swimming in place to circulate water. Finally, if water flow ceases and conditions become unfavorable the alevins begin undirected (kinetic) movements that tend to move them to more favorable conditions.

These adaptations for respiration are important because the embryos need oxygen and are initially close to each other. Fast (1987) reported oxygen consumption rates of about 1500 mg per gram of alevins per hour (using dry weight of alevin tissue, without yolk, at 8°C). As expected, oxygen consumption rates increased with temperature and decreased on a per-weight basis with development but increased on an absolute scale as the alevins got larger. Given these requirements, and the variation in DO among egg pockets, it is not surprising that the alevins have a series of adaptations related to breathing. The DO requirements of embryos may be one reason why females deposit their eggs in a series of pockets rather than all together.

At the end of the alevin period there is a phase of upward (geotactic) movement combined with orientation into the water current. At this phase the alevins are more tolerant of light than they are at earlier stages, but light still inhibits emergence and most fry emerge from the gravel at night (reviewed by Godin 1982). For example, Heard (1964) found that 84% of the sockeye fry emerged during the night from redds in the Brooks River, Alaska. This site is close to 60° north latitude, and the peak of emergence was around June 15, near the summer solstice. At that time of year the days are more than 19 hours long, so the night constitutes only about 20% of the 24-hour period. Godin (1980) also reported that the great majority of pink salmon emerged from simulated gravel redds at night but that the proportion emerging during the day increased over the course of the emergence period, similar to the results of Mason (1976a) for coho salmon fry. The subgravel phase ends when the fry emerge and first fill their swim bladder, enabling them to maintain neutral buoyancy.

Alevin development

The date of emergence depends on the process of yolk absorption, which depends on temperature and varies among species. At high temperatures metabolism is less efficient and the fry from a given sized egg are smaller. Heming (1982) incubated chinook salmon alevins at 12°, 10°, 8°, and 6°C and found progressive reductions in yolk conversion efficiency from 68% to 49%, in length from 42.2 to 38.3 mm, and in weight from 691 to 604 mg over that range of temperatures. These differences may seem small but they are more than enough to affect the outcome of territorial disputes or predator attacks on free-swimming fry. In addition to the effect of temperature on yolk conversion and fry size, there are other subtle but important factors. Bams (1969) presented a conceptual model to describe the energetics of alevin development (fig. 9-3). We can think of the process as a transformation of yolk into salmon. The fertilized egg is negligible in size and the yolk is large (salmon eggs weigh about 0.1 to 0.4 g before water hardening, and trout eggs are a bit smaller). Until the alevin begins to

FIGURE 9-3. Conceptual model of the transfer of mass from yolk to fry during development and the reduction of fry size with increased alevin activity (from Bams 1969).

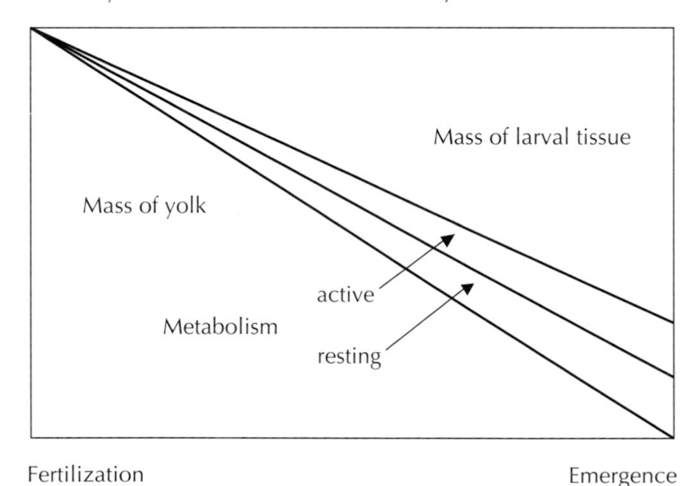

feed (which it may do while still taking nourishment from the yolk), the sole source of energy comes from the yolk. Thus the yolk gets smaller (both before and after hatching) as the embryo gets larger. However, there cannot be a perfect transfer of mass from yolk to fish because some energy is used in metabolism rather than growth. If environmental conditions are good (meaning that it is dark, with a good flow rate of oxygenated water), the alevins are quite inactive, and their metabolic rate is low. If they are active, more yolk is metabolized for movement and less is available for growth. This seemingly obscure fact turns out to be quite important.

Fry incubated in hatcheries are often smaller than wild fry, even if the eggs were the same size. How can this be? Alevins have a "righting response," meaning that they like to remain upright, with their yolk below them. Snuggled in the interstitial spaces with gravel all around them, they tend to move very little. However, if they hatch and find themselves in an artificial surface such as the trays commonly used in hatcheries, they fall over on their side. Their struggle to right themselves results in considerable activity, especially if they are in a group and the activity of one alevin disturbs others. Mead and Woodall (1968) incubated sockeye salmon from Weaver Creek, Cultus Lake, and the upper Pitt River (all in British Columbia) under hatchery conditions (flat troughs), in artificial spawning channels with gravel, and under natural conditions. For all three populations, the hatchery-incubated fry emerged 1–2 weeks earlier and were significantly smaller (25% less dry body weight) than the fry from the natural channels. The hatchery fry used their yolk more rapidly and less efficiently and so emerged smaller and earlier. The reduction in size of fry resulting from incubation under artificial conditions was sufficient to reduce their swimming performance and increase vulnerability to predators (Bams 1967).

As the alevins reach the final stages of yolk absorption, the yolk sac that once dominated their thin bodies is reduced to an ellipse of orange, like a slim football, on their ventral surface. When this is finally enveloped by their body wall they are said to be "buttoned-up." It is common to see newly emerged fry with a bit of yolk showing on the belly, especially early in the emergence period. Late-stage alevins still nourished from the yolk can feed as well (e.g., Zimmerman and Mosegaard 1992; García de Leániz et al. 2000), even while still in the gravel. During this period, the yolk does not seem sufficient to meet the energetic demands of the large alevin, and the alevin actually loses weight by the time the yolk is fully absorbed. For example, chinook salmon incubated at 6°C had a dry weight of 20 mg tissue and 140 mg yolk when they hatched (total = 160 mg), reached a maximum tissue weight of 112 mg with 9 mg yolk (total = 121 mg), and weighed only 75 mg when the yolk was completely gone (Heming 1982). During the late stages of alevin development, alevins would normally begin feeding opportunistically below the surface of the gravel and so have overlap between yolk and external food sources.

The combination of intrinsic individual variation in developmental rate, variation in activity, and in environmental conditions (temperature, flow, and oxygen) experienced by embryos and alevins can result in considerable variation in emergence from a single redd. Mason (1976a) put alevin siblings into simulated redds 10 days after hatching. Fry emergence spanned 20 days, though more than 75% of the fry emerged during the median

10-day period. As we will see in chapter 11, these differences in emergence date can have considerable consequences for the fry.

Initial migration of fry

Upon emergence from the gravel, juveniles of some species (notably chum and pink salmon) migrate directly downstream. We will return to them in chapter 12 but at this point only indicate that their migration is usually at night. The initial movements of stream-rearing fry such as coho and rainbow trout are not well known. Most probably stay near the redd and try to find territories or disperse downstream; the density would be so high that many must move away. Detailed analysis of the movements of fry from specific redds would be of considerable interest, and this might be generated from either genetic analysis of naturally spawned fry or collection of fry marked (e.g., thermal marking of otoliths: Volk et al. 1994) as embryos and planted at known locations.

Downstream migration is at night, but some populations migrate upstream and they do so in the day (Godin 1982). Upstream migration is essential for fish migrating to a lake to rear (typically sockeye salmon but also trout). For example, sockeye salmon fry migrating up the Chilko River to Chilko Lake (in the Fraser River system) do so in the day (Brannon 1972), as do fry migrating up the Babine River to Babine Lake in the Skeena River system, British Columbia (Clarke and Smith 1972). How do newly emerged fry decide which way to go? Their siblings are as inexperienced as they are, and no parents or older juveniles are available to show the way. Similarly, many trout populations exist above waterfalls. Downstream migration by fry might lead them to favorable feeding areas below the falls but they could never get back, so the population can only persist if the juveniles hold position against the current or move upstream.

Studies, primarily on sockeye salmon and cutthroat trout, have revealed a genetic basis for the tendency of newly emerged fry to swim upstream or downstream. For example, sockeye fry emerging in the Chilko River migrate upstream to rear in Chilko Lake whereas fry emerging in the Stellako River migrate downstream to rear in Fraser Lake (also in the Fraser River system). Brannon (1972) designed an artificial stream and tested fry by placing them in the middle and seeing how many moved to traps at the upstream and downstream ends. When tested in the field, Chilko River fry swam upstream and Stellako River fry swam downstream. To tease apart the roles of genetic and environmental factors, Brannon took eggs and milt from adults and incubated embryos for laboratory experiments. In addition to pure Chilko and Stellako fry, he produced hybrids between the two and also used fry from Cultus Lake, where sockeye spawn on beaches and so the fry need not migrate to reach their lake. In experiments with about 5000 fry per group, 91% of the Chilko fry migrated upstream, only 12% of the Stellako fry migrated upstream, and their hybrid offspring showed an intermediate response (49% upstream), demonstrating genetic control over the trait. Cultus Lake fry showed an indifferent response to current (53% upstream). Differences in orientation between inlet and outlet populations have been obtained with other salmonid species and other sockeye populations (fig. 9-4).

The fry from outlet populations tend to swim upstream, but the orientation is considerably more complex than this. Brannon (1972) pointed out that the upstream

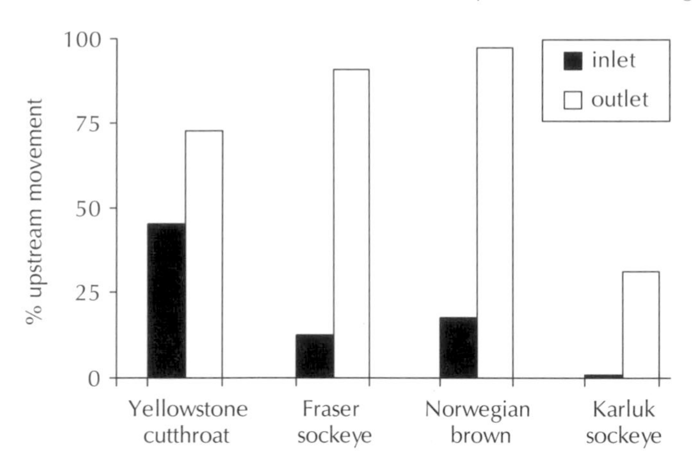

FIGURE 9-4. Percentage of salmonid fry from lake inlet and outlet populations that moved upstream in experimental tanks (Yellowstone Lake cutthroat trout from Bowler 1975; Fraser River system sockeye salmon [Stellako and Chilko rivers] from Brannon 1972; Norwegian brown trout from Jonsson et al. 1994; and Karluk Lake sockeye salmon from Raleigh 1967).

migration toward the lake may bring fry into contact with water from a tributary of the river, and migration up that water source would lead to progressively less suitable habitat for the many fry. For example, Chilko River fry must bypass Madison Creek on their way upriver to Chilko Lake. Tested in the experimental arena, 84% of the fry moved upstream (as expected) in water from the river draining the lake, but only 5% migrated upstream when tested in water from Madison Creek. This suggested that some clue indicated the presence of the lake and triggered upstream swimming. Subsequent experiments indicated that fry are more attracted to the water source in which they have been incubated than to unfamiliar water. This response would assist fry in swimming upstream toward a lake because those odors would be experienced by alevins during incubation. Moreover, sockeye fry also seem to have an innate attraction to lake water. Chilko Lake fry, incubated on either creek or lake water at Cultus Lake lab, were more attracted to lake water than creek water. Subsequent experiments with Lake Washington sockeye fry supported these findings (Bodznick 1978). Perhaps the presence or odors of zooplankton, or the odors of juvenile sockeye salmon or some other chemical trace, indicate the presence of a lake and stimulate upstream swimming. Given the need for the millions of sockeye salmon fry to find food, and their morphological and behavioral adaptations for life in lakes, poor growth and predation provide strong selection against fry failing to find the lake (photo 9-1).

Summary

After the alevins hatch they are incompetent to swim in streams and would find harsh conditions if they emerged. Their initial responses are to wriggle downward through the interstitial spaces in the gravel of the stream or lake and then to disperse. They orient toward flowing water, away from light, and toward higher concentrations of dissolved

9-1 School of newly emerged sockeye salmon fry in the Cedar River, Washington. Photograph by Julie Hall, City of Seattle.

oxygen. Otherwise, alevins are inactive, making most efficient use of the yolk provided by their mothers. As the yolk is metabolized, the organ systems of the alevins develop, and there is a period when some yolk remains but the digestive tract is functional. Fry can emerge at this stage, and premature emergence may be triggered by stressful conditions. The fry typically emerge from the gravel at night and take up residence in the stream or migrate from it. There is a high degree of genetic control over the migratory behavior of newly emerged fry and their responses to flowing water, and this has been best studied in sockeye salmon.

10

Sockeye Salmon and Trout in Lakes

Most juvenile sockeye salmon rear in lakes for the first year or two of their lives before migrating to sea (though there are also river-type and ocean-type sockeye). Lakes are also typical habitat for many species of trout and char, and coho salmon may inhabit lakes in winter (e.g., Quinn and Peterson 1996) or as a primary rearing area (e.g., Ruggerone and Rogers 1992). In addition, some populations of pink, chum, and chinook migrate through lakes on their way to the ocean. It is not possible to do justice to the myriad patterns of all these species, so this chapter focuses on sockeye and, to a lesser extent, trout.

Research on sockeye salmon in lakes has emphasized three related processes: spatial distribution and movements (horizontal and vertical) within the lake, growth, and predator avoidance. Sockeye could grow faster in the ocean than in a lake, but they emerge from the gravel so small that the mortality they would experience at sea apparently counters the higher growth potential. Sockeye are the smallest, on average, of the North American salmon as eggs (Beacham and Murray 1993), and fry average 28.4 mm (based on the average of thirty-four populations, ranging from 23.5 to 31.4 mm: Ruggerone 1989; photo 10-1). So, sockeye stay in the lake for a year or two, grow slowly, try to avoid being eaten, and then take their chances at sea when they are bigger.

Distribution and horizontal movements

Sockeye emerge, often at high densities, and migrate to lakes that are usually very oligotrophic (unproductive). Sockeye fry generally feed on aquatic insects and crustacean zooplankton in the nearshore (littoral) zones of lakes and on zooplankton species in the offshore (limnetic) zone. How might sockeye fry orient their movements upon entering their lake? Do they move at random, move in response to proximate stimuli such as

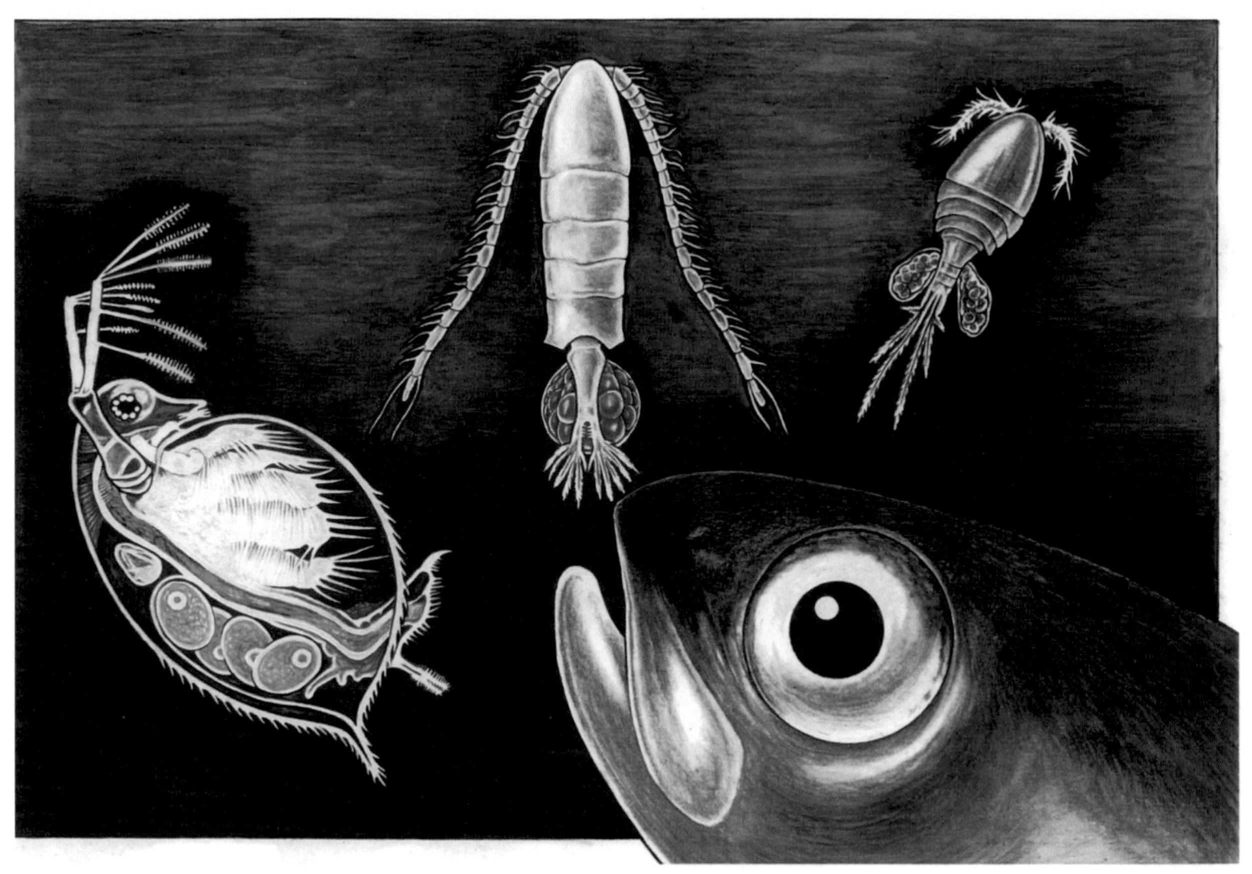

PLATE 13. Close-up of the head of a juvenile sockeye salmon and three of its common crustacean zooplankton prey: *Daphnia* (on the left; note the prominent eye and developing embryos), *Diaptomus* (in the center; note the cylindrical body, long antennae and single egg cluster), and *Cyclops* (on the right; note the more tear-drop shaped body, shorter antennae, and two clusters of eggs). Copyright: Charles D. Wood, Ph.D.

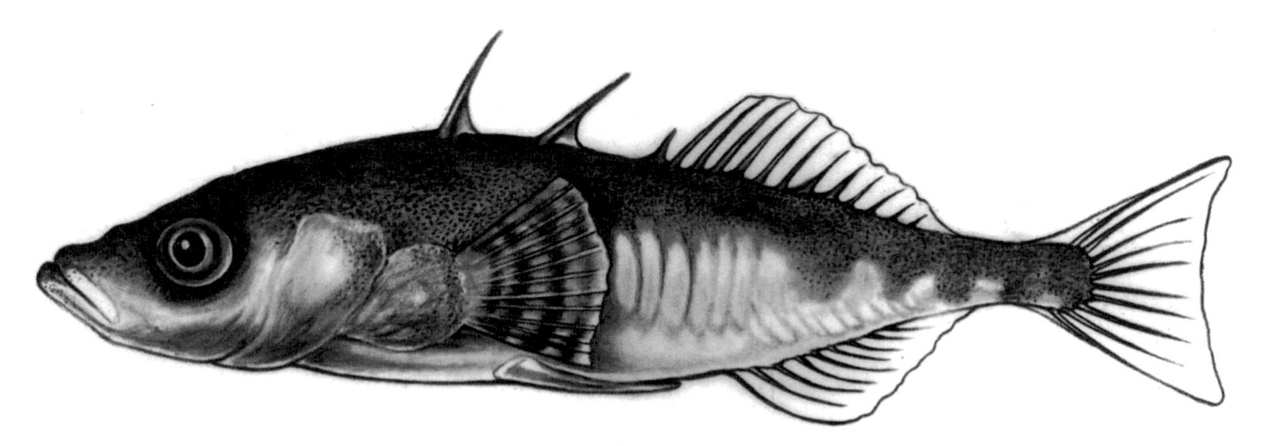

PLATE 14. Illustration of a threespine stickleback, *Gasterosteus aculeatus,* a common competitor with juvenile sockeye salmon in lakes. The development of lateral plates and spines varies among populations, and the colors change in the breeding season. Copyright: Charles D. Wood, Ph.D.

10-1 Newly emerged sockeye salmon fry. Photograph by Richard Bell, University of Washington.

food and competition, or move in a directed manner that has evolved to facilitate growth and predator avoidance? Extensive sampling in Lake Aleknagik, Alaska, revealed that newly emerged sockeye salmon (about 26–28 mm long) occupy the littoral zone from early June through mid-July, growing to about 50 mm long as the waters of the lake warm (Rogers 1973). They then move offshore and remain in the open water of the lake until they leave as smolts the following spring. It is not clear whether the fry move offshore after food becomes scarce in the littoral zone or if they move in anticipation of better feeding offshore, but they tend to move offshore earlier in years when the ice leaves early and the lake is warm.

Some populations show highly directed migrations. For example, sockeye fry enter Babine Lake (a very long, narrow lake in the Skeena River system, British Columbia) from a number of tributaries, including the Fulton River, near the lake's middle. Fulton River fry enter in late May and early June and have largely left the littoral zone by mid-July. McDonald (1969) found a very strong tendency for fry to move southward, along the shore and in open water, rather than to the north. Simulations by Simms and Larkin (1977) also indicated a highly oriented, southward migration for the first few months. The movements were more random from late in the first summer until the following spring, when a highly directed northward migration took the smolts to the outlet and down to the ocean.

Foraging opportunities (and perhaps also predator avoidance) may provide the ecological motivation for migration, but how do the fish know which way to go? In the 1960s there was a blossoming of research on the orientation abilities of animals, including homing pigeons, a variety of migratory birds, fishes, honeybees, reptiles, intertidal invertebrates, and others. In many cases, migratory animals released in the middle of radially symmetrical areas (round tanks or platforms, cages, or multi-armed apparatus) will move in the direction that is appropriate for their migration in open space. Cognizant of this work and that of Groot (1965) on sockeye salmon smolts leaving Babine Lake, Brannon (1972) released sockeye salmon fry from the Chilko River population in the middle of tanks with six arms and recorded the directions of movement. The fry

TABLE 10-1. Numbers of Chilko River sockeye salmon fry trapped in the four cardinal compass directions when the sky was clear and visible or obscured by clouds or covers, and in the normal magnetic field of the earth or in a field in which south was shifted to the east (Quinn 1980).

Sky	*Field*	*North*	*South*	*East*	*West*	*Mean direction*
Visible + obscured	normal	488	916	662	617	155°
Visible	S in the E	83	229	162	154	177°
Obscured	S in the E	166	147	231	151	77°

tended to move southward, as would be appropriate for migration because they enter at the north end of this long, narrow lake.

I built smaller, four-armed versions of Brannon's arenas, captured fry migrating to lakes, released them in the center of the tanks, and then counted the numbers trapped in the north, south, east, and west arms after 45-minute trials (Quinn 1980). Some tests were conducted with a view of the sky (clear or cloudy, as the weather determined), others under covers that allowed light but obscured the image of the sun (during the day), or were fully opaque (at night). In addition, a direct current of electricity running through coils of wire around the tanks rotated the horizontal component of the earth's magnetic field 90° counterclockwise (moving north to the west, south to the east, etc.).

The fry were denied the water current and odors that so strongly control migration to the lake and were in a highly artificial environment, so the absence of directional movement would not have been surprising. However, Lake Washington fry (in Washington State) tended to move northward when tested at night, and this is the compass direction that they migrate from the spawning grounds to the lake and in the lake itself (Quinn 1980). Chilko River fry, tested in the day as appropriate for their migration, oriented to the south, as they would in the lake. They showed this response on both clear and cloudy days and when the tanks were covered. An alteration in the magnetic field did not affect their orientation when they could see the sky but when tested under cloudy skies or covers, their orientation shifted from south-southeast to east-northeast, as predicted (table 10-1). These results indicated that the primary orientation mechanism was visual, probably the position of the sun or polarized light patterns, but that in the absence of such clues the earth's magnetic field gave the salmon directional information.

These orientation abilities are impressive but not unique. Many kinds of aquatic and terrestrial animals migrate in fixed compass directions by linking their circadian rhythm (internal clock) to the apparent movement of the sun across the sky (at 15° of arc per hour). The animals shift their direction of movement, relative to the sun, to compensate for the changing position of the sun in order to move in a fixed direction. Many kinds of animals, including salmonids, can also detect and orient to the plane of polarized light (Parkyn et al. 2003), and this can indicate the position of the sun when it is obscured by clouds or is low on the horizon. The ability to detect earth-strength magnetic field has been documented in many animals, including nudibranchs, bees, salamanders, sea

10-2 Threespine stickleback, a common competitor for food with sockeye salmon in lakes. Photograph by Richard Bell, University of Washington.

turtles, birds, and tuna, as well as salmon. The actual mechanism by which organisms detect the field has been a subject of much research and controversy over several decades. Single-domain sized crystals of magnetite are apparently an essential component of this sense (e.g., research on rainbow trout by Walker et al. 1997; Diebel et al. 2000).

Feeding and growth

Once the sockeye fry enter the lake, regardless of the mechanisms orienting their movements, they are occupied with two goals: to grow and to avoid being eaten. These two goals are not entirely compatible, as we will see shortly. The lakes that they rear in are generally oligotrophic, meaning that primary production rates are low, relative to most lakes. These lakes often have rather simple communities (i.e., few species) of fishes. Sockeye salmon are often the most abundant fish species feeding on crustacean zooplankton in the open water of the lake, though there may be sticklebacks (threespine: *Gasterosteus aculeatus*, photo 10-2; ninespine: *Pungitius pungitius*), smelt (e.g., pond smelt, *Hypomesus olidus*, or longfin smelt, *Spirinchus thaleichthys*), and whitefish (e.g., least cisco, *Coregonus sardinella*) competing for the zooplankton as well. Some lakes have freshwater shrimp (e.g., *Neomysis mercedis* or *Mysis relicta*) that may be prey for the sockeye salmon but can also compete with them for zooplankton. In addition to these species in the open waters of the lake, there are typically sculpins (e.g., *Cottus asper*) on the bottom and various invertebrates including insects near the edges. Larger salmonids are the main predatory fishes, foraging in nearshore and open-water areas.

Growth is fastest in late spring and summer and very slow in winter (fig. 10-1). Growth results from the interactions of food and temperature, because water temperature controls fish metabolism. Brett et al. (1969) conducted controlled feeding trials of sockeye salmon at different temperatures and demonstrated clearly that the maximum scope for growth is at about 15°C. That is, if fed a high or excess ration, this is the temperature at which they grow most rapidly. At lower temperatures, food is digested too slowly for maximum growth, and at higher temperatures the metabolic

FIGURE 10-1. Growth of sockeye salmon in Lake Washington, Washington (from Woodey 1972), and Babine Lake, British Columbia (from McDonald 1969).

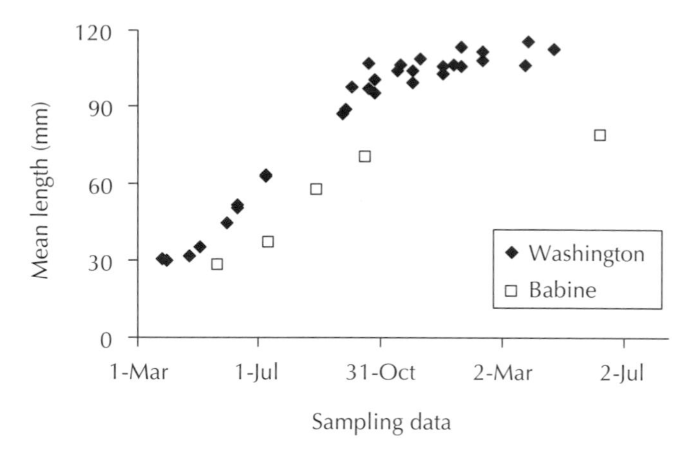

rate is so high that much of the food is required to maintain the fish and little remains for growth (fig. 10-2). Food conversion was relatively efficient from 5–17°. Such curves are typical of fishes, but the optimal temperature varies greatly among species. However, if the fish are fed a reduced ration (as might occur in oligotrophic lakes), the optimal temperature for growth shifts downward because at 15° the metabolic demand is higher than at 10°, for example, and if the fish is not satiated then some of the

FIGURE 10-2. Growth of juvenile sockeye salmon as a function of food ration (percentage of body weight per day) and temperature (from Brett et al. 1969). Note that growth peaks at intermediate temperatures, and as ration is reduced, the optimal temperature for growth is lower.

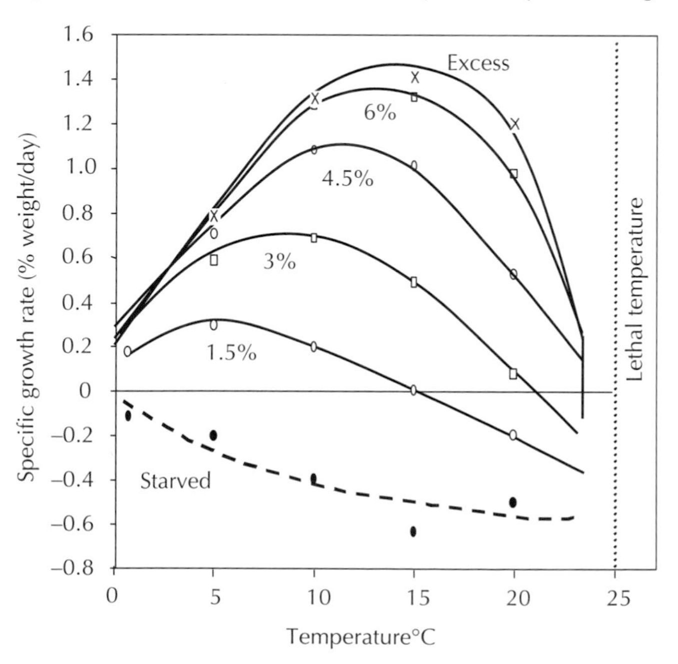

FIGURE 10-3. Average length and weight of age-1 sockeye salmon smolts from thirty-six lakes in Alaska differing in zooplankton density (Edmundson and Mazumder 2001).

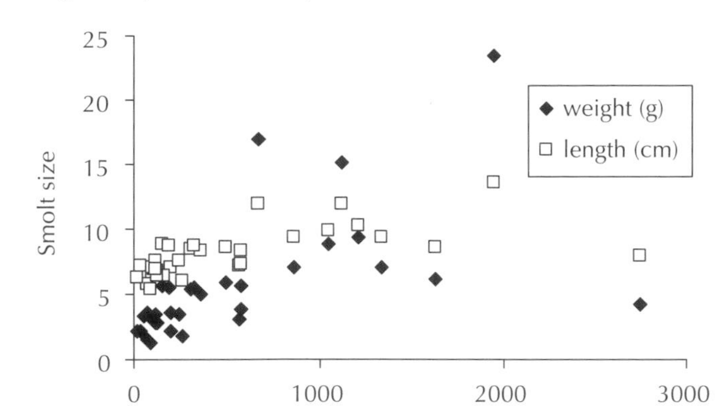

higher metabolism comes at a cost to growth. This can be understood by the extreme case; if fish do not feed at all, they lose weight most slowly in the coldest water because their metabolism is so slow.

Sockeye salmon occupy lakes with a wide range of thermal regimes and other physical attributes, including coastal lakes in Washington that stratify in summer with surface temperatures near 20° and that remain mixed without freezing in winter, to lakes in the interior and in northern latitudes that are covered by ice for at least half the year and have summer surface temperatures barely above 10°. These lakes range in elevation from essentially sea level to 2000 m, in area from 1 to 2600 km^2, and include coastal lakes from Washington to Alaska and lakes in the interior of the Columbia, Fraser, and Skeena river systems. The mean length of age-1 smolts varies among lakes from about 50 to 140 mm (a more than fifteenfold range in weight), but the mode is between 60 and 90 mm, based on data from Burgner (1991), Gustafson et al. (1997), Edmundson and Mazumder (2001), and K. D. Hyatt (personal communication).

What controls the variation in growth among lakes? Edmundson and Mazumder (2001) found that water temperature alone explained 24% of the variation in smolt length among thirty-six Alaskan lakes (warmer water being linked to bigger fish), consistent with the physiological role of temperature. However, most (52%) of the variation in smolt size was explained by the density of zooplankton, especially at densities up to 1000 mg/m^2 (fig. 10-3). Beyond that point, lakes with more zooplankton did not produce larger smolts. The density of juvenile sockeye salmon explained little of the variation in smolt size. Hyatt and Stockner (1987) also reported that the size of sockeye smolts in coastal British Columbia lakes was related to the concentration of zooplankton per fish (including sockeye and their main competitor, threespine sticklebacks).

The zooplankton that the sockeye salmon eat are crustaceans, primarily calanoid and cyclopoid copepods and cladocerans. They vary considerably in size, visibility, and mobility (including escape responses), and these all affect their ease of capture and desirability for sockeye. Zooplankton are not taken by filter feeding but rather are located, pursued, and consumed one at a time, and the stomach of a juvenile sockeye might

FIGURE 10-4. Monthly mean surface water temperature, density of the primary zooplankton species eaten by sockeye salmon (*Daphnia* and *Diaptomus;* data from Daniel Schindler, University of Washington) and the relative abundance of sockeye salmon fry entering Lake Washington (from Seiler 1995 and Seiler and Kishimoto 1997).

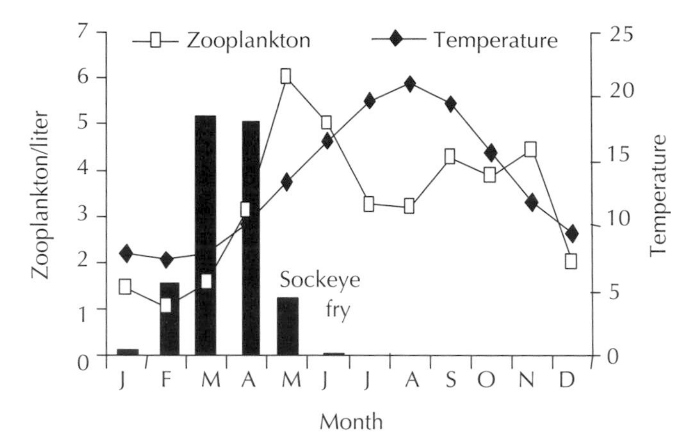

contain dozens, hundreds, or even more than a thousand such individual prey items, eaten within a day. Planktivores, including sockeye, tend to eat large prey more often than would occur by chance (Eggers 1982; O'Neill and Hyatt 1987).

All other things being equal, larger prey will be easier to see (think about trying to spot individual zooplankters 1 mm long or smaller, at twilight, 10 m below the surface of a murky or tea-colored lake). However, encounter rate is not the only factor affecting diet. Larger prey provide the predator with a bigger meal than smaller prey, assuming it does not take much longer to catch, subdue, and consume the larger ones. Thus under the principles of optimal foraging theory, a predator should eat the larger of two prey items available to it. The optimal size of prey also increases as the predator gets larger because the gape of its mouth allows it to handle larger prey. The morphology, presence of eggs, and other features of zooplankton make some more easily caught than others. However, prey do not passively accept their fate but are more or less actively trying to evade the fish. Eggers (1982) concluded that the diets of Lake Washington sockeye indicated an active preference (i.e., more than just differential encounter rate) for large, nonevasive prey and a selective avoidance of small prey. However, he pointed out that zooplankton communities are dynamic, thus sockeye diets shift over the seasons with the relative abundance of more or less desirable prey.

The biology of the zooplankton species varies greatly but in general they are herbivores, grazing on phytoplankton. Phytoplankton abundance depends on light, hence there is a marked seasonal pattern in primary production in lakes. The zooplankton often pass the winter as eggs or in some other resting stage and are thus quite scarce as prey to salmon. In spring, as days lengthen and the lake thaws, wind mixes the water column and brings inorganic nutrients to the surface where photosynthesis takes place once the lake warms up and begins to develop thermal stratification. At this time, a phytoplankton bloom usually occurs, followed by an increase in zooplankton density. The emergence of sockeye salmon and their entry into lakes can precede the peak density of zooplankton (fig. 10-4). Zooplankton are preyed upon by the sockeye, and the

TABLE 10-2. Responses of coastal British Columbia lakes to fertilization in primary production (mg carbon per square meter of lake area per day), zooplankton biomass (mg ash-free dry weight/m^3), and sockeye salmon smolt weight (g) based on average August and September values in years before and after treatment; see Stockner and MacIsaac (1996) for details.

Variable	*Before treatment*	*After treatment*	*% increase*
Primary production	68	167	146
Zooplankton biomass	5	18	260
Sockeye smolt weight	2.12	3.5	68

abundance, size distribution, and species composition of zooplankton can be strongly affected by the selective feeding of sockeye and sticklebacks (O'Neill and Hyatt 1987).

Growth depends on the quality as well as the quantity of food, and there are substantial differences among zooplankton in the highly unsaturated fatty acids that are important for growth. For example, Ballantyne et al. (2003) concluded that sockeye growth in Lake Washington was more constrained by food quality (as measured by the fatty acid DHA) than quantity or temperature. Nutritional quality varies among zooplankton species, and it might also vary among lakes, depending on the phytoplankton they eat.

The abundance of phytoplankton, combined with temperature, thus fuels the growth of sockeye. Phytoplankton production depends largely on inorganic nutrients, chiefly nitrogen and phosphorus, and it has long been recognized that the lakes inhabited by sockeye salmon are often nutrient-poor (Stockner 1987). Therefore, the growth of sockeye, especially in some of the coastal lakes of British Columbia and southeast Alaska, is exceptionally poor (barely reaching 2–3 g in a year). Experiments to enhance primary production, hence zooplankton, hence sockeye salmon growth (and hence, it was hoped, survival at sea and adult returns) were initiated more than 50 years ago (P. R. Nelson 1959). A much larger-scale program was undertaken in British Columbia, starting with Great Central Lake in 1969 and growing to include many control and treated lakes (Stockner 1987; Hyatt and Stockner 1987; Stockner and MacIsaac 1996). Considerable research went into determining the appropriate ratio of ammonium polyphosphate and urea ammonium nitrate to encourage the growth of phytoplankton edible by zooplankton (rather than the often inedible cyanobacteria) for each lake, the appropriate concentration, timing of release, and so on (Stockner and MacIsaac 1996). In most cases, application of fertilizers from airplanes to large lakes resulted in striking changes that propagated through the trophic links to juvenile sockeye salmon (table 10-2).

Thus the variation in sockeye growth among lakes is primarily related to the availability of food, which is a consequence of the intrinsic productivity of the lake (i.e., flushing rate and inorganic nutrients from local geology) and the lake's temperature. However, growth also varies considerably within lakes among years, as temperature and food availability vary. In northern lakes, the earlier the ice leaves the lake and the warmer the spring conditions, the larger the fry at the end of the summer (e.g., Iliamna Lake, Alaska; Quinn, unpublished data). This probably reflects both the energetics of the salmon and the ecology of the lake itself. In addition, the density of fry can reduce zooplankton

FIGURE 10-5. Relationship between number of adult sockeye salmon spawning in tributaries of Lake Aleknagik, Alaska, and the mean length of their fry on September 1 of the next year (Schindler et al., forthcoming).

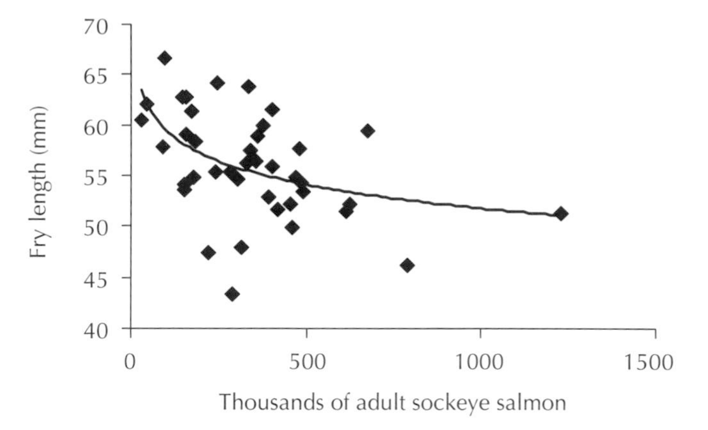

density in some lakes, and growth is often reduced at high fish densities (fig. 10-5). Results reported by Mazumder and Edmundson (2002) based on extensive experimental fertilization and stocking in Packers Lake, Alaska, confirm the connections between food, density, and growth. Before fertilization, sockeye were scarce and grew slowly, and *Daphnia* were also scarce and small. Fertilization increased the density and size of *Daphnia*, and the growth of sockeye, but high densities of sockeye resulting from stocking reversed these effects. These relationships are true for kokanee as well as sockeye, as shown by the positive effects of lake productivity and negative effects of density on growth in Idaho lakes and reservoirs (Rieman and Myers 1992).

For anadromous sockeye, the negative effect of fry density on growth may be offset, to some extent, by the possible positive, fertilizing effect of the carcasses of the parents that spawned them (Donaldson 1967; Schmidt et al. 1998; Finney et al. 2000). In general, however, large escapements tend to give rise to numerous but slow-growing fry. Clear evidence of the effects of density was provided by Koenings and Burkett (1987), reporting on the experimental stocking of Leisure Lake, Alaska. As they increased the density of fry, the growth rate (to smolt stage) decreased dramatically. The proportion leaving the lake after a single year decreased and more fish remained for 2 full years (table 10-3), and the combination of poorer growth and mortality reduced the overall biomass almost threefold after a fourfold increase in number of fry stocked. In general this effect seems to be restricted to a single year (that is, the growth of fry in one year is affected by their own density). However, in cases of exceptionally large escapements, as occurred in 1989 when concern about salmon contaminated by oil from the *Exxon Valdez* tanker spill greatly reduced commercial fisheries, the competition seems to affect growth in the following year as well (Ruggerone and Rogers 2003).

Sockeye salmon not only compete with each other but sometimes also with threespine and ninespine sticklebacks (Rogers 1973), as these fishes also feed on zooplankton (O'Neill and Hyatt 1987). Sticklebacks breed in spring in the lake's littoral zone or in slow-moving streams, and the males guard nests containing fertilized eggs. The larval sticklebacks are too small to effectively compete with sockeye fry but later in the summer they may be

TABLE 10-3. Relationship between the number of sockeye salmon fry stocked in Leisure Lake (otherwise devoid of sockeye) and the average size (g), age, and biomass (kg) of smolts (Koenings and Burkett 1987).

Fry stocked	*Age-1 smolt weight*	*Age-2 smolt weight*	*% age-1 smolts*	*Smolt biomass*
0.5 million	8	13.2	97	2009
1 million	4	7	77	1894
1.5 million	2.2	3.6	87	888
2 million	1.8	3.4	58	771

large enough to compete, and age-1 sticklebacks certainly overlap in diet with sockeye. In some lakes the sticklebacks primarily inhabit the littoral zone and so do not compete extensively with the sockeye offshore. However, in other lakes the sticklebacks are common offshore and more competition may occur. Because they are spawned in spring, stickleback abundance may be related to different factors than juvenile sockeye, such as perhaps the date of ice breakup or lake level. Overall, growth of sockeye (and other fishes) in lakes is most directly controlled by temperature and food, and these reflect a series of site-specific features including the lake's location, morphology, and nutrient inputs, climatic factors, and levels of competition (fig. 10-6).

Diel vertical migration

The zooplankton consumed by sockeye concentrate near the surface of lakes, feeding on phytoplankton. The sockeye might be expected to spend all their time near the surface, where there is enough light to see and catch the zooplankton. However, in most lakes the sockeye spend the day in deep water, ascending toward the surface near dusk and back down around dawn (e.g., Narver 1970; reviewed by Levy 1987). Rhythmic behavior patterns such as this, displayed every day, are termed "diel," hence this is a diel vertical migration (DVM). DVM is a very common phenomenon among organisms in lakes and oceans, and it has intrigued biologists for decades. In Babine Lake, for example, sockeye are found at 20–40 m depth during the day in summer. They move to the upper 5–10 m at dusk, down to the thermocline at night, briefly up to surface at dawn, and then down again during the day. In Lake Washington, they stay deep during the day, followed by a brief ascent to about 10 m below the surface at dusk (Eggers 1978). They go back down during night and do not ascend again at dawn. In Great Central Lake, they are 70–120 m deep during the day, ascend to about 10–20 m at night, stay in this upper range all night, and descend at dawn (Levy 1987). In Lake Tustumena (a glacially turbid lake in Alaska), they show the reverse pattern: near the surface in the day and deep at night (Thorne 1983).

How, then, do we explain the general pattern and the variations among lakes? Three main hypotheses have been proposed, related to prey capture, bioenergetic efficiency, and predator avoidance. From the perspective of prey capture, perhaps the sockeye vertically migrate because the zooplankton also migrate. Although this is plausible and may occur in some cases, it does not seem to be a general pattern. Levy (1990) examined

FIGURE 10-6. Conceptual representation of the factors with positive (+) and negative (-) effects on each other, and eventually on the two factors (food and temperature) most strongly affecting growth of salmonids in lakes, with an effort to distinguish factors that primarily vary among lakes from factors that primarily vary from year to year.

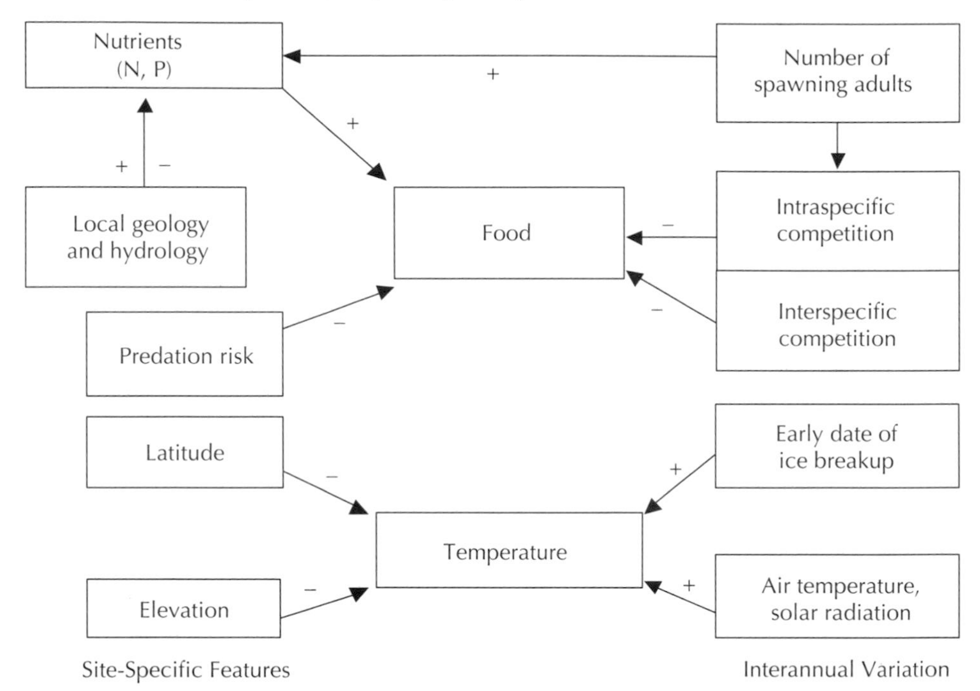

data from four lakes where sockeye vertically migrate (Cultus, Babine, Quesnel, and Shuswap) and in no case was there a clear vertical migration by the prey. In contrast, zooplankton showed vertical migration in Nimpkish Lake where the sockeye and sticklebacks did not consistently migrate. Levy (1990) concluded that this reciprocal pattern of migration by prey or predator but not both was inconsistent with the hypothesis that sockeye migrate simply to follow their prey. Moreover, the sockeye in Great Central Lake spend the day far below the zone of maximum food density. Thus though the distribution of sockeye and their food must sometimes overlap, prey movements are not sufficient to explain sockeye vertical migrations.

Another hypothesis was presented by Brett (1971), who developed the relationship between growth, temperature, and ration size (Brett et al. 1969; see fig. 10-2). In the summer, if sockeye were to feed at the surface at dusk and remain there all night (when it is too dark to forage successfully), the water would be too warm for optimal digestion efficiency. If the fish moved below the lake's thermocline, they could digest the food more efficiently and then either rise to the surface again at dawn or descend directly to the deep water during the day. Laboratory feeding trials with live zooplankton showed that at moderate rations (similar to what might be experienced in lakes), sockeye grew faster under a temperature regime simulating diel vertical migration than under constant temperature regimes (Biette and Geen 1980). Modeling based on observed patterns of temperature and zooplankton experienced by kokanee in a North Carolina reservoir also indicated a significant increase in growth for vertically migrating sockeye when the optimal temperature is below the depth where prey are concentrated (Bevelhimer and Adams 1993). However, Steinhart and Wurtsbaugh (1999) documented diel vertical migrations among three populations of *O. nerka* during the winter. The 3–5°C isothermal environment under the ice led them to propose that foraging was more important than temperature in structuring the observed vertical distribution patterns.

Energetic efficiency might explain why juvenile sockeye do not remain at the surface all night (though Levy 1987 pointed out that there is an energetic cost to migrating up and down, and also some cost to maintaining neutral buoyancy during the migration), but what determines the depth to which they go in the day? The most plausible explanation is that they are trying to avoid predation from birds (near the surface) and from larger fishes that feed by sight. The safest place is at the bottom of the lake but there is little to eat there, and the surface is the best place to feed but it is hazardous. Eggers (1978) reported that sockeye in Lake Washington seem to avoid predators by forming schools in darker, deeper water by day and only venturing up to feed in the late afternoon and dusk, when the schools dispersed. There is sufficient food in this lake (based on both zooplankton density and sockeye growth) for the sockeye to minimize vulnerability to predation by feeding to satiation at dusk and then descending until the next evening, rather than reascending to the surface again at dawn to feed again as occurs elsewhere.

Clark and Levy (1988) sought to unify the different hypotheses and proposed that each population strikes an appropriate trade-off between maximizing the opportunity for growth (based on food and temperature) and the risk of predation (based on light levels), given the prevailing local conditions. The sockeye should feed near the surface during an "antipredation window" when it is light enough for them to feed but dark enough to reduce predation risk. Levy (1990) gathered data from a number of lakes in support of this hypothesis. He concluded that visual range, affected by water clarity and light level, affects predator efficiency. Daytime depth is regulated by water clarity; sockeye go as deep as they have to in order to reduce light levels, minimizing risk of predation. The fish ascend to feed during the crepuscular periods but at night it is too dark for them to feed or for the predators to eat them. Thus sockeye move up to feed in the evening in most lakes but can also do so during the day in highly turbid water. The higher the productivity or higher the predation risk, the less time the sockeye should spend near the surface. At night they are found at or near the thermocline (in summer), at temperatures where they can efficiently process their food. Thus the nighttime depth is regulated by temperature, and the vertical pattern varies with the season. Therefore, the conditions of water clarity, predation pressure, food availability, and temperature seem to control variation in vertical migration among lakes and among seasons at a given lake. Recently, Scheuerell and Schindler (2003) provided strong empirical evidence supporting this integrated model to explain DVM. The vertical movement of sockeye was tightly linked to light levels; the predators were suspended in the water column, feeding on the passing sockeye each day; and the sockeye went deeper in the lake than would be needed for energetic efficiency (fig. 10-7). Thus the model by Clark and Levy (1988) explains a great deal about the DVM patterns of sockeye, though sockeye movements in winter are much less well known than those in summer, and the winter patterns are not always consistent among lakes (Steinhart and Wurtsbaugh 1999).

FIGURE 10-7. Vertical distribution of sockeye salmon fry (scattered dots) and predatory fish (the box indicates the depth range of 50% of the predators and the line within the box indicates the median depth), determined by hydroacoustic surveys, and light levels (continuous line: $\log_{10}$[W m^{-2}]) from evening to morning in July in Little Togiak Lake, Alaska (modified from Scheuerell and Schindler 2003).

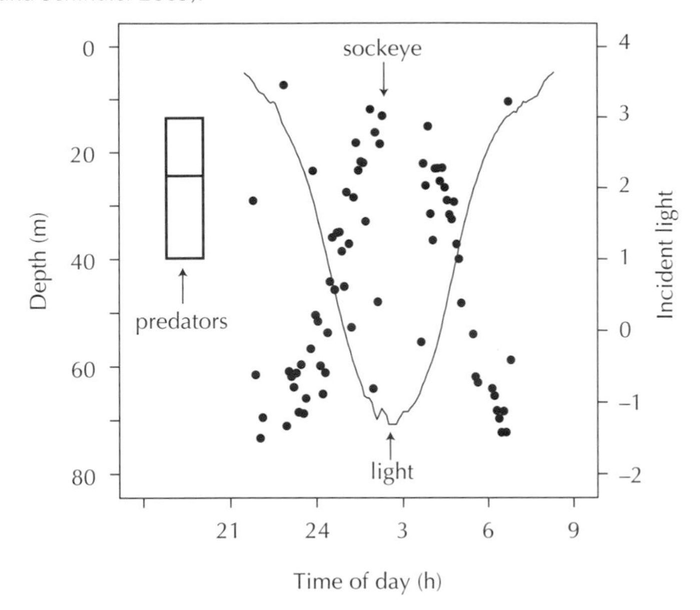

Predation, mortality, and trout in lakes

The subject of predation raises the issue of mortality. Among six populations (Chilko, Babine, Lakelse, Karluk, Washington, and Port John lakes), the average survival of sockeye fry to the smolt stage was 25.8% (see table 15-1). This is probably much higher than they would experience in a year if they went to sea as fry, but the losses call for some explanation. West and Larkin (1987) determined that the mortality in Babine Lake was size selective. They developed a relationship between the lengths of sockeye fry and the radius of their otolith (ear bone). They then measured the otoliths of juveniles at the end of their first summer in the lake, and of smolts leaving the lake, and estimated how large the fish had been when they entered the lake by measuring the mark on the otoliths made when the fry entered the lake. Only 27.2% of the fry from the lower half of the length-frequency distribution survived the year, compared to 43.3% for the larger half of the population. McGurk (1999) proposed that the mortality rate of sockeye and kokanee in lakes decreases as they grow, and he estimated the relationship as annual mortality = 1.38 x weight $^{-0.19}$. This is the equivalent to mortality decreasing from 90% to 80, 70, 60, and 50% as fish grow from 1.2 g to 2.3, 4.7, 10.6, and 27.8 g.

In most sockeye lakes, the primary predators seem to be other salmonids, especially trout and char. With the exception of lake trout (that spawn in lakes) and juvenile coho salmon rearing in the lake before migrating to sea, these species are generally represented by adfluvial populations. This term refers to fish that were spawned in streams, reared there (often for a few years), and then migrated to the lake to feed. Adfluvial trout may remain in the lake during the whole year, leaving only to spawn in streams, but some

10-3 Arctic tern, Lake Aleknagik Alaska, a predator on sockeye salmon. Photograph by Gregory Buck, University of Washington.

populations also make feeding migrations to streams to feed on insects, salmon eggs, and other prey (e.g., Eastman 1996; Meka et al. 2003). However, sockeye are also eaten by birds, including Arctic terns (photo 10-3), loons, mergansers, gulls, and other species.

Predation is presumably responsible for most of the mortality of sockeye in lakes, and of salmonids in general, though we may be too quick to ignore the role of disease. Pathogens of various sorts can weaken or even kill young salmon. For example, a cestode, *Eubothrium salvelini*, infects young sockeye salmon when they eat copepods, especially *Cyclops*, that contain the parasite. Smaller sockeye (26–35 mm) were more than three times as likely to get infected as larger fish (56–65 mm) in experiments (Boyce 1974), and there was indirect evidence that the parasite might reduce growth and survival. However, the proximate cause of mortality is usually predation. The lakes inhabited by sockeye often contain one or more species of predatory salmonid, including cutthroat or rainbow trout, Dolly Varden, Arctic char, lake trout, and bull trout. Ruggerone and Rogers (1992) conducted an especially thorough study of predation by juvenile coho salmon on sockeye in Chignik Lake, Alaska, and estimated that the coho ate 24–78 million sockeye, or about 59% of the sockeye, per year.

Sockeye salmon nursery lakes from the Columbia River to the Nass River, British Columbia, may also contain a large, piscivorous cyprinid, the northern pikeminnow

10-4 Rainbow trout in Iliamna Lake, Alaska, known for its large adfluvial trout. Photograph by Gregory Ruggerone, Natural Resources Consultants, Inc.

(*Ptychocheilus oregonensis*). Foerster's (1968) review of the Cultus Lake and Lakelse Lake predation studies indicated that pikeminnow (formerly known as squawfish) were the most numerous piscivore in the lake. They consumed considerable quantities of sockeye, but their per capita predation was lower than that of salmonids. By comparison, the per capita predatory effects of cutthroat trout, coho salmon, and Dolly Varden were estimated to be fivefold, fourfold, and threefold greater than those of pikeminnow, respectively. Similarly, Beauchamp et al. (1995) estimated that each pikeminnow ate about 5.6 fry per year, compared to 138.9 eaten by each cutthroat trout in Lake Ozette, Washington. Though they do not seem to be heavy predators in these lakes, pikeminnow are blamed for much of the juvenile salmonid mortality in the Columbia River system reservoirs (see chapter 12).

The biology of salmonids with adfluvial rather than anadromous life cycles (photo 10-4) is diverse but they tend to be generalist predators, foraging in the littoral zone on emerging insects, benthic invertebrates, and small fishes such as sculpins; and in the limnetic zone on fishes, including smaller salmonids, sticklebacks, smelt and other planktivores, and on zooplankton. These habitats may be used daily on an opportunistic basis but there may also be seasonal or ontogenetic habitat shifts. For example, Lake Washington's cutthroat trout typically spend 2 years in streams before migrating to the lake at a length of about 150 mm (fig. 10-8). They remain in the littoral zone, feeding mainly on insects until they reach about 250 mm. They then move to the limnetic zone and feed much more heavily on fishes (Nowak and Quinn 2002; Nowak et al. 2004).

FIGURE 10-8. Fork length distributions of cutthroat trout sampled in a tributary of Lake Washington (Quinn, unpublished data) and in the lake's littoral and limnetic zones (Nowak et al. 2004). The timing of sampling in the creek results in underrepresentation of young-of-the year trout, which would be less than 50 mm in spring.

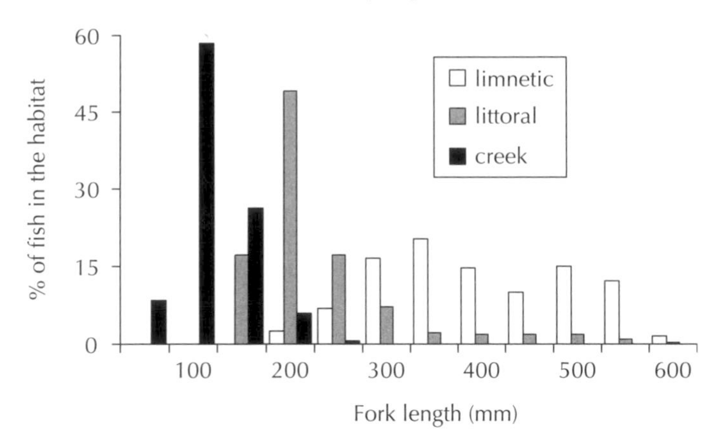

They do so by remaining just below the thermocline in summer, apparently moving slowly and intercepting the sockeye salmon and longfin smelt (*Spirinchus thaleichthys*) that vertically migrate (e.g., see fig. 10-7). However, individual trout move throughout the lake, tending to forage in the littoral zone most often in spring. Adfluvial bull trout also prey heavily on fish (almost exclusively so in Flathead Lake in Montana; Fraley and Shepard 1989), including both benthic and pelagic species.

Like the sockeye, trout seek depths that balance physiologically efficient temperatures, sufficient prey, and light to catch them. They are visual predators, and their encounter rates with prey depend on prey density, water clarity, and ambient light conditions (Beauchamp et al. 1999). Henderson and Northcote (1985) found that Dolly Varden reacted to prey from a greater distance at low light levels, but at higher light levels cutthroat trout showed a greater reactive distance. This is consistent with the higher density of cones in the retina of cutthroat trout (Henderson and Northcote 1988) and the tendency for char to reside deeper in the water column than trout, especially when they are sympatric. In addition to light, the vertical distributions may be constrained by other physical factors, notably temperature and dissolved oxygen (e.g., Rowe and Chisnall 1995).

Besides these physical influences on trout, there is considerable evidence that their distributions reflect interspecific competition. Nilsson and Northcote (1981) obtained information on rainbow and cutthroat trout distributions and diets in a series of lakes, including some containing both species (sympatry) and others where the species existed alone (allopatry). Both species were broadly distributed and fed on a wide variety of prey but differences between the species seemed to be magnified in sympatry. Cutthroat trout were more piscivorous than rainbows, especially when they were found together. Most striking were the patterns of growth. In allopatry, rainbow grew faster than cutthroat but when they were found together the cutthroat grew much faster. Further research on competition in lacustrine trout populations revealed that cutthroat trout tended to dominate Dolly Varden and exclude them from productive littoral feeding areas (though cutthroat were dominated by rainbow; Hindar et al. 1988).

Summary

For juvenile sockeye, the 1 or 2 years spent in the lake are a period of slow growth and low mortality compared to the years at sea. The sockeye often show diel vertical migrations, spending the day in deep water and moving up near the surface at dusk to feed. Such movements seem to balance the need to feed near the surface with the risk of predation there and the energetic efficiency of digesting their meal at moderate temperatures. Sockeye feed on insects in the littoral zone and in the limnetic zone they feed selectively on the largest crustacean zooplankton available (often *Daphnia* or large copepods). Growth is influenced by temperature or length of the growing season and the availability of food. Food is, in turn, influenced by intrinsic features of the lake (chemistry, depth, temperature, etc.) and by the density of planktivores, including both sockeye salmon and other fish species. Adfluvial trout migrate to lakes, often after rearing for a few years in the natal stream. They commonly feed on insects and other invertebrates in the littoral zone and on zooplankton and fish in the limnetic zone. Trout may move back and forth between the lake and streams as food becomes available (e.g., in fall when they eat eggs from spawning salmon). Their diet is increasingly composed of fish as they grow, and they are often the most important predators on young sockeye salmon.

11

Juvenile Salmonids in Streams

Most salmonid species spend at least some time in streams, and many populations spend their entire lives there. The volume of scientific literature on salmon and trout in riverine habitats is staggering, and there are many books on the subject. In an effort to present something meaningful but not overwhelming, I will first briefly consider the physical and biological attributes of streams relevant to salmonids. From rainbow trout in streams of Mexico, to coho salmon and steelhead in the coastal rainforests of Washington, British Columbia, and southeast Alaska, to chinook salmon in the mountains of Idaho or the upper reaches of the Yukon River, Pacific salmon inhabit a wide range of rivers. Therefore, the consideration of rivers and their communities must be brief but some information on the physical processes and biological communities is needed. I will then take two approaches to this period in salmonids' lives: one at the level of whole populations and the other at the level of individual fish. These approaches yield complementary insights by posing population-level questions: What determines the productivity of streams for salmonids among streams and within streams among years? And by posing individual-level questions: What determines the distribution, movements, growth, and survival of individual fish?

Physical characteristics of streams

Throughout the world, streams share certain similar features, derived from basic physical principles, and the resident fishes and other organisms have adapted to these features. Temperature is one such feature, and it plays a very strong role in the distribution and performance of fishes in streams, through the physiological effects explained in the previous chapter and through its effects on the rest of the stream community, including other fishes and prey. Water gains and loses heat slower than air, so stream temperatures tend to follow local air temperatures but streams fluctuate less than the air (Beschta et al. 1987). Thus streams will not get as warm nor as cold as the air around them and will lag behind air temperature changes. The shallower the stream, the more closely it will track air temperature. Thus as one moves downstream from very shallow trickles in mountainous headwaters to large rivers at low elevations, the average temperature tends to increase and the variation (daily and seasonal) tends to decrease. In addition to air temperature, the main factor affecting stream temperatures is solar radiation. Streams exposed to the sun heat up faster but they may also lose heat more rapidly at night without a canopy of trees. Removal of trees can thus affect stream temperatures both by affecting local air temperatures and by increasing incident radiation and heat loss. The heavy canopy over small streams can reduce the magnitude of temperature fluctuations that the shallow water would otherwise experience.

Streams thus differ fundamentally from lakes in the problems and options they present to fishes. Lake temperatures change gradually over the year and have negligible changes from day to night but are often stratified vertically. Streams are generally mixed vertically but change more rapidly, and this can present fish with stressful conditions. Intolerable or undesirable stream temperatures can be mitigated by the influence of groundwater. This water, either entering the stream from the adjacent ground below the surface or flowing through gravel bars or islands, tends to equilibrate with the ground, which loses and gains heat even slower than water. Groundwater temperatures approximate the area's mean annual air temperatures, being about 10°C in coastal Washington, warmer to the south, and cooler to the north (see Meisner et al. 1988 for detailed maps). Consequently, groundwater is usually cooler than river water in summer and warmer in winter and can provide a critical refuge for adult and juvenile salmonids. Matthews and Berg (1997) reported that rainbow trout moved into such thermal refuges in southern California, but the cooler water was also lower in dissolved oxygen, or DO (commonly the case with groundwater), so the fish had to balance the stresses from high temperature and low DO in the different habitats.

Perhaps the most fundamental feature of a stream is its flow regime: the seasonal pattern in average discharge and the level of variation around the average. Much of the range of salmon can be divided into coastal streams, which are mostly rain-fed and have peak flows in winter, and streams in the interior and far north whose water is primarily from melting snow and so have greater flows in late spring and early summer. There are, of course, streams draining areas with intermediate elevations, with flows showing mixtures of winter rain and spring-summer snowmelt effects. Flow regimes strongly influence the dynamics and movement of salmonids, but the patterns are complex and depend heavily on the local backdrop of conditions and life-history stage of the fish. Low flows in summer may limit some populations in dry areas (see examples in this chapter) but snowmelt floods in the interior can also reduce densities (Latterell et al. 1998). In temperate coastal streams, high winter flows may drive fish from the main stems of rivers but allow them to access wetland refuge areas inaccessible in summer (e.g., Peterson 1982a). The streams inhabited by salmon and trout at different life-history stages are part of riverine networks that come from primary water sources up in the mountains, combining to form progressively larger rivers at lower elevations, all eventually reaching the estuarine interface with the ocean.

As water flows from the mountain ridge to the ocean it typically experiences a reduction in gradient with the inevitable reduction in elevation. The gradient and volume of water, in the context of the overall landscape, will determine the shape of the stream channel. At steep gradients and high elevations, the channel is often confined by a narrow valley, characterized by boulders and large woody debris, and flows as a cascade. As the gradient lessens somewhat, the stream typically sequentially changes shape, adopting step-pool, plane bed, pool-riffle, and regime morphologies (Montgomery and Buffington 1997). High-gradient streams are characterized by erosion, while those of intermediate and low gradient have progressively more transport and deposition, until the depositional zone in the estuary is reached. Larger rocks are deposited first, smaller particles are transported farther, and so the substrate size varies with gradient. Gradient can exert a crucial role in salmonid distributions because even a short, steep reach can render the rest of the stream inaccessible to anadromous species. The resident trout and char populations above these barriers may be impervious to straying and so may be genetically isolated (e.g., Latterell 2001). Farther downstream, gradients will produce stream habitats for spawning and rearing that are not equally suitable for all species.

When water flows down a slope it tends to meander rather than flow in a straight line, and its path is affected by the structural elements in the stream, notably boulders and large woody debris (often abbreviated LWD, alternatively, large organic debris or coarse woody debris). Large trees characterize the native vegetation in the coastal region, where most salmon are produced, and areas of the interior, though there are certainly areas with salmon but few trees. Streams have a natural tendency to meander and will create pools and riffles in the absence of structure, but the LWD found in many streams increases the complexity of stream habitats by creating areas with different depths, velocities, substrate types, and amounts of cover. Thus streams with more LWD tend to have more pools, and larger pools, and have more diverse physical conditions than streams with more limited woody material. However, the influence of the wood is mediated in part by the size of the stream and the type of channel (Bilby and Ward 1989; Montgomery et al. 1995). Because salmonid species and age groups differ in habitat preferences, the more complex the stream the higher the abundance and diversity one typically finds, though changes in conditions that benefit one species may render the habitat less suitable to another.

There are two perspectives that one may take regarding stream habitats, and both are very useful. The first views the physical and biotic attributes of streams as consequences of location along the continuum from the headwaters to the sea. Streams are categorized by order; first-order streams are the primary channels, joining to form second-order streams, and so on in an open-ended system. In their classic paper, "The River Continuum Concept," Vannote et al. (1980) proposed that intrinsic physical processes related to gradient and elevation control primary production, consumers, and communities (table 11-1).

In addition to this basin-scale perspective on fish distributions, there is also benefit to considering finer spatial scales. Fish are not uniformly distributed throughout a stream

TABLE 11-1. Generalized attributes of river systems along a continuum from small (low-order) to large (high-order) streams (derived from Vannote et al. 1980).

Attribute	*Progressive changes with increasing stream order*
Elevation and gradient	High to low
Stream size	Narrow, shallow, and fast to wider, deeper, and slower
Temperature	Low and variable to warmer with less variation
Substrate	Coarse, with erosion, to finer, with more transport and deposition
Tree canopy	Decreasing percent of the stream under canopy, more sunlight
Primary production	Shift from allochthonous material (leaves and needles) to autochthonous material (algae on rocks and phytoplankton)
Insect community	Transition from insects specialized to shred leaves, to species grazing on algae, and finally to species collecting or filtering fine organic material
Salmonid community	Non-anadromous cutthroat trout to steelhead and coho salmon, to chinook, pink, and finally chum salmon in the lowest elevations

of a given order and particular characteristics. Rather, differences in density and species composition are commonly observed among the three primary types of habitat units: riffles, glides, and pools. At still finer spatial scales, within these habitat units we can measure microhabitat features (e.g., depth, velocity, and substrate) at the specific location where an individual fish is found and observe variation with fish species and size. The fish community may differ as one moves upstream, but the locations where individuals establish foraging stations or breeding sites may also be explained by local features. However, there is no reason to see the broad-scale and fine-scale perspectives on fish distribution as being at odds with each other. The features of a given reach of stream are controlled by their positions in the stream network and by the features of the landscape in which the stream flows. Thus we have watersheds (or catchments, as they tend to be known in Europe), with valley segments, and discrete types of channel reaches within the segments, habitat units such as pools and riffles within the reaches, and microhabitats at the specific locations where individual fish are found (see reviews of the various stream classification systems by Naiman et al. 1992 and by Bisson and Montgomery 1996).

Salmonid populations in streams

Salmonids are saddled with what might be called the egg-fry conflict. Cold, well-oxygenated, sterile streams are ideal environments for incubating embryos, but fry need food and warmer temperatures for digestion, so the ideal environment for embryos is not ideal for fry. Chum and pink salmon migrate to sea, sockeye migrate to lakes, but most salmonids make do with stream habitats for a few months or years before migrating to sea or to a lake, or they spend their whole lives in riverine habitats. The capacity of a stream for incubating embryos is far greater than its capacity for sustaining juveniles, so the stream-rearing species are far less numerous than are pink, sockeye, and chum.

11-1 Chinook salmon fry. Photograph by Richard Bell, University of Washington.

Thus there are generally low densities of spawning adults, high survival rates of eggs to emerging fry, and considerable thinning of the population in the stream after emergence. At a population level, this typically means that the rearing space and food for juveniles, rather than space for spawning, limits the production of smolts. At an individual level, the fish compete for food by establishing and maintaining feeding territories that are large enough for the fish to intercept enough food falling onto the stream's surface or drifting in the current to allow them to thrive and grow, albeit slowly. I will first take up the population-level patterns, then discuss the perspective of individual fish, and then unite the two perspectives.

The distribution patterns of juveniles are determined by broad, species-specific habitat requirements for breeding and rearing, by preferences for physical habitats, and by competition with other fishes, especially other salmonids. Juvenile ocean-type chinook salmon (photo 11-1) are usually found in the lower reaches of medium-to-large rivers from the southern end of their distribution north to about 56° latitude (E. B. Taylor 1990a). At higher elevations and in the northern areas, the stream-type dominates. In the regions where the two types overlap (e.g., the Columbia River), stream-type adults tend to spawn in smaller streams and the juveniles rear there, whereas ocean-type tend to spawn in larger rivers (J. M. Myers et al. 1998). The ocean-type tends to emerge earlier in the spring and grow for a few months in the river or while gradually moving downstream. The stream-type establishes feeding territories (Everest and Chapman 1972) and remains there throughout the winter, during periods of low temperatures and flows. Steelhead may be found from coastal areas to the interior, and they typically spend 1–3 years in streams (Busby et al. 1996).

Coho salmon (photo 11-2) are often the most numerous salmonid in streams where they occur and are generally found spawning in smaller streams than chinook, and at

11-2 Coho salmon fry. Photograph by Richard Bell, University of Washington.

higher gradients. From California to British Columbia, most coho spend a year in freshwater prior to seaward migration, but more often 2 years in Alaskan streams (Weitkamp et al. 1995). Coho commonly co-occur with steelhead and cutthroat trout and are more numerous than the trout. For example, smolt trapping at Big Beef Creek, a tributary of Hood Canal, Washington, yielded 93% coho salmon, 5% steelhead, and 2% cutthroat trout (David Seiler, Washington Department of Fish and Wildlife, unpublished data). Surveys of fourteen streams in coastal Oregon revealed an average of 66% young-of-the-year coho, 31% age-1 steelhead, and 3% age-1 cutthroat (not counting the chinook found in only three of the creeks; Reeves et al. 1993). However, eight of the Oregon stream communities were overwhelmingly (more than 80%) coho, four had mixed communities (from 63% coho and 36% steelhead to 30% coho and 66% steelhead), and two streams had greater than 90% steelhead. The relative abundance patterns change along the coast of California, where coho salmon are often less numerous and more restricted in range than steelhead (e.g., Shapovalov and Taft 1954). Hicks and Hall (2003) sampled salmonids in ten streams along the Oregon coast and found predominately coho salmon in lower-gradient streams whereas higher-gradient streams had lower densities of coho and higher densities of steelhead and cutthroat. The low-gradient streams tended to be formed from sandstone and had many pools, whereas the high-gradient streams were primarily in basalt, and riffle habitats predominated. Thus the relative abundance of the species reflected interactions among geology, gradient, and habitat type (table 11-2).

Within a given stream, chinook tend to be in the lower reaches, the coho at intermediate elevations and distances upriver, and the steelhead and cutthroat still farther up. Snorkel surveys in the Sitkum River, a tributary of the Calawah River in the Quileute River drainage on the coast of Washington, illustrated this pattern. All four species

TABLE 11-2. Density of juvenile salmonids in six high-gradient and four low-gradient streams along the Oregon coast (Hicks and Hall 2003). The fish were separated into those in their first year of life (age 0+), at which age the trout could not be distinguished to species, and those in their second year of life (age 1+).

		Density (fish/m²)			
	Gradient (m/km)	*0+ coho*	*0+ trout*	*1+ steelhead*	*1+ cutthroat*
High gradient	24.15	0.030	0.492	0.101	0.029
Low gradient	10.8	0.459	0.143	0.026	0.008

occurred within the watershed but chinook and coho reached 50% of their cumulative distributions farther downstream than the trout (fig. 11-1).

The coho salmon are essentially all anadromous whereas the steelhead/rainbow trout and cutthroat trout may exist as anadromous, resident (photo 11-3), or mixed populations, and this complicates comparisons somewhat. Farther south we find nonanadromous trout such as the Mexican golden, Apache, and Gila trout (Behnke 1992). The other common native, stream-resident salmonids in the Pacific Northwest are char: Dolly Varden and bull trout. These outwardly similar species are now recognized as different on the basis of genetic and morphological traits (Haas and McPhail 1991).

Better data are available on coho salmon than the other species, so we will consider them for the moment. The first question might be, what controls fish abundance, which might be measured as the number of smolts produced by different streams? Marshall and Britton (1990) surveyed the literature and found a linear relationship between the number of smolts produced and the length of the stream. One might think that stream area would be more important than length but probably in larger rivers there is unsuitable habitat in the middle, so the productive habitat along the edges is better estimated

FIGURE 11-1. Cumulative distributions of young-of-the-year chinook salmon, coho salmon, trout (not identifiable to species), and yearling cutthroat trout and steelhead, measured upstream from the mouth of the Sitkum River, in the Quileute River drainage on the coast of Washington. Data provided by John McMillan and James Starr of the Wild Salmon Center.

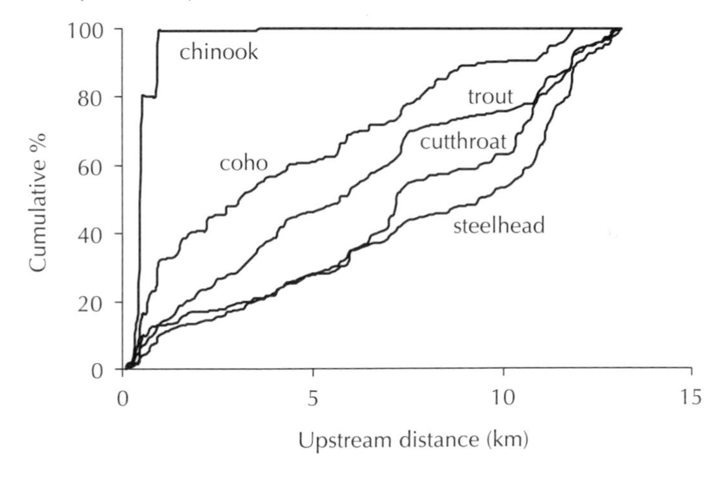

11-3 Resident rainbow trout, Little Lost River, Idaho. Photograph by Bart Gamett, U.S. Forest Service.

by length. Bradford et al. (1997) built on this work and increased the sample to ninety-eight rivers from Alaska to California (but mostly in British Columbia and Washington). The results again showed the relationship between length and number of smolts, with an average of 1952 smolts/km (fig. 11-2).

It is hardly surprising that larger streams produce more fish than smaller streams, just as larger lakes usually produce more sockeye smolts than smaller lakes, but what determines productivity besides size? Sharma and Hilborn (2001) obtained data on the number of coho smolts produced by different rivers, as well as data on various habitat

FIGURE 11-2. Number of coho salmon smolts produced from streams of different lengths from Alaska to Oregon, but primarily in British Columbia and Washington (from Bradford et al. 1997).

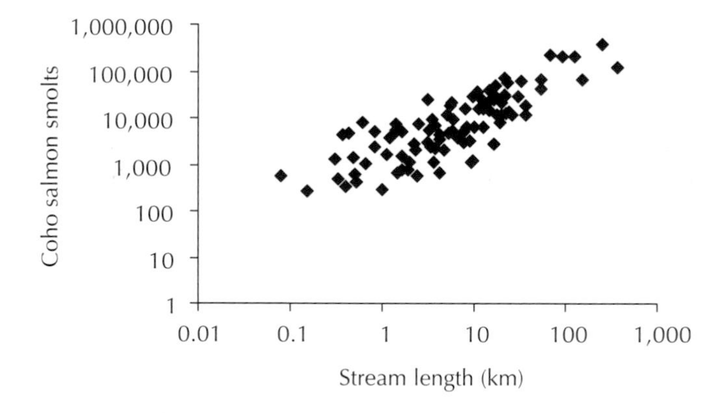

FIGURE 11-3. Relationship between the density of pools (m²/km) and the number of smolts produced per kilometer in a series of streams in western Washington (Sharma and Hilborn 2001).

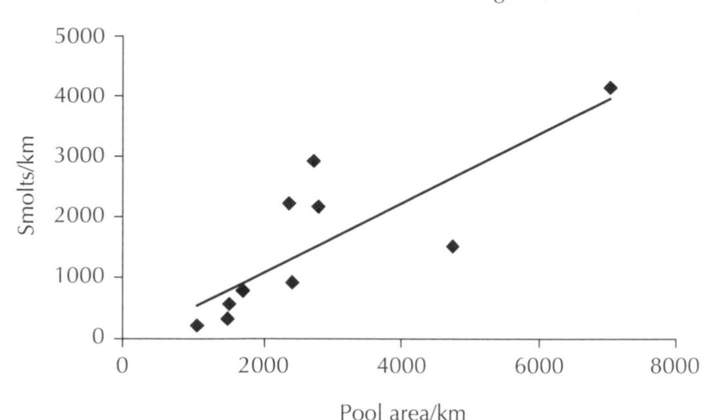

features hypothesized to affect production. The two clearest patterns pertained to gradient and slow-water habitat. There was a decline in number of smolts produced per kilometer with an increased gradient of the stream and slope of the valley, and an increase in smolt production per kilometer with an increased area of pools and ponds associated with the stream (fig. 11-3). These attributes, of course, are not independent. Streams with low gradient, in broad valleys, are likely to have more pools in the stream and more ponds connected to it than high-gradient streams in steep valleys.

The variation in coho smolt production among streams is related to stream size and habitat quality, but what controls the variation among years within a single stream? The obvious answer is the number of spawning adults. Clearly, if there are no adults there will be no smolts, and at low densities there may be a linear relationship between the numbers of parents and smolts. The slope of this line is referred to as the population's productivity, and this varies among streams for the reasons mentioned above. However, at some point the carrying capacity of the stream is reached, also a function of its size and habitat features, and no more smolts are produced from additional adults (see Mobrand et al. 1997 for a discussion of capacity and productivity in salmon restoration efforts). Data from Snow Creek and Big Beef Creek, Washington, show this pattern (fig. 11-4). The carrying capacity of Snow Creek is less than 500 adults whereas Big Beef Creek seems to produce more smolts until about 1000 adults spawn.

Once the carrying capacity has been reached, the number of smolts produced per year can still vary considerably. So, what controls the number of smolts produced, once a sufficient number of adults return to spawn? Decades ago, Smoker (1955) pointed out that the commercial catch of western Washington coho salmon in one season was positively correlated with the flow of water in the region's rivers during the summer that those coho would have been rearing in freshwater, 2 years earlier. He reasoned that the limiting factor was rearing space and that low flows would shrink it, resulting in poor growth or survival of juveniles (fig. 11-5). As the proportion of hatchery fish in the region increased, the relationship weakened but research by Seiler, Volkhardt, Neuhauser et al. (2002) on Bingham and Big Beef creeks clearly shows the importance of flow during the summer for coho smolt production (e.g., fig. 11-6). This does not imply that all

FIGURE 11-4. Numbers of spawning adult coho salmon and smolts produced by them in Snow and Big Beef creeks, in western Washington. The lines are based on Beverton-Holt models (Beverton and Holt 1957). Snow Creek data are from Lestelle et al. (1993) and Thom Johnson, Washington Department of Fish and Wildlife, unpublished data; Big Beef Creek data are from Seiler, Volkhardt, Neuhauser, et al. (2002).

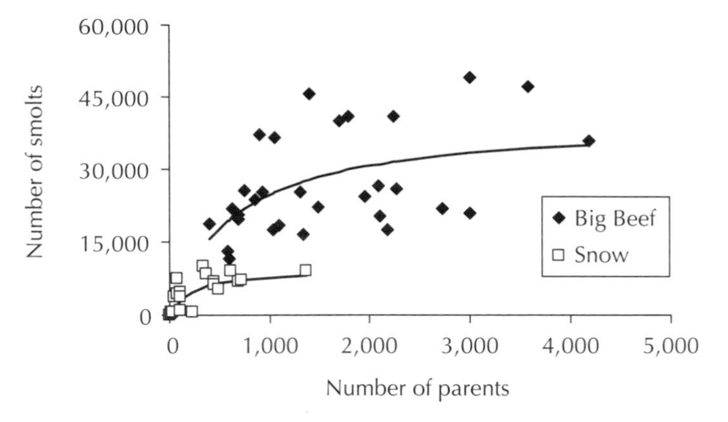

FIGURE 11-5. Relationship between the catch of adult coho salmon and an index of stream runoff in western Washington during the summer 2 years earlier (i.e., when those fish would have been rearing in streams; from Smoker 1955).

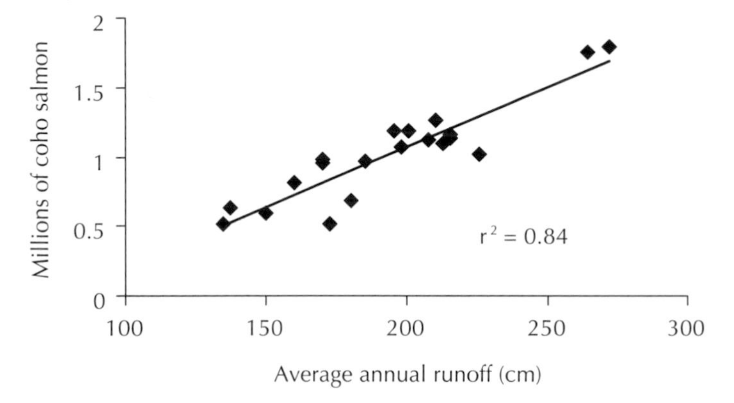

FIGURE 11-6. Number of coho salmon smolts leaving Big Beef Creek, Washington, as a function of the average flow during the 45-day period the previous summer with lowest flows. Smolt counts from Seiler, Volkhardt, Neuhauser et al. (2002); flow data from USGS records.

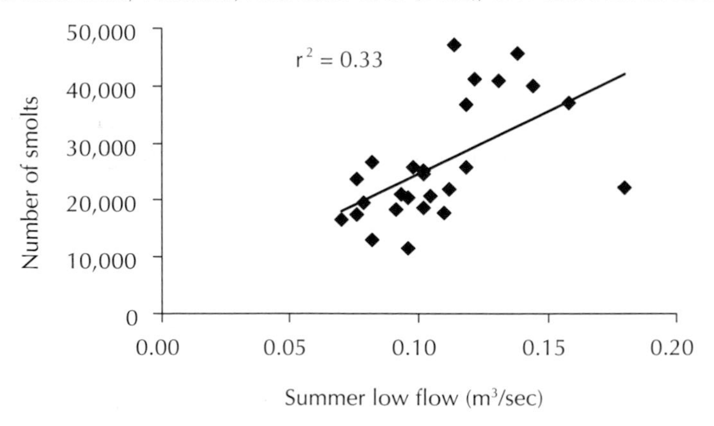

coho salmon populations are limited by summer low flows. Cederholm et al. (1997) presented evidence that high flows in winter reduced abundance, and experimental manipulation of winter rearing habitat in Oregon increased the production of coho salmon, steelhead, and cutthroat trout (Solazzi et al. 2000). Interestingly, the populations of trout in Big Beef Creek were not as strongly affected by summer low-flow conditions as were the coho. Seiler's data indicated at most a very weak relationship between the low flow in the first summer of life and the number of steelhead smolts leaving the creek, and no relationship whatsoever for cutthroat trout.

Stepping back from the details of what controls the abundance or density of a given species or a given population, we may consider what level of production is typical of salmonids in general. Bisson and Bilby (1998) compiled ninety-two estimates of salmonid production, measured as grams of fish per square meter per year, from seven species in their native and introduced habitats. Most studies reported low levels of production (60% were no more than 6 g/m^2) but some were well over 20 g/m^2. At this broad scale, production was determined primarily by high growth rates of fish rather than by high density. Bisson and Bilby found that "productive sites tend to possess hard waters with relatively high inorganic nutrient concentrations; moderate temperatures, especially in spring-fed streams where temperatures are buffered by groundwater inputs year-round; relatively low vegetative canopy coverage allowing ample sunlight to reach the streams; and abundant macrophytes and mosses, or dense growths of filamentous algae" (391). This is not a description of typical coho salmon or coastal cutthroat trout habitat but rather of a limestone-buffered stream in England, Pennsylvania, or New Zealand. Not only does fish production seem to depend on food resources but the nature of the controlling factors varies between seasons. Summer production is most strongly affected by autochthonous pathways (i.e., algal growth in the stream itself, feeding the insects that feed the fish), whereas in winter allochthonous pathways (leaf litter, needles, and salmon carcasses) are the important source of nutrients.

Foraging, territorial behavior, and growth

To understand the underlying relationships between habitat attributes and salmonid density at the population level, it is useful to consider the foraging ecology and behavior of salmonids in streams as individuals. The fish need food and space, and the behavior of the fish shifts seasonally with the changing availability of these two linked requirements. In early spring, newly emerged fry may school in the margins of the stream, either to avoid predatory fish in the deeper water or to avoid high flows, and the fry grow rapidly. Growth rates slow in the early summer (fig. 11-7), when territorial behavior becomes more important, and densities seem to decline rapidly. In the late summer, streams along the coast commonly reach their lowest flows, limiting space, and the high temperatures and reduced food can result in slow growth or even weight loss. During this period, coho salmon may abandon their territories (Brian Fransen, Weyerhaeuser Company, personal communication). These factors may explain the correlation between summer low flows and smolt production mentioned above. When fall rains increase flows, coho often move from the main channel of the stream to off-channel ponds, sloughs, and wetlands that might not have been even accessible during the summer. Space is probably not limiting because area is increasing and the population has been

FIGURE 11-7. Average weights of juvenile coho salmon from emergence to smolt transformation in Huckleberry Creek, Washington (samples taken from several years; from Fransen et al. 1993).

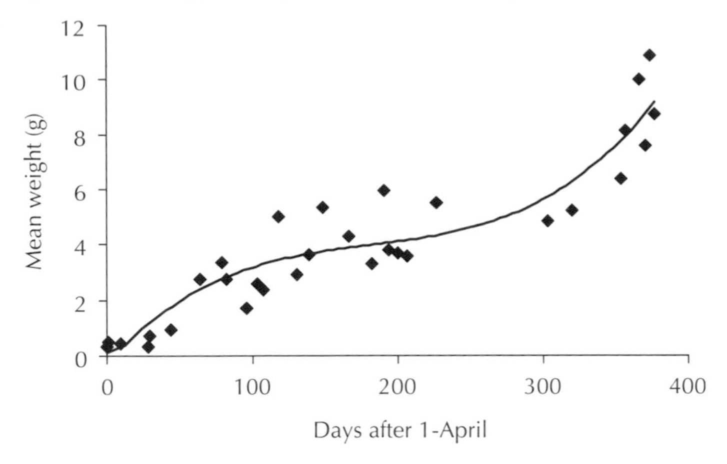

thinned during the summer. In some cases growth can be good during the winter and accelerates rapidly in spring prior to seaward migration (fig. 11-7).

In general, juvenile salmonids trade off the need to obtain food against the cost of acquiring it. The primary food organisms are drifting aquatic insects and the larval stages of terrestrial insects, though some adult stages also are taken at the surface. For example, Martin (1985) studied the diets of cutthroat trout in their first, second, and third years of life and found that aquatic insects made up 87–90% of the diet by number and 71–93% by weight. There was a tendency for the larger fish to eat larger prey but there was considerable diet overlap (hence, presumably, competition) among the different ages (fig. 11-8). The availability of terrestrial and aquatic invertebrates varies seasonally; Nakano and Murakami (2001) reported peaks of terrestrial insects in streams from July to September, whereas aquatic insect biomass peaked from December to July. Consistent with this pattern, Wipfli (1997) reported greater reliance on terrestrial invertebrates from late spring until fall than in the other seasons (and also greater use of terrestrial invertebrates in young-growth compared to old-growth forests).

FIGURE 11-8. Variation in prey size among different age groups of cutthroat trout in streams in the Olympic Peninsula, Washington (from Martin 1985).

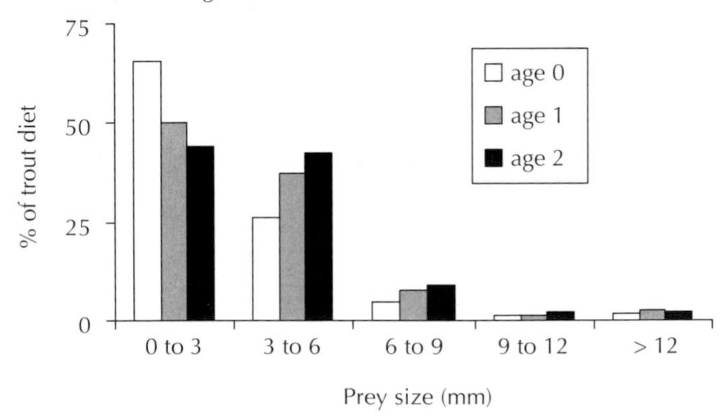

Insects drift in the current, so the fish orient upstream and hold positions from which to dart out and take food. Insects are sufficiently scarce that juvenile salmonids establish and defend feeding territories in the stream in order to intercept enough food to maintain themselves and grow. Territoriality is characteristic of stream-resident salmonids, though the behavior is not displayed by all individuals or at all times of the year. Feeding territories allow the resident access to the insects dropping on the surface or drifting into it. If the territory is too large, insects may drift by at the periphery and be unobtainable. Moreover, defending a large territory requires energy, reduces the amount of time available for foraging, and might also increase the risk of predation. Thus the territory's size tends to shrink when food is plentiful (Mason 1976b; Dill et al. 1981), and territorial behavior can be eliminated in free-living salmon by supplemental feeding (Brian Fransen, Weyerhaeuser Company, and Peter Bisson, U.S. Forest Service, personal communication). However, when food is limited and space is defensible, territorial behavior is common. As the fish grow larger they need more food, hence larger territories, and there is positive relationship between the size of juvenile salmonids and their territories. For example, salmonids 3, 5, 10, and 20 cm long occupy territories about 0.02, 0.08, 1.0, and 3.0 m^2, respectively (Keeley and Grant 1995). As Grant et al. (1998) pointed out, many fry emerge from redds. Their density is reduced by mortality but growth increases the food requirements of the survivors. If density is adjusted for the size and territorial requirements of the fish, the "percent habitat saturation" may remain largely constant over the seasons in streams.

Contests for territories are settled by stereotyped displays, emphasizing the black and white margins on the fins and the size of the fish, though outright attacks also occur. The dominant fish maintains its position in the water column and the submissive fish lowers its fins, darkens its colors, and retreats. Not surprisingly, bigger fish tend to dominate smaller fish, and a size disparity of only 5% in body weight confers significant advantage (Abbott et al. 1985). However, the advantage of size can be offset by territorial possession; residents are able to resist eviction by larger intruders (Rhodes and Quinn 1998). Thus the parents affect the prospects of their offspring in several ways. Fry from large eggs are larger than those from smaller eggs. When reared in conditions of abundant food, such as in hatcheries, the size advantage of the larger individuals is not maintained and the small fish catch up (e.g., Heath et al. 1999). However, recent experiments on brown trout by Einum and Fleming (1999) revealed that under food-limited conditions fry from large eggs had higher growth and survival rates than siblings from smaller eggs. In addition, females that spawn early produce fry that will emerge earlier in the spring and so possess territories when the offspring of later-spawning females emerge. The early fry will also be larger, having grown before the others emerge (fig. 11-9). These processes all favor early spawning, but the first fry to emerge may suffer heavy losses if predators are waiting for them and may grow slowly if food is scarce or temperatures cold. Thus there are conflicting pressures on emergence date related to competition, feeding, and predator avoidance (Brännäs 1995). Field studies, with DNA parentage analysis, can reveal the extent to which these processes operate under fully natural conditions (Seamons et al. 2004).

Recent evidence indicates that large size per se is not the cause of dominance but rather that dominance results from a high metabolic rate (e.g., masu salmon: Yamamoto

FIGURE 11-9. Diagrammatic representation of some of the factors influencing movement, growth, and survival of juvenile salmonids in streams.

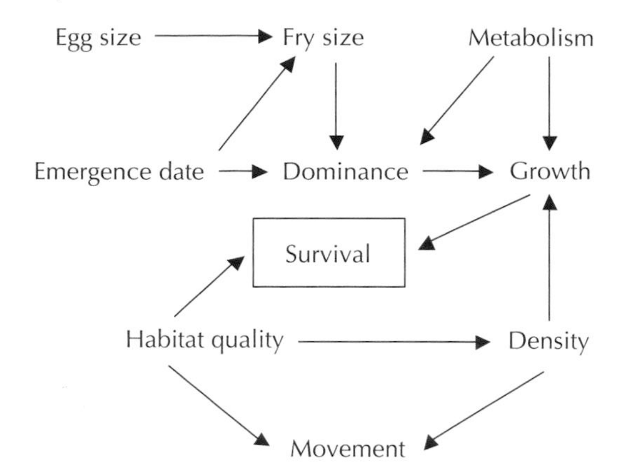

et al. 1998), and the combination of metabolic rate and ability to dominate access to food allows those individuals to attain large size. Experience also affects competition; fish that have recently lost a competitive encounter are more likely to lose again than fish of the same size that have recently won (Abbott et al. 1985; Rhodes and Quinn 1998). If these patterns were not complex enough, young salmonids can also distinguish their siblings from other members of their population, based on odors alone, even if all fish are the same age, fed the same food, and held in the same water (Quinn and Busack 1985). This sibling recognition ability seems to be widespread among stream-rearing salmonids (reviewed by Brown and Brown 1996). Fish tend to aggregate with relatives (at least in seminatural arenas) and are more tolerant of them than they are of nonrelatives. The hypothesis is that they spend less time fighting their relatives and can devote more time to foraging. It is not clear to what extent siblings continue to rear together and whether kin recognition affects growth and survival under natural conditions.

In any case, if earlier-emerging, larger fish obtain and hold territories, what happens to the offspring of females that spawned late, had small eggs, or both? The conventional wisdom is that they either forage without establishing territories or are displaced downstream. Chapman (1966) presented experimental evidence that these "nomads" form territories if given the chance, and he hypothesized that they are unable to do so because they are competitively inferior. Nielsen (1992) pursued this line of investigation with a detailed study of individually marked coho salmon in a small stream. She classified the fish in the foraging hierarchy as either dominant or subdominant and also observed "floaters" that were not part of the hierarchy. The fish displaying these behavior patterns differed in initial size, rates of foraging, aggressive and submissive displays, and growth rates as did water velocity where they fed and ambient food levels available to them (table 11-3). Martel (1996) supported these results, showing that floaters grew slower than territorial fish, even if they were initially the same size.

Competition for food and space has long been thought to cause some individuals to abandon territorial behavior, and to displace others downstream where they may establish territories or keep moving until they reach the sea. Chapman (1966) presented

TABLE 11-3. Attributes of juvenile coho salmon displaying one of three behavioral patterns, sampled in June (Nielsen 1992). Ambient food levels (g/m^3/h) and growth rates were averaged over four types of pools.

Attribute	*Dominants*	*Subdominants*	*Floaters*
Initial length (mm)	64	55	50
Focal point water velocity (m/sec)	0.14	0.12	0.01
Foraging rate (number/min)	2.7	1.8	1.7
Aggressive acts (number/min)	0.77	0.42	0.16
Submissive acts (number/min)	0.05	0.21	0.33
Ambient drift food available	0.29	0.22	0.06
Ambient drop food available	0.14	0.13	0.15
Growth rate (mg/d)	5.5	3.2	2.1

evidence that the production of downstream migrating fry is a function of the number of females, whereas the number of smolts is stable over a wide range of escapements (e.g., see fig. 11-4). He concluded that migrant fry are in some sense "surplus" to the population and constitute evidence that the carrying capacity of the stream for fry has been exceeded. Consistent with this concept, Hartman et al. (1982) presented data indicating that the abundance of coho fry in Carnation Creek, British Columbia, in late May and early June ranged widely among years, from 15.6 to 46.6 thousand, but by late September the range was from 9.1 to 13.2 thousand. During these years, tens of thousands of fry migrated downstream, often associated with flood events. In large rivers, these migrant fry may find suitable habitat and will rear successfully in the lower reaches of their stream, but in small coastal streams like Carnation Creek the fry will enter saltwater and may not thrive. Such fry are found in estuaries, and their fate is a fascinating subject needing far more work.

Does the presence of migrating fry indicate that the carrying capacity of the stream has been reached, or do they represent an alternative life-history pattern? This is an important and fascinating issue in the ecology of these fishes. Recent observations of individually marked coho salmon in streams indicated that movement reflects habitat choice rather than failure to compete (Kahler et al. 2001). In this study, the fry that moved were not, on average, initially smaller than those that remained in the same place, nor did they grow slower after they moved. Indeed, they tended to grow somewhat faster than fish that remained, indicating that they were not "losers" at all. Movement was related to habitat quality more than competition, as indicated by density. Fish in small, shallow pools tended to move whereas those in larger, deeper pools were more likely to stay. These larger pools often held higher densities, slowing the growth rates. Perhaps the fish found the smaller pools too risky, and so they left, despite the fact that their growth had been adequate. On the other hand, experiments by Keeley (2001) with juvenile steelhead indicated that increased competition (higher density of fish or lower density of food) caused increased mortality, lower and more variable growth rates, and emigration of smaller fish.

FIGURE 11-10. Relationship between density of juvenile coho salmon and average weight at the end of the summer at sites in western Washington (from Fransen et al. 1993).

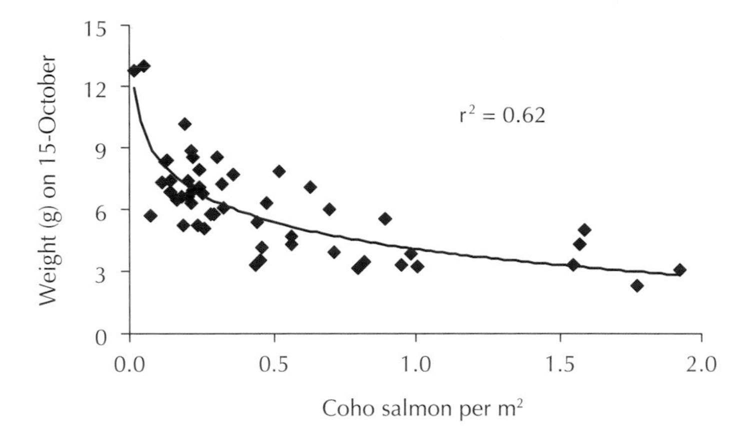

There is certainly evidence that growth of salmonids in streams is density dependent. For example, Fransen et al. (1993) sampled different reaches of eight western Washington streams and found that the average size of coho salmon on October 15 ranged from as much as 12 g under very low densities (more than 10 m^2 per fish) to about 3 g at high densities of 1–2 fish per m^2 (fig. 11-10). The modal density at that time of year was about 0.1–0.3 fish per m^2, or 1–3 fish per 10 m^2, and a biomass of 1–3 g per m^2. In addition to these social aspects, the ability of fish to grow in streams is a function of the amount of food, itself a combination of the production of food and the competition for it, and the temperature for processing the food. Growth is often food limited in streams. Age-1 cutthroat trout increased their body weight by only 0.022% per day on ambient food but grew 1.73% per day when the food was supplemented (Boss and Richardson 2002). The density of insect drift may be greatest in spring (e.g., Fransen et al. 1993) when water temperatures are cool, making digestion inefficient. In summer, when the water is warmer and digestion more efficient, the low density of food may reduce growth. Fransen et al. (1993) reported that the peaks of coho salmon growth were in April–June of their first year of life, and again in the April–June period just prior to seaward migration (see fig. 11-7). Consistent with these findings, Martin (1985) showed that cutthroat trout food consumption peaked with temperature in July and August but that growth was most rapid in April and May (fig. 11-11). The declining growth rates in summer may be an ecological reason for migration to sea, where growth opportunities are better. In addition to these seasonal patterns, the growth rate of fish, expressed as a proportion of their size, is greatest for small fish. For example, the August growth rates of Martin's (1985) cutthroat trout were 14.3, 2.4, 1.1, and 0.4 mg/g/d for fish in their first, second, third, and fourth years of life, respectively.

The foregoing paragraphs have pointed to the importance of food and space for stream-dwelling salmonids such as coho salmon. Mason (1976b) conducted a particularly revealing study on the interaction between these factors, augmenting the food available to free-living coho salmon by feeding them krill (a marine crustacean). This reduced both territory size and density-related migration from the study areas and increased both growth rate and lipid storage. However, Mason concluded that "the 6-fold increase

FIGURE 11-11. Rates of food consumption and growth and ambient temperature experienced by age-1 cutthroat trout in Bear Creek, Olympic Peninsula, Washington (Martin 1985).

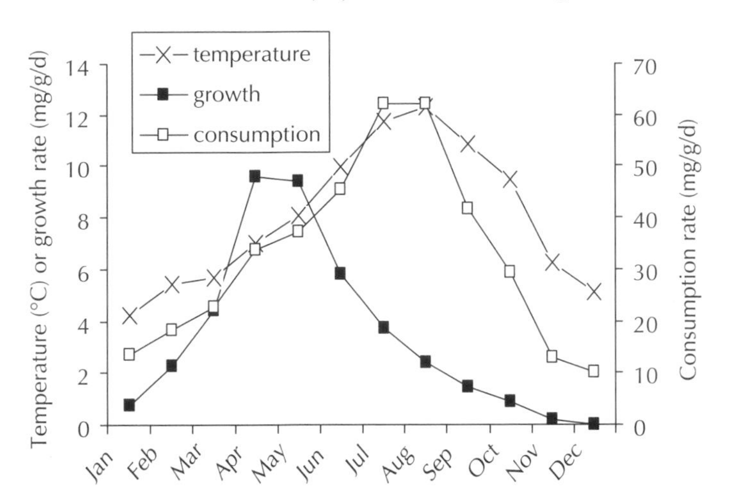

in potential smolt yield induced by supplemental feeding during the summer was nullified by the natural carrying capacity of the stream during winter" (775), so both fed and unfed populations produced about 1 smolt per 4 m^2 of stream area.

Distribution, habitat preferences, and movements

Habitat attributes affect fish distribution, density, and movements over a range of spatial scales. The upriver distribution of anadromous salmonids may be limited by gradients above 4–5% in many cases, so they will be absent from the upper reaches of streams. There are also more local "habitat-unit" scales, such as the relative densities of coho salmon, steelhead, and cutthroat trout in pools compared to the adjacent riffles. Finally, at so-called microhabitat scales, one may measure the depth and velocity at the specific point in the pool where an individual coho has established its feeding territory and consider the benefits (in terms of access to drifting food) and costs (in predation risk and energy expenditure) associated with that location.

Not surprisingly, juvenile salmonids have definable habitat preferences (reviewed by Bjornn and Reiser 1991). That is, they are more likely to occur in locations with certain combinations of depth, velocity, and other physical features than would occur by chance and that occur less often in other areas. For example, Healy and Lonzarich (2000) showed that coho salmon selectively used the deeper, slower water characteristic of pools rather than shallower, faster moving water (fig. 11-12). Second, there are ontogenetic shifts in habitat preference; generally towards deeper, faster water as fish become larger, probably related to foraging or predator avoidance. Coastal cutthroat trout occupy very shallow habitats in the margins of streams after emergence, and their density may be determined by the quantity of such lateral habitat (e.g., Moore and Gregory 1988), but they occupy deeper water as they get older (fig. 11-13). These shifts in habitat use might result from predator avoidance. Small fish may occupy shallow water to avoid predation from bigger fish, and the bigger fish may occupy deeper water to avoid predation from

FIGURE 11-12. Habitat use patterns by coho salmon in Wisconsin in relation to the available habitat (from Healy and Lonzarich 2000).

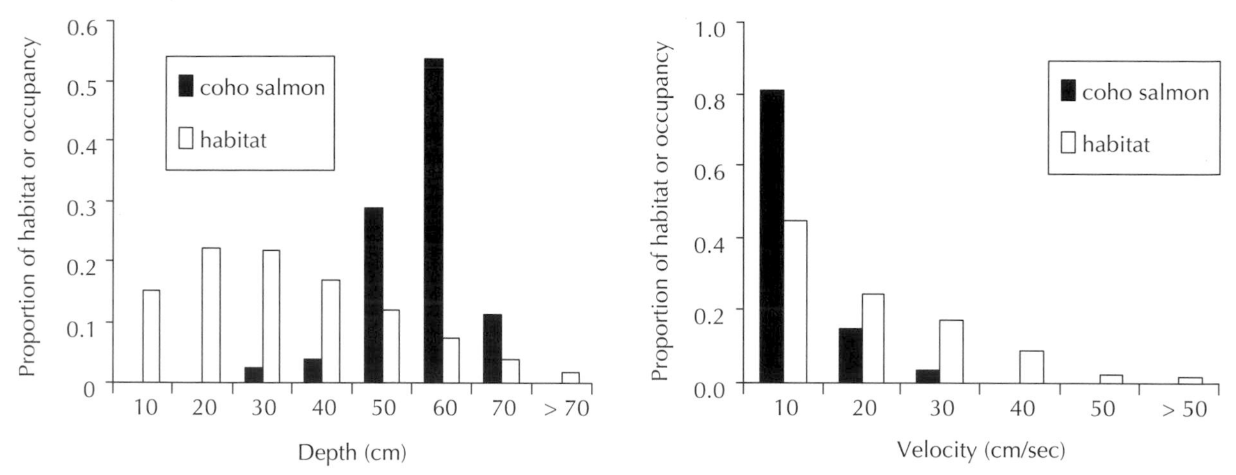

birds. However, experiments by Rosenfeld and Boss (2001) indicated that age-0 trout occupied pools when given access to them and could grow in both pools and riffles. In contrast, larger trout were unable to grow in riffles, owing to their higher food requirements. Thus smaller trout can exploit habitats that are simply unsuitable for larger fish.

In addition to ontogenetic shifts (Bisson et al. 1988; Dolloff and Reeves 1990), there are also species-specific differences is habitat use. These may reflect differences related to body size, which is controlled by egg size and emergence date. For example, Lister and Genoe (1970) found that chinook fry in the Big Qualicum River, British Columbia, were both larger at emergence than coho (about 42 vs. 38 mm) and also emerged earlier. However, at a given size, both species seemed to prefer the same habitats. After emergence all fry sought the margins of the creek and areas with cover, but they moved to progressively faster water as they grew, and so the chinook were found in different areas than the coho.

FIGURE 11-13. Water depths where age 0, 1, and 2 cutthroat trout (estimated from length frequency) were sampled in April in a reach of the east fork of Issaquah Creek, Washington (Quinn, unpublished data).

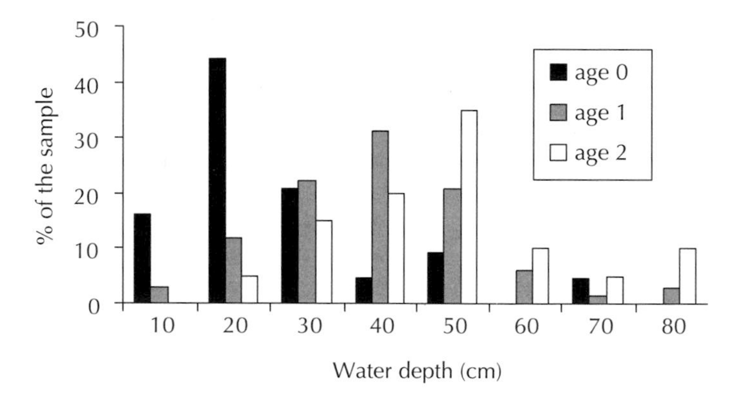

However, there are species-specific patterns of habitat use in addition to those differences related to size and emergence. For example, juvenile steelhead in Idaho occupied faster, shallower water than chinook salmon (Everest and Chapman 1972), and Dolly Varden and coho salmon were segregated in Alaskan streams (Dolloff and Reeves 1990). Bisson et al. (1988) surveyed fifteen sites in nine Washington creeks and found that coho salmon were most likely to be found in pools, steelhead in shallower, faster water, and cutthroat in intermediate conditions. They also noted subtle differences in body morphology that seemed consistent with the salmons' habitat-use patterns. The coho salmon had deep, laterally compressed bodies and long fins, the steelhead had wider and more cylindrical bodies, and the cutthroat were elongate like steelhead but compressed like coho. Similarly, Roni and Quinn (2001) reported that stream-restoration projects that increased the quantity of LWD and pools tended to increase coho density but decrease that of steelhead.

Coho salmon, being spawned in fall, emerge before cutthroat and steelhead trout, and coho generally have larger eggs as well, so how much of the habitat segregation results from intrinsic preferences and how much from competition? The diets of these species overlap considerably and their basic foraging patterns and territorial behavior are similar. Glova (1987) found that in when the species were sympatric the cutthroat tended to occupy riffles and the coho occupied pools. However, in the absence of coho, cutthroat tended to occupy pools. He concluded that competition resulted in habitat segregation, with the competitively inferior cutthroat occupying the less favorable habitat when dominated by coho. Sabo and Pauley (1997) revisited this hypothesis and found that when cutthroat and coho were matched for size and experience, the trout were equal to or competitively superior to the salmon. They concluded that the advantages of size and experience allow coho to dominate and that trout populations that have evolved with coho are less competitive than those that have evolved in isolation.

Besides the variation in habitat-use patterns among fish of different species and sizes, there are also temporal patterns. In general, there seems to be an initial movement of fry immediately after emergence, but this period is poorly known. Then, movement is comparatively slight during summer, followed by significant downstream movement in fall and winter. This fall movement has been best studied in coho salmon in coastal streams that experience high flows during this time. The salmon move from inhospitable main channel areas and occupy flooded wetlands, beaver ponds, tributaries and a variety of off-channel habitats, sometimes moving up to 30 km downstream when flows increase (e.g., Peterson 1982a). This seasonal habitat shift also takes place in the interior, where ponds provide a thermal refuge (e.g., Swales and Levings 1989). Coho salmon seem particularly inclined to use such habitats, and other salmonids such as chinook salmon, steelhead, and Dolly Varden may make much less use of off-channel habitats. Such seasonal shifts are not limited to anadromous populations; Brown and Mackay (1995b) showed that cutthroat trout in Alberta moved from shallow streams to deeper water in winter, as low water and ice made shallow habitats unsuitable. Juvenile chinook salmon may move in and out of streams prior to seaward migration (e.g., Murray and Rosenau 1989; Scrivener et al. 1994), but in general their movements in large rivers are much less well known than are those of coho salmon and trout in small streams.

All of these patterns of distribution and movement show considerable variation, as streams flood or dewater, freeze, or become warm depending on the region and elevation. Layered on these seasonal patterns are dramatic diel (day-night) differences in behavior and microhabitat, which change from summer to winter. Daytime snorkeling is an accurate way to assess abundance in summer but in winter the nighttime counts typically far exceed those in the day (e.g., 92 vs. 2 for cutthroat and steelhead trout and 177 vs. 39 for coho salmon: Roni and Fayram 2000), indicating that the fish hide and are inactive in the day in winter. However, the seasonal shift from diurnal to nocturnal activity differs between interior and coastal streams, as Grunbaum (1996) showed for Oregon steelhead. In summer, divers counted similar numbers of trout in the day and at night in both coastal and interior streams. In winter, trout in coastal streams seemed to reduce their daytime activity, as the daytime counts were only 30% of the total counts. In the interior, trout were entirely absent from the open water in the day and only emerged from gravel hiding places at night. The difference in behavior between regions indicates that it is not merely the shorter days but the lower temperatures that affect the fish, and this is corroborated by lab studies.

Stream fish communities

The scientific literature on Pacific salmon and trout tends to ignore other fish species or to consider them only in the context of their interactions with salmon (i.e., predators, competitors, and prey). This book will do nothing to reverse this regrettable pattern because there is insufficient time to do justice to the region's fish diversity. Still, a few words are called for. The region inhabited by Pacific salmon has fewer freshwater fish species than central and eastern North America. Indeed, the structure of the fish communities in the West is more similar to those of western Europe than eastern North America (Moyle and Herbold 1987). The most common primary freshwater fishes (species that evolved in freshwater) are cyprinids (dace, shiners, and chubs) and catostomids (suckers). The dominant secondary freshwater fishes (species whose lineages evolved at sea but presently live entirely in freshwater) are the cottids or sculpins (Wydoski and Whitney 2003 listed eleven species in Washington; photo 11-4). In addition to salmonids there are other anadromous fishes including the Acipenseridae (sturgeons), Petromyzontidae (lamprey), Osmeridae (smelt), and Gasterosteidae (sticklebacks), but most of these have limited presence in small streams where juvenile salmonids rear. There are also many nonnative species in the western United States and southern Canada, often introduced for recreational fishing. These tend to be large-bodied, temperate, or warm-water species that are most common in lakes and large rivers rather than small streams, including centrarchids (sunfish and bass), ictalurids (catfishes), percids (yellow perch and walleye), and cyprinids (carp) in freshwater; and two anadromous species, *Alosa sapidissima* (American shad) and *Morone saxatilis* (striped bass), in some large rivers.

The dominant factor shaping the region's fish diversity (or lack thereof) is the relatively recent glacial retreat, only about 10,000–15,000 years ago (Pielou 1991). As indicated in the introduction, the major river systems like the Columbia and Fraser have 52 and 39 native freshwater fish species, respectively, and Wydoski and Whitney (2003)

11-4 Shorthead sculpin, *Cottus confusus*. Photograph by Richard Bell, University of Washington.

listed 88 species of freshwater (including euryhaline) fishes in Washington (55 native to the state or the region). G. C. Carl et al. (1977) listed 56 native freshwater fishes in British Columbia, though several have very restricted distributions in the province. Not withstanding this apparent diversity, most salmonids exist in streams with very simple communities. In coastal streams, there might be 2 or 3 species of sculpins, 3 species of salmonids, lamprey, and perhaps 1 or 2 species of salamanders. Roni (2002) sampled thirty forested streams in coastal Oregon and Washington and most streams had a subset of the following species: coho salmon, steelhead and cutthroat trout, reticulate and torrent sculpins, larval lampreys, and Pacific giant salamander. The densities of sculpins approached those of salmonids in pools and exceeded them in riffles, so we might refer to these as "sculpin streams" rather than "salmon" or "trout" streams. The basic ecology and population dynamics of sculpins and other non-salmonid species are poorly known. This is unfortunate in itself, and it limits our understanding of salmonid ecology as well.

The few experiments on interactions between salmonids and non-salmonids suggest fascinating, complex patterns. Reeves et al. (1987) studied competition between juvenile steelhead and redside shiner (*Richardsonius balteatus*). In the field, steelhead occupied cooler water than redside shiner, even in the same river. In experimental tanks, steelhead dominated the shiners at low temperatures (12–15°C) by aggressive, territorial behavior (so-called interference competition), but at higher temperatures (19–22°C) the shiners were more active and outcompeted the trout for food in "exploitative competition." Reese and Harvey (2002) reported somewhat similar results with a different cyprinid species, the Sacramento pikeminnow (*Ptychocheilus grandis*). In warm water (20–23°C), pikeminnow had the same per capita effect on the growth of dominant juvenile steelhead as did other steelhead, indicating similar effectiveness in competition for food. In cooler water (15–18°C), the growth of dominant steelhead was unaffected by the presence of pikeminnow. Subdominant steelhead were more affected by intraspecific competition than interspecific competition, regardless of temperature.

Survival and predation

Direct studies on the patterns and causal agents of mortality in stream-resident salmonids are not nearly as well known as one might guess from all the research efforts. Studies on mortality tend to focus on one of three aspects: (1) the rate of mortality at discrete life history stages, (2) the attributes of the fish or its habitat that are associated with mortality (or survival), and (3) the causal agents. In coho salmon (the best-studied species), in-stream mortality rates are about 80–90% in summer and 50–75% the following winter (reviewed by Quinn and Peterson 1996). For example, Crone and Bond (1976) studied coho in Sashin Creek, Alaska, for 3 years and estimated that the average survival from egg deposition was 20.9% to just before emergence, 7.1% to late June or early July of the first summer, 2.21% to late August of the first summer, 0.68% to late June or early July of the second summer, and 0.39% to late August of the second summer.

Except in unusual cases, one does not find dead juvenile salmon washed up at the margins of streams, so it is assumed that they are all lost to predation. Undoubtedly, they get eaten, but the role of disease in natural populations has not received nearly as much attention as it should. There are certainly diseases and parasites that afflict the fish, and debilitated fish may be more likely to get eaten, so disease-related mortality may masquerade as predation. There are acute pathogens like *Ceratomyxa shasta* that can cause considerable mortality during outbreaks, but populations tend to evolve resistance (Buchanan et al. 1983). There are also parasites that do not cause acute problems but may weaken the fish, at least temporarily. For example, *Nanophyetus salmincola*, a trematode, can reduce the swimming performance of juvenile salmonids for some days while the parasites move through the muscle tissue (Butler and Millemann 1971), and this might increase susceptibility to predation.

Notwithstanding such diseases, in summer the proximate cause of mortality is probably predation, though various factors might contribute to it, such as small size or isolation in pools when portions of the stream are dewatered. In winter, low temperatures may place the fish under energetic stress and contribute to mortality. Four lines of evidence indicate that mortality is most severe on small individuals. First, population-specific survival rates of Carnation Creek coho salmon were higher in years when the fish were large at the end of the summer (Hartman et al. 1987). Second, coho salmon that were large at the end of the summer were more likely to survive the winter than smaller ones (Quinn and Peterson 1996). Third, larger steelhead stocked in the Englishman River, British Columbia, had lower mortality rates than smaller ones (from 9.6% per month for 6 g fish to 15.4% per month for 0.2 g fish; Hume and Parkinson 1988). Fourth, Smith and Griffith (1994) found that small steelhead in Idaho had lower survival rates than larger ones, even when protected from predation, and especially at the coldest sites. However, Connolly and Petersen (2003) studied winter stress under milder conditions and found that larger steelhead (among the age-0 fish) encountered greater energetic stress than small ones when the water was warmer, but that smaller fish fared less well

TABLE 11-4. Percent of juvenile salmonids of different species that survived exposure to predatory torrent sculpins in stream aquarium experiments (Patten 1975).

	coho	*chinook*	*steelhead*	*sockeye*	*chum*	*pink*
size (mm)	36–39	38–42	29–31	41–43	35–38	35–37
% survival	76	52	23	20	5	2

when it was cold. Depending on the region, the winter may be a time of low or high water, the thermal stress may be severe or mild, and ice can also play a role (see Cunjak 1996 for a good review of winter habitat needs). Within the stream, survival varies considerably among habitats, being higher (especially for coho) in side-channels, pools, and areas with complex woody debris than in simpler habitats and main-channel areas (Peterson 1982b; Tschaplinski and Hartman 1983; Bell et al. 2001). In Big Beef Creek the coho salmon that enjoyed the highest rates of growth and survival were those that apparently spent the winter in a lake, even though they were comparatively small at the end of the fall (Quinn and Peterson 1996).

Juvenile salmonids in streams experience predation primarily from fishes, including larger salmonids, from birds, and from mammals. I do not know of a comprehensive study that evaluated the relative mortality attributable to these different kinds of predators in one system, so inferences have to be drawn from field and lab studies in different settings. In general, predation from fishes is size selective, with the smaller individuals being more likely to be eaten than larger ones because the limitation is the gape of the predator's mouth. However, salmonids may fall prey to fish not too much larger than themselves. For example, Pearsons and Fritts (1999) reported that juvenile coho could eat chinook that were more than 40% of their length (a 140 mm coho ate a 64 mm chinook).

Besides salmonids, the other main predatory fishes are sculpins. Patten (1977) exposed juvenile coho salmon to predation from torrent sculpins (*Cottus rhotheus*) and found that sculpins 60, 80, 100, and 120 mm long could eat salmon 50, 60, 70, and 80 mm long, respectively. In addition to size, some species of salmonids seem to be more vulnerable to predators than others. Patten (1975) exposed six salmonid species to torrent sculpins about 90 mm long in tanks. The highest mortality was suffered by pink, chum, and sockeye salmon, despite the fact that they were about the same size as the coho (the least vulnerable species) and were larger than the steelhead (table 11-4). Probably the species that typically reside in streams (coho, chinook, and steelhead) are better adapted to avoid sculpin predation by virtue of morphology, color, swimming performance, or behavior than the species typically migrating from streams at once.

In addition to fishes, a variety of birds (e.g., kingfishers, herons, and mergansers) routinely prey on stream fishes, including salmonids. Wood's (1987b) detailed study on the Big Qualicum River, British Columbia, indicated that merganser broods consumed about 82,000 to 131,000 juvenile coho salmon between June 10 and August 25. This was the equivalent to 24–65% of the wild smolt production from the river, assuming the fish would have otherwise experienced typical survival rates until the following spring. A number of mammals, but especially river otters, also prey on juvenile salmon. Dolloff

(1993) examined the scats from two adult otters and their two young on a section of the Kadashan River, Alaska. During a 6-week period in late spring, the otters ate at least 3300 juvenile coho salmon and Dolly Varden (and a smaller number of coast range sculpins). Birds and mammals may take larger than average fish, as they might be easier to see and more rewarding to eat. This shift in enemies may contribute to shifts toward deeper habitats as fish grow, though as indicated above this may also reflect foraging needs of the juveniles.

Summary

Streams are the initial rearing habitat for most salmonid species. As a type of habitat, streams offer juveniles the opportunity to grow prior to migration to lakes, larger rivers, or to the ocean. These alternative habitats may promise better growing opportunities but they also have a high risk of size-selective mortality. Throughout their lives in streams, salmonids experience relatively slow growth and more or less continual predation risk. Individual fish must compromise the need to feed, including both actual pursuit of prey and defense of the feeding territory, with the need to avoid being eaten. They shift their habitat seasonally and from day to night to avoid peril or to take advantage of favorable opportunities, including movement to off-channel habitat in the winter along the coast, and into the shelter of the gravel. Complex interactions between size, growth rate, territorial behavior, movement, and survival of fish have yet to be unraveled. In addition to these views of the stream from the perspective of the individual, we can also view streams from a holistic perspective. Streams have a finite amount of space and limited food, and limiting the number and size of migrants or adults that can be produced. The quality and quantity of space in a stream largely determines these limits, though environmental conditions can cause fluctuations in production. Streams are inherently variable, as a consequence of seasonal and short-term changes in flow, temperature, and other factors, and salmonids have evolved behavioral and demographic responses to the disturbances that episodically occur in streams. Floods and movement of debris may seem devastating in the short run but in the long run may create the diversity of habitats needed to accommodate the different species and age groups of salmonids and non-salmonid species. Eventually, though, many salmonids leave the stream and migrate to the ocean, and anadromy is a defining trait of this family.

12

Downstream Migration: To Sea or Not to Sea?

The physiology of salmonids prepares them for seaward migration every spring, and the critical life-history decision, to go to sea or remain in freshwater, must be made annually. Some salmonid populations are truly landlocked, persisting above barriers to upstream migration only because individuals do not migrate downstream. As indicated in chapter 9, there are strong genetic controls over such behavior, though the resident populations were probably founded by anadromous fish in the past. However, there are many salmonid populations in rivers and lakes that are accessible to anadromous fishes, yet they remain in freshwater. In some cases these populations even co-occur with anadromous ones of the same species. Before taking up the behavior of the salmon that migrate to sea, it is important to consider those that do not.

Gross (1987) proposed that the best approach to the puzzle of the two life-history patterns within and among salmonid species is to consider the fitness of the alternative patterns. He defined fitness as "the lifetime summation of an individual's probability of surviving to reproduce at any age, multiplied by its fecundity (or male fertility) and breeding success at that age" (17). Put simply, fish can remain in freshwater and have a comparatively high survival rate but a low growth rate and so be likely to survive to breed but produce few eggs (in the case of females), or they can go to sea and have a lower probability of surviving but a higher growth rate (hence, fecundity) if they survive.

Extensive studies have been conducted on the biology of kokanee, the nonanadromous form of sockeye salmon. The two forms can be sympatric, not only rearing in the same lake but spawning in the same river. However, there are also many lakes with only sockeye, and a few that are accessible to anadromous fishes but have only kokanee. What is the relationship between sockeye and kokanee, how can the two forms persist, and how might they have evolved? The traditional view of such evolutionary events has been that species evolve in spatial isolation (e.g., on separate islands or in lakes). Eventually, the patterns of life history, morphology, color, behavior, and so on of the allopatric populations are sufficient to prevent interbreeding among adults and to prevent hybrid offspring from thriving even if adults do interbreed. Alternatively, one species can diverge into two or more species while still in each other's presence if variations in life history cause them to develop distinct forms that might tend to breed with each other rather than with the alternative form, and this is termed "sympatric divergence."

Consistent with allopatric divergence, one might hypothesize that all kokanee are part of a single "race," found in many lake systems, so kokanee are more closely related to each other than they are to the "race" of sockeye. However, patterns of genetic variation showed that the sockeye and kokanee within lakes are usually more closely related to each other than they are to members of their form in other lakes (Foote et al. 1989; E. B. Taylor et al. 1996). Thus it seems that, in each lake, sockeye colonized after the last glacial period and gave rise to kokanee, implying independent evolution of kokanee on many occasions. How might this occur, and how might the forms remain distinct?

Separation between forms results in part from assortative mating. Kokanee are much smaller than sockeye, and male sockeye do not court female kokanee, instead courting the larger and more fecund sockeye (with ten to twenty times more eggs; Foote and Larkin 1988). Male kokanee sometimes attempt to sneak fertilizations with the sockeye but prefer to court female kokanee. Thus there is considerable assortative mating isolating the forms. However, the proportion of sockeye eggs fertilized by male kokanee was too great to permit the observed level of genetic isolation, so there seems to be some reduced performance of the hybrids between forms. The hybrids are viable, and both pure kokanee and hybrids are able to tolerate the transition to seawater. However, there are subtle but significant differences in a series of traits, such as embryo development, swimming performance, color, and especially size threshold for sexual maturity that reinforce the separation between forms by making the hybrids less viable than either pure form (reviewed by Wood and Foote 1996).

If assortative mating and differential survival keep the forms separate at present, how did the separation begin? The link between these two rather different forms seems to be what Ricker (1938) termed "residuals"—the nonanadromous offspring of anadromous parents. It is hypothesized (Foote et al. 1989; Wood 1995; E. B. Taylor et al. 1996; Wood and Foote 1996) that lake systems were invaded by sockeye after glacial recession. Most individuals continued their anadromous life history but a few did not leave the lake. These residuals were much smaller at maturity than the anadromous fish because growing conditions in the lakes are generally poorer than those at sea (the residuals are also less red in color; photo 12-1). Residual females were only courted by male residuals, though residual males courted with both residual and anadromous females. Assuming some heritable tendency to remain in the lake rather than migrate to sea, this assortative mating might promote genetic divergence between forms as they specialize for their freshwater and marine habitats. These genetic differences would be reinforced by size-specific preferences for breeding sites, accompanied by the evolution of isolating mechanisms to reduce interbreeding between forms.

Gross's (1987) fitness approach may be used to examine the life-history patterns of sockeye and kokanee. I assembled table 12-1 with generalized estimates of growth,

12-1 Two male sockeye salmon from Iliamna Lake, Alaska, one (above) a jack and the other apparently a residual (nonanadromous) fish, based on the smaller size and markedly less red color on the side. Photograph by Thomas Quinn, University of Washington.

TABLE 12-1. Model of growth, survival and reproductive potential of sockeye salmon and kokanee. The model began with 500 kokanee eggs and 3000 sockeye eggs and applied fictitious but plausible survival rates. Fecundity was estimated from empirical relationships with length but was only hypothetical for the very small fish. Fitness, calculated as the probability of survival times the expected fecundity at that age, must exceed 2.0 for the life-history pattern to replace itself.

Age	*Form*	*Length*	*Survival*	*Number*	*Fecundity*	*Fitness*
1st spring	kokanee	28	0.1	50		
	sockeye	28	0.1	300		
1st fall	kokanee	60	0.4	20	3	0.12
	sockeye	60	0.4	120	3	0.12
2nd spring	kokanee	80	0.5	10		
	sockeye	80	0.5	60		
2nd fall	kokanee	120	0.8	8	24	0.38
	sockeye	180	0.3	18	85	0.51
3rd fall	kokanee	180	0.6	4.8	85	0.82
	sockeye	360	0.4	7.2	700	1.68
4th fall	kokanee	300	0.8	3.84	500	3.84
	sockeye	560	0.8	5.76	3000	5.76

survival, fecundity, and fitness (cumulative survival multiplied times fecundity) for the two forms. I assumed similar survival and growth rates of the two forms when they are both in freshwater, but sockeye at sea experience faster growth and higher mortality than kokanee in the lake. These factors compensate for each other to some extent and the sockeye actually produce more surplus fish in this model (which assumes populations below their carrying capacity and so more than two adults are produced per female). Note that in this model, maturity at any age below 4 does not result in replacement, so the risk of mortality during the fourth year is more than balanced by the increased size-related fecundity.

Research on sockeye and kokanee indicates a relatively sharp division between anadromous and nonanadromous forms. That is, populations are basically all one form or the other, with strong genetic control over the traits associated with the forms, both sexes showing the same patterns, and few residuals. However, salmonids as a group vary in the patterns of anadromy and residency. Nonanadromy is often largely or entirely a male trait. In masu salmon, many populations (especially those in the northern end of the range) are dominated by anadromous individuals of both sexes whereas toward the middle of the range most smolts are females; most males remain in freshwater and mature as parr. Some of them survive after spawning (Ivankov et al. 1977) and migrate to sea, to return as jacks or fully grown males (Tsiger et al. 1994). Mature male parr also occur in some stream-type chinook populations, whereas all females go to sea (E. B. Taylor 1989; Mullan et al. 1992). The phenomenon of mature male parr has received little attention in chinook salmon, unlike Atlantic salmon, where this life-history pattern is both common and well documented (Fleming 1998).

The tendency of males rather than females to mature as parr reflects the alternative life-history approach of sneaking fertilizations. Female reproduction is controlled primarily by fecundity and in these cases the lifetime mortality is so great that only large, anadromous females can produce enough eggs to replace themselves. Having said this, it is not clear why there shouldn't be as many populations of nonanadromous coho salmon or many mature coho parr, for example, as there are in cutthroat, rainbow, and sockeye. Foerster and Ricker (1953) described nonanadromous coho in Cultus Lake, British Columbia. In this population, apparently residuals produced by sea-run parents, both males and females occurred. However, such residual coho salmon seem to be a rare phenomenon.

In addition to species like chinook salmon that have mature male parr, there are other salmonids with both anadromous and nonanadromous populations, notably cutthroat trout and rainbow/steelhead within *Oncorhynchus*, and Atlantic salmon, brown trout, Arctic char, brook trout, Dolly Varden, and others. There is no reason why all species must have the same degree of genetic control over nonanadromy, but some separation between forms seems to be the general rule. In their study of the genetic basis for anadromy in rainbow/steelhead trout, Zimmerman and Reeves (2000) took advantage of the fact that seawater has a higher ratio of strontium to calcium than freshwater. The elemental concentrations in a female are passed to her offspring via the yolk of the egg, and the composition of the otolith can reveal whether the fish had an anadromous or nonanadromous mother. The rainbow trout in the Deschutes River, Oregon, all had nonanadromous mothers and all steelhead had anadromous

TABLE 12-2. Characteristics of redds made by rainbow trout and steelhead on the Deschutes River, Oregon (Zimmerman and Reeves 2000).

	Redd site			*Redd dimensions*	
Form	*Depth (cm)*	*Velocity (cm/sec)*	*Gravel (mm)*	*Length (m)*	*Width (m)*
steelhead	54.1	71.4	32.5	2.1	1.2
rainbow	42.6	63.4	25.1	1.5	0.8

mothers, though the separation was less clear in the Babine River, British Columbia. Consistent with differences in body size (about 20–35 cm for rainbow vs. 61–69 for steelhead), the fish tended to occupy different breeding sites (table 12-2). Moreover, the median spawning by steelhead was completed by the fifteenth week of the year, at which time only 1.3% of the rainbow trout spawning had taken place. The median date for rainbow spawning was week 24 and it extended to week 32. This separation in spawning date is consistent with the sockeye-kokanee pattern reported by Wood and Foote (1996) and with the general tendency for larger individuals within populations to breed earlier than smaller ones.

The focus on anadromy versus nonanadromy may have distracted us from the more basic phenomenon of migration versus residency. As Northcote (1992) and Jonsson and Jonsson (1993) have pointed out, there is a great range in behavior among salmonids, including maintenance of a very restricted home range throughout the entire life or migration from a small stream to a large river, to a lake, to an estuary, or to the open ocean. Given the limited carrying capacity of streams to grow fish, the volitional migration or forced eviction of many individuals is hardly surprising. In some ways, trout that spend 2 years in a stream before migrating to a lake to feed until they mature are equivalent to anadromous fish. Admittedly, the shift from fresh- to saltwater has great consequences for the physiology of the fish, but from an ecological standpoint it is the migration (i.e., the habitat change) that is important.

Timing of migration

The decision to go to downstream (e.g., to sea) is thus relatively hard-wired in some species and probably more flexible in others, with different degrees of flexibility between males and females. Assuming the fish will go to sea (or to the lake, etc.) at all, the next decision is, "this year or next year?" As table 12-3 shows, salmonids vary greatly in age at seaward migration. Pink and chum salmon all migrate in their first year of life but the other species are more flexible, and the proportion of older smolts tends to increase toward the northern end of the range. This pattern, exemplified by steelhead (table 12-4), is also seen in sockeye and coho salmon. Chinook salmon migrate to sea in their first or second year of life, with the ocean-type predominating in the southern end of the range and in the more productive, coastal rivers and the stream-type found in the interior and the north. E. B. Taylor (1990a) reported that chinook salmon populations north of about 56° north latitude (i.e., the Skeena and Nass rivers) were exclusively stream

TABLE 12-3. Characteristic (++) and less common (+) ages at seaward migration of different salmonids, from Randall et al. (1987); masu salmon added based on Kato (1991).

	Age								
Species	*0*	*1*	*2*	*3*	*4*	*5*	*6*	*7*	*8*
pink salmon	++								
chum salmon	++								
chinook salmon	++	++	+						
coho salmon	+	++	++	+					
sockeye salmon	+	++	++	+	+				
masu salmon		++	++	+					
steelhead trout		+	++	++	+				
brown trout		+	++	++	+	+	+		
cutthroat trout		+	++	++	++	+	+		
brook trout		+	++	++	++	+	+	+	
Atlantic salmon		+	++	++	++	+	+	+	+
Dolly Varden			+	++	++	+	+		
Arctic char		+	+	++	++	++	++	+	+

type, though the Situk River in south-central Alaska (59°30'N) is an exception to this rule, producing almost exclusively ocean-type juveniles (S. W. Johnson et al. 1992).

The age composition of smolts depends on growing conditions and presumably reflects the trade-off between growth and mortality regimes in freshwater and marine environments. Large size increases the odds of survival at sea, but there will be some probability of mortality for every year spent in freshwater. Thus sockeye salmon at the southern end of their range, for example, grow sufficiently large after 1 year in the lake to greatly enhance their survival rate at sea, relative to fry. However, the sockeye might only grow a bit more in a second year in the lake (as a new cohort of fry enters and competes with them) and would have some finite chance of dying during that year. The survival benefit at sea of that additional increment of growth is apparently not sufficient to offset the additional risk of mortality in freshwater, so they all go to sea after 1 year. In less

TABLE 12-4. Percent of steelhead migrating to sea after different numbers of years in freshwater (i.e., smolt age), based on averages of populations from different regions of North America (sample size in parentheses; reported by Busby et al. 1996).

Smolt age	*Alaska (6)*	*British Columbia (15)*	*Washington (9)*	*Columbia River (18)*	*Oregon (4)*	*California (8)*
1	0.0	0.7	6.3	3.8	3.8	23.3
2	15.2	34.9	85.7	79.0	79.3	69.0
3	73.8	60.5	8.0	16.6	16.3	7.5
4	11.0	3.9	0.0	0.6	0.8	0.3

FIGURE 12-1. Average population-specific lengths of age 1 smolts (117 coho salmon populations, data from Bradford et al. 1997, Weitkamp et al. 1995, and Crone and Bond 1976; 85 sockeye salmon populations, data from Burgner 1991, Gustafson et al. 1997, Edmundson and Mazumder 2001, and Kim D. Hyatt, unpublished data).

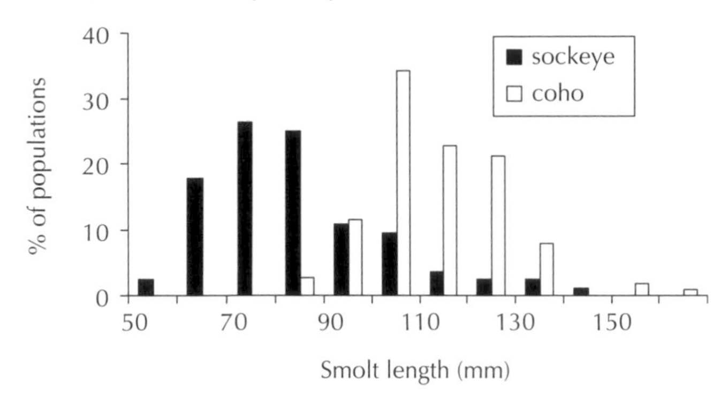

productive lakes, growth is so poor that the sockeye may face severe size-selective mortality at sea (see chapter 15) if they leave after only 1 year. Mortality rates in the lake may be low and so the fish stay a second year, accepting the modest risk of mortality, to increase their survival chances at sea. Obviously, there is no thought process involved; the evolutionary pressures have culled from the population the fish that make the wrong "choices." In years with good growing conditions (warm water, low density of fish, etc.) there may be a higher proportion of age-1 smolts than in years with poorer conditions. The effects of density on the age composition of smolts (see table 10-3) illustrate this. However, it is important to note that each population has its own "set point" for the decision to go or stay. In some lakes the sockeye go to sea after 1 year but at a small size, whereas in other lakes such small fish would stay for a second year.

Data from a large number of sockeye and coho salmon populations illustrate the range of sizes of age-1 smolts (as well as the generally larger size of coho than sockeye after a year in freshwater; fig. 12-1). Data from four lakes within the Bristol Bay region of Alaska (table 12-5) further illustrate the variation in smolt size and age patterns. Sockeye

TABLE 12-5. Average age composition, size, and marine survival of sockeye salmon smolts from four lake systems in Bristol Bay, Alaska (Alaska Department of Fish and Game data, complied and summarized by Gregory Ruggerone, National Resources Consultants, Inc.).

Lake system	*Years*	*Smolt age*	*Survival*	*Weight*	*% at age*
Iliamna	1976–93	1	7.5	5.6	53
		2	11.9	9.4	47
Becharof	1982–93	1	18.4	9.8	29
		2	28.6	14.3	71
Ugashik	1982–93	1	4.9	7.1	51
		2	10.0	11.5	49
Wood River	1975–90	1	6.0	7.0	93
		2	6.9	9.3	7

12-2 Sockeye salmon smolts migrating down the Wood River lake system, Alaska. Photograph by Gregory Ruggerone, Natural Resources Consultants, Inc.

smolts from the Wood River lakes average 7.0 g in weight after 1 year (photo 12-2) and 93% leave at this age, whereas smolts from Ugashik are as large and those from Becharof are even larger, but the smolts from these latter systems more commonly stay for 2 years in the lake. Compared to the other systems, the Wood River sockeye seem to grow less in their second year in the lake, and the age-2 smolts have less advantage in marine survival. In the other lakes, the age-2 smolts are much larger than the age-1 smolts, and enjoy a greater survival advantage at sea. Perhaps the mortality rate in the Wood River lakes during the second year is greater than the survival advantage at sea, so they migrate, despite the fact that they are similar in size to fish in other systems that do not migrate.

On an absolute basis, age-2 smolts are generally larger than age-1 smolts (with some overlap, of course), but the paradox is that the larger, age-2 smolts, actually represent the slower growing component of the population. They stayed in freshwater for another year because they were too small to face the size-selective mortality at sea. Thus the population as a whole and individuals within the population seem to strike a balance between growth and survival in freshwater and mortality at sea. In some cases, older smolts are often not much larger than younger individuals. Data on Columbia River steelhead smolts illustrate this point (fig. 12-2). Most steelhead sampled at Rock Island Dam were aged 2 and 3 but they ranged from 1 to 7. However, over this range of ages the average size only increased slightly, indicating a threshold of size rather than a common age for seaward migration.

Having decided to go to sea in a given year, the next question is, when? Broadly speaking, all salmon migrate to sea in spring, but there is considerable variation among populations over the latitudinal range of the species, as the local concept of "spring" differs greatly from California to Alaska. In the north the migration is generally later than it is farther south (e.g., sockeye: Burgner 1991), reflecting the growing conditions in freshwater and marine environments. As illustrated by the data on juvenile salmonids in streams, the

FIGURE 12-2. Average length at age and age composition of steelhead smolts sampled at Rock Island Dam on the Columbia River (Peven et al. 1994).

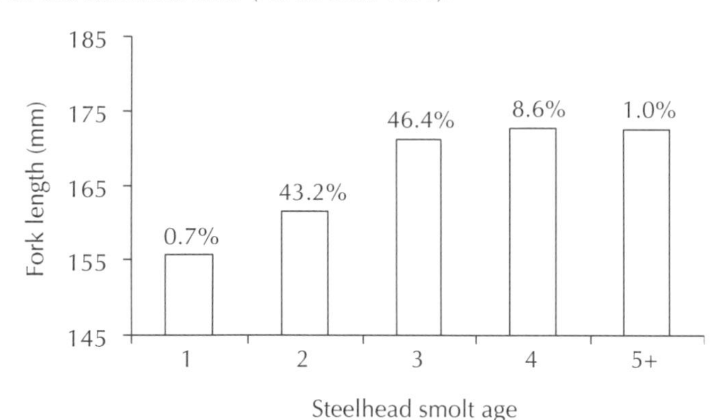

summer is not necessarily the period of greatest growth in freshwater, so the salmon migration timing has presumably evolved to best balance the growth and survival opportunities in the freshwater and marine environments.

Is the timing of smolt migration an adaptation to the marine environment where the smolts are heading or to the freshwater one that they are leaving? Obviously, migration timing results from a bit of both, but data on sockeye salmon leaving Wenatchee and Osoyoos lakes in the Columbia River system suggest a large role for freshwater factors. The smolts of these populations originate in different lakes but they share a journey of hundreds of kilometers to the ocean, and presumably factors along the route, in the estuary, and at sea would select for similar timing. However, the Wenatchee Lake fish migrated much earlier than the Osoyoos Lake fish (Peven 1987). Even within a single lake system there can be bimodal patterns of migration timing, as illustrated by Babine Lake, British Columbia. In this very long lake, there was a distinct early run with fish from the North and Morrison arms of the lake, followed by a run of smolts from the main lake (Groot 1965).

Within a given river, older smolts tend to leave before younger ones. For example, age-2 smolts constituted a declining percentage of the coho smolts leaving the Keogh River, British Columbia, over the run (April: 17.6%, May: 9.8%, June: 4.0%; Irvine and Ward 1989). Data on sockeye salmon leaving Iliamna Lake, Alaska, not only show this pattern but also show that earlier migrants tend to be larger than later migrants of the same age, despite the fact that the later fish have an extra month in which to grow (fig. 12-3). Like the other patterns of size, age, and migration, this tendency also presumably reflects the need for small individuals that are going to leave in a given year to grow a bit more in order to improve their odds of survival when they go to sea. These late, small fish may sacrifice the superior growing conditions at sea for that small size advantage, and this may have repercussions for their marine ecology that merit investigation.

Layered under the broad species-specific patterns, latitudinal clines, and aspects of timing related to age and size, there is also interannual variation in timing that is often related to temperature. In has long been reported that salmon migrate earlier after or during a mild spring than a cold one (e.g., Burgner 1962). Data on chinook salmon migrating from two tributaries of the Umpqua River, Oregon (Roper and Scarnecchia 1999), show this, as do data on sockeye salmon migrating from Iliamna Lake(fig. 12-4).

FIGURE 12-3. Percentage of age-2 (rather than age 1) sockeye salmon smolts (bars) and average length of age 2 smolts (line and diamonds) over the 1990 migration period from Iliamna Lake, Alaska (Crawford et al. 1992).

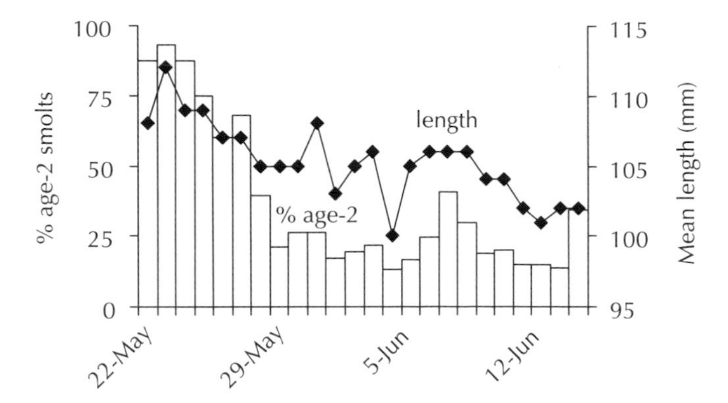

Finally, the individual fish need to decide what time of day or night to move. In general, the downstream migration is a nocturnal phenomenon, whether the fish are moving to a rearing habitat or to the sea. Data presented by McDonald (1960) on pink, coho, and sockeye salmon fry in the Lakelse River system in British Columbia provide an example. There was a clear tendency for all species to migrate at night (fig. 12-5). The minor differences in the timing among species may be genuine or they might have resulted from the varied distances between the trap and the areas where the fish emerged. However, in northern areas there is very little nighttime in late May and early June when smolts are leaving, so the tendency to migrate at night breaks down to some extent.

The parr-to-smolt transformation

The interannual variation in smolt timing in response to environmental conditions is really the fine-tuning of a fundamental internal sequence of changes in morphology, physiology, and behavior needed for transition to saltwater. The salmon are preparing for a migration to a distant and totally new environment, whose conditions they cannot know until they get there. The best predictor for seasonal processes is day length, and

FIGURE 12-4. Relationship between spring water temperature and interannual variation in migration timing of (left) chinook salmon leaving tributaries of the Umpqua River, Oregon (Roper and Scarnecchia 1999), and (right) sockeye salmon leaving Iliamna Lake, Alaska (Alaska Department of Fish and Game annual reports).

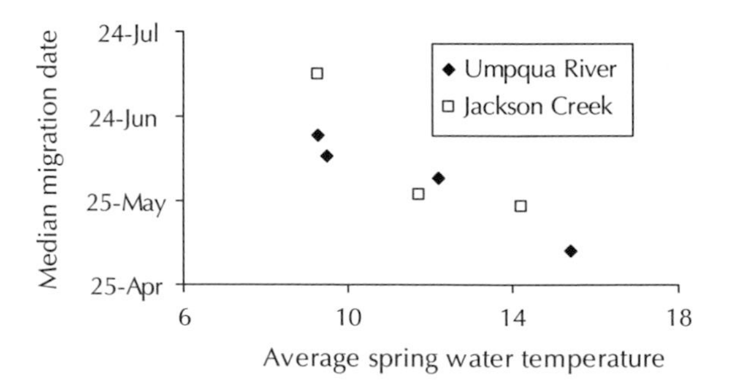

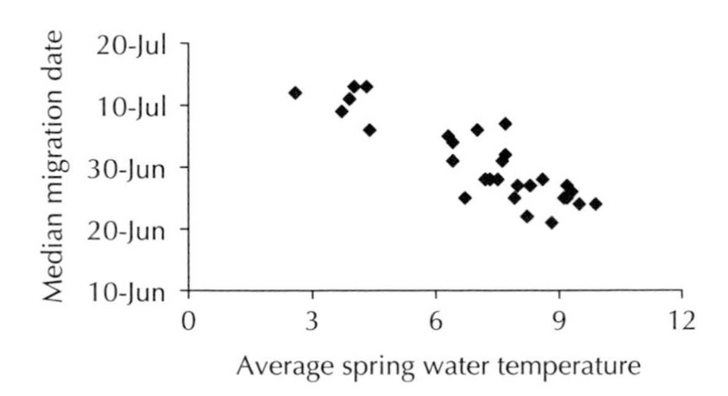

FIGURE 12-5. Nocturnal downstream migration of juvenile salmon in the Lakelse River system, British Columbia (from McDonald 1960).

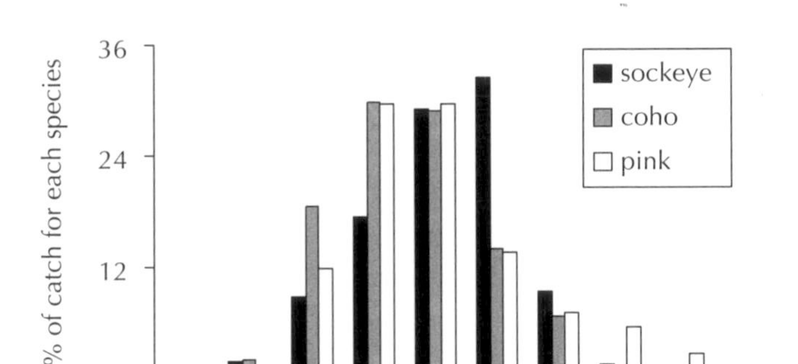

this strongly influences parr-smolt transformation. However, there are internal (circa-annual) rhythms of hormones that prepare the fish for this suite of profound changes. These rhythms are synchronized by day length and also affected by temperature. Some discussion of this parr-smolt transformation is needed, though this is not the place for a review of this fascinating and complex process. Much more detailed information is available in the reviews by Hoar (1976), Wedemeyer et al. (1980), Dickhoff and Sullivan (1987), and Clarke and Hirano (1995). In brief, however, the transformation of salmon parr into smolts is the preparation for the radical change in environments required of anadromous fishes. There are several components to this transformation, including changes in shape and color, preparation of the body's osmoregulation (salt balance) system, energy storage, and mobilization; and the endocrine system plays a key role by responding to external stimuli (photoperiod and temperature) and driving the internal changes (photos 12-3 and 12-4).

Perhaps the most obvious changes are in color. The fish change their appearance, from the green-brown backs and vertical parr marks that camouflage them in streams to the blue-green backs, silver sides, and white bellies that are typical of so many pelagic marine fishes. These colors best hide them in the unstructured ocean against the color of the water when seen from above, from the side, and from below by predators. This is accomplished by gradual deposition of guanine over the skin pigments of the parr during the transformation process. The fins also darken as melanin is deposited. There are changes in morphology at this time, though the salmon body plan suits them as generalists in both environments. The body gets slimmer, however, especially by elongation of the caudal peduncle (Winans and Nishioka 1987), and teeth develop on gums and tongue to permit the fish to catch larger and different prey.

Beneath the surface of the salmon's body, however, much more profound changes are taking place. Freshwater is virtually devoid of salts, the internal fluids of salmon are at about 10 ppt salt, and saltwater is about 33 ppt. Thus in freshwater, fish must keep from losing salts and being flooded with water. To accomplish this, they refrain from drinking, excrete dilute urine, and actively transport ions inward at the gills, where the huge surface area would otherwise lead to ion loss. In saltwater, the fish

12-3 Coho salmon undergoing the parr-smolt transformation process. Photograph by Richard Bell, University of Washington.

drink but they cannot produce urine more concentrated than their body's fluids so ion excretion at the gills is very important. Thus, substantial changes in the kidneys and gills are required to permit the fish to migrate to sea. In coho salmon, for example, the concentration of sodium (Na+) ions in the blood is about 120 milliequivalents in freshwater. This increases to about 160–200 in the first day after introduction to saltwater, then goes down to 140–150 within 2 more days, as the fish's system regains control over its salt balance.

Smolting is not just a matter of ion regulation in a new environment but an overall set of endocrine and metabolic adjustments. The transformation process seems to be very energetically demanding. In the late winter, insulin levels increase, and the body stores lipids and carbohydrates, later mobilized to fuel the changes from parr to smolt. Mobilization of stored energy is stimulated by growth hormone from the pituitary gland, thyroxine from the thyroid gland, and corticosteroids (especially cortisol) from interrenal gland. Cortisol plays an important role in gill Na+K+ ATPase production, allowing the gills to regulate ions adequately when fish enter seawater. However, cortisol also inhibits immune function, so smolts may be more vulnerable to diseases than parr. Prolactin is involved in ion regulation in freshwater; growth hormone and cortisol stimulate seawater tolerance.

Thus every spring the young salmon are prepared by internal rhythms for the transition to seawater, and the increasing day length synchronizes the body's rhythm to local conditions. Increasing temperatures play an important secondary role. The decision to go or stay is associated with size; larger parr are more likely to become smolts in a given year and to smolt earlier in the year than smaller ones. However, rapid growth rate is more important than absolute size for proper synchrony of the physiological processes associated with smolt transformation (Dickhoff et al. 1997; Beckman et al. 1998). There

12-4 Chinook salmon undergoing the parr-smolt transformation process. Photograph by Richard Bell, University of Washington.

seems to be a window of time in spring when the fish either begin the process of smolt transformation, including energy storage, mobilization, thyroid hormone surges, changes in color, shape, ion regulation, and behavior, or they do not (fig. 12-6). In species and populations with sexually mature male parr, it is typically the largest or fattest males that mature, the fish of intermediate size that leave as smolts, and the smallest ones that remain for another year (Thorpe 1987). Later in the summer, the fish lose the ability to thrive in seawater and the tendency to migrate.

External stimuli thus trigger a cascade of events, including energy storage and mobilization, changes in behavior and morphology, and readiness to enter seawater. However, these are not like a row of dominoes that are lined up and then fall in a rigid

FIGURE 12-6. Generalized sequence of plasma hormone levels (IGF-I is "insulin-like growth factor I") and body size during the critical period of smolt transformation. Conceptually, a lower growth rate and lower levels of hormones might lead to unsuccessful transformation (modified from Dickhoff et al. 1997).

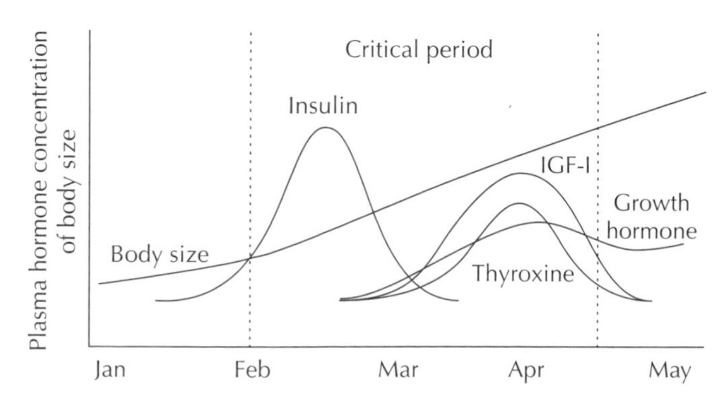

FIGURE 12-7. Levels of Na+K+ ATPase activity in the gill tissue of wild steelhead caught migrating down the Snake River, steelhead released from Dworshak Hatchery and caught during their migration, and steelhead held in the hatchery (Zaugg et al. 1985).

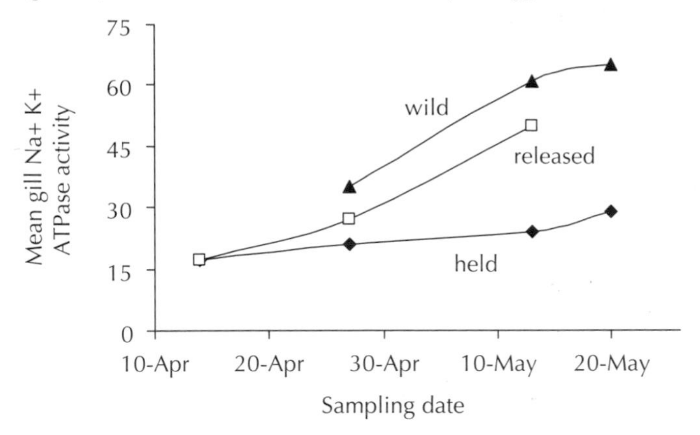

sequence. Just as migration seems to be an important part of the process of olfactory imprinting (see chapter 5), Zaugg et al. (1985) showed positive feedbacks between seaward migration and development of the physiological traits needed for successful entry into seawater. Steelhead smolts held at Dworshak National Fish Hatchery had lower gill Na+K+ ATPase activity levels than hatchery fish migrating downstream, and wild fish sampled at the same locations and dates had still higher levels (fig. 12-7).

Downstream migration behavior

Besides the complex internal changes involved in smolt transformation, perhaps the most critical change is the behavioral one: freshwater residents migrate to sea. This very obvious point attracted attention decades ago, and Hoar (1958) noted that migration necessitates changes in several behavior patterns, including hiding, schooling, territoriality, predator avoidance, and feeding. These changes in behavior are associated with changes in response to "directing" or "guiding" factors such as light, temperature, water current, salinity, and structure. Thus sockeye change from diel vertical migrations to directed, horizontal migrations to the lake outlet and downstream orientation, and the stream-dwelling species change from upstream orientation and territorial behavior to downstream orientation and schooling.

Pioneering work was conducted on sockeye salmon migrating from Babine Lake by Groot (1965, 1972). Direct observations and sonar showed the fish swimming about 24–30 cm per second. The fish (about 8 cm long) swam at least 3 body lengths/sec, virtually their maximum sustainable speed. They did not swim at this rate all day, however, but primarily after sunset and before sunrise, and their net travel rate was about 7.5 km per day. This is the equivalent in body lengths of adult salmon (e.g., 600 mm long) swimming 56 km/d. Groot (1965) hypothesized and then demonstrated that smolts can use the sun and polarized light patterns to orient in experimental arenas. He also found that sockeye smolts could orient without a view of the sky, and he called this type-X orientation because it could not be explained. Subsequent experiments identified the earth's magnetic field as the source of this mysterious orientation ability (Quinn and Brannon

FIGURE 12-8. Percentage of juvenile ocean-type chinook salmon showing downstream movement in experimental tanks (left panel) and preferred salinity, based on distribution of fish in vertical salinity gradients (right panel; all data from Whitman 1987).

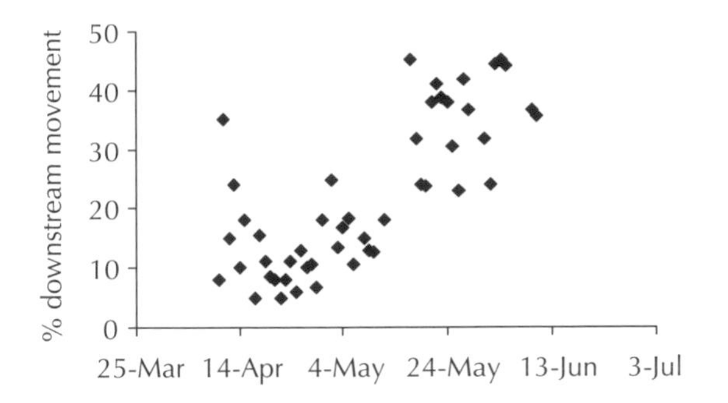

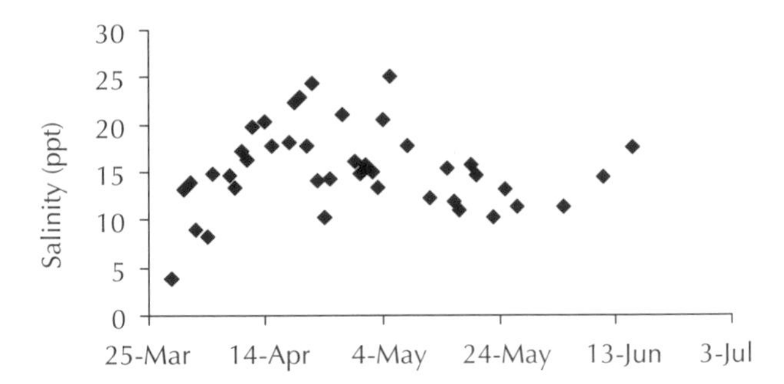

1982). Smolts migrating from Babine Lake were caught and placed in circular tanks with water upwelling into the center of the tank and flowing out via eight ports, equally spaced around the edge of the tank. During the day the smolts showed no inclination to leave, just as the trap in the lake's outlet caught virtually nothing. However, at dusk, when smolts in the lake were migrating, the fish in the tanks moved downstream out of the tanks and left via the ports in the compass direction that they had been swimming in the lake. Alteration of the magnetic field by coils around the tanks redirected the smolt movements.

For the most part, though, smolts migrate downstream in creeks and rivers of increasing size. Neave (1955) observed that migration was not accomplished by passive displacement or drift but by active, downstream swimming. In the British Columbia pink and chum populations he studied, the migration was exclusively nocturnal, and the fish that did not reach saltwater during their first night moved back into gravel and hid during the day, resuming migration the next night. Movement is typically in the fastest water available, and even minor changes in the position of a trap in a river can result in great differences in catch rate. Radio tracking suggests that migration is not continuous but is interspersed by periods of holding. Coho salmon, migrating down the Chehalis River, Washington, showed a net travel rate of 1.2 km per hour, but during periods of active movement they traveled 3.1 km/h, implying that about they spent about 40% of the time moving and 60% holding (Moser et al. 1991). Much of their movement took place in the day but they also held in back-eddies and other off-channel habitats, consistent with observations by McMahon and Holtby (1992) on the affinity of migrating coho smolts for woody debris and cover.

Experiments by Whitman (1987) on ocean-type chinook salmon revealed increased downstream orientation in fish held in captivity from mid-April through mid-June. He also documented an increase in preferred salinity, determined by the position of the fish in a vertical salinity gradient (fig. 12-8). Curiously, the highest salinities were preferred in late April, well before the maximum display of downstream orientation. Under natural conditions, a combination of downstream orientation and preference for increasing salinity facilitates migration downstream, through an estuary and out to sea.

Considerable information on patterns of downstream migration comes from the Columbia River system, where there is intense interest in both the movements of fish and of water. Smolt migrations take place primarily in spring and early summer but

TABLE 12-6. Counts of juvenile salmonids trapped migrating downstream in a bypass channel at Bonneville Dam, 1950–1953, expressed as a percentage of the annual total number trapped (coho: 2287, chinook: 158,520, sockeye: 5555, steelhead: 15,247; data from U.S. Army Corps of Engineers, annual fish passage reports).

Species	*Jan*	*Feb*	*Mar*	*April*	*May*	*June*	*July*	*Aug*	*Sept*	*Oct*	*Nov*	*Dec*
coho	5.9	6.4	8.4	18.1	46.0	4.0	0.9	0.1	0.3	7.1	1.9	0.9
chinook	1.2	8.6	22.1	32.2	22.5	11.2	1.5	0.4	0.1	0.2	0.1	0.1
sockeye	0.4	1.7	5.0	19.1	70.1	1.9	0.8	0.2	0.4	0.4	0.0	0.0
steelhead	0.2	0.2	0.6	33.2	62.0	2.9	0.1	0.0	0.1	0.4	0.1	0.1

data from a small bypass channel at Bonneville Dam, sampled in the early 1950s, are worth noting. This time period was before the greatest increase in hatchery production and modification of the river's flow regime, but considerable reduction in habitat integrity and population diversity had already taken place. Nevertheless, during a 4-year period, at least some chinook, coho, sockeye, and steelhead were taken in every month of the year (table 12-6). The trapping efficiency probably varied with discharge and so the data are useful only in a qualitative way but they still show interesting patterns. Sockeye and steelhead showed the narrowest timing distribution (primarily May). Significant numbers of coho salmon (fry were not distinguished from smolts) migrated from January onward, reaching a peak in May but there was another smaller peak in October, probably fry rather than post-smolts. Chinook showed the broadest distribution pattern, consistent with work on the species elsewhere.

The peak of migration coincided with the Columbia River's peak flows, powered by the melting of snow in the upper reaches of the basin, and this would be true of many of the larger rivers in the region. Modification of the river's flow regime to store water in summer for release in winter to produce hydroelectric power has greatly reduced the discharge during the peak period. Flows in June were about 14,000 m^3/sec in the 1950s but only about 6000–8000 m^3/sec since the 1990s. In addition, the river has become a series of reservoirs and dams, so the patterns of water velocity have also been changed. The Columbia River's discharge (i.e., volume per unit of time: m^3/sec) affects the water velocity (m/sec) at a given point, but the relationship is not linear (a small increase in discharge causes a greater increase in velocity at low discharges than at high ones). Berggren and Filardo (1993) reviewed a large number of marking studies and concluded that travel rates of smolts were approximately equal to the rate of travel of water. However, in some cases the fish were moving faster than the estimated travel rate of water. This suggests that the fish were actively swimming downstream, especially if they did not move continuously but stopped for some time each day. The relationship between discharge and fish travel rate is probably tempered by different responses of the fish to different velocities and shears that they experience. At low velocities, they may swim downstream and exceed the speed of the water but at higher velocities they may back downstream or move discontinuously.

However, there are many factors that affect fish travel rate besides water speed, including date, location where the fish commence migration, fish size, and extent of parr-smolt

FIGURE 12-9. Travel rate of coho salmon smolts released at Ice Harbor Dam on the Snake River and captured at Jones Beach, 462 km downstream in the lower Columbia River (from Dawley et al. 1986).

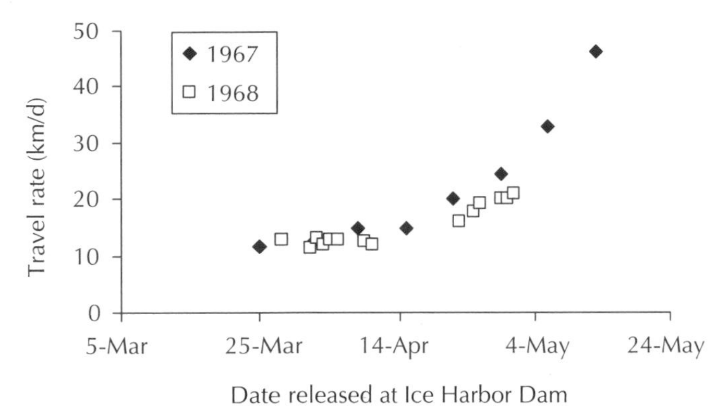

transformation. Extensive sampling in the upper Columbia River estuary in the late 1960s to early 1980s by the National Marine Fisheries Service (Dawley et al. 1986) provided many fascinating insights into smolt migrations. The focal area was Jones Beach, Oregon, about 75 km upriver from the mouth and about 50 km above the normal intrusion of saltwater. One finding was that fish released later in the season moved much faster than those released earlier (e.g., coho salmon: fig. 12-9). Giorgi et al. (1997) reported similar results for sockeye, steelhead, and yearling chinook salmon but not for age-0 chinook. Fish length affected migration speed but results varied among species. Age-0 chinook salmon (relatively small to begin with) showed the greatest positive correlation between length and speed. Larger sockeye and yearling chinook also migrated faster than smaller ones but the relationship was weaker, and in steelhead (typically the largest smolts) larger fish migrated slower than smaller ones (Giorgi et al. 1997). Finally, fish released farther from the estuary seem to migrate faster than those originating farther downriver (fig. 12-10), possibly reflecting a genetic adaptation to the distance that populations would migrate. Not only do fish from hatcheries farther upriver migrate

FIGURE 12-10. Relationship between downstream travel rate of age-0 chinook salmon in 1968 and 1969 and the distance from their hatchery to the capture site (Jones Beach, in the lower Columbia River; from Dawley et al. 1986).

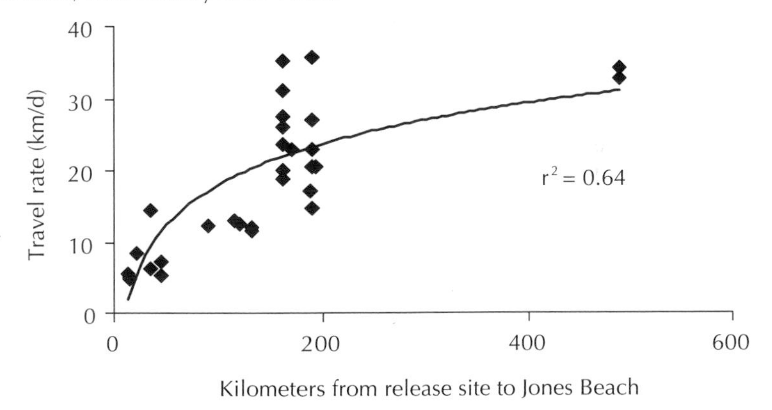

12-5 Rock Island Dam, mid-Columbia River. Photograph provided by Charles Peven, Chelan County Public Utility District.

faster than those from downriver sites, but they also have a greater tendency to migrate in open water (consistent with rapid migration) rather than in the margins, as reflected by the relative catches in purse seines and beach seines (Zaugg and Mahnken 1991).

Fish size and date are generally viewed (e.g., by Berggren and Filardo 1993 and Giorgi et al. 1997) as surrogates for the state of smolt transformation. However, growth rate rather than absolute size may be critical in migratory behavior (Beckman et al. 1998) and in the general development of smolt transformation (Beckman and Dickhoff 1998). There are other indications that the state of parr-smolt transformation affects migratory behavior. For example, the chinook salmon guided into bypass channels at Lower Granite Dam differed in gill Na+K+ ATPase levels from those passing through the turbines (Giorgi et al. 1988), though it is unclear what connection between physiology and behavior would lead fish to pass the dam one way or the other.

Mortality during downstream migration

The reduction in the Columbia River's discharge over the past decades, the connection with smolt travel rate, and the overall decline in the river's salmon runs all stimulated interest in the possible correlation between travel rate and survival. A positive relationship might justify releases of water from upriver reservoirs to speed the fish safely to sea. However, water is so valuable for hydroelectric power production that there is great interest in information on the relationship between flow and survival (photo 12-5). Sims and Ossiander (1981) reported that not only travel rates but also survival rates of chinook and steelhead were positively related to river flows, using annual estimates from 1973 to 1979, for fish migrating between Ice Harbor Dam on the Snake River to The Dalles Dam on the lower Columbia River. These and other data pertaining to flow and salmon survival have been subjected to intense scrutiny in the years since this publication. Bickford and Skalski (2000) examined fifty-three survival studies conducted over 25 years and concluded that about 87–88% of the juvenile salmon survived passage through each "project" (dam and associated reservoir). The authors noted the consistency in estimates among salmon species and among dams and the absence of a clear relationship between survival and flow. There have been changes in the operations of the dams and conditions in the river since the 1970s, but research under present conditions (e.g., Skalski 1998; Muir et al. 2001; S. G. Smith et al. 2002) indicated that while travel rate depends on flow, survival seems to be an almost constant function of distance; the farther the fish have to go, the less likely they are to survive the trip. Survival rates of 88% per project and 99.8% per kilometer may seem benign until one remembers that some salmonids migrate through as many as eight or nine projects and about 700–800 km on the main stems of the Snake and Columbia rivers, not including migration in tributaries. For example, if 100,000 smolts experienced 90% survival through each of eight projects, only 43% would enter the estuary. A survival rate of 85% per project would reduce the number of survivors to 27% of the initial cohort.

If the 10–15% of the salmon smolts dying per project are killed outright by the physical forces of turbines and pressure changes, then it is fundamentally an engineering problem (albeit a big one). However, if the fish are eaten by predators, then it is an ecological problem, or an interaction between engineering and ecology. Ruggerone (1986) reported that gulls (chiefly ring-billed gulls) consumed about 2% of the smolts at Wanapum Dam in 1982, though some of these were fish that might have been mortally wounded by physical damage at the dam. Wires are set up at dams to hinder gull predation but most research has been directed toward predation on smolts by fishes, especially by the northern pikeminnow (referred to in earlier literature as northern squawfish). Stomach content analysis indicated that the pikeminnow eat smolts but estimates of overall predation depended on a number of factors, which were determined in a thorough series of field and laboratory studies reviewed by Beamesderfer et al. (1996). To estimate predation we need to know the relationships between pikeminnow size and tendency to eat smolts of different sizes (varying among species and between hatchery and wild fish), smolt consumption and water temperature (which varies over the season, as does the proportion of smolts of each species), the abundance and size distribution of predators and prey, the proportion of fish eaten that were already killed by the physical stress of the turbines or bypass systems, and so on. In summary, it was estimated that predatory fishes consumed between 1.9 and 3.3 million of the 19 million salmonids migrating annually through the John Day reservoir. Pikeminnow accounted for 78% of the mortality, with lesser contributions from three nonnative fishes: walleye (*Stizostedion vitreum*), channel catfish (*Ictalurus punctatus*), and smallmouth bass (*Micropterus dolomieui*). Underyearling chinook salmon suffered the heaviest predation rate because they are smaller than the other species and tend to migrate later, when the water is warmer and the predators' appetite and swimming speed are greater. Expansion of these estimates and additional sampling indicated that there may be about 1.765 million pikeminnow, each consuming an average of 0.06 live smolts per day, or 16.4 million smolts per year in the entire main stem Snake and Columbia River system (thirteen reservoirs and the reach from Bonneville Dam to the estuary).

Predation has been most intensively studied on the Columbia River, where many changes in the river resulting from the dams may have contributed to predation, in addition to the direct mortality from the dams themselves. The increased temperatures increase the metabolic demands of the predators, and Petersen and Kitchell (2001) estimated that pikeminnow predation was about 26–31% higher during warm than during cool periods of the twentieth century, and as much as 68–96% higher in the warmest compared to the coolest year on record. The dams have also increased water clarity, and this may make smolts easier for predators to catch. However, it is important to remember that survival rate before dams were constructed would not have been 100%. Pikeminnow are native fishes, and some predation from them, other fishes, and from birds surely took place. Interpretation of mortality in the Columbia River would greatly benefit from research on the Fraser River, as it is in many ways comparable but lacks any dams along the main stem. Many studies have indicated significant predation in wild populations migrating down unregulated rivers. Hunter (1959) studied predation by sculpins and coho salmon on chum and pink salmon fry for 10 years at Hooknose Creek, British Columbia. Sculpins at least 6 cm long ate fry, and the number of fry in the stomachs of sculpins increased steadily from 0.5 per 6 cm sculpin to more than 5.5 for those 13 cm and larger. The number of fry eaten per year increased with fry abundance but the predators seemed unable to eat more than about 0.5 million fry; this saturation was reached when there were about 1 million or more fry present. Predation ranged from about 80% of the fry at low densities to about 20% at high densities. Ruggerone and Rogers (1984) found that Arctic char ate 5–40% of the sockeye smolts leaving Little Togiak Lake, Alaska, under low salmon densities. Wood's (1987a) research also indicated that predation by common mergansers was density dependent but he estimated maximum mortality rates of less than 10% in Vancouver Island, British Columbia, streams.

Summary

Every year, salmonids are faced with the choice of migrating to sea or remaining in freshwater. This critical life-history transition is triggered by size, growth, or some other factor related to the individual's condition. Large individuals in good condition tend to migrate whereas smaller individuals are more likely to remain in freshwater for another year. In some species, all or a fraction of the individuals in some populations do not migrate to sea at all. These fish sacrifice the growing opportunities at sea for the relative safety of freshwater, and males are more inclined to remain than females. This difference is related to the fact that reproductive success in females is linked to fecundity, hence large size, whereas small males can fertilize many eggs by sneaking rather than fighting.

The timing of migration is determined by internal seasonal rhythms of physiology, synchronized by the increasing day length in spring and modulated by temperature. Larger and older individuals tend to migrate earlier in the season than smaller ones, and the downstream migration itself generally takes place at night. Many studies have investigated the effects of hydroelectric dams and impoundments on the rate of downstream migration and the survival of smolts, though there is still some controversy on the subject and uncertainty about the levels of mortality in free-flowing systems.

13

Estuarine Residence and Migration

As salmon migrate to sea they encounter estuaries, the ecotone habitat between fresh- and saltwaters. Estuaries extend from the lower reaches of rivers, where flow and water level are influenced by tides but the water is fresh, through regions with intermediate salinities (mixed or stratified vertically, depending on site-specific features) and increasing tidal influence, to coastal waters with salinities more characteristic of the ocean. Estuaries are thus not a single type of habitat but the term encompasses a range of habitats whose physical and biological characteristics are exceedingly complex. The habitats progress (not always smoothly) from riverine to marine and include edge, bottom, and open-water environments. Not only is there great spatial variation in estuaries, but the physicochemical attributes of the water such as depth, salinity, temperature, turbidity and velocity, vary over complex temporal scales including seasonal, lunar, and tidal periods.

From a physical standpoint, estuaries are the meeting place of two hydraulic forces: the river and the tides. Winter flows are low in snowmelt-dominated rivers and high but variable in rain-dominated coastal rivers, and spring (when most salmon migrate to sea) is often a period of varying river discharges, followed by lower and more stable summer flows. Thus the river's power to push back the ocean is changing over the period when salmon migrate to sea. In addition to the seasonal variation in river flow, the semidiurnal (twice a day) tidal cycle of the ocean causes a given site to experience very different regimes of salinity, velocity, temperature, and other conditions from hour to hour. The magnitude of these tides also varies with the stage of the moon, on a 28-day period, further complicating the temporal variation. Likewise, at a given point in time there is a great deal of spatial variation, both horizontally and vertically (saltier as one goes seaward and deeper in the water column). Freshwater is less dense than saltwater, and it starts out above it in the river, but this stratification breaks down as the water

FIGURE 13-1. Salinity-depth profiles at high and low tides at three locations (marked with solid diamonds) in the lower Chehalis River (entering from the right) and Grays Harbor (to the left), Washington, during the period of salmon smolt migration (May 14–24, 1989; Quinn 1990b).

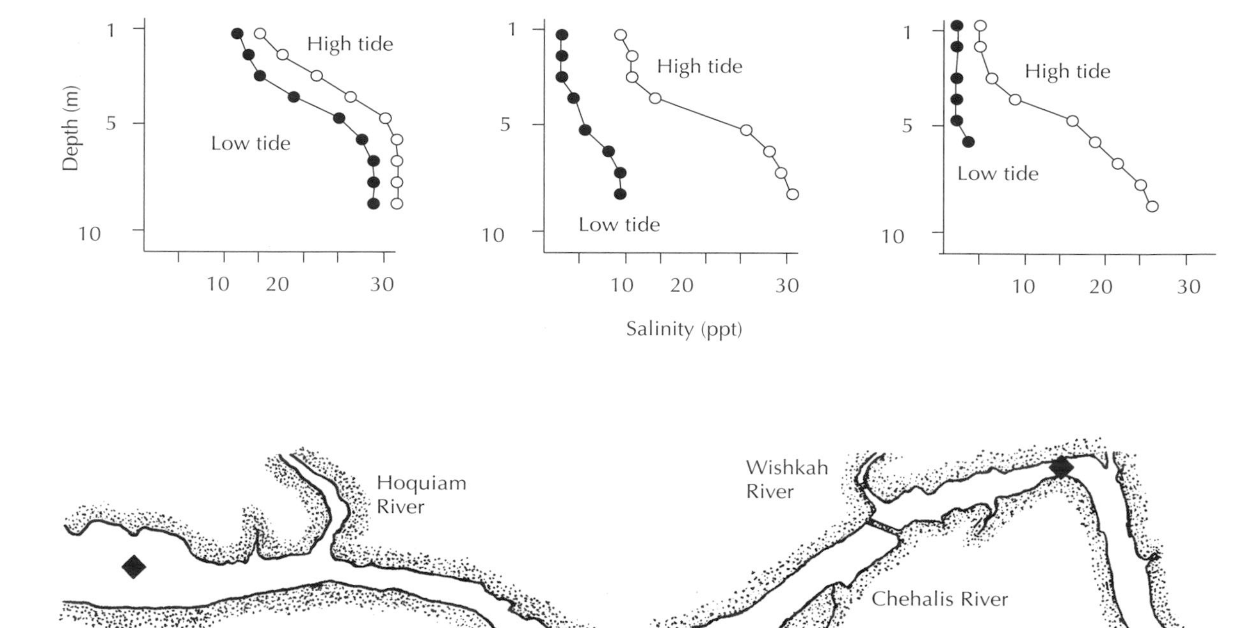

becomes mixed from wind and tide action. Figure 13-1 provides an example of the three main sources of variation in salinity: from the surface of the water to the bottom, from low to high tide, and from one site to another moving from the river towards the ocean.

Temperature has much less effect than salinity on the density of water, so the relationship between temperature and depth differs from that in lakes, where only temperature affects density. In late winter or early spring the rivers are often colder than the North Pacific Ocean, by midspring the temperatures may be similar, and in summer the rivers are normally warmer. Thus the fish may experience a conflict in their efforts to find optimal temperature and salinity conditions. In addition to the conditions of salinity and temperature, estuaries are also areas of sediment deposition, as fine material settles out of the water column. However, resuspension of sediment from interactions between riverine and tidal flows and wind can result in very turbid water. This turbidity likely affects foraging and predator avoidance in complex ways. Water flow and velocity also have complex patterns in estuaries. The overall flow is seaward but at any point in space and time the actual flow is a combination of river discharge and tide, itself determined by the phase of the moon (strongest on new and full moons) and the point in the tidal cycle. The velocity experienced by a fish is also related to its distance from the bottom, from the shoreline, and from structure, as well as the dominant flow at that moment and location.

In terms of the salmon's life history, the estuary lies between the breeding and incubation habitat in freshwater, and the ocean, where most growth occurs. The salmon go through the estuary on their way to the sea, and the estuarine environment may play one or more roles for them, depending on the species and population of salmon and the particular estuary in question. The estuary may be a relatively neutral highway through which salmon pass quickly on their way out to sea, a haven where they adjust their bodies to the change in osmotic environments, a rich pasture where they feed before the levels of competition or the seasonal shifts in estuarine productivity make it less beneficial than the coastal waters and ocean, or a maze fraught with predators that they must escape as quickly as possible. Estuaries are inherently difficult environments in which to conduct behavioral and ecological research, given the spatial and temporal variation described above, and their open-ended nature. Some are large and not very amenable to manipulation or experimentation, though the restoration of estuarine habitats in smaller systems or parts of larger estuaries presents opportunities in this regard. In addition, estuaries are less numerous and more different from one another than streams and so are less amenable to comparative research approaches. Because of these difficulties there is still considerable uncertainty regarding the roles that estuaries play in the lives of salmon. This chapter considers the patterns of estuarine use among species and the evidence for the importance of estuaries for salmon osmoregulation, feeding, and predator avoidance.

Patterns of entry and use

Simenstad et al. (1982) reviewed information on the timing of arrival and patterns of estuarine residence by salmon in Puget Sound, as did Healey (1982a) for southern British Columbia and Bottom et al. (2001) for the Columbia River. We may think of the patterns in terms of arrival, duration of residence, and habitat occupied by the population as a whole, and duration of residence by individuals. The general patterns are more or less as follows. Pink and chum fry (photos 13-1, 13-2) enter from early February to late May, with the peak typically in late March to early May. Sockeye and coho salmon smolts enter in mid-April to mid-June, peaking typically in mid-May, but coho salmon fry may migrate earlier and over a longer period. Some ocean-type chinook salmon migrate downstream as fry shortly after emergence or after growing in rivers for a few months, and the timing of yearling stream-type chinook is similar to that of coho salmon. For example, Myers and Horton (1982) reported that in Yaquina Bay, Oregon, the juvenile chum salmon arrived and apparently moved through the estuary in March and April, coho smolts in April and May, and chinook salmon were present from April to November but their peak in abundance followed after the other species. In addition, coho salmon fry migrate to sea from many coastal rivers (e.g., Hartman et al. 1982). They have generally been regarded as "surplus" to the carrying capacity of the stream and have been

13-1. Juvenile pink salmon at the stage when they migrate downstream and enter estuaries. In the background there is a juvenile chum salmon (with visible parr marks) as well as another pink salmon. Photo by Richard Bell, University of Washington.

assumed to perish at sea. I suspect that these assumptions may not be entirely justified, and the behavior and ecology of these coho fry is a fruitful topic for further research. They may leave the estuary for the ocean in their first year of life or reenter freshwater and leave as smolts the next spring.

The duration of estuarine residence varies greatly among salmon. In general, the species arriving in the estuary at a larger size spend less time there than do species that

13-2. Juvenile chum salmon at the stage when they migrate downstream and enter estuaries. Photo by Richard Bell, University of Washington.

TABLE 13-1. Catches of juvenile salmonids in 1968 at Jones Beach, Oregon, 75 km upstream from the mouth of the Columbia River (from Dawley et al. 1986).

		Catch per set			
Gear	*Month*	*age 0 chinook*	*age 1 chinook*	*coho smolts*	*steelhead smolts*
Beach seine	May	177.6	2.1	25.1	1.1
	June	164.4	0.1	0.6	0.0
	July	497.0	0.0	0.2	0.0
Purse seine	May	15.7	12.1	61.3	31.3
	June	24.9	0.4	1.5	1.4
	July	14.1	0.1	0.2	0.1

are smaller when they enter. Specifically, steelhead, coho, sockeye, and yearling chinook seem to move through estuaries rapidly whereas underyearling chinook, underyearling sockeye, and chum salmon stay there longer. Pink salmon are anomalous in this regard, as they are the smallest when they arrive in the estuary but tend to move through faster than the larger chum and ocean-type chinook salmon. The species not only vary in duration of residence but also in habitat use within the estuarine system.

Extensive work in the Columbia River estuary by Dawley et al. (1986) revealed much about the patterns of arrival and use by salmon. They sampled the littoral zone with beach seines and the offshore region with purse seines. One cannot directly compare the numbers of fish caught with these different types of gear, and they might have caught more ocean-type fish had they sampled peripheral channels more extensively (Charles A. Simenstad, University of Washington, personal communication). Nevertheless, the relative catches revealed several patterns. Coho, steelhead, and yearling chinook smolts were caught in May but catches then dropped off rapidly, whereas many ocean-type chinook were caught into the summer. The yearling chinook, coho, and especially the steelhead were caught primarily offshore (with purse seines), whereas the ocean-type chinook were caught primarily with beach seines (table 13-1). In open water the chinook tended to be near the surface. At each of three different sites, no fewer than 95% of the ocean-type chinook were caught in the upper 3 m of the water column, as opposed to 3–6 m (2–5%) or below 6 m (less than 0.3%).

McCabe et al. (1986) reported that juvenile chinook salmon were caught in the Columbia River estuary during every month of the year, though 97% were caught in March through September. There was a progressive shift over the season from peak concentrations in littoral areas in the upper estuary early in the season, to pelagic habitats in the upper estuary and littoral habitats in the lower estuary later, and finally to pelagic habitats in the lower estuary (table 13-2). The chinook caught in pelagic areas were larger than those in intertidal sites at a given period of time, and the average fish size increased over the season.

Healey (1980, 1982a) showed similar patterns for chum and ocean-type chinook salmon in the Nanaimo River, British Columbia, estuary: increasing average size from March through July, a shift from catches in the inner to the outer part of the estuary, and a

TABLE 13-2. Catches of juvenile ocean-type chinook salmon in upper and lower regions of the Columbia River estuary, in intertidal (littoral) and pelagic (offshore) habitat in 1980 and 1981 (McCabe et al. 1986).

Month	*Upper intertidal*	*Lower intertidal*	*Upper pelagic*	*Lower pelagic*
January	18	0	0	0
February	26	5	0	0
March	171	25	3	7
April	260	55	38	7
May	379	297	657	257
June	484	602	380	441
July	406	380	587	598
August	111	28	316	215
September	145	25	177	116
October	20	10	17	29
November	3	1	9	12
December	6	0	24	13

tendency for larger fish to be caught offshore at each time period (fig. 13-2). Thus, as individuals grow they seem to move progressively downstream and offshore. Chum salmon probably grow little in many of the very small coastal streams where they are spawned, so increasing sizes of fish in the estuary suggest estuarine growth. However, ocean-type chinook salmon may reside in rivers for varying periods of time, often several months or more before migrating to sea, so the large fish caught in estuaries late in the summer did not necessarily grow there but may have only recently left the river. Carl and Healey (1984) presented morphological and genetic evidence indicating three life-history types of chinook

FIGURE 13-2. Sizes of chum salmon fry caught in the inner and outer areas of the Nanaimo River, British Columbia, estuary (Healey 1982).

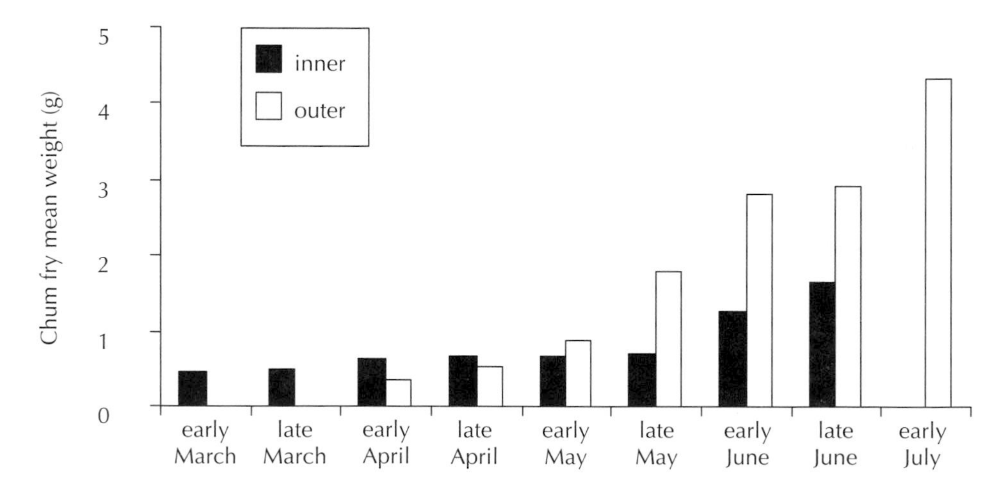

in the Nanaimo River system: those migrating as newly emerged fry, fingerlings, and yearling smolts; and Reimers (1971) reported five different life-history types of chinook in the Sixes River, Oregon, based on patterns of rearing in the river and the estuary.

Although salmon as species and populations occupy estuaries for months, individuals tend to remain for shorter periods. Estuary size and other physical attributes affect residence patterns, but Shreffler et al. (1990) compiled data from several studies indicating average individual residence times of about 1 week for chum salmon and 1–3 or more weeks for chinook. Such data are typically obtained by marking a number of fish and then repeatedly sampling for them. Declining catch rates are taken to indicate dispersal, though mortality occurs too. The observation that chum salmon repeatedly occupy tidal channels that dewater on each cycle (Mason 1974; Levy and Northcote 1982) provides additional evidence for active retention in estuaries rather than direct migration to sea or even passive displacement. However, residence time may depend on fish size. Pearcy et al. (1989) marked chum salmon in Netarts Bay, Oregon, and reported residence half-life values (days for the catch rate of marked fish to decline by half) of 7.4, 4.9, and less than 2 days for fish weighing 1.0, 2.2, and 6.5 g, respectively.

Importance of estuaries

It is widely believed that estuaries are an important habitat for young salmon, despite the fact that salmon only reside there seasonally. Thorpe (1994) reviewed the three main hypotheses regarding the functional significance of estuaries: feeding, transition to seawater, and predator avoidance. As indicated above, the size distributions of juvenile salmon in estuaries often increase over time, and this may reflect some combination of growth in the estuary and the arrival of fish that had grown in the river prior to entering the estuary. Marking studies can provide direct information on growth, though they are weakened by the inability to measure the fish that died (probably the smaller fish) and those that had already left (typically the largest fish). Ocean-type chinook salmon grow about 0.5 to 1 mm per day in estuaries (reviewed by Bottom et al. 2001). As reviewed by Healey (1982a), Simenstad et al. (1982), and Pearcy et al. (1989), the growth rates of juvenile chum salmon (measured as a percentage increase in body weight per day) in estuaries range from about 2% in Netarts Bay, Oregon, 3.5% in Nitinat Lake, 4.2% in the Fraser River and Gulf Islands, 5.7% in the Nanaimo River (these latter sites in B.C.), 6% in the Skagit River, and 8.6% in Hood Canal, both in Washington. Chum salmon held in seawater tanks and fed live zooplankton by LeBrasseur (1969) showed a maximum growth rate of about 5.5% on an excess ration (greater than 16% of their weight in food per day). These figures are very close to Healey's (1979) estimates of 6% growth at a ration of 15% of their weight per day, suggesting that chum fry may be feeding at close to their maximum ration in some estuaries. This indication that estuaries present excellent growing conditions is consistent with evidence presented by Kjelson et al. (1982) that juvenile chinook salmon grew faster in the estuary of the Sacramento–San Joaquin river system than they did in the Sacramento River. However, MacFarlane and Norton (2002) reported slow growth in the San Francisco estuary but much more rapid growth once the salmon entered the coastal waters, so there is certainly more work to be done on comparative growth in different environments along the migratory route from river to the ocean.

Estuaries are often very productive environments, and young salmon feed on different types of organisms, depending on the salmons' size, species, and habitat (e.g., Macdonald et al. 1987). These variations reflect both the differences in food available in different habitats and different times of the year, the preferential use of habitats by different species, and the downstream and offshore shifts in fish distribution as the fish get larger. Estuaries may contain such habitats as (1) tidally influenced but fully freshwater sloughs with rich production of such insects as chironomid (midge) larvae, preyed upon by ocean-type chinook and coho fry for protracted periods and by chum fry during their seaward migration; (2) brackish marshes with emergent vegetation providing insect larvae, mysids, and epibenthic amphipods for chinook and chum fry; (3) sandy beaches, eelgrass beds, tidal mudflats, and channels with epibenthic copepods, benthic amphipods, and cumaceans for chum and pink fry; (4) gravel and cobble beaches with amphipods and isopods for coho smolts; and (5) open-water habitats with drifting insects, zooplankton such as crab larvae, pelagic copepods, and larval fish for chinook and coho smolts and larger chums and pinks.

Several studies have pointed out the especially large contribution of harpacticoid copepods in the diets of juvenile salmon in estuaries, especially chum and pink salmon (e.g., Kaczynski et al. 1973; Mason 1974; Mayama and Ishida 2003). These copepods (e.g., *Harpacticus uniremis*) seem to be much more common in fish diets than their abundance in the environment would suggest. The fish may seek them out, though it is difficult to accurately assess the relative abundance of epibenthic prey (Simenstad et al. 1980). Regardless of these sampling issues, the harpacticoid copepods feed on bacterial flora associated with the organic detritus that is the base of the food web in many estuaries (Sibert et al. 1977; Simenstad et al. 1990). The detritus comes from the breakdown of several types of plant material, including *Zostera* (eelgrass), *Carex* (a sedge found in intertidal marshes), intertidal algae, and phytoplankton from freshwater.

If the growth in estuaries is so good, why do the fish leave? Sibert (1979) reported that the juvenile chum salmon consumed *Harpacticus uniremis* out of proportion to its relative abundance in the Nanaimo River estuary. Chum salmon apparently consumed nearly the entire production of this copepod, leading Sibert to conclude that the emigration of chum salmon may have been related to limited food resources in the estuary. Likewise, Reimers (1971) concluded that density might have limited the growth of juvenile chinook salmon in the estuary of the Sixes River, and Wissmar and Simenstad (1988) reported that the densities of harpacticoid copepods declined at the end of the residence period of chum salmon. Thus the seasonal patterns of estuarine use by salmon may reflect the inherent changes in estuarine productivity, the densities of salmon (in numbers and biomass, as they arrive, grow, and depart or die), and the relative growth opportunities at sea in nearshore and oceanic habitats.

In evaluating the role of estuaries in providing food for salmon, it is important to remember that salmon are not the only members of the community. McCabe et al. (1983) caught thirty-nine different fish species in the Columbia River estuary from March through September 1980. They classified thirteen (including chinook, coho, and steelhead) as common in at least some habitats and time periods. In the pelagic region of the lower estuary, the catches were dominated by marine fishes, especially Pacific herring (*Clupea pallasi*) and northern anchovy (*Engraulis mordax*). The upper estuary's intertidal areas were also dominated by non-salmonids, especially euryhaline species (i.e., tolerant of a range of salinities) such as starry flounder (*Platichthys stellatus*), threespine sticklebacks (*Gasterosteus aculeatus*), peamouth chub (*Mylocheilus caurinus*), and one nonnative species, the anadromous American shad (*Alosa sapidissima*). The lower intertidal areas were also dominated by euryhaline non-salmonids, including starry flounder, threespine sticklebacks, shiner perch (*Cymatogaster aggregata*), staghorn sculpin (*Leptocottus armatus*), and also marine species, notably Pacific herring and surf smelt (*Hypomesus pretiosus*). Bottom and Jones (1990) characterized the fish community of the Columbia River estuary as typical of coastal estuaries; they noted that this reflects the dominance of marine species from the continental shelf and euryhaline species with broad distributions, though they also noted that larger estuaries tended to contain more species. High abundance of non-salmonids is common to many estuaries. For example, Gray et al. (2002) reported tenfold higher densities of staghorn sculpins than chinook salmon in the Salmon River, Oregon, estuary.

Examination of the diets of these non-salmonids and the juvenile chinook salmon indicated very little predation on salmonids by fishes and varying degrees of diet overlap (hence, potential for competition; McCabe et al. 1983; Bottom and Jones 1990). In intertidal habitats, diet overlap for ocean-type chinook salmon was greatest with starry flounder, and in pelagic habitats overlap was greatest with shad, sticklebacks, herring, shiner perch, and longfin smelt, in addition to competition with other salmonids. Amphipods, especially the tube-dwelling *Corophium salmonis* and *C. spinicorne*, were heavily preyed on by salmonids and non-salmonids alike, and the freshwater cladoceran *Daphnia* that is favored by sockeye in lakes was also prey to many species. The salmon diets in the Columbia River differed from those in other estuaries, though the general patterns of habitat use and distribution of euryhaline fishes were similar. Many factors influence the diets of fishes, including the relative abundance and behavior of the predators and prey, physical structure in the habitat, temperature, salinity, turbidity, and so on, and estuaries are no exception in this regard (see review by Levings 1994).

In addition to the role of estuaries in feeding and growth, estuaries are widely viewed as providing a gradual transition from fresh- to saltwater, easing this stressful period for the fish. In estuaries, salmon can regulate the salinity around them by moving up and down only a few meters if they are near the interface between fresh surface water and saltier water below, as Iwata and Komatsu (1984) observed chum salmon doing after release from a hatchery. The progressive shift toward more offshore water seen in chinook salmon (e.g., Healey 1980; McCabe et al. 1986; Levings et al. 1986) over the course of the season is also consistent with progressive tolerance for higher salinities. Salmon seem to prefer intermediate salinities as they reside in and move through estuaries, and the tendency of chinook to occupy lower salinities than coho (e.g., less than 5 vs. greater than 10 ppt in the Campbell River estuary: Macdonald et al. 1987) is consistent with the idea that coho are more ready for the ocean and move out more rapidly than chinook. Moreover, Campbell River chinook salmon taken from freshwater and put directly into cages in seawater had much higher mortality rates than those moved to riverine or estuarine sites (Macdonald et al. 1988).

McInerney (1965) conducted laboratory experiments to measure salinity preference of juvenile salmon over time. Chum salmon, held in freshwater, shifted their preferred salinity from freshwater in May to 3 ppt in June, 6 ppt in July, 8 ppt in August, and 10 ppt in October. At no time did the chums or any other species of salmon prefer to make the abrupt shift from fresh- to full-strength seawater (about 33 ppt), and McInerney concluded that the progressive shift in salinity preference might facilitate migration through estuaries. Under normal conditions, the fish would seek increasingly salty water as they became able to regulate their internal salt concentration. However, some rivers have such small estuaries that juvenile salmonids must make a rather abrupt transition to seawater. They can do so, though it may not be their preference.

The water in estuaries is commonly quite turbid, as fine material drifting downriver meets a "null zone" of velocity, and shifting riverine and tidal currents keep it suspended. This turbid water may provide a refuge from predators, and this is the third role that estuaries are commonly hypothesized to play in the lives of juvenile salmon (e.g., Simenstad et al. 1982; Thorpe 1994). However, it is difficult to find hard evidence for this function because the magnitude and agents of mortality in estuaries are so poorly known and may be difficult to separate from marine mortality. Estuaries may have cutthroat trout and other piscivorous salmonids, benthic species such as staghorn sculpins that could eat salmon, and other potential fish predators. McCabe et al. (1983) reported little apparent predation by fishes in the Columbia River estuary but this may not be representative of the many much smaller estuaries and wetlands used by salmon as migratory corridors or temporary rearing areas.

Estuaries often have high densities of piscivorous resident and migratory birds, including wading birds such as great blue herons and also swimming and diving birds such as mergansers, kingfishers, cormorants, gulls, and terns. Thus estuaries may not be very safe places for salmon, and evidence for predation by birds in the Columbia River's estuary has come to light in recent years. Material dredged to maintain the navigational channel in the lower 168 km of the river and estuary has been deposited to form several small, artificial islands. For example, 120-ha Rice Island was created in 1962 at river km 34. By 1987 there were colonies of Caspian terns (*Sterna caspia*) and double-crested cormorants (*Phalacrocorax auritus*) and by 1997 there were more than 9000 terns and 1000 cormorants (Ryan et al. 2001). Examination of the islands using electronic detectors revealed large and increasing numbers of passive integrated transponder (PIT) tags that had been inserted into juvenile salmon. In 1999–2001, about 4% of the tags placed on salmon in the Columbia River system were recovered from bird colonies, implying a minimal predation rate of greater than 4% (Ryan et al. 2001, 2003; Brad Ryan, National Marine Fisheries Service, personal communication). Interestingly, the predation varied markedly among salmonid species; sampling for PIT tags on bird colonies produced 11.5% of the steelhead, 4.6% of the coho, and only 2.6% of the yearling chinook salmon detected migrating past Bonneville Dam (Ryan et al. 2003). It is unclear what factors (fish size, timing of migration, proximity to the surface or shoreline, etc.) make steelhead so vulnerable, and chinook much less so.

These data reveal the capacity of birds to prey on salmon but they should not be taken as characteristic of either the Columbia or other rivers. The open-ended nature of estuaries makes it possible to count the fish going in but very difficult to count those leaving, and so makes it difficult to determine the rates and especially the agents of mortality. Given this uncertainty, is there evidence that estuaries play a critical role in

FIGURE 13-3. Estimated lengths of juvenile chinook salmon of life history-type 3 leaving the Sixes River (left panel) and estimated growth in the estuary (right panel) of the juveniles as a whole (sample = 74) and those fish surviving to adulthood (sample = 145). Note that the fish surviving to adulthood are overrepresented among the larger individuals (from Reimers 1971).

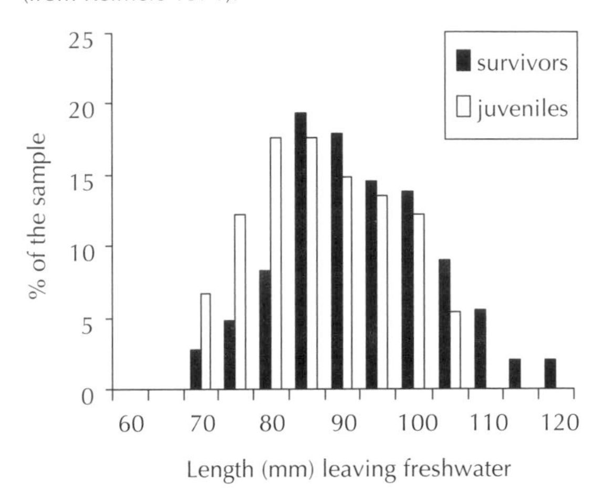

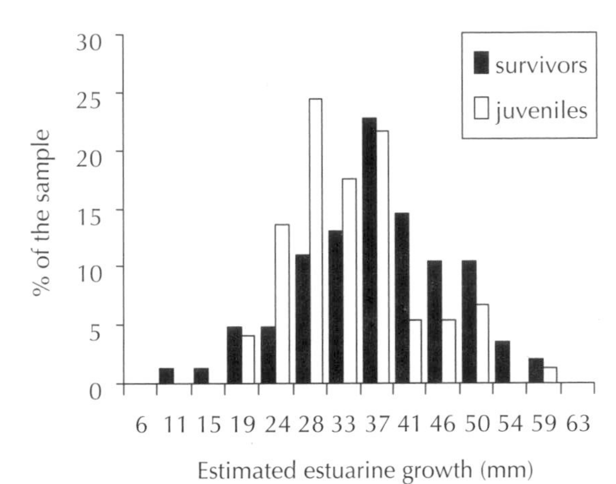

the lives of salmon? Work by Reimers (1971) on the Sixes River provided important though indirect evidence that residence and growth in the river and estuary increased the survival rates of juvenile chinook salmon. He measured juvenile salmon in the river and its estuary, carefully examined the number and spacing of circuli (rings) on their scales, and identified five life-history types of juveniles, varying in use of the river and estuary for growth. Type-1 fish migrated through the river and out to sea as newly emerged fry; type-2 fish reared in the river or its tributaries until early summer, reared for a short time in the estuary, and moved out to sea in midsummer; type-3 fish reared in freshwater until early summer and in the estuary until fall; type-4 fish stayed in freshwater until fall and then migrated out to sea without significant use of the estuary; and type-5 fish spent a year rearing in freshwater and migrated directly to sea as yearling smolts. There was no quantitative estimate of the proportions of these types of juveniles but Reimers ranked the types from most to least abundant as: 2, 3, 4, 5, 1. However, the vast majority of adults had scale patterns indicating that they were of type 3, suggesting higher survival of this group.

Reimers also compared the counts of circuli on scales of type-3 juveniles in one year to those of adults of this type in the subsequent years when they returned. The circuli counts (indicating fish size) showed that the surviving adults had grown more in the river and also in the estuary than the average for the juveniles of their type (fig. 13-3). Specifically, the surviving adults were significantly larger than the population's overall average when they left the river (92.4 vs. 86.7 mm long), and the survivors grew more in the estuary as well. This difference may seem slight but much smaller differences can affect survival at sea (see chapter 15).

It seems very likely that salmon, especially ocean-type chinook and chum, derive significant benefits from estuarine residence, though many important aspects need further work. Evidence supporting this idea was provided by Magnusson and Hilborn (2003), who showed that the survival of ocean-type chinook (but not coho) released from hatcheries was positively correlated with the percentage of the estuary in natural condition. It should be noted that in parts of the range of salmon there have been dramatic and in some cases irreversible changes in estuaries. Maser and Sedell (1994) pointed out that the lower portions of rivers and estuaries on the forested coastline had dense concentrations of large woody debris but much of the material was removed in the late nineteenth and early twentieth centuries. This reduction in wood and the removal of mature trees from the riparian zones of forests (that might have eventually fallen and drifted into the estuary) has resulted in structurally simpler estuaries, and this may have had consequences for foraging and predator avoidance by salmon.

In addition to such losses of woody material from coastal estuaries, some estuaries have been permanently altered or lost to human development in the form of ports, industrial areas, farms, and residences. Hutchinson (1988) estimated the area of all intertidal marshes greater than 50 ha in the Puget Trough at the time of European settlement and at the present. The data indicated a reduction from 4453 to 3636 ha in the Strait of Georgia and from 9360 to 3887 in Puget Sound, for an overall loss of 45%. Bottom et al. (2001) conducted an especially thorough review of the history of development of the Columbia River estuary, with special emphasis on chinook salmon. This system has not only been greatly modified by dikes, industrial development, farming, port facilities, and cities, as with many estuaries, but it has also been affected by the great reductions in peak discharge as the river was harnessed for hydroelectric power production. The human activities in estuaries are so numerous and complex that it proves difficult to ascribe losses of salmon abundance to one specific activity or another. However, Bottom et al. (2001) concluded that the complexity of the predevelopment estuary was important for the diverse life-history patterns of chinook salmon and that the simplification of the estuary has contributed to the simplification of life-history patterns seen today. In recognition of these losses, there have been efforts to understand the functional roles of estuaries and to use this information to design and evaluate wetlands created to mitigate losses (e.g., Shreffler et al. 1990; Miller and Simenstad 1997).

Summary

Estuaries are an important transitional habitat, used to different extents by different species and populations. In general, yearling and older smolts (sockeye, coho, steelhead, and chinook) make less use of estuaries than young-of-the-year chum, coho, and especially ocean-type chinook salmon. Growth is rapid in estuaries, and analysis of scale patterns of smolts and surviving adults indicates that larger individuals are more likely to survive at sea than smaller ones. However, the vast majority of salmon vacate estuaries by mid to late summer; the density of salmon and seasonal changes in food availability may provide the ecological pressure to leave. Estuaries may also benefit the smolts by easing the transition from fresh- to saltwater, and the fish tend to move downstream and offshore as they grow and as the season progresses. With these changes in habitat and fish size come changes in diet, and the smolts feed on a combination of insects, benthic and planktonic crustaceans, and small fish. The predator-prey relations in estuaries are not well known. The turbid water in some estuaries may facilitate predator avoidance, but estuaries also abound with birds and other predators on salmon.

14

Marine Migration Patterns

The International North Pacific Fisheries Commission (INPFC) was responsible for much of the early work on the movements of juvenile salmonids at sea. This work, reported in detail by Hartt and Dell (1986), involved extensive field sampling along the North American coast and in offshore waters and relied chiefly on catch rates to infer migration routes from the changing patterns of apparent abundance (though mark-recovery work contributed considerably to the emerging picture). From July to September, there seems to be a particularly dense concentration of salmonids from about Cape Flattery, Washington, to the eastern Aleutians. The band is narrow (30–40 km wide) off British Columbia and southern Alaska and widens with the continental shelf in the Gulf of Alaska. For example, twenty-nine purse seine sets were made off Baranof and Chichagof islands in southeast Alaska in August and September 1964 and 1965. The average catch rates of salmonids in their first year of marine life were 434 fish per set between 7 and 18 km from shore, 267 fish per set 20–30 km offshore, and only 1.3 fish per set 42–50 km offshore. If we make some assumptions about the efficiency of the gear and other factors, a catch rate of 350 fish per set correspond to about 0.0015 salmon per square meter, or 681 m^2 per salmon. In the area sampled by Hartt and Dell, the dominant species were pink (49.4%), sockeye (26.7%), and chum (13.0%), followed by coho (10.5%), chinook (0.3%), and steelhead (less than 0.1%). The relative abundance varied considerably among areas; more sockeye were caught in the Bering Sea, more pinks in southeast Alaska and British Columbia (table 14-1), and chums were rather evenly distributed.

Sampling in the 1980s (Jaenicke and Celewycz 1994) in an area of southeast Alaska that overlapped with the southern area sampled by Hartt and Dell (1986) also produced primarily pink salmon and a higher relative abundance of coho than was found farther up the coast. Jaenicke and Celewycz, however, caught significant numbers of juvenile

TABLE 14-1. Catches of juvenile salmon in the summer by purse seines along the coast of North America: 1964–1968 (Hartt and Dell 1986, indicated by H&D), 1983 and 1984 (Jaenicke and Celewycz 1994, indicated by J&C), and 1981–1985 (Pearcy and Fisher 1990, indicated by P&F). The nets were not identical, so only relative catch rates should be compared among the studies. Cutthroat were the remaining 1.59% of the catch by Pearcy and Fisher.

	% of the catch in each area						
Area (source)	*sockeye*	*chum*	*pink*	*coho*	*chinook*	*steel-head*	*catch/set*
Bering Sea (H&D)	91.7	7.0	0.5	0.8	< 0.1	0.00	154.3
Northern Gulf of Alaska (H&D)	38.3	5.1	48.6	7.8	0.2	0.00	82.6
South side, Alaska Peninsula (H&D)	45.6	12.7	36.8	4.9	0.00	0.00	127.0
Yakutat, Alaska to Cape Flattery, Washington (H&D)	13.9	14.4	58.8	12.5	0.4	< 0.1	229.4
Southeast Alaska (J&C)	7.07	10.32	70.37	11.65	0.59	no data	38.3
Washington and Oregon (P&F)	1.1	6.6	4.1	64.3	21.0	1.3	11.5

salmon 74 km offshore (as far as they sampled) in August and concluded that the fish were distributed farther offshore than was indicated by Hartt and Dell. Still farther south, summer sampling by Pearcy and Fisher (1990) off Washington and Oregon revealed mainly coho and, to a lesser extent, chinook, and very few sockeye, chum, or pink salmon.

Jaenicke and Celewycz (1994) examined their data on catches of juvenile salmon off southeast Alaska for evidence of aggregation within species. They compared the distribution of catches per set (that is, how many sets caught no pink salmon and various numbers of pinks) to what would have been expected by chance given the overall catch rate. The results indicated that all five species were aggregated rather than randomly distributed and that the pink salmon were most highly aggregated, followed by chum. Sockeye and coho were least strongly aggregated; too few chinook were caught to perform such analyses. Patterns of association among species, inferred by looking at the relative catches of different species in different sets, indicated that pink and chum salmon were most often associated with each other, followed closely by pink-sockeye and then sockeye-chum. Coho were not associated with any of the other three species.

In addition to the general existence of this "stream of salmon," Hartt and Dell (1986) drew some inferences about movement patterns from paired purse seine sets made in opposite directions. There were nineteen pairs of sets between Cape Flattery, on the northern coast of Washington, and Yakutat, in south-central Alaska, in July to September from 1964 to 1967. The vast majority (83%) of the fish were taken in southeast facing sets (i.e., the fish were tending to swim northwest), and only 17% were taken in the sets with the net open to the northwest. This pattern did not arise from a few unrepresentative sets; the southeast-facing set caught more fish than the northwest-facing set in sixteen of the nineteen pairs. Estimates of migration rate vary greatly but most are on the order of 1–10 km per day (Brodeur et al. 2003).

The invention and widespread use of coded wire tags to mark hatchery-produced (and, to a much lesser extent, wild) salmon, made it possible to know the origin of fish caught at sea. Most tags are recovered in commercial and recreational fisheries, so little is known about salmons' first few months at sea compared to information on the distribution of maturing fish. Moreover, the coded wire tags have been most widely applied to coho and chinook salmon, so information on the other species is more limited. In recent years techniques have been developed to induce banding patterns on the otoliths of salmon embryos and alevins prior to release in hatcheries, so the entire output of a hatchery can be marked, improving our understanding of the marine ecology of salmon. Recent reviews have summarized the research on early marine life of salmon conducted by Canadian (Beamish et al. 2003), Japanese (Mayama and Ishida 2003), Russian (Karpenko 2003), and American scientists (Brodeur et al. 2003), and these papers provide a wealth of details, references, and thoughtful suggestions for future research directions.

The migration patterns of salmon and trout at sea are as varied as any other aspect of their biology, but research suggests that these patterns can be broadly put into four groups: (1) sockeye, pink, and chum, (2) coho, chinook, and masu, (3) steelhead, and (4) cutthroat and Dolly Varden. These patterns will be described, with emphasis on the initial migration to sea and the subsequent movements that lead to the locations from which the maturing salmon began their journey in chapter 2.

Sockeye, pink, and chum salmon

These three species, comprising the vast majority of salmon, enter the eastern Pacific Ocean from Oregon northward (mostly from Puget Sound and southern British Columbia to western Alaska). Chum often spend time in estuaries, but the migration of all species is characterized by relatively directed, rapid movement northward through inside passages and along the continental shelf during these fishes' first summer at sea, and then by distribution in the offshore waters of the Bering Sea, Gulf of Alaska, and open North Pacific Ocean until they mature. These species are seldom found in coastal waters such as Puget Sound, the Strait of Georgia, or the inside waters of southeastern Alaska after their first fall at sea until they return to spawn. Sockeye salmon from Bristol Bay (Alaska) and the Fraser River and pink salmon from southeast Alaska will serve as examples.

Juvenile sockeye salmon enter Bristol Bay from five main areas: Togiak, Nushagak, Naknek-Kvichak, Egegik, and Ugashik (see fig. 3-1). In general, age-2 smolts precede age-1 smolts, and smolts from complex, multilake systems like the Wood River, part of the Nushagak River system, tend to migrate later than those from lakes closer to the ocean (Rogers 1988). Thus there is some initial segregation of populations and age groups, but all seem to migrate within about 48 km of the north side of the Alaska Peninsula, in the upper 5 m of the water column (Straty and Jaenicke 1980). The Naknek-Kvichak, Egegik, and Ugashik populations swim along the peninsula and the populations on the northwestern side (mostly from the Nushagak) migrate southeast to join them, along a gradient of increasing salinity. These routes keep the young salmon in water that is more turbid and lower in salinity than the water farther offshore, where the adults are migrating homeward. Food is scarce and little feeding seems to occur in June, but by July the young salmon have largely moved into clearer, more oceanic and productive waters in outer Bristol Bay. They then seem to move primarily south into the Gulf of Alaska and North Pacific Ocean, but some remain in the Bering Sea.

Extensive sampling of Fraser River populations by Groot and Cooke (1987) also indicated a directed migration. Aware of the variable proportion of adults that return to the Fraser River via the outside of Vancouver Island and the San Juan Islands versus via Johnstone Strait and the Strait of Georgia (described in chapter 3), Groot and Cooke wanted to determine if the migration of adults reflected the route they took as smolts on the outward journey. Contrary to this hypothesis, virtually all smolts appeared to migrate through the northern route. Sockeye smolts started to enter the Strait of Georgia in the last 2 weeks of April. By early to mid-May, two centers of distribution were evident, one northwest of the river on the British Columbia mainland and the other southwest of the river, in the Gulf Islands. By early June, no smolts were found off the mouth of the river. The smolts that had been on the mainland coast north of the river had moved up into Johnstone Strait, and the southern mode had moved north through the islands and crossed the Strait of Georgia to the mainland side. Groot et al. (1989) concluded that the fish, moving about 6–7 km/d, had a northward orientation, and this explains the group migrating directly north along the mainland. However, the net movement of fish is the resultant of two vectors: the speed and direction of their swimming and those of the water mass in which they are moving. For small fish in coastal waters, the water's movement can be a large or even dominant component of travel until the fish grow larger and move into slower-moving water masses. Groot and Cooke (1987) noted that the Fraser River's flow, combined with an ebb tide, might displace some of the fish westward, where they would "regroup" and then head north to join the others migrating along the mainland coast. Simulations by Peterman et al. (1994) supported their hypothesis of strong northwesterly orientation by the fish but concluded that the water movement (and hence displacement of fish) resulted more from wind patterns than river flow and tides.

A final example of the orientation of juvenile salmon is that of pinks migrating through the maze of inlets and islands of southeast Alaska. Jaenicke et al. (1984) set paired beach seines in opposite directions along stretches of shoreline in Chatham Strait, west of Juneau. The pink salmon showed a strong northward tendency; nets open to the south caught 82% of the 77,828 salmon in 1981 and 95% of 40,977 salmon in 1982 in False Bay. The fish were not drifting with prevailing currents but swimming actively. In 1982, the predominant direction was against the current in twenty-two of thirty-two sets. More equal proportions of the fish were moving north and south elsewhere, especially at a site with more complex currents, and the authors concluded that in some areas the fish mill around and at other sites they migrate with a strong northward orientation.

Thus, pink, sockeye, and chum salmon enter the ocean from thousands of streams along the coast and, after some delay in the estuary by the chums, they migrate northward during their first summer at sea. By the fall they seem to be concentrated in the northern Gulf of Alaska and then move offshore. The winter distribution of salmon at sea is very poorly known, as the conditions are inhospitable for researchers. The typical models (e.g., French et al. 1976; Healey and Groot 1987) hypothesize that the salmon move southward in the winter and then northward the following summer. Pink salmon would then migrate homeward that summer and fall, but the sockeye and chum would stay at sea for another year or more, moving westward. However, it is also possible that the salmon might disperse and even move north in the winter rather than south. This seems counterintuitive because we tend to regard cold as undesirable, but if food is scarce the fish may be better off in cold water, where their metabolism is slow, than in warmer water where they would need more energy to maintain themselves. Fundamentally, we still know very little about the movements of immature salmon at sea, and especially their behavior and ecology in winter.

In addition to the migrations of North American salmon, there are also large populations from Asia, especially chum salmon from hatcheries in Japan and wild salmon from Russia and in particular from the Kamchatka Peninsula. Their migrations take them into the open ocean, where they overlap with North American salmon (e.g., see fig. 2-1). There is a general tendency for Asian salmon to migrate farther toward North America than the reverse, as indicated in the INPFC studies described in chapter 2. Presumably, the migration patterns of species and populations have evolved to optimize the use of the "ocean pasture." Areas of the ocean differ in productivity and temperature, but if all the salmon went to the best places then food might be scarce. The extent to which the migrations of salmon shift in response to short-term and long-term patterns of competition, food resources, and temperature are poorly known but would seem to be a suitable topic for future fieldwork and models.

Chinook, coho, and masu salmon

The relatively rapid, northward and then offshore migrations of pink, chum, and sockeye salmon differ from those of coho and chinook (and masu in Asia). These species travel more slowly at sea and have a much greater tendency to remain along the continental shelf than the other three. To be sure, there are coho salmon in the open ocean, and chinook as well. Healey (1983) inferred from various sources of information that stream-type chinook salmon use the open ocean to a much greater extent than ocean-type fish. However, this does not seem to be an absolute pattern, as some stream-type chinook (e.g., from the Columbia River) are caught in coastal fisheries from Oregon to Alaska. The tendency for ocean-type fish to have a coastal distribution may result from their concentration in the southern end of the range, where offshore waters are too warm and unproductive for them. These populations, therefore, distribute along the coast whereas stream-type populations, the form that predominates in the north, can move offshore to rear. Likewise, is it unclear why some coho salmon use the open ocean while most are found near the coast. Putting this puzzle aside, we will focus on the coastal migrations, as we know them best.

Off the coast of Oregon, Pearcy and Fisher (1988) caught 98% of their juvenile coho salmon in purse seine sets facing south, indicating that the fish had been swimming northward. Similarly, D. R. Miller et al. (1983) reported that 76% of the coho and 80% of the chinook were caught in sets facing south in May and June. Pearcy and Fisher (1988) also inferred movement patterns by recovery of young salmon with coded wire tags indicating their origin, and this presented a more complex picture. Columbia River coho were exclusively (25 of 25) caught south of the river in May, primarily (30 of 40) south of

TABLE 14-2. Relationship between the origin of coho salmon and the coastal area where they were caught. Data represent the averages from 2 hatcheries in Alaska, 10 in British Columbia, 33 in Washington, 16 in Oregon, and 4 in California (from Weitkamp and Neely 2002).

	Catch area				
Origin	*Alaska*	*British Columbia*	*Washington*	*Oregon*	*California*
Alaska	98.8	1.3	0.0	0.0	0.0
British Columbia	5.6	90.1	4.1	0.2	0.0
Washington	0.1	37.9	42.3	17.8	1.8
Oregon	0.0	4.0	14.8	58.5	22.7
California	0.0	0.0	0.3	17.3	82.3

the river in June and July, but exclusively north of the river (16 of 16) by September. The authors concluded that the fish were swimming northward but that early in the season the winds generate strong southerly currents that displaced the fish to the south. Later in the summer, the currents weakened and the fish grew (and hence became stronger swimmers), and their distribution shifted northward.

Recoveries of coho salmon with coded wire tags in coastal fisheries provide further insights into movement patterns. Weitkamp and Neely (2002) showed a marked association between the locations of sixty-five hatcheries from Alaska to California and the area where the maturing coho were caught. Most coho were caught in the broad region of the coast where they originated (table 14-2), but individual populations varied (fig. 14-1). Some salmon moved long distances along the coast but most moved shorter distances to the south or north. However, it is unclear where the fish were before they were

FIGURE 14-1. Catch distributions of coho salmon originating from 65 hatcheries, arranged along the x-axis in order from north to south, from Alaska (AK), British Columbia (BC), Washington (WA), Oregon (OR) and California (CA) (from Weitkamp and Neely 2002).

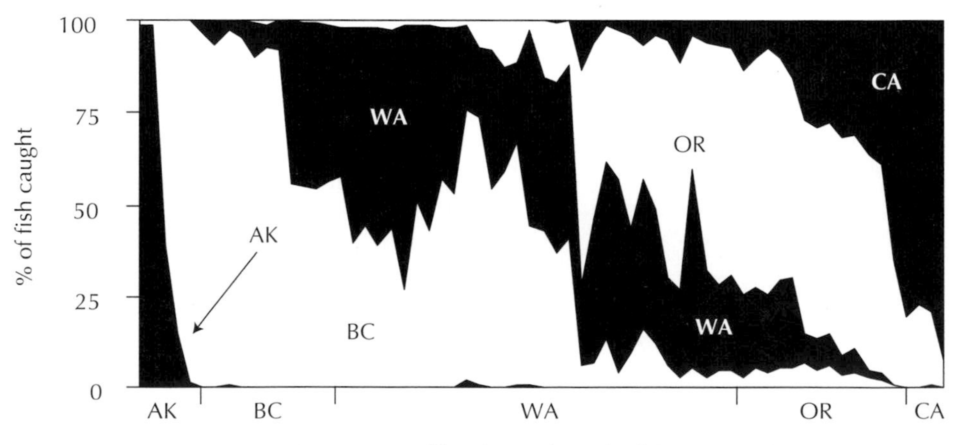

caught. It is not safe to assume that they were between their origin and capture location, as they may have been caught on their way home, having been feeding still farther away.

The marine distribution of chinook also varies with the latitude where they entered the ocean. Healey and Groot (1987) showed that most juveniles rearing in southeastern Alaska had originated in Alaska, British Columbia, and Washington; most fish rearing in British Columbia had come from British Columbia and, to a lesser extent, from Washington, with few from Oregon; most rearing off Washington had originated there; fish rearing off Oregon had come from California, Oregon, and Washington; and most fish rearing off California had originated there but some had migrated south from Oregon and Washington.

At a finer scale, Nicholas and Hankin (1989) summarized a wealth of information on the life history of Oregon coastal chinook salmon populations, including the proportions recovered in fisheries from California to Alaska. The fishing effort is not uniform, so these data, like the coho data reported by Weitkamp and Neely (2002), reflect relative rather than absolute abundance. Nevertheless, they reveal two main migration patterns. Chinook over much of the coast tended to migrate north, with most of the adults caught off British Columbia (56%) and Alaska (25%). However, south of Cape Blanco, Oregon, the populations tended to migrate in a more southerly direction, with an average of 49% each off Oregon and California (fig. 14-2). The apparent division at Cape Blanco is interesting because coho populations originating south of Cape Blanco also showed much more southerly distributions than those north of it (Weitkamp and Neely 2002).

In addition to overall spatial distribution, we can infer movements over time using recoveries of salmon with coded wire tags. These data indicate the origin of the salmon and can be adjusted for the fraction of the salmon catch that gets sampled. However, it is more difficult to adjust for differences in effort among areas and years, as regulation, gear, and numbers of vessels change in complex ways. With this caveat in mind, there are some interesting patterns. As an example, I have looked at the coho salmon returning to the Soos Creek Hatchery, in central Puget Sound. As shown in table 14-3, catches were very heavy off the southwest coast of Vancouver Island in July, decreasing in August

Figure 14-2. Catch distribution of chinook salmon originating from Oregon hatcheries (from Nicholas and Hankin 1989).

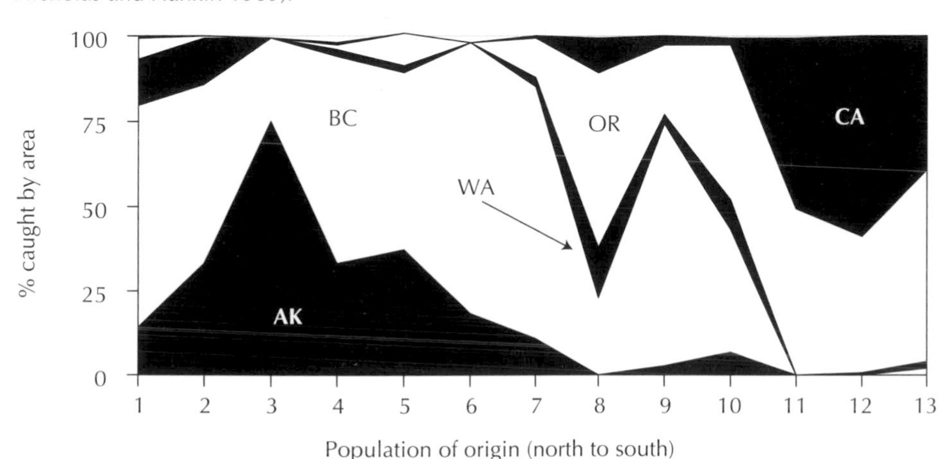

TABLE 14-3. Estimated catches of coho salmon from Soos Creek Hatchery, Puget Sound, Washington, based on coded wire tag data. The data have been adjusted for incomplete sampling of the catch but have not been adjusted for effort and so should not be taken to indicate densities, only general patterns, which are highlighted in bold.

Regions	*June*	*July*	*Aug*	*Sept*	*Oct*	*Nov*	*Dec*
SW Vancouver Island	984	**14,236**	**6557**	4148	135	0	0
Strait of Juan de Fuca	260	904	**4023**	**5746**	754	27	11
Puget Sound	207	379	650	**29,240**	**15,804**	**14,581**	1207
N Washington coast	271	**2423**	**2434**	376	0	0	0
S Washington coast	408	**2046**	926	158	5	0	0

and September, and negligible thereafter. The fish were also caught in July off the southern coast of Washington and off the northern coast in July and August, but were largely gone from these areas by September. In August and September they were caught in the Strait of Juan de Fuca, as would be expected for fish migrating from the coastal areas into Puget Sound. The catches were not very high in Puget Sound until September (despite considerable effort earlier in the summer, as indicated by catches of chinook salmon). Significant numbers were caught until November, when the fish were essentially all in their natal river. Similarly, Machidori and Kato (1984) concluded that masu salmon migrate within a restricted area. Anadromous individuals (predominately females in many populations) tend to spend one winter at sea, like the coho, and they are distributed primarily in the Sea of Japan and Sea of Okhotsk.

In addition to these migratory patterns, some coho and chinook salmon are caught in the vicinity of their natal stream outside of the spawning season. For example, some chinook salmon, known as residents or blackmouth, are caught in Puget Sound in the winter. These may be immature salmon that never left the area but it is also possible that there are seasonal migrations by immature as well as mature salmon. Coded wire tagging data on the University of Washington (UW) population showed that there were very few UW chinook in Puget Sound in spring and early summer. In fall, there was the expected migration of maturing adults, but in late fall and winter there were still UW chinook in central Puget Sound (Brannon and Setter 1989). These fish must have been immature, and Brannon and Setter hypothesized that a fraction of the population migrates south in Puget Sound each season, then back north in the spring and summer.

Steelhead

Steelhead are much less abundant than the semelparous salmon species, so data on their marine distribution are harder to obtain. Inferences about steelhead migrations are further complicated by the extremely broad period (essentially year-round) when maturing adults may be returning to coastal locations and by the presence of repeat spawners, as well as "maiden" fish returning for the first time, and kelts (post-spawning adults returning to sea). One might assume that steelhead, being the anadromous form of a species with many nonanadromous populations, might have a limited distribution at

TABLE 14-4. Purse seine catches per set of juvenile salmonids off Washington and Oregon in 1980, indicating patterns of relative abundance (Miller et al. 1983).

Species	*27 May–7 June*	*4–15 July*	*28 Aug–8 Sept*
chinook	6.67	2.36	6.38
coho	12.26	0.73	2.80
steelhead	5.20	0.10	0.00

sea. However, the North American populations tend to migrate through estuaries and quickly leave the coastal waters soon after entering the ocean. For example, purse seine sets off Washington and Oregon caught significant numbers of chinook and coho from late May through early September but the steelhead were essentially gone by July (D. R. Miller et al. 1983; table 14-4).

The distribution of North American steelhead has been demonstrated by recovery at sea of juveniles tagged in freshwater and by recovery of adults that had been tagged at sea (Burgner et al. 1992). In addition, many of the steelhead from the U.S. Pacific Northwest (Washington south to northern California, including Idaho) are infected by a parasitic trematode, *Nanophyetus salmincola*, but this parasite is absent in steelhead from Asia or elsewhere in North America. The parasite's range is controlled by that of its obligatory first intermediate host, snails of the genus *Juga*. The combination of tagging data and parasite prevalence demonstrated that North American steelhead travel almost as far west as 160° east longitude, south of the Kamchatka Peninsula. Notably, a steelhead tagged in the Quinault River, Washington, was recovered on the ocean 5370 km to the west. Steelhead are rare in the Bering Sea, however, because northerly populations of the species are seldom anadromous.

Not all steelhead undertake such lengthy migrations, and there are two specific exceptions to be recognized. First, in some northern California and southern Oregon rivers, a significant fraction of the population returns to freshwater after only a summer at sea (Kesner and Barnhart 1972). These fish are usually not sexually mature but spend the fall and winter in freshwater (feeding heavily, unlike mature fish that seldom feed) and leave again the following spring. The local term "half-pounders" refers to their approximate weight when they first return. The ecological pressures that give rise to this pattern are not known. In addition, steelhead in some Asian populations spend only the summer at sea and stay in coastal waters, returning at a small size when they first mature.

Cutthroat trout and Dolly Varden

If the scarcity of steelhead hinders our development of a clear picture of their marine migrations, the problem is even greater for cutthroat trout. However, their migrations seem distinctly different from those of steelhead. Despite the close relationship between these two species and many similarities in their freshwater ecology, cutthroat migrations at sea are more similar to those of Dolly Varden char. Cutthroat trout generally spend only the summer at sea rather than a year or more, as is the case with steelhead. In the Washington–Oregon area, cutthroat return to large rivers in the late summer and

TABLE 14.5. Catches per set of cutthroat and steelhead and numbers of purse seine sets off the Washington and Oregon coasts in 1981–1985, indicating patterns of relative abundance (Pearcy and Fisher 1990).

	May	*June*	*July*	*August*	*September*
cutthroat	0.18	0.18	0.34	0.21	0
steelhead	0.38	0.14	0.08	0.02	0
sets	180	327	130	66	152
% cutthroat	32.7	55.7	81.5	93.3	

early fall (chiefly September and October). In small streams that flow directly into saltwater, the cutthroat return in midwinter, having spent more time at sea (J. M. Johnston 1982). This difference in timing is probably related to the difficulty in accessing some small streams in early fall if water levels are low, but availability of prey resources might also play a role.

Purse seine sets off the Washington and Oregon coasts (Pearcy et al. 1990) caught a small but relatively steady number of cutthroat from May to August, but the fish were apparently gone (perhaps toward streams to overwinter) by September. In contrast to the cutthroat but consistent with the patterns seen by D. R. Miller et al. (1983), the steelhead catches dropped off dramatically after May, presumably because the fish had moved to the open ocean (table 14-5). Neither species was common near shore (less than 9 km), but most cutthroat were caught at intermediate distances (about 9–45 km offshore). Steelhead were more common, relative to cutthroat, farther offshore, consistent with the idea that the species have different distributions. J. M. Johnston (1982) hypothesized that the tendency of cutthroat to be found several kilometers offshore along the Washington coast may result from the exposed and inhospitable nature of the shoreline there. In contrast, in the more protected waters of Puget Sound cutthroat are found (and caught by anglers) very close to shore. The movements of cutthroat from inland waters and large estuaries are poorly known, but the fish may stay in protected waters all summer.

The movements of anadromous Dolly Varden at sea are as poorly known as those of cutthroat. Armstrong (1965, 1971) documented the movements of these species from and back to Eva Lake, in southeastern Alaska. Both species tended to leave in May and return in September, though the cutthroat trout were a bit later than the Dolly Varden (fig. 14-3). Armstrong (1984, 559) described the movements of Dolly Varden as "a manager's nightmare" because they were so complex. The Dolly Varden leaving rivers in spring included smolts going to sea for the first time, immature fish that had spent the winter in freshwater, and post-spawners. The fish returned to freshwater in the fall but not always to their home stream. If their home stream had a lake and they were going to spawn, they would spawn and then spend the winter in the lake. If they were mature but their natal stream did not have a lake, they would home to their natal river, spawn there, but then leave to spend the winter in a system with a lake. If they were not going to spawn, they would spend the winter in any system with a lake. More recent work by Bernard et al. (1995) supported many of these conclusions but added the further complexity that some spend the winter at sea, and DeCicco (1992) reported long-distance

FIGURE 14-3. Seasonal movements of anadromous Dolly Varden (Armstrong 1965) and cutthroat trout (Armstrong 1971) leaving and returning to Eva Lake, southeast Alaska.

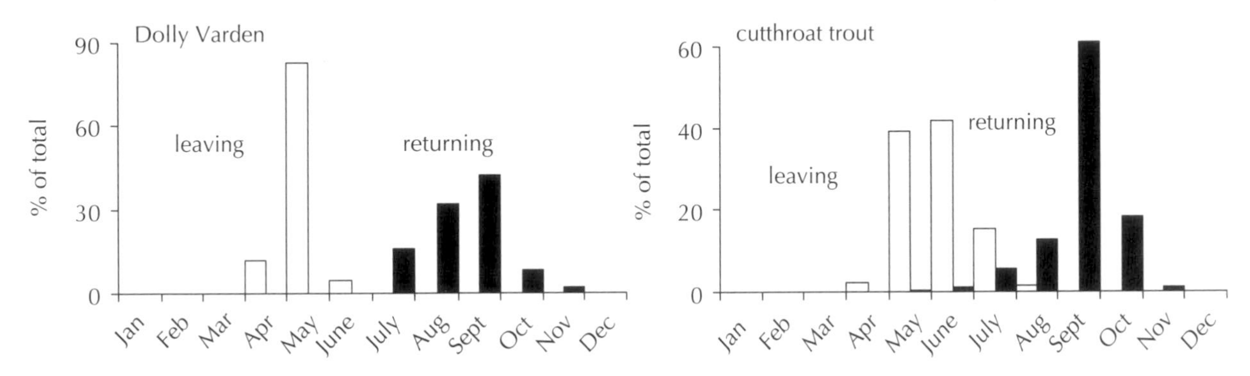

movements by some individuals. One remarkable char was tagged in the Wulik River, Alaska, in August and recovered 540 km up the Anadyr River, a total of 1560 km away, only 2 months later! Arctic char also seem to spend only the summer at sea, stay close to their natal river, and remain near shore (L. Johnson 1980), though there may be exceptions here as well.

In addition to the species-specific differences in distribution and migration, and differences arising from the location where the fish enter the ocean, there are genetic dispositions to migrate in certain areas. One example of this was documented in an early application of coded wire tags. Cowlitz and Kalama Falls spring (stream-type) chinook from the 1970 and 1971 brood years showed a more northerly distribution than Snake River fish (table 14-6). Such differences in distribution can occur in populations that are even closer together (geographically). The Toutle River is the main tributary to the Cowlitz River, which enters the Columbia River below Bonneville Dam. Hager and Hopley (1981) reported that the Cowlitz coho had a more northerly distribution than those from the

TABLE 14-6. Distribution (%) of catches of spring chinook released from different parts of the Columbia River system (Wahle et al. 1981), coho released from the Toutle and Cowlitz rivers in the lower Columbia River system (Hager and Hopley 1981), and chinook from the University of Washington (UW) and Elwha River populations (including hybrids), released from the UW hatchery, Puget Sound (Brannon and Hershberger 1984).

Species	*Origin*	*Alaska*	*British Columbia*	*Coastal Washington*	*Puget Sound*	*Oregon*	*California*
chinook	Cowlitz/Kalama	2.9	21.7	72.8	0	2.5	0.1
	Snake	1.1	9.0	50.0	0	27.7	12.2
coho	Toutle	0	0	33.5	0	55.7	10.8
	Cowlitz	0	0	70.1	0	29.1	0.8
chinook	UW x UW	0	43	6	50	1	0
	UW x Elwha	0	41	0	59	0	0
	Elwha x UW	0	31	0	69	0	0

Toutle (table 14-6). Additional evidence for the genetic control over distribution comes from the catches of different populations of salmon released from the same hatchery. Brannon and Hershberger (1984) spawned, reared, and released chinook salmon from the University of Washington (UW) hatchery that included representatives of the UW population, the Elwha River (on the Strait of Juan de Fuca), and their hybrids. The spatial distributions, inferred from catches, differed between the populations, and those of the hybrids were intermediate.

Summary

The migrations of salmon at sea remain poorly known, as does their marine ecology in general. Their distribution patterns show general species-specific tendencies: sockeye, chum, and pink salmon tend to migrate to the open ocean; coho and chinook use both coastal areas and the open ocean; steelhead generally migrate to distant marine areas and spent several years at sea before returning, but their relatives, the cutthroat trout (photo 14-1), seem to remain near the natal river and spend only a summer at sea, as do Dolly Varden and Arctic char. In addition, there is evidence that populations have genetically determined patterns of migration and distribution but that they are also affected by varying oceanic conditions. After reviewing some 50 years of research by their respective nations, Beamish et al. (2003; Canada), Brodeur et al. (2003; United States), Karpenko (2003; Russia), and Mayama and Ishida (2003; Japan) concluded that large-scale, holistic, ecosystem-oriented work linking physical processes, prey, competition, and predation is needed, rather than isolated, small studies. The role of estuaries and the transition from estuaries to the ocean is poorly understood, as is the relationship between distribution, growth, exposure to predators, and survival. The offshore ecology is largely a mystery, especially the distribution, movements, feeding, and mortality agents in winter. The effects of changing environmental conditions on migration, growth, and survival are recognized but not understood. There are significant problems of logistics and study design associated with ocean work but such research is critical if we are to truly advance our understanding of salmon behavior and ecology.

14-1. Anadromous cutthroat trout. Photograph by Michael McHenry, Lower Elwha Tribal Fisheries Office.

15

Survival in the Marine Environment

Most of the lifetime mortality of salmonids typically takes place in the gravel (see chapter 8), before salmon are even free-swimming, and many salmon perish while they are residing in freshwater or migrating to sea. However, the great majority of salmonids that migrate to sea do not return, and this chapter examines mortality (or, more optimistically, survival) at sea. The thrust of research on marine ecology is quite different from that conducted in freshwater. We have studied the ways in which human activities (e.g., mining, logging, and urbanization) have degraded or altered stream habitats and lakes, and we have studied ways to restore or otherwise modify these habitats. This focus on active management in freshwater systems contrasts with that in marine systems, where we are unable to modify the environment itself and so we primarily seek to understand the processes. With this understanding, we try to predict the survival rates of salmon at sea to better manage the fisheries that depend on their return to coastal waters and rivers after the natural mortality has taken place. We also try to better prepare the fish for the ocean by manipulating their size or condition when released from hatcheries. This chapter first considers the average proportions of salmonids of different species that survive at sea and then examines the patterns, processes, and sources of variation.

I obtained data on survival of wild salmon populations from a wide variety of sources and periods of record (including but not limited to those reviewed in Groot and Margolis 1991, Bradford 1995, and McGurk 1996), and I then averaged these population-specific values (table 15-1). The methods and periods of record varied among studies, and few reported more than one life-history stage for the population. In some cases the data are sparse, but despite these drawbacks, several patterns are evident. First, the species that migrate to sea at a larger size tend to have higher marine survival rates. Steelhead and cutthroat tend to spend 2 or more years in freshwater and go to sea at a large size; coho and

TABLE 15-1. Average stage-specific survival rates for different species of salmon, calculated from 215 published and unpublished estimates for wild or naturally rearing populations. Female fork length and fecundity are based on the average of population-specific average values. Sizes of eggs (Beacham and Murray 1993), fry (Beacham and Murray 1990), and smolts are averages of reported values or general figures but obviously vary among populations. For coho, chinook, steelhead, and sockeye, some studies reported egg–smolt survival, so separate overall estimates were used combining egg–fry and fry–smolt and using just egg–smolt estimates. No studies were found for cutthroat egg–fry, fry–smolt, or egg–smolt survival.

Life-history stage	*pink*	*chum*	*sockeye*	*coho*	*chinook*	*steelhead*	*cutthroat*
Female length (mm)	522	683	553	643	871	721	413
Fecundity	1648	2876	3654	2878	5401	4923	1197
Egg size (mg)	190	290	130	220	300	150	110
Egg-to-fry survival	0.115	0.129	0.127	0.253	0.380	0.293	
Fry size (mm)	32	34	28	33	35	28	25
Fry-to-smolt survival			0.258	0.165	0.101	0.135	
Smolt size (mm)	32	40	80	105	60–120	200	190
Smolt-to-adult survival	0.028	0.014	0.131	0.104	0.031	0.130	0.198
Adults per female	5.2	5.1	15.7	12.5	6.4	25.5	
Egg-to-smolt survival			0.021	0.018	0.104	0.014	
Adults per female			10.2	5.4	17.5	9.2	

sockeye tend to spend 1 or 2 years in freshwater and are somewhat smaller; chinook vary in size as smolts; and pinks and chums go to sea at the smallest size. Second, there seems to be an effect of the length of time spent at sea: chinook spend more time at sea than pinks and so, despite being larger, have similar survival rates. In the broader context of overall mortality, we can see that the species that spawn at high densities (sockeye, chum, and pink) tend to have lower survival rates during incubation than the species whose densities are lower. If we include average fecundity data for each species, we can construct a crude life table. This does not include any density-dependent effects, and the values for most populations were recorded under conditions of fishing, so a typical female's eggs are more than sufficient for replacement (i.e., greater than 2.0 to return a female and a male).

For pink and chum salmon, the table results in plausible levels of lifetime productivity (2 adults per female would replace the population, so values around 5 would support fishing levels of 60%). The value for coho based on egg-to-smolt survival seems reasonable, as does the chinook value using all the stage-specific data. The overall productivity estimates for sockeye and steelhead, and some of the estimates for coho and chinook, seem implausibly high. This may result from unrepresentative periods of record, small sample sizes, biased methods, or other problems, though some of the best data are available for sockeye so it is not clear why these might be erroneous.

Size, timing, and survival

Most of the research on marine mortality has fallen into three categories: (1) attributes of the fish, such as their size, physiological condition, or migration date, that are associated

TABLE 15-2. Average marine survival (%) for sockeye salmon smolts as a function of size (small: < 85 mm, medium: 85–114 mm, large: > 115 mm) and latitude (southern: ≤ 55° N, central: 55°–60° N, and northern: > 60° N; Koenings et al. 1993).

Smolt size	*Southern*	*Central*	*Northern*
Small	5.9	16.7	17.7
Medium	12.6	25.1	32.0
Large	17.1	37.0	39.1

with survival; (2) correlations between environmental conditions and survival; and (3) studies devoted to particular species whose predation on salmon might be especially significant. It has long been recognized, in ecological studies of many organisms, that size is related to survival. As animals grow, they are better able to escape from potential predators, and they also become either too big to be eaten or big enough to defend themselves. Salmon are not well equipped to fight for survival, but swimming speed increases with size and they eventually grow to exceed the mouth gape of many predators. McGurk (1996) combined data from different species and populations showing that larger smolts have higher survival rates than smaller ones. Some of the best data on wild salmon are for sockeye because the populations are large and valuable enough to warrant systematic monitoring. Koenings et al. (1993) reported a strong relationship between size and survival among populations, though northern populations had higher survival rates than southern ones (table 15-2). The effect of size was largely driven by the nearly twofold greater survival of age-2 compared to age-1 smolts (see also table 12-5). However, the relationship was not linear; above about 100 mm there seemed to be no increase in survival rate and there was even some evidence of lower survival rates among the largest smolts.

In addition to evidence at the population level, there is also evidence for the increased survival of large individuals within a cohort. For example, older steelhead smolts from California had much higher marine survival rates than younger ones (table 15-3). One might think that the modal age of smolts should match the age conferring maximum survival but this is not the case. The reason is that there is also mortality in freshwater. So, the fish have to balance the survival advantage at sea associated with large size against the risk of mortality during the additional year(s) in freshwater needed to grow that

TABLE 15-3. The proportion of steelhead smolts aged 1–4, their modal lengths, and survival rate to adulthood in Waddell Creek, California (Shapovalov and Taft 1954).

Smolt age	*% of smolts*	*Modal length (mm)*	*% marine survival*
1	59.8	100	2.4
2	37.3	160	5.8
3	2.9	220	18.1
4	0.03	270	16.7

large. Thus the majority of steelhead smolts left after only 1 year in the creek, even though their odds of surviving at sea would have more than doubled after an additional year and would have increased more than sevenfold after 2 more years.

One clever method for studying selective mortality is to measure the size of individual smolts and the radius of their scales, and then to examine the scales of adult salmon. A mark is apparent on the scale, indicating the stage when an adult was a smolt. By measuring the distance from the center of the scale to this mark, one can estimate how large that fish was as a smolt. This technique, known as back-calculating, allows comparison of the size distribution of all smolts with the distribution of the survivors (i.e., those measured as adults). Such studies have shown a size advantage for survival of chinook (Reimers 1971), chum (Healey 1982b), steelhead (Ward et al. 1989), coho (Holtby et al. 1990), sockeye (Henderson and Cass 1991), and white spotted char (Yamamoto et al. 1999).

Kaeriyama (1999) reported higher marine survival rates in years when chum salmon released from Japanese hatcheries were larger, and positive relationships between size and survival have also been reported for cutthroat trout (Tipping and Blankenship 1993) and masu salmon (Miyakoshi et al. 2001). These results, combined with the common observation of size-biased survival within cohorts, would suggest that variation in survival among years within wild populations might vary with average size. However, neither Henderson and Cass (1991) nor Holtby et al. (1990) found a relationship between average smolt size and survival among years within wild populations of sockeye and coho, respectively. Presumably, other more important factors affected survival among years for those populations.

The link between size and survival is intuitive, and for many years hatchery managers have attempted to increase the survival of the smolts they release by maximizing size. One way to produce large smolts is to grow them longer into the spring, but this confounds the effects of size and date, and prolonged rearing may be costly and even deleterious. Bilton initiated a series of studies (e.g., Bilton, et al. 1982; Bilton, et al. 1984; Morley et al. 1988) on the interactions between the smolt size, date of release, and survival. The basic design was to release small, medium, and large fish on four different dates. The results for coho salmon released from Quinsam River Hatchery, on Vancouver Island, were both consistent between years and remarkable in their outcome (fig. 15-1). There was little variation in survival among the size groups released on a given date but great variation in survival among dates. Survival was low on April 20, intermediate on May 10, high on May 30, and as low on June 19 as it had been in April. Thus the date had a very strong effect but there was little effect of size within dates, so any overall pattern of survival with respect to size was explainable by covariation with date. Whitman's (1987) experiment on University of Washington hatchery chinook salmon was consistent with Bilton's coho studies. Chinook of three size groups were released in early April, late April, and mid-May, with only about a twofold effect of size on survival compared to a ten- to twentyfold effect of release date. Survival rates increased from one release to the next, but it is unclear if even later releases would have resulted in low survival, as was seen in the coho studies. S. G. Taylor's (1980) experiment on Auke Creek, Alaska, pink salmon provided still more evidence of the importance of timing; the survival rate of early hatchery fry was 0.17%, compared to 1.46% for fish migrating 35 days later.

FIGURE 15-1. Survival of coho salmon smolts from three size groups, differing from each other by about 5 g, released on four dates from Quinsam River Hatchery, Vancouver Island. Each point represents the mean of three replicate groups, in 1980 or 1981 (data from Bilton et al. 1984; Morley et al. 1988).

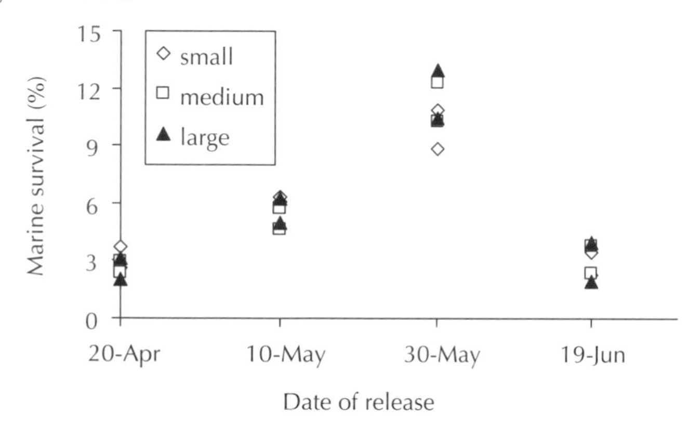

The application of these findings to wild populations is complicated by the tendency for larger smolts to leave freshwater earlier than smaller ones, and there are many open questions regarding this subject. Do the smolts have higher survival rates because they are large or because they leave early? Are the smaller juveniles spending another few weeks in freshwater in order to grow a bit more, at a sacrifice in terms of the date of entry into the ocean? What is the agent of mortality, and why is timing so important? Might the arrival date of salmon into the coastal ocean, relative to the period of peak density of predators in the vicinity, control survival rates? The migration patterns of a particular predator may be related to water temperature and so vary somewhat among years, so the ideal date for salmon to go to sea is not fixed. Moreover, the production of zooplankton and other food resources for salmon in coastal waters shows both a dramatic seasonal pattern (see chapter 16) and also variation among years. Growth of the salmon probably gives them a substantial survival advantage, so it is important to enter the ocean when food is available. Walters et al. (1978) created a simulation model to examine the interactions between date of entry into the ocean, food resources, migration, growth, and size-biased survival for British Columbia salmon. The researchers concluded that absolute food limitation (i.e., starvation) is highly unlikely, but the synchrony of arrival timing in coastal waters and the availability of food may be critical for determining the survival rates of different cohorts. Recent marking studies on pink salmon from Auke Creek by Mortensen et al. (2000) supported this model, showing significant variation in survival among groups entering the ocean on different dates, and also a significant survival benefit for rapid early growth. The next chapter provides further details on the linkages between the abundance of prey for salmon and interactions with other species (notably walleye pollock, *Theragra chalcogramma*, and Pacific herring, *Clupea pallasi*, in Prince William Sound) that can both compete with salmon and prey on them (Willette et al. 2001).

Our understanding of the processes leading to mortality is hamstrung because we know so little about the behavior and other attributes of salmon that may be associated with survival. Experiments in which salmon from different families were given distinctive marks

FIGURE 15-2. Survival of wild and hatchery-produced stream-type chinook salmon smolts from the time they passed their first dam on the Snake River to their return as adults (from Raymond 1988).

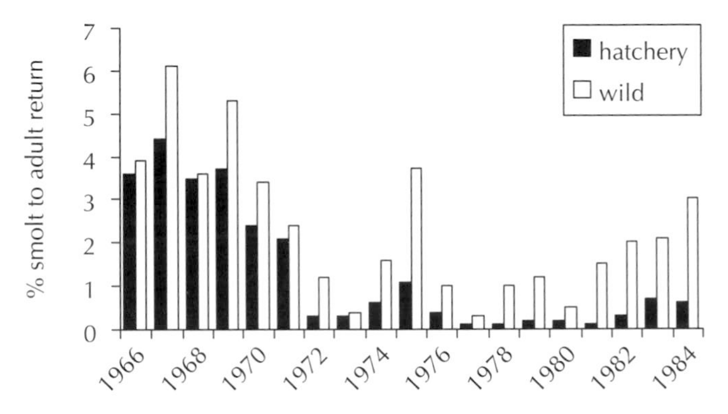

revealed significant variation in survival at sea. Some of this variation was attributable to differences in size among families, but there was strong evidence that genetic factors other than size affected survival of pink (Geiger et al. 1997) and chinook salmon (Unwin et al. 2003). Obviously, natural selection favors traits associated with high survival, but survival is the result of many traits interacting with environmental conditions. Growth may lead to large size, hence higher survival, but bold foraging may also pose higher risks of predation. A more cautious fish may grow slower but be less likely to get eaten (for example, by feeding farther from the surface, later in the evening, less often, etc.).

In addition to body size, timing of migration, and family attributes, one other factor has received attention regarding survival at sea: the rearing history of the fish. It is commonly observed that wild smolts have higher survival rates than conspecifics produced in hatcheries. Stream-type chinook from the Snake River (fig. 15-2) provide an example. There is certainly mortality during the migration to the ocean, and interceptions in distant fisheries are not included. Still, the hatchery fish had consistently lower rates of return than the wild fish (averaging 1.3 vs. 2.3%). The return rates of the two types tracked closely among years, indicating some factors influencing both groups in common. Raymond (1988) also reported similar results for steelhead (1.6% return for hatchery fish vs. 3.0% for wild ones in the same years), and many other studies have shown lower survival of hatchery compared to wild fish, even if the hatchery fish were larger and so should have had an advantage. The reasons for the disability of hatchery fish are complex and seem to result from at least three processes: (1) short-term effects of culture (such as pale colors) that make them vulnerable to predators immediately after release; (2) longer-term behavioral effects of culture such as reduced predator avoidance and tendency to feed at the surface (Maynard et al. 1995; Olla et al. 1998); (3) and genetic effects of selection for life in the hatchery environment as opposed to a natural environment (Reisenbichler and Rubin 1999).

When does most mortality occur?

Knowing when most salmonid mortality takes place would help us determine the causal agents and perhaps also help predict variation among years, species, and populations.

FIGURE 15-3. Average weight and number of Karluk Lake, Alaska, sockeye salmon (starting with fifty smolts), based on Ricker's (1976; table 7) monthly mortality and growth-rate estimates.

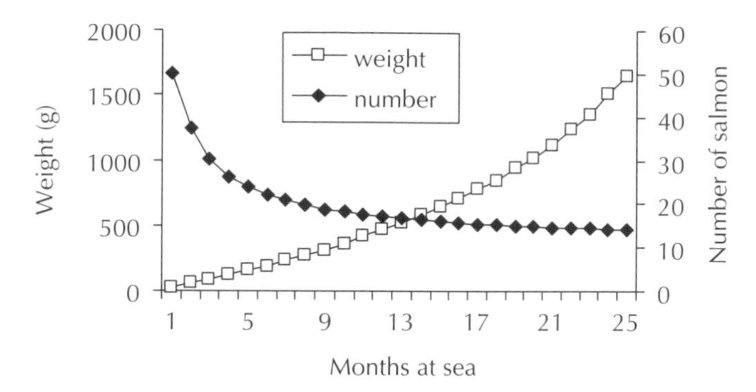

Several lines of evidence suggest that most mortality occurs early in the lives of salmon at sea, discussed in detail by Ricker (1976). There is good evidence, as well as logic, indicating that mortality rate diminishes as size increases, so most of the mortality should take place in the first few months (fig. 15-3). Several tagging studies have indicated that the early period at sea is marked by higher mortality per day than occurs later. Parker (1968) conducted an extensive, 3-year marking project with pink salmon leaving the Bella Coola River and migrating through Burke Channel, on the central coast of British Columbia. He concluded that about 59–77% of the juveniles died in their first 40 days at sea, or about 1.5–2% per day. During the remaining period of about 410 days at sea, about 78–95% of those surviving the first 40 days were estimated to die (i.e., about 0.2% per day). This mortality rate may initially seem high but in fact it results in a total marine survival of about 5%, and this is enough to not only sustain the population but also to support a heavy fishing rate, as table 15-4 shows.

Another marking study, by Bax (1983) on chum salmon in Hood Canal, Washington, indicated that mortality averaged 31–46% per day over the first 2–4 days. More recently, Fukuwaka and Suzuki (2002) marked chum salmon in hatcheries and recovered them in the coastal waters in the Sea of Japan. They estimated mortality rates of about 9–15% per

TABLE 15-4. Hypothetical dynamics of Bella Coola River, British Columbia, pink salmon, based on marine mortality estimates from Parker (1965, 1968). The initial number of fry, arbitrarily set at 60 million, is within the range for the population, and the fecundity and egg–fry mortality levels are assumed but plausible for the species. The fishing rate reflects only the level calculated for replacement and is not based on any actual records.

Process	*Loss rate*	*Survivors*	*Stage*
initial fry production		60,000,000	fry
59–77% mortality in first 40 days	65%	21,000,000	early marine
78–95% mortality in the next 410 days	85%	3,150,000	adults
85% fishing rate	85%	472,500	spawners
% females in the population	50%	236,250	females
fecundity @ 2000 eggs/female		472,500,000	eggs
egg–fry mortality	87.3%	60,000,000	fry

day (depending on the group of fish and the method of data analysis) in the 14–43 days after release. Even these rates, well below those reported by Bax, could not be sustained for long before the population would go below the replacement level. These and all such marking experiments are based on the assumptions that the loss of marked fish reflects mortality and that the marked fish fairly represent the population as a whole. However, movement of marked fish from the study area, shifting vulnerability to the sampling gear, and elevated mortality rates from handling can all bias such work.

Notwithstanding the concerns about marking experiments, there are other reasons to believe that most mortality occurs early in the lives of salmon at sea. Time-series analysis identified a 2-year lag between winter conditions and sockeye salmon abundance, consistent with mortality during the early period at sea (Francis and Hare 1994). Another line of evidence is the relationship between the number of jacks observed in one year and the number of older fish in the following year(s). Assuming the population has some intrinsic rate of producing jacks (whether based on genetic or environmental factors), the observed ratio of jacks to adults would vary greatly if the mortality took place after the time when the jacks had returned. For coho salmon, the smolts go to sea April through June and the jacks return in October or November, so the survival of the cohort seems to be determined in the first summer. Moreover, Fisher and Pearcy (1988) reported a close correlation between catches of post-smolt coho salmon off Oregon in June and the return of jacks that fall, and they concluded that the survival of the cohort may be largely determined within the first month or two at sea.

Additional evidence for the importance of the first few days or weeks in determining overall survival patterns comes from the efforts to increase the survival of smolts from commercial ocean ranches and public hatcheries by transporting them in barges and releasing them offshore rather than from the hatchery. For example, Pearcy (1992) reported data on fourteen releases of coho salmon from a commercial facility at Yaquina Bay, Oregon. On average, the overall marine survival rate of barged coho was twice that of fish released from shore, though there was a great deal of variation, and in five of the fourteen cases the offshore fish fared less well (Frank R. Severson, unpublished data cited in Pearcy 1992, 77). Solazzi et al. (1991) reported similar results, though the effects of transportation on survival were slight. Finally, extensive sampling at sea off the coast of Oregon allowed Pearcy and Fisher (1988, 1990, summarized in Pearcy 1992) to estimate mortality rates. They had an estimate of the number of coho salmon smolts entering the ocean from 1982 through 1985. They used their net sampling at sea to estimate density about 7 and about 19 weeks after the salmon had entered the ocean, and they obtained an estimate of the number surviving to maturity the following fall. These estimates depend on the coho remaining within the sampling area and on the absence of other coho salmon in the area, and both of these were not entirely true. Still, an average of the estimates over the 4 years presents plausible (though not necessarily accurate) estimates of abundance and declining mortality rate (on a per day basis) as time went by (table 15-5).

It should be noted that Beamish et al. (2000) used similar serial sampling techniques in the Strait of Georgia, British Columbia, and concluded that much of the mortality occurred during the first winter rather than the first summer at sea. Subsequent work in the Strait of Georgia indicated that coho salmon that grew slowly during their first

TABLE 15-5. Estimated numbers of coho salmon released in the Oregon Production Index area annually from 1982 through 1985 and their abundance (based on purse seine sampling) that June and September and the following fall, when they were caught or returned to spawn (Pearcy 1992, 41; data provided by W. G. Pearcy and J. Fisher, Oregon State University, Corvallis, Oregon). The release date of May 1 is a generalization, not an absolute average, and many of the smolts were released into the Columbia River, so some of the mortality probably took place before they entered the ocean itself.

Date	*Days at sea*	*Millions of salmon*	*% mortality per day*	*% mortality per period*
1-May	0	35.243		
17-Jun	47	8.496	2.9204	75.9
11-Sep	133	3.865	0.9113	54.5
1-Sep	488	1.111	0.3506	71.3

summer at sea were less likely to survive the late fall and winter than were faster-growing coho (Beamish et al. 2004). Further indirect evidence for mortality after the first summer comes from the negative correlation between the abundance of Asian pink salmon and the survival of Bristol Bay, Alaska, sockeye salmon (Ruggerone et al. 2003). These sockeye would not encounter Asian pink salmon until their second growing season at sea, and the density of pink salmon depressed both growth and survival during that season. The issue of when mortality takes place is thus not fully resolved, and it is one of many critical uncertainties regarding the marine ecology of salmon.

Environmental correlates of survival and long-term trends in abundance

For many decades, fisheries scientists have tried to predict the number of adult salmon that will come back to a given area so that fisheries can be better managed to provide appropriate levels of catch and escapement. In many cases there has been a plausible connection between the physical variable and survival (or, more typically, the total run or catch of adult salmon). In chapter 11 we saw the relationship between flow and abundance of coho salmon derived by Smoker (1955), based on the premise that marine survival varied little and production of smolts from freshwater determined the number of adults. All such correlations should be viewed with caution, especially if the causal mechanism is unclear. Future patterns may not be predictable from past experience, especially if the predictive relationship is based on a limited range of environmental data. A good example of the strengths and weaknesses of the correlation approach to salmon survival may be found in Vernon's (1958) report that the adult run of Fraser River pink salmon was strongly correlated with the temperature in the Strait of Georgia in the spring and summer (April to August) when the fry left the river and entered the strait. Over the first 9 years (1937 to 1953, odd years only), 69% of the variation in adult catch could be explained by this simple physical variable: higher return (and presumably survival) rates were associated with colder water. After Vernon's report was published, the relationship deteriorated and by 1961 explained only 16% of the variation in return. However, records through 1999 reveal an apparently dome-shaped relationship (fig. 15-4). Returns have

FIGURE 15-4. Relationship between the catch of Fraser River pink salmon in a given year and the mean April to August water temperature in Departure Bay, Vancouver Island, in the previous year (i.e., the year when those fish went to sea). The solid squares and the line represent the data reported by Vernon (1958), showing a decrease in survival with increasing temperature; and the open squares show the more recent data (from Pacific Salmon Commission annual reports and Department of Fisheries and Oceans, Canada), indicating a dome-shaped relationship.

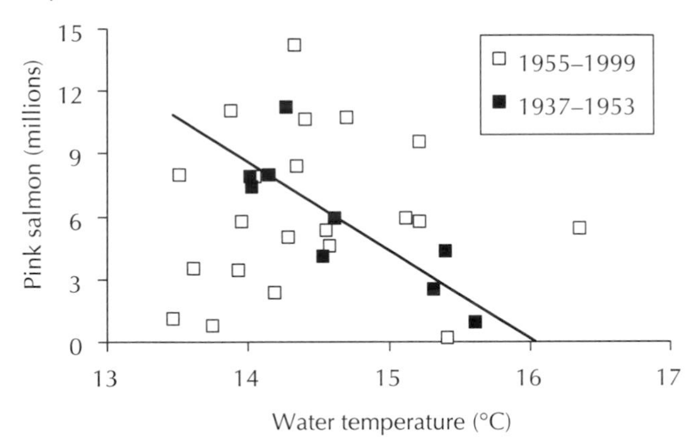

been weak after both cold and warm springs (averaging 4.77 million below 14°C and 4.31 million above 15°C) but higher (7.54 million) at intermediate temperatures. These temperature differences are too small to have a direct effect on survival. Do they indicate a complex ecological process affecting survival that is correlated with temperature, or is the entire pattern an illusion? If there is an underlying relationship, is it driven by marine processes or does the temperature in the estuary really reflect the temperature and flow in the Fraser River itself, and therefore a process affecting fry production or survival in the river? Correlations are intriguing but can be misleading.

The power of marine conditions to affect salmon survival was demonstrated in the summer of 1983, when the climate generated exceptionally warm temperatures and weak upwelling along the coast of Oregon and warm temperatures farther north as well. As mentioned in chapter 2, the El Niño conditions that year were associated with a record proportion of sockeye salmon migrating to the Fraser River via the northern route. Returns of Oregon coho and lower Columbia River chinook were only 42% and 46% of the predicted runs, respectively, and salmon growth was also very poor (S. L. Johnson 1984). The mortality apparently took place in the final summer at sea, unlike the usual pattern of mortality in the first summer. This delayed mortality was inferred from the fact that the number of adults surviving to maturity in 1983 was very small in proportion to the number of jacks in 1982.

The magnitude of the 1983 El Niño event and its biological effects caused scientists to look back in time and find correlations between physical processes and marine survival of salmon. Nickelson (1986) reported higher survival of coho salmon off the Oregon coast in years with stronger upwelling. In this case, upwelling is plausibly related to salmon growth (hence perhaps survival) because nutrients are brought from deep water to the surface where they can fuel primary production and zooplankton for the salmon to eat in their important first summer at sea. Several subsequent studies (e.g., Koslow et

FIGURE 15-5. Left panel: relationship between catches of pink salmon from central and southeast Alaska and winter (November through March) air temperatures in Sitka, southeast Alaska. Right panel: relationship between catches of central and southeast Alaska pink salmon and coho off Washington, Oregon, and California (data from Francis and Sibley 1991, updated by Robert Francis, personal communication).

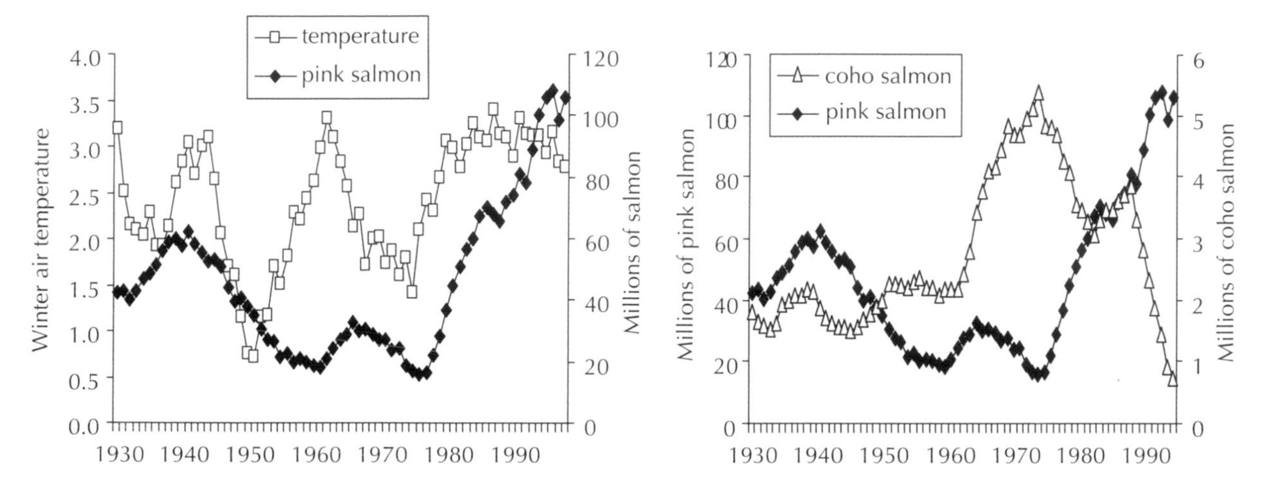

al. 2002) confirmed and elaborated on the earlier work; survival of coho salmon off Oregon was high in years with a suite of linked meteorological and oceanographic conditions including strong winds, upwelling, and cool water. The authors concluded that their findings "support the hypothesis that coho marine survival is regulated by marine productivity and food availability during their periods of early marine residence and prior to their return migration" (Koslow et al. 2002, 76), but they acknowledged that we still do not understand the connection between climate and food for salmon, and the predators that are actually responsible for the mortality.

Another major breakthrough in the study of salmon survival came when Francis and Sibley (1991) pointed out the correlations between long-term trends in climate and the abundance of pink and coho salmon during the twentieth century. The researchers first showed that the abundance of pink salmon in the Gulf of Alaska (as indicated by catches in central and southeast Alaska) tracked the regional air temperatures (indicated by records from Sitka, in southeast Alaska). The pink salmon catches were high from the 1920s to about 1950, then were low until the 1970s, and have been high again since then (fig. 15-5, left). Moreover, the trends in pink salmon catches were opposite to those for coho salmon from Washington, Oregon, and California (fig. 15-5, right). Subsequent studies have confirmed the pattern of opposite survival rates in northern and southern populations of salmon (e.g., Hare et al. 1999; Mueter et al. 2002).

During the 1990s, several scientists pointed out that the relative abundances of salmon species from the northern end of the range have been more or less synchronous during the twentieth century, coincident with climate and ocean conditions (e.g., Beamish and Bouillon 1993; Francis and Hare 1994; Hare et al. 1999). Individual populations, of course, have been affected by local conditions, both natural and human related, that influence survival; but the coherence at broad spatial scales indicates common influences. By analogy, the abundance of each individual population is like a tile

FIGURE 15-6. Illustration of the positive and negative phases of the Pacific Decadal Oscillation (PDO). Colors show the deviation of winter sea surface temperature from the long-term average. The direction and size of the arrows indicate surface wind anomalies. The contour lines in the ocean indicate sea level pressure. The positive phase of the PDO was associated with high survival of salmon in the northern part of their range and lower survival rates in the south, and the opposite effect was detected during the negative phase. Figure modified from Mantua et al. (1997) by Steven Hare, International Pacific Halibut Commission.

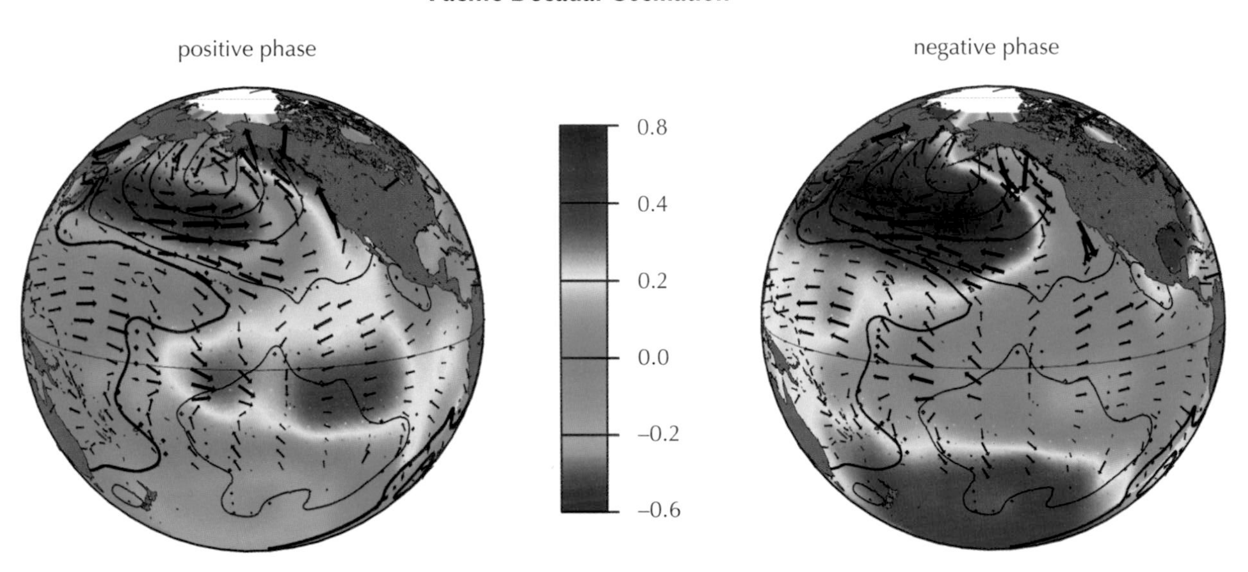

in a mosaic whose design is only clear at a distance. It appears that the locations and strengths of two atmospheric pressure cells, one high and one low, strongly affect the patterns of wind and circulation in the eastern North Pacific during winter. These circulation patterns continue to affect temperature later in the year and have a profound direct or (more likely) indirect effect on the survival and growth of salmon and other components of the ecosystem. Generally speaking, the period from 1977 to 1998 was characterized by relatively strong Aleutian low-pressure systems and weak subtropical high-pressure systems. The winds resulting from this combination tend to direct more of the subarctic current toward the north when it encounters the coast of North America, resulting in relatively warm water along the coast and cool water offshore in the Gulf of Alaska (fig. 15-6). This condition is associated with favorable survival and growth of northern salmon populations. However, the weaker coastal currents that it brings reduce the normal upwelling of cold, productive water off the coast farther south, hence the inverse relationship between survival rates in the two regions. When the weather patterns are in the other configuration, as they tended to be from 1947 to 1976, the winds and current along the Washington, Oregon, and California coasts increase the upwelling of cold, nutrient-rich water, and both survival and growth rates improve. The important point is not that the weather varies from year to year but rather that there are "regimes" lasting about 2–3 decades with somewhat similar climate and ocean conditions, with important consequences for the ecosystem. These conditions have been quantified by composite indices such as the Pacific (inter) Decadal

Oscillation (PDO: Mantua et al. 1997; fig. 15-6). At the time of this writing, it is unclear whether the regime that favors northern populations at the expense of southern ones has changed, but there is great interest in this issue.

Predators of salmon

The absolute magnitude of the shifts in physical conditions at sea between years is not great. The water is not so warm or cold that the fish are killed outright. Indeed, the differences in survival seem disproportionately large for the environmental changes that are correlated with them. This fact highlights the poverty of our knowledge concerning the actual agents of mortality. One seldom finds sick or starving salmon, so the proximate cause is presumably predation. There have been many efforts to identify the important predators, and some of these species will be considered here as examples.

Parker (1968) emphasized the importance of coho salmon smolts as predators of young pink and chum salmon, and his experiments showed that smaller fry were more vulnerable than larger ones. These findings were supported by later experiments by Hargreaves and LeBrasseur (1986), who reported that 120 mm coho smolts ate more small than large chum fry, over a range of 43–63 mm. The coho smolts also preyed more heavily on pink fry than chums, even when the chums were smaller and more abundant in the enclosures (Hargreaves and LeBrasseur 1985). It is unclear what behavioral or other attributes made the pink salmon more vulnerable, and this line of investigation might be profitably pursued in the future.

There have been several studies on possible salmon predators in the Strait of Georgia. Beamish et al. (1992) concluded that the per capita consumption of salmon by spiny dogfish (*Squalus acanthias*), a small shark, was slight. However, if their estimate of some 35 million dogfish was correct, the overall predatory effect of dogfish could be very significant. Another potential predator is lamprey. Larval lamprey (known as "ammocoetes") reside below the surface of streams, feeding on microorganisms. However, lamprey transform into adults, and in some species they attack fishes, using an oral sucking disk ringed with teeth to rasp through the skin of the fish. Newly transformed river lamprey (*Lampetra ayresi*) leave the Fraser River from April to July, coincident with salmon smolt migrations, and seem to feed almost exclusively on Pacific herring and, to a lesser extent, smolts in the Strait of Georgia. Beamish and Neville (1995) estimated that lamprey killed 20 and 18 million chinook salmon, and 2 and 10 million coho salmon smolts, in 1990 and 1991. These losses were estimated to be a large fraction of the total marine mortality of these salmon species, and there were losses of other species as well. Recent reviews of predatory fishes and birds affecting Asian salmon have been published by Karpenko (2003) and Mayama and Ishida (2003), and they include white spotted char (*Salvelinus leucomaenis*), Arctic char (*S. alpinus*), Arctic smelt (*Osmerus mordax*), and spiny dogfish. However, the magnitude of predation from these fishes has not been quantified and may be less significant than predation from seabirds.

Estimates of the impact of a predator species on its prey depend on reliable estimates of both the per capita consumption rate (itself affected by many factors) and the abundance of the predator. Marine fish populations are difficult to enumerate but mammals can be easier to count. Olesiuk's (1993) study on harbor seals (*Phoca vitulina*) in the Strait of Georgia reveals the kinds of data that are needed for a solid estimate of predation. First, the seal population was estimated at 12,990 prior to pupping and 15,810 after pupping. Laboratory observations were then used to estimate that seals eat about 4.3% of their body weight per day, or an average of 1.9 kg per day. Then, extensive analysis of scat samples (which may be biased) indicated that from April to November the seals ate mainly Pacific hake (*Merluccius productus*), and from December to March they ate mostly Pacific herring (*Clupea pallasi*). These two species comprised about 75% of the overall annual diet. Olesiuk estimated that the seal population ate about 9892 metric tons of fish per year and that about 398 tons (4%) were salmon. Thus salmon were not a critical component of seal diets as a whole, though in some cases seals might kill a significant fraction of the salmon and local depletion might occur. Considered in a broader ecological context, the seal data pose as many questions as they answer. How do we balance the predation on salmon with the ecological interactions with hake and herring? Hake may eat juvenile salmon. Herring can prey on pink and chum fry (e.g., Willette et al. 2001), compete with salmon for

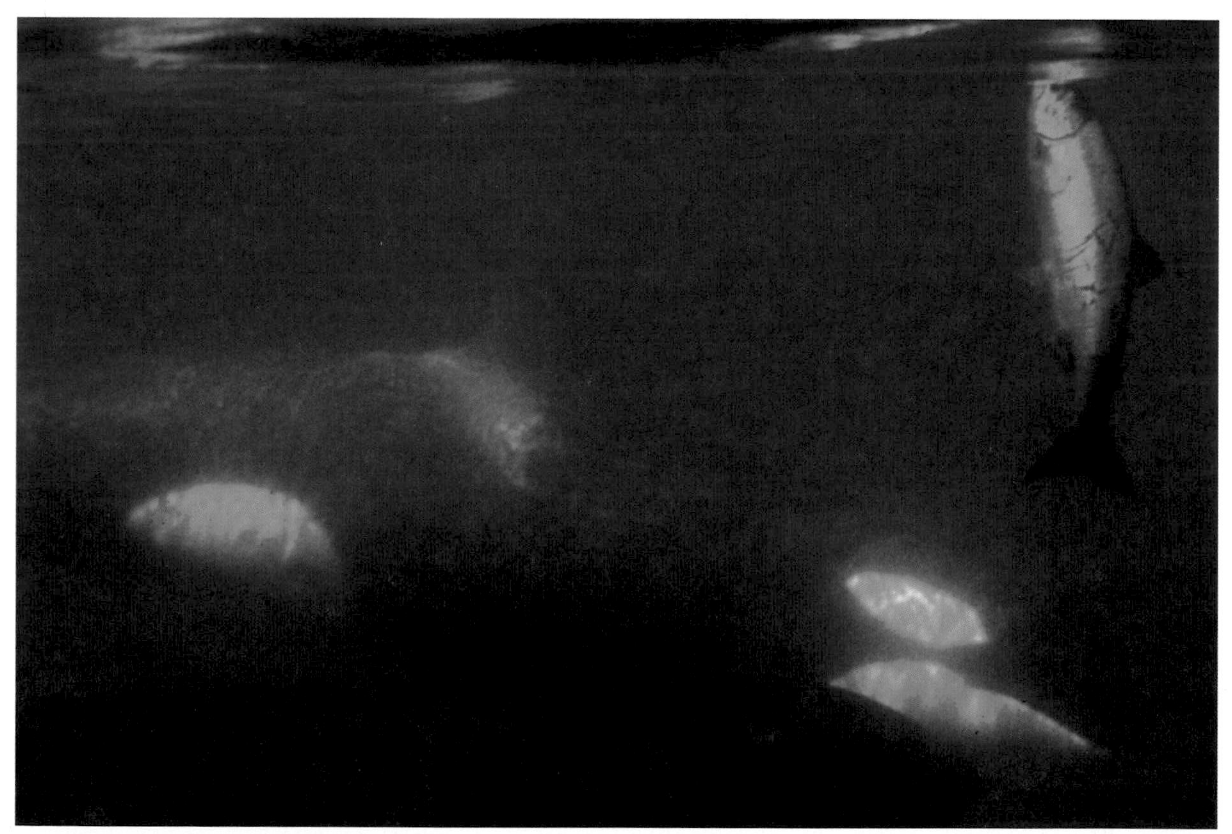

15-1 These juvenile killer whales attacked this adult pink salmon in Johnstone Strait, British Columbia. The fish sought refuge under the photographer's small boat, but eventually the whales killed and ate it. Photograph by John Ford, Department of Fisheries and Oceans, Canada.

food, buffer salmon against predation from seabirds (Scheel and Hough 1997), and can be eaten by adult salmon.

The ambiguous ecological relationship between seals and salmon is mirrored by the relationship between killer whales (*Orcinus orca*) and adult salmon. Killer whales exist as two ecophenotypes: "transients" that prey on marine mammals (including harbor seals) and "residents" that feed on fish (Ford et al. 1998). The arrival of residents in coastal waters such as Johnstone Strait and the San Juan Islands of Puget Sound tends to coincide with the summer salmon migrations. Nichol and Shackelton (1996) reported significant correlations between the occurrence of certain pods and the estimated abundance of pink (fig. 15.1) and sockeye salmon, and between the occurrence of one pod with chum salmon, but they were unable to distinguish salmon abundance from other factors that might be associated with the presence of the whales. Analysis of whale diets, based on field observations and stomach contents of stranded whales, showed a distinct tendency to eat salmon rather than other marine fishes (Ford et al. 1998). There was also a strikingly high proportion of chinook salmon (mostly adults aged 3 and 4) in the diet. Not only are chinook the least abundant species of salmon but they also tend to swim relatively deep in the water column, so the diet would seem to reflect active preference for the larger prey rather than encounter rate.

Finally, there can be mortality of salmon from pathogens and parasites as well as predators. Such mortality goes largely undetected but occasionally conditions allow the parasite access to the salmon and also allow us to document the event. Salmon at sea are commonly afflicted by parasitic copepods known as sea lice (*Lepeophtheirus salmonis*). These ectoparasites do not ordinarily seem to be a serious problem at sea and they drop off soon after the salmon enter freshwater. However, S. C. Johnson et al. (1996) reported heavy infestation of sockeye salmon with sea lice in Alberni Inlet, British Columbia. High water temperatures and low dissolved oxygen levels in the rivers caused the sockeye to delay in the inlet much longer than they normally do, and they were literally eaten alive by the copepods. Many salmon died in the inlet, and many of those that entered freshwater died prior to spawning.

Summary

Studies on the mortality of salmon at sea have emphasized the attributes of the fish that may affect survival, correlations between mortality and conditions at sea, and the roles of specific predators in overall mortality (photo 15-2). Survival at sea is generally size-selective, so salmon must enter the ocean at a large size or grow rapidly if they are to survive. However, the date when they enter the ocean is also important, probably through connections with the abundance of food and predators. Still, the ideal entry date may vary from year to year as ocean temperatures, currents, and other correlates of productivity and mortality vary. Ocean conditions (e.g., El Niño) that bring unusual growing conditions such as warm water and weak upwelling may also bring unusual associations between juvenile salmon and predators. Overall, there have been very substantial changes in the abundance of Pacific salmon over broad areas during the twentieth century, marked by inverse patterns of abundance of salmon in Alaska

15-2. Coho salmon smolt, showing the size of a coded wire tag at the tip of the tweezers, and an insert showing the tag. This invention played a crucial role in advancing our knowledge of salmon survival and migrations. Photograph by Andrew Dittman, National Marine Fisheries Service.

and northern British Columbia versus Washington, Oregon, and the Columbia River. From the late 1970s through the 1990s, salmon were comparatively abundant in the north and less so in the south, whereas from the 1940s to 1970s the reverse was true. Earlier in the twentieth century, salmon were abundant in the north as well. These seem to be related to broad-scale climate shifts that affect ocean conditions, but the direct links to salmon survival are unclear. However, information on the specific agents of mortality is very limited in most cases. Even the timing of mortality at sea is open to question, though most information points to a high mortality rate during the first summer at sea and declining rates as the fish grow.

16

Feeding and Growth at Sea

As described in chapter 14, salmon (and steelhead) feed in the coastal waters of both continents and the open ocean, where they may increase their weight a hundredfold, a thousandfold, or even more. This range of habitats and sizes of salmon complicates a review of their feeding ecology at sea. The subarctic domain (see fig. 2-3) that they occupy is not a static feature but rather a region defined oceanographically, whose exact boundary moves with the seasons and also varies from year to year.

Salmon in the marine food web

The top level of the "trophic pyramid" in northern waters is occupied by marine mammals, such as fur seals and sea lions, and by sharks, including the salmon shark (*Lamna ditropis*). In the offshore waters, salmon are among the most abundant epipelagic (near-surface) fishes, though in coastal waters there are other abundant species. There are several species of fishes and squid that co-occur with salmon, and they may be predators, competitors or prey, depending on the relative sizes of the salmon, the fish, and the squid. For example, in northern waters small salmon may compete with northern lampfish (*Stenobrachius leucopsarus*) and juvenile Atka mackerel (*Pleurogrammus monopterygius*). Squid are often a very important part of the diet of salmon but salmon also prey on zooplankton species, larval crabs, amphipods, polychaetes, krill (e.g., *Euphausia pacifica* and *Thysanoessa* spp.), and other crustaceans. Large salmon eat fishes such as herring (*Clupea pallasi*), sand lance (*Ammodytes hexapterus*), and eulachon (*Thaleichthys pacificus*), though these species may be competitors or even predators when the salmon are small. Farther south, in the warmer, more saline, and less productive waters of the transition and subtropical domains, the top predators are

TABLE 16-1. Catches of fishes and squids in gillnets at a range of latitudes along 155° west longitude by the *Oshoro maru* on July 15–30, 1984 (from Pearcy 1991). Some species caught at low abundance have been deleted. Species are not equally vulnerable to the gillnets, so abundance comparisons among species are not reliable. The data primarily illustrate the shifting relative abundance of species with latitude.

	Latitude											
	Subarctic ridge domain						*Transition domain*					*Sub-tropic domain*
Species	*55°*	*54°*	*53°*	*52°*	*51°*	*50°*	*49°*	*48°*	*47°*	*44°*	*41°*	*38–30°*
sockeye	263	357	171	169	88	52	33	50				
chum	294	316	209	183	42	121	117	38	2			
pink	313	110	134	105	104	158	118	5				
coho	146	27	16	15	31	33	16	5	2			
chinook	3	1	1	1	1		1	1				
steelhead	42	24	33	16	11	9	10	6	4			
eight-arm squid, *Gonatopsis borealis*	4	10	2	7	9	5	6	6				
Pacific pomfret, *Brama japonica*				9	12	14	177	1040	72	94	2	
nail squid, *Onychoteuthis borealijaponicus*				8	12	24	15	1				
Pacific saury, *Cololabis saira*										1	349	6
blue shark, *Prionace glauca*										4	128	
albacore, *Thunnus alalunga*										23	3	6
neon flying squid, *Ommastrephes bartrami*										26	167	112
skipjack tuna, *Euthynnus pelamis*												108
yellowtail, *Seriola dorsalis*												64
bigeye tuna, *Thunnus obesus*												7
frigate mackerel, *Auxis thazard*												6

blue sharks, tunas, and billfish, and salmon give way to mackerel and pomfret, and different species of squid.

Although these oceanographic domains are not fixed, they nevertheless reflect major differences in the aquatic communities. Table 16-1 displays the catches of different fishes and squids by gillnets set by the Japanese research ship *Oshoro maru* in late July on a transect from north to south. The net is not equally efficient at catching different species, so the data should not be used for comparisons of abundance among species. Nevertheless, the data show a clear preponderance of salmon in the north and a relatively abrupt shift to other distinct communities farther south.

FIGURE 16-1. Relationship between the sizes of salmonids and their prey, separated into fishes (all environments combined) and invertebrates (streams and lakes combined, and ocean). Data from Keeley and Grant (2001).

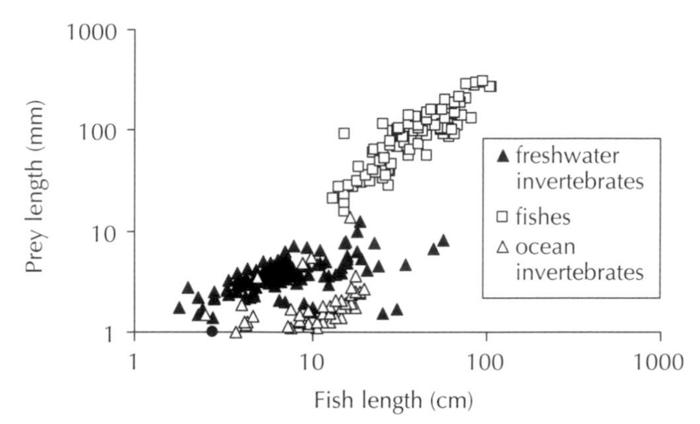

Even within the range of salmon, the quantity and nature of available prey vary among regions, seasons, and years. It is therefore difficult to state their food habits categorically because salmon cannot eat prey that are not available to them. However, salmon are trophic generalists without highly specialized mouthparts, tactics, or other features that would constrain their diet. They are well designed for swimming and, once they leave the littoral zones of estuaries, feed in physically unstructured environments. They rely chiefly on sight to locate prey and on speed and maneuverability to catch them. As with most predators, the size of their prey increases as the predators grow. This is both because they can catch larger prey and also because small prey require more energy to catch than they yield as food. Keeley and Grant (2001) reviewed a large number of studies of salmonid diets that included data on fish size, prey size, and habitat. The invertebrates (typically insects) eaten in streams were a bit larger than the invertebrates (mostly zooplankton) eaten in lakes and at sea. However, in all habitats there was a clear shift to diets dominated by fish (and larger fish) as the salmonids became larger than about 15–20 cm (fig. 16-1).

The salmon leave freshwater "anticipating" conditions at sea. The productivity of the ocean is obviously controlled by complex physical and biological processes, with great variation in space and time, among and within years, and this variation affects the growth and survival of salmon. Research in Prince William Sound, Alaska, provides an excellent example of these principles. Pink salmon originating from the nearly 1000 small streams and rivers (and four major hatcheries) left freshwater and entered the sound primarily in early to mid-May. Most reared in nearshore areas as they migrated slowly southward to staging areas prior to leaving for the open Gulf of Alaska in late July. Their timing matched the peak in biomass of large calanoid copepods in the surface waters (two species of *Neocalanus*, and *Calanus marshallae*; fig. 16-2). Salmon released from the hatchery during the time of the bloom had higher survival rates than those released after the bloom (average over 4 years: 6.95% vs. 3.35%), even though the fish released later were somewhat larger (0.32 vs. 0.26 g; Cooney et al. 1995). These findings are consistent with the links drawn by Mortensen et al. (2000) between food, growth, and survival of pink salmon in southeast Alaska.

FIGURE 16-2. Synchronous downstream migration of pink salmon fry (daily counts in six creeks in Prince William Sound, Alaska, averaged between 1990 and 1991) and density of zooplankton (7-day running average of samples collected episodically from 1981 to 1991 near the Armin F. Koernig Hatchery) (from Cooney et al. 1995).

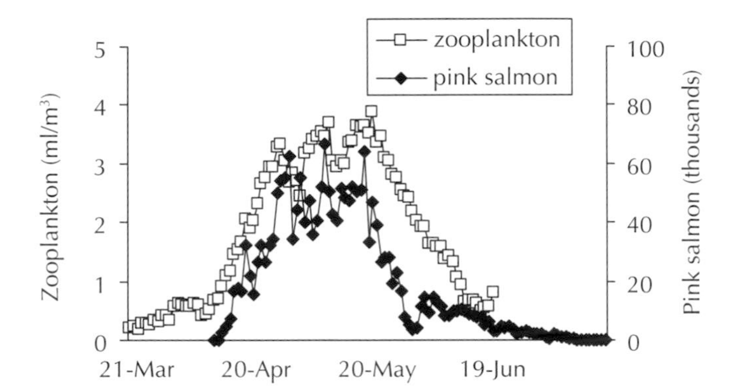

An extensive series of studies revealed fascinating interactions between food availability, growth, and survival of pink salmon in Prince William Sound (Cooney et al. 2001; Willette 2001; Willette et al. 2001). First and most obviously, abundant prey allowed the fry to grow rapidly and thus limited their vulnerability to two important predators: walleye pollock and Pacific herring. However, pollock and herring are facultative planktivores, and they fed almost exclusively on *Neocalanus* when copepod densities exceeded about 500 per cubic meter. At lower copepod densities, herring and pollock searched for alternative prey and frequently ate pink salmon fry. Approximately 75% of the fry entering the sound from 1995 through 1997 were lost to predation in the first 45 days at sea. Not only did the scarcity of copepods intensify predation on pink salmon in the nearshore area, but when copepod densities were low the salmon tended to leave the relative safety of the coast to forage offshore, where their risk of predation increased about fivefold. Thus the density of copepods affected not only salmon growth but also their movements and vulnerability to predators in complex ways.

Although the fish migrate at the time of year when food is most likely to be available, there is considerable variation from year to year in the timing of migration (affected by spawning date and incubation temperature) and the timing of food production. Thus the ideal date will vary somewhat from year to year but will tend to stabilize the population around the long-term optimum. In cases where the food availability is highly unpredictable or nearly uniform, we might see broader migration timing than in cases where food abundance shows a sharp and predictable seasonal peak.

Farther offshore, there are also seasonal patterns in food availability, as indicated by the repeated sampling at Ocean Station P (50° north latitude, 145° west longitude, in the Gulf of Alaska) by Canada. As summarized by Brodeur et al. (1996), these data (fig. 16-3) showed a pronounced peak in May and June, though there was also considerable variation among years, and these data do not take into account the variation in composition or desirability of the zooplankton species for salmon. Still, it is noteworthy that the

FIGURE 16-3. Annual cycle of average zooplankton density at Ocean Station P (mean and standard error; from Brodeur et al. 1996).

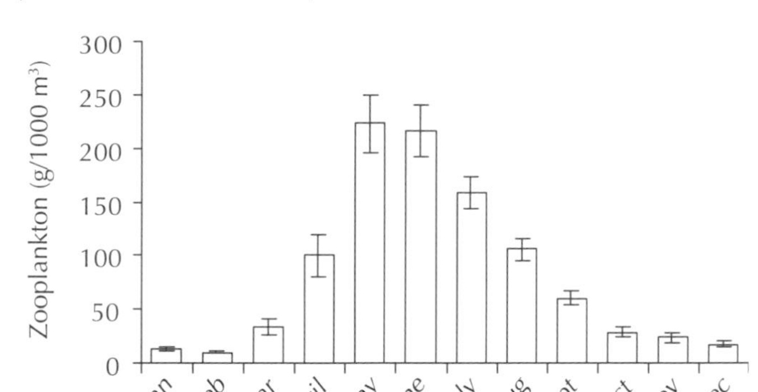

abundance of food peaks in late spring, before the water temperatures peak. Thus the fundamental relationship between temperature and ration size in determining growth (see fig. 10-2) is relevant here too.

Brodeur et al. (1996) also examined data over a broader spatial area (including much of the range of salmon at sea) and reported much lower zooplankton densities during the 1956–1962 period than the 1980–1989 period (average: 93.9 vs. 226.5 g/1000 m^3). This result is not a consequence of the scarcity of predators to eat the zooplankton. Quite the opposite; as seen in the previous chapter, the biomass of salmon was higher in the 1980s than the earlier period. Brodeur et al. (1996) estimated that the biomass of species likely to eat the zooplankton (chiefly salmon, squid, and Pacific pomfret) from 1980 through 1989 was more than double that from 1955 through 1958. Apparently, there were fundamental changes in the marine ecosystem with positive effects at several trophic levels, lending further support for the idea of a climate-driven "regime shift" in the late 1970s.

As indicated in previous chapters, the water masses in which salmon feed at sea are slowly moving, and the salmon are swimming in a relatively undirected manner, with modest rates of net travel compared to the rapid rates shown by maturing salmon migrating homeward. The actual distribution of salmon at sea is not well understood at a fine scale. They can be found in schools, but these are not the tightly packed schools that might characterize such fishes as herring. Rather, the schooling by salmon might be described as facultative rather than obligate, and catches in research gillnets and purse seines are sufficiently spotty that we may conclude that many salmon are not in sight of other salmon (e.g., Pearcy and Fisher 1990). Nero and Huster (1996) used low-frequency acoustic imaging to survey salmon (or, technically, targets likely to be salmon) in the Gulf of Alaska and concluded that most salmon were alone or in groups of two to four individuals. Their estimated density, 114 salmon per square kilometer, was not too different from that estimated from fisheries statistics (160 fish/km^2). These values should not be taken as precise, and of course there is variation in space and time, but they still give a useful perspective.

There is also variation among species in the depth of foraging, and this may affect their diet. Catch rates on commercial trolling gear reported by Orsi and Wertheimer (1995) indicated that juvenile coho were closer to the surface than chinook. Coho catches were relatively even down to about 25 m and then decreased with depth, whereas chinook catch rates were low from the surface to 15 m and increased to the deepest depth fished—37.5 m. The depth distribution of chinook increased dramatically with age but was always deeper than coho. These patterns were consistent with ultrasonic tracking at sea by Ogura and Ishida (1992, 1995), who reported average depths as follows: chinook at 29.1 m, chum at 10.2 m, sockeye at 9.9 m, coho at 9.2, and pink at 5.9 m. These studies indicated that chinook travel or feed deeper than the other species, and this is consistent with other lines of evidence. Archival tags on chinook salmon off southeast Alaska revealed that chinook routinely traveled to depths of 100 m or more (Murphy and Heard 2001). Chinook salmon have also been taken as bycatch, especially in winter, in commercial bottom trawl fisheries along the California, Oregon, and Washington coasts, with many records between 100 and 300 m, and some as deep as 482 m (Erickson and Pikitch 1994). In addition to the differences in average depth among species, most salmon species seem to move closer to the surface at night but Murphy and Heard (2001) reported chinook moving downward during the night.

Feeding patterns

Although salmon are trophic generalists, there is still considerable variation in diet. Some of this is a consequence of variation in the prey available to them, but there are also underlying patterns related to fish size and species. For example, Beacham (1986) reported on the diets of adult salmon caught by research trolling in the Strait of Juan de Fuca, British Columbia. One notable pattern was that a significant fraction (about 30–50%) of the fish had empty stomachs. Of those with prey, chinook had the greatest tendency to eat fish (especially sand lance), followed by coho, then pink, and then sockeye (no chum or steelhead data were presented). Reliance on crustaceans showed the reverse order (from sockeye to chinook), with euphausids being the dominant prey, followed by amphipods. In addition, fish comprised a greater fraction of the diets of larger individuals than smaller ones.

There have been many other studies describing the diets of salmon species caught in a given area and time period. I refer readers to Brodeur's (1990) synthesis and compilation of this extensive literature and give only a few representative examples here. LeBrasseur (1966) sampled in the northeastern Pacific Ocean in summer and reported that pink salmon primarily ate amphipods and fish; immature sockeye ate small crustaceans; maturing sockeye ate larger crustaceans (euphausiids) and squid; coho ate euphausids, squid, and fish; and steelhead ate fish and squid. Manzer (1968) reported that winter samples from the Gulf of Alaska showed that the sockeye ate primarily fish (especially lantern fish, Myctophidae) and squid, whereas the pink salmon ate almost exclusively amphipods. More extensive sampling by Pearcy et al. (1988) revealed variation in diet among areas but they found generally similar patterns of diet among species. Analysis of stable isotopes of nitrogen and carbon allowed Welch and Parsons (1993) to infer that chinook salmon fed highest on the food chain, followed by coho, then sockeye, and then pink salmon. Broadly speaking, this is consistent with the results of direct studies of diet, and steelhead seem to eat rather high on the food chain as well.

In these and other studies, chum salmon had a rather distinctive diet, relying heavily on soft-bodied organisms such as jellyfish and ctenophores that are seldom eaten by other salmonids. Azuma (1992) showed that not only do chum have an unusual diet but that their digestive tract has an unusually low pH and distinctive morphology. The volume of the digestive tract is large, its wall is very thick, and the pyloric caeca (fingerlike extensions of the digestive tract) are numerous. Azuma contrasted sockeye with chum and concluded that sockeye are adapted for eating highly nutritious foods whereas chum are specialized for less nutritious, soft-bodied items. Tadokoro et al. (1996) reported that the diets of chum salmon varied, depending on the relative abundance of pink salmon (the most numerous species). When pink salmon were abundant, chums tended to eat gelatinous prey; but when pinks were less abundant the chums ate more crustaceans and had a generally more varied diet.

Unfortunately, many studies provide little or no information on the relative abundance of food available to the salmon, so it is difficult to infer much about the behavior of the fish from their dietary patterns. However, some studies, such as those by Brodeur (1989) off the coast of Washington and Oregon and by Landingham et al. (1998) off northern British Columbia and southeastern Alaska, explicitly compared diets of different salmon species to the available prey. These studies indicated that the diets were not a random sample of the available prey, that the species differed in their use of the available prey, and that they tended to eat larger prey relative to what was available. The level of feeding, however, often varies over the day-night period. For example, Pearcy et al. (1984) reported that gillnet catch rates of sockeye, pink, chum, and coho in the Gulf of Alaska were greatest, and also closest to the surface, at night. Sockeye, pink, and coho fed on euphausids at night and ate mostly fishes, squid, and amphipods in the day, whereas chum showed no feeding transition and ate mostly salps.

One complication in these and all diet studies is how best to characterize the diet. Some prey might be numerous but small, so they would be a large fraction of the individual items eaten but a modest fraction by weight or volume. On the other hand, a large item eaten only rarely would be less representative of the feeding behavior than a small one eaten more often. It is also important to know what fraction of the individual salmon ate a particular item. If a single salmon ate a huge number of a certain prey, this might be less indicative of overall behavior than if a larger proportion of the salmon ate at least some. To help address this issue, Pinkas et al. (1971) proposed the Index of Relative Importance (IRI), a formula that incorporates the percent that a prey species contributes by volume and by number among all the predators sampled, and the proportion of the predators that consumed at least one such prey. This IRI is commonly used in studies of many kinds of fishes, including salmon.

Growth of salmon at sea

One general conclusion from the many diet studies is that salmon feed actively on mobile, nutritious prey. The food items that salmon eat at sea are not only nutritious in

TABLE 16-2. Approximate fork lengths of juvenile salmon entering the ocean and their lengths in August of that year. Hecate Strait, British Columbia data are averaged from 1986 and 1987 (Healey 1991b); southeast Alaska data are from 1983 and 1984 from Jaenicke and Celewycz (1994), except the chinook lengths, which were in September 1986 and 1987, estimated from figure 3 of Orsi and Jaenicke (1996).

Salmon species	*Entry size (mm)*	*Southeast Alaska*	*Hecate Strait*
pink	32	149	141
chum	40	153	148
sockeye	80	158	147
coho	105	243	
chinook (ocean-type)	70	180	
chinook (stream-type)	100	280	

terms of caloric content, fat, and protein, but they are also rich in the highly unsaturated fatty acids. In this regard they seem to be superior to the common prey of salmon in freshwater, and this may contribute to the rapid growth of salmon at sea (see Higgs et al. 1995 for a thorough review of the feeding habits and growth of salmon at different life-history stages and habitats, and the nutritional content of their various prey). In general, salmon have high metabolic rates, feed heavily, and grow fast, compared to other fishes in their range. Salmon can double their length and increase their weight more than tenfold in their first summer at sea. Sampling at the end of their first summer provides some examples of these kinds of growth patterns. As table 16-2 shows, pink and chum salmon have virtually caught up to sockeye, despite having been much smaller (especially the pinks) upon entering the ocean. Coho are somewhat larger upon entering the ocean than sockeye (see fig. 12-1) and continue their more rapid growth while at sea. Another good example is the growth rates of juvenile chinook salmon caught off southeast Alaska (Orsi and Jaenicke 1996). Stream-type chinook salmon were larger when

FIGURE 16-4. Growth rates at sea estimated for stream- and ocean-type chinook salmon caught off southeast Alaska (Orsi and Jaenicke 1996), beginning in September of their first year at sea.

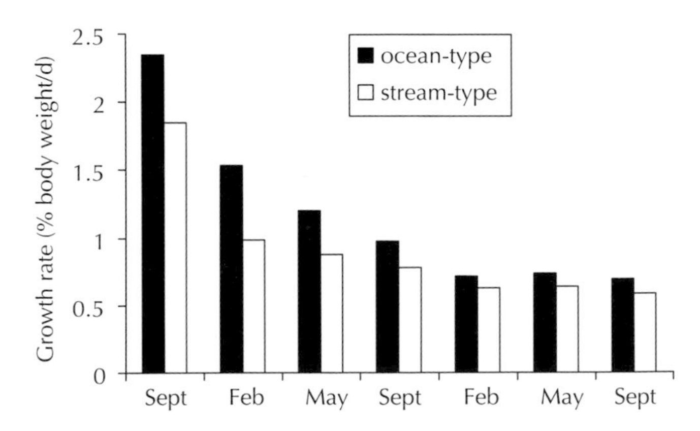

TABLE 16-3. Generalized weights of salmon as they enter the ocean (averaged from McGurk 1996 and Groot and Margolis 1991) and adult weights (averaged from the 1975 and 1993 values presented by Bigler et al. 1996). The steelhead value is for fish caught at sea after 2 years (Burgner et al. 1992); masu data are from Kato (1991). Numbers of years spent at sea represent typical values for the species.

	pink	*chum*	*chinook*	*sockeye*	*coho*	*steelhead*	*masu*
Smolt weight (g)	0.22	0.4	5 to 18	10	18	50	18
Adult weight (kg)	1.63	3.73	7.22	2.69	3.02	3.48	1.45
Full years at sea	1	2, 3, 4	2, 3, 4	2, 3	1	2	1

they entered the sea than ocean-type and were larger in September of their first year at sea, but the ocean-type fish grew faster and caught up to the stream-type fish by September of their third year at sea (fig. 16-4). However, the measurement and reporting of growth rates can be deceptive because larger fish grow faster, on an absolute basis, than smaller fish (in grams per day) whereas smaller fish grow faster in proportion to their size (in percent increase in body weight per day). Thus, relative growth rate declines with size and age whereas absolute growth tends to increase.

Overall, more than 98% of the final weight of most salmon is achieved at sea (table 16-3). The species enter the ocean at different sizes, and they achieve different growth rates. Sockeye grow slowly, being smaller after 2–3 years at sea than coho are after 1 year. Pink salmon are about half the weight of coho after a similar period at sea, though the coho are nearly 100 times heavier than pinks when they enter the sea. A pink salmon entering the sea at 0.2 g and returning at 2 kg has increased its weight ten thousandfold! Masu salmon also grow slowly, being similar to coho as smolts but about the same size as pinks at maturity. One must evaluate these data with some caution, however. Needless to say, there is a great deal of variation in size at maturity within and among populations, and among years (more on this shortly), so the species-specific values are not precise. In addition, the timing of return migration varies greatly. For example, coho salmon in Washington commonly leave the ocean in September or October, whereas masu leave in spring, so the latter have a much shorter growing season, even though both species might be considered to have spent a year at sea. As indicated earlier, timing variation can be extreme in steelhead populations, making it difficult to compare growth rates at sea.

The rapid growth of salmon results from feeding rate and metabolism. The metabolic rates are a product of both the salmons' intrinsic physiology and the ambient temperatures. There are regions of the ocean where salmon are scarce or entirely absent, even though the temperatures are by no means lethal. The salmon seem to seek a compromise between the food resources and the temperatures at which to digest them. The thermal regimes occupied tend to increase over the summer. For example, Burgner et al. (1992) related the catch rates of steelhead at sea (with research gear) to the ambient thermal regime (fig. 16-5). The great majority of steelhead were caught over a 3°C range of temperatures, increasing from 6–9° in May to 7–10° in June and 9–12° in July, but some were caught over a range of at least 7° in a given month.

FIGURE 16-5. Catch rates of steelhead at sea as a function of sea surface temperature (Burgner et al. 1992).

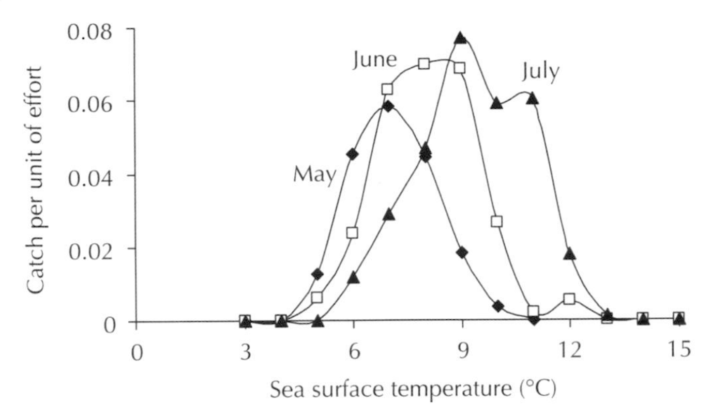

There is much less information on salmon at sea in the winter than in the summer, and this greatly hampers our understanding of their migrations and feeding ecology. However, catch rates in surface gillnets from January through March 1967 reported by French and McAlister (1970) provide some insights. Sockeye were caught regularly at surface temperatures from 4 to 6°C, but fewer were caught above 6°. Chum and coho were less abundant overall and were caught almost exclusively in water from 5 to 6°. The authors concluded that the sockeye moved south from their summer range. Doing so would increase the ambient temperature, and this would facilitate growth if food were available. It also appeared that maturing and immature sockeye had different winter distributions.

Regardless of where salmon spend the winter, this is a period of reduced prey base and little if any growth, as indicated, for example, by data on pink and chum salmon at sea (Mayama and Ishida 2003). Reduction in total lipid levels and changes in fatty acid profiles caused Nomura et al. (2000, 349) to conclude that "the low lipid content in the muscle in the winter suggests that chum and pink salmon have inadequate food at that season. Such a low lipid content jeopardizes survival of salmon in high-seas during winter." Further indirect evidence of energetic stress in winter was provided by an unusual data set on the fecundity of Asian pink salmon reported by Grachev (1971). Salmon were sampled from August of their first year at sea to the following August, when they were maturing. The assumption that the population was uniform over this time period might be questioned. Notwithstanding this issue, the samples showed a much higher number of oocytes (undeveloped eggs) in females from August through February than were found at maturity. However, the late winter and early spring samples were marked by a sharp reduction in oocyte counts, followed by stable counts during the final summer at sea (fig. 16-6). These oocytes are energy-rich, and salmon seem to produce more than are needed at maturity and then to reduce the number during periods of poor growth. The final size of eggs is probably determined quite late in the maturation process, reflecting the number of remaining eggs and the energetic demands of migration (Kinnison et al. 2001).

FIGURE 16-6. Average number of oocytes found in Asian pink salmon sampled monthly at sea by Grachev (1971).

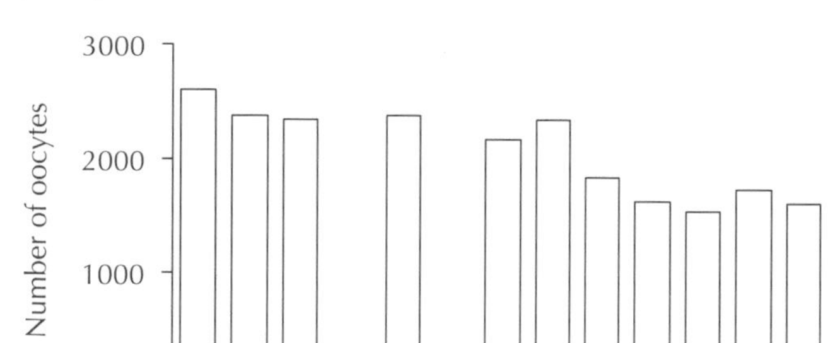

Variation in growth among regions and years

Notwithstanding the generalizations above regarding growth, there is great variation. Some of this will be taken up in the next chapter, but two themes will be mentioned here. First, there are regional patterns of body size. For example, both pink and chum salmon are larger at a given age toward the southern end of their North American ranges (fig. 16-7). These patterns might result from faster growth rates at sea, but they are complicated by the fact that populations in southern areas often enter the ocean earlier in their first year and also leave the ocean later in their last year, so these southern populations are able to take more advantage of these two summers of growth.

FIGURE 16-7. Average sizes of mature pink and chum salmon along the North American coast. Weights of pink salmon were from odd-year runs, 1963–1971 (Takagi et al. 1981). Chum salmon fork lengths were from age-4 fish (Salo 1991).

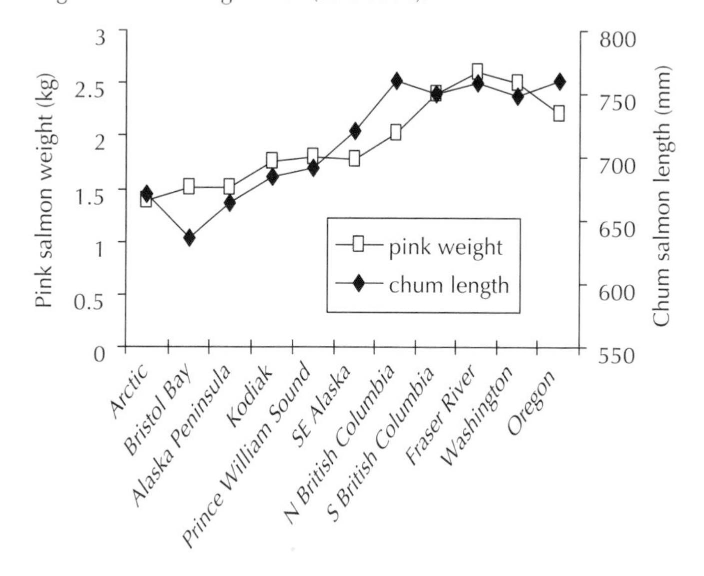

FIGURE 16-8. Mean length (mideye to fork of tail, males and females combined, for fish spending 2 years at sea) of adult Bristol Bay sockeye salmon as a function of the number of adults returning to spawn that year. The line represents the best linear fit to the early years (1958–1976, solid diamonds). From 1977 to 1997 the salmon were larger, for a given density, than in the earlier period. The most recent 5 years seem more similar to the earlier pattern (Alaska Department of Fish and Game data, complied and analyzed by Donald Rogers, University of Washington).

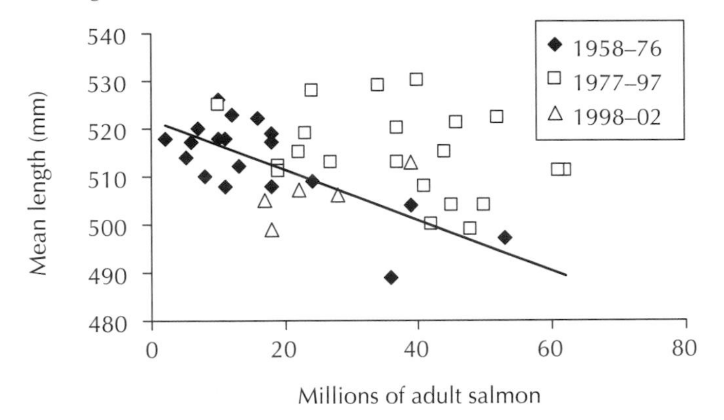

In addition to regional and population-specific variation in growth (discussed in the next chapter), there is also considerable variation from year to year. In a pioneering study, Rogers (1980) presented evidence that the size at a given age of Bristol Bay, Alaska, sockeye was density dependent. That is, in years (from 1958 to 1973) when many salmon returned to the region, the fish were smaller. This suggested that, despite the comparatively low density of salmon at sea, they were in competition for a limited prey base. Rogers continued to track this relationship and noticed that after 1977 the salmon were larger, for a given level of abundance, than the earlier years. Rogers and Ruggerone (1993) pointed out that two processes seem to be at work. First and foremost, growth was depressed at high densities, and densities of Alaskan salmon tended to be higher since the late 1970s than in the 2 previous decades. The secondary effect was improved growth at warmer temperatures, and temperatures have tended to be warmer since the late 1970s. Thus the effects have been offsetting. The annual average lengths of Bristol Bay sockeye were virtually identical in the 1958–1976 and 1977–2002 periods (fish spending 2 years at sea: 513.1 vs. 513.1 mm, 3 years at sea: 573.0 vs. 570.9 mm). Figure 16-8 displays an update of the information through 2002, showing the difference in patterns in the early years and those since the climate-driven regime shift in 1977. The most recent 5 years (1998–2002) seem more similar to the earlier period, and oceanographic conditions may have changed back to the earlier type of regime at about this time.

Pyper and Peterman (1999) confirmed the earlier findings that density depressed growth in sockeye salmon from Alaska and British Columbia. However, in contrast to the earlier work, they found that the secondary effect of elevated temperature was to decrease growth. They suggested that the difference between their findings and those of Rogers and Ruggerone (1993) may have arisen from the different places where temperatures were recorded for the analysis. The nature of the physical oceanography of the

FIGURE 16-9. Mean fork length of Hokkaido, Japan, chum salmon (eleven populations) as a function of the number of adult chum salmon returning to Hokkaido (catch + escapement) that year (from Kaeriyama 1998).

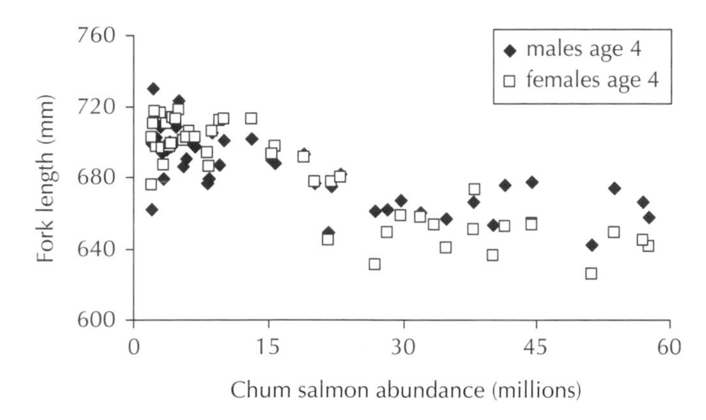

region is such that years are not marked by higher temperatures in all areas; indeed, opposite patterns may occur in coastal and offshore areas.

Density-dependent growth is also evident in the decline in size at a given age of Japanese chum salmon as the number of fish released from and returning to their hatcheries has increased (Kaeriyama 1998; fig. 16-9). Peterman (1987) took such analyses one step farther when he combined the zooplankton data from Ocean Station P with estimates of annual abundance of Fraser River pink salmon to show that salmon growth declined as the density of fish increased, relative to the amount of zooplankton. Thus it is clear that some of the large population complexes show density-dependent growth, but is there competition among species, and does the competition affect survival? Rogers and Ruggerone (1993) reported no evidence that abundance of Asian chum salmon affected growth of Bristol Bay sockeye, perhaps because their diets differ. However, Ruggerone et al. (2003) found that the abundance of Asian pink salmon reduced not only the growth of Bristol Bay sockeye (during their second and third years at sea, when spatial overlap occurs) but also their survival. Further research along these lines is certainly needed, but these results indicate important ecological interactions among salmon at sea and highlight the connection between growth and survival, in the context of changing physical conditions at sea.

Summary

Pacific salmon enter the ocean, mainly in spring and early summer, and grow rapidly (photo 16-1) on a diet of small fishes, crustaceans, and squid. The salmon feed in the epipelagic region, where they are the dominant fishes. Their diet varies among species, and the salmon tend to eat larger prey and rely more heavily on fish and larger squid as they grow. Salmon grow primarily in the summer, and growth rate is related to prey availability and temperature. These environmental features vary in a somewhat predictable seasonal and spatial manner but also vary from year to year. The

16-1 Adult chinook salmon in salt water. Photograph by Andrew Hendry, McGill University.

variation among years has some apparently random components but there also seem to have been general regimes of conditions from 1947 through 1976, 1977 through 1998, with perhaps a new one starting in 1998. These regimes have been characterized by physical and biological processes at sea that affect salmon growth and survival. The period from the late 1970s to 1990s was favorable for growth of northern populations, though the benefits were offset to some extent by increased density.

17

Age and Size at Maturity

Before considering the biological aspects of age at maturity, I must describe the two common systems for designating ages of anadromous salmon. They both indicate the numbers of years spent in freshwater and at sea, not merely the total age. This is critical because fish of the same age that have spent different periods in freshwater and at sea will differ in size, and will have gone to sea in different years, and so will have experienced different conditions as smolts. These two designation systems are part of the salmon jargon and the formats differ entirely. One system was developed for Atlantic salmon (sometimes known as the European system) and was applied to Pacific salmon by Koo (1962a). The designation consists of two numerals, separated by a decimal point. The first numeral indicates the number of full years spent in freshwater as a free-living fish, and the second indicates the number of full years spent at sea. Thus a sockeye that was spawned in October of 2000, emerged from the gravel in April of 2001, left the lake in May of 2002 after 1 full year, and returned to spawn in October of 2004 after 2 full years at sea would be designated 1.2 (spoken "one, two" or "one point two"), for a total age of 4 years. The months in the gravel from fertilization to emergence are not counted in the designation, nor are the months of the last summer at sea. Thus 1 plus 2 equals 4 years. The other designation system, called the Gilbert-Rich system, uses subscripts (Koo 1962a). For the same fish, the first numeral would be a 4, indicating the total age of the fish at maturity. The subscript indicates the year of its life when it went to sea. Thus the 1.2 sockeye would be a 4_2 (spoken "four sub two") because its total age was 4 years and it went to sea in its second year of life (i.e., after 1 full year in freshwater). A fish that spent another year at sea would be a 1.3 with a total age of 5, also designated 5_2. Typical coho salmon in Washington and Oregon, for example, would be aged 1.1 or 3_2, and jacks would be 1.0 or 2_2.

TABLE 17-1. Percentage of adult sockeye salmon with different combinations of years in freshwater and at sea in the Chignik Lake–Black Lake system, 1926–1966 (Dahlberg 1968).

	Years in freshwater			
Years at sea	*1*	*2*	*3*	*Total*
1	0.05	0.05	0.003	0.103
2	5.58	10.68	0.32	16.58
3	46.10	36.27	0.26	82.63
4	0.53	0.17	0	0.70
Total	52.26	47.17	0.583	

Sockeye salmon have an especially large number of age groups compared to other semelparous species. For example, sockeye from the Chignik Lake–Black Lake system on the Alaska Peninsula included eleven different combinations of freshwater age (1–3) and marine age (1–4; Dahlberg 1968; table 17-1). However, the great majority of adults (82.6%) spent 3 years at sea, and 16.6% spent 2 years at sea. Roughly equal proportions spent 1 and 2 years in freshwater (52.6 and 47.2%), so most of the fish were in two of the eleven age groups, 1.3 and 2.3. Including populations of ocean-type sockeye (those going to sea in their first year of life), there are fourteen age groups of sockeye (the eleven reported by Dahlberg plus 0.2, 0.3, and 0.4).

People working on Atlantic salmon have somewhat different terminology related to age at maturity. They use the term "grilse" for salmon that spent 1 full year (i.e., from one spring to the next spring or summer) at sea. Some populations are almost exclusively grilse whereas in other populations the grilse are mostly males and the females are "multi–sea winter" (MSW) salmon. MSW salmon are also indicated by the number of winters spent at sea (2SW, 3SW, etc.) but these terms are not used for Pacific salmon.

The systems for designating ages get complicated by the many variations in life history. First, the range in dates of seaward and homeward migration can make it hard to compare the numbers of "years" spent at sea by fish from different populations. We often refer to the number of winters spent at sea, and these are indicated on the scales or otoliths (see "Age determination," below). However, chinook salmon returning in March have had different growing opportunities from those returning in September. The second complexity is the tendency for some salmonids to return to freshwater after a period at sea but not to spawn. These are the "half-pounders" in steelhead, but many cutthroat trout, Dolly Varden, and Arctic char also show this migration pattern. Finally, the life histories of the iteroparous fish need special designations to show the number of times that they have spawned. In some char species these life-history patterns get especially bewildering, as fish migrate back and forth repeatedly, spawning on some years but not others. In these cases, the total age of the fish may convey little information about its history of migration and reproduction, or its size. The ages of iteroparous, anadromous salmonids can be designated as follows (e.g., Loch and Miller 1988). A numeral followed by a period indicates the number of years spent in freshwater prior to seaward migration. After the period, "+" signs indicate journeys to sea of less than a year's duration, and "F" and "S" symbols designate feeding and spawning migrations to freshwater, respectively. Thus a fish designated 2.+F+S+S+ would have spent 2 full years in freshwater, migrated to sea for a summer, returned but not spawned, then gone to sea twice more—spawning after each trip—and then would have gone to sea once more.

Age determination

Unless they were marked as juveniles, the ages of salmonids are determined by examination of the patterns of rings on their scales and otoliths. Alevins have no scales but they develop them after emergence. For example, sockeye salmon fry emerge about 28 mm long but do not develop scales until they are at least 38 mm (Koo 1962b). Interestingly, the scales do not form simultaneously over the entire body. Rather, the first scales are formed above the lateral line and behind the dorsal fin. The body is then enveloped progressively from each side until it is fully scaled. The scales grow as the fish grows and have concentric ridges known as "circuli." During periods of rapid growth, the circuli are widely spaced, but when growth is slow (typically during late winter) they are more closely spaced and give the appearance of a ring. These rings of tightly spaced circuli are called "annuli." When the salmon go to sea the annuli are spaced farther apart, reflecting the more rapid growth. Experienced "scale readers" can consistently and accurately count the annuli corresponding to winters in freshwater and at sea. The transitions from the ocean to freshwater are detectable, as are the reproductive seasons in iteroparous species.

Scales are very useful for age determination, in part because they can be quickly removed without sacrificing the fish. However, fish lose some scales during their life, especially during the smolt period, and the replacement scales cannot reveal the annuli of the lost years. In addition, as the salmon become fully mature, the skin thickens and scales deteriorate, making them no longer reliable for ageing. However, growth increments are also recorded on the saggital otoliths (one of three pairs of ear bones in fishes). They are formed prior to hatching, and experienced observers can identify marks indicating hatching and emergence from the gravel, and they can count daily rings laid down during the first few weeks or months. Later, the otoliths record the winter annuli in freshwater and at sea. Otoliths are normally used for determining the ages of salmon recovered dead on spawning grounds.

In addition to recording the age and reproduction of the fish, scales and otoliths can provide further insights into the fish's history. Variation in growth, as may occur in sockeye from different lakes, can provide patterns of circuli counts and spacing, so experienced observers can determine the likely origin of fish in samples with salmon from many populations (Davis et al. 1990). These differences in growth can not only distinguish among some wild populations but can reveal whether a fish was reared in a hatchery or a river (e.g., chinook salmon: Unwin and Lucas 1993; steelhead: Bernard and Myers 1997). Scales and otoliths also incorporate aspects of the chemistry of the water in which the fish are living and thus preserve a record of fishes' environments. Seawater has a higher ratio of strontium to calcium than freshwater, so otoliths record whether a fish has been to sea or not. These differences in chemistry are even passed from a female to her eggs, so embryos or fry can be assayed to determine whether their mother was anadromous or not (e.g., Rieman et al. 1994; Zimmerman and Reeves 2000).

Variation in age and size at maturity

Before discussing biological issues related to size and age, I must issue a note of caution. Many behavioral ecologists study adult salmon in streams and rivers as though these fish represent their populations. As we stand on the stream's bank this seems reasonable enough, but this is only because commercial and recreational fisheries are out of sight. Almost all populations are subjected to some degree of fishing, and in some cases the majority of the adults that survive all natural mortality are caught. Not only does this obviously affect the density-dependent processes observed on spawning grounds, but it can also greatly affect the age, size, and even sex distribution of spawning salmon.

The extent of these effects depends on the variation in age and size in the population, and the degree of selectivity of the fisheries. Jacks are often below the size limit in recreational fisheries and so are overrepresented on spawning grounds relative to their true abundance. Chinook salmon in coastal waters may be vulnerable to fishing over most of their lives, so individuals with a tendency to mature at an old age are more likely to get caught before they mature and are underrepresented in the escapement. Some forms of commercial fishing gear, notably gillnets, are highly selective. Small fish tend to pass through the nets and, depending on the size of the mesh, some fish may be large enough to avoid capture as well. Data on sockeye salmon in the gillnet fishery in upper Cook Inlet, Alaska, provide an example. Sampling the fishery would indicate a population dominated by fish that spent 3 years at sea, whereas sampling on the spawning grounds would indicate a more even mix of 2-ocean and 3-ocean fish and a higher (but still small) percentage of jacks (table 17-2). Gillnets often catch more males than females, for a given size, because males' longer jaws and teeth become entangled in the net. The size difference between males and females and the differences in shape (e.g., dorsal hump) as the fish mature can affect selection of salmon by gillnets as well.

The age composition and size of salmon at maturity reflect not only variation in growth rate, but also also the different thresholds of size or growth rate that trigger the transition from growth at sea to the onset of sexual maturation, homeward migration, and the events described at the beginning of the book. As indicated in the previous two chapters, our knowledge of the ecology and behavior of salmon at sea is still quite limited, and many conclusions are based on indirect evidence. The patterns related to maturation apply to nonanadromous populations as well, and this chapter describes and attempts to explain some of the variation.

The first pattern of variation is among species. Based on data from a wide range of sources, average lengths (in millimeters, from tip of snout to fork of tail) of female

TABLE 17-2. Percentages of sockeye salmon in the upper Cook Inlet, Alaska, gillnet catches and in the escapement, as a function of the number of years the sockeye spent at sea (Waltemyer 1994).

Years at sea	*1*	*2*	*3*	*4*
Catch	0.3	38.7	60.7	0.3
Escapement	0.7	53.5	45.6	0.2

17-1 Philip Roni with a male chinook salmon from the Kitsumkalum River, British Columbia, known for its very large chinook. Photographer unknown.

salmon are as follows: chinook: 871, steelhead: 721, chum: 683, coho: 643, sockeye: 553, pink: 522, masu: 506, and cutthroat: 413 (see fig. 6-4). These values represent the average of population-specific averages and thus estimate the central tendency for each species, integrating the variation in age at maturity and length at a given age. However, the values mask many interesting and important variations between males and females, among populations, and among individuals within populations.

One of the most obvious (but seldom studied) sources of variation in age and size at maturity is between males and females. Males vary more in size at maturity than females and (except in pinks) more in age at maturity too. Roni (1992) obtained data on the average length at maturity in 108 chinook populations (photo 17-1). The population-specific maximum sizes of males and females were similar, but males showed a much greater range of sizes than females (fig. 17-1), and for most populations the average male was smaller than the average female. This results from the fact that some males mature before the youngest females of the population, pulling down the average length of males. These may be anadromous individuals (jacks) or males that mature without going to sea, sometimes referred to as "precocious parr." I could find no systematic study of the incidence of jacks and precocious parr in Pacific salmonids, but jacks seem to be more prevalent in coho and

FIGURE 17-1. Mean lengths of male and female chinook salmon from 108 populations throughout the range in North America (from Roni 1992). The measurement, from the back of the eye to the end of the spinal column, is not biased by the greater development of jaws in males compared to females at maturity, so the sexes can be properly compared.

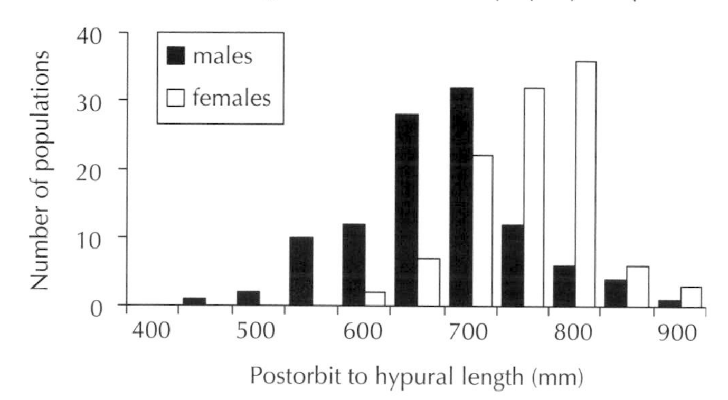

chinook than sockeye, virtually absent in chum, and absent in pinks. The rarity of jacks in chum salmon is curious because chum do vary in age at maturity. Jacks do not seem to be numerous in masu but there are many nonanadromous males.

Sexually mature nonanadromous males are, to my knowledge, unknown in chum and pink salmon (in their natural range), rare in sockeye and coho, present in some chinook salmon populations, more abundant in steelhead and cutthroat trout, and very common in masu. Some masu that mature as parr survive after spawning, later migrate to sea, and spawn again (Kato 1991; Tsiger et al. 1994), though the sea-run fish die after spawning. In chinook salmon, mature parr seem to occur primarily, if not exclusively, in stream-type populations (E. B. Taylor 1989; Mullan et al. 1992). Robertson (1957) reported that such mature parr were able to survive for up to 5 months after the spawning period, and work on chinook salmon in New Zealand showed that under controlled conditions males that matured as parr could survive, grow, and spawn again the next year, and even in a third year (Unwin et al. 1999). Under normal conditions the demands of gonad development and reproduction would leave these parr with insufficient energy to survive the unproductive fall and winter seasons. However, it is fascinating that even semelparity, which seems like such a fundamental trait of Pacific salmon, is more like a point along a continuum of life-history traits than an absolute state.

One complication in the study of alternative male life-history patterns is that such males are difficult to accurately census in wild populations. Spawning-ground crews are much more likely to see and recover carcasses from large fish than small ones. For example, Zhou (2002) found that age-2 male chinook (jacks) were underestimated by 75% relative to their actual abundance and age-6 males were overestimated by 21% on the Salmon River, Oregon. Mature parr would be even more difficult to sample accurately. On the other hand, if all salmon can be counted in the river (for example, at a trap or a hatchery), the apparent proportion of jacks will be an overestimate because they are much less vulnerable to fishing than the larger and older males.

Notwithstanding such methodological issues, the various male life-history patterns are a fascinating feature of salmonids (photo 17-2). Even populations with entirely freshwater life cycles may have alternative male forms. For example, Yamamoto et al. (2000) described three different male life-history types in the nonanadromous masu of Lake

17-2 Sexually mature masu salmon parr from the Shiribetsu River, Japan. Photograph by T. Aoyama, Hokkaido Fish Hatchery.

Toya, Japan. The "standard" males reared for a year in a stream after emergence, then underwent the appearance of a smolt transformation, migrated to the lake, spent a full year there, and returned to spawn at age 3. There were also precocious parr that matured without leaving the stream. Finally, there were so-called "puerile" males that migrated from the creek to the lake but retained parr marks even at sexual maturity. These three forms ranged in average size from standard (41.2 cm) to puerile (25.1 cm) to mature parr (11.3 cm). All the females were the standard form, averaging 49.8 cm.

Jacks and mature parr reflect the alternative pathways to reproductive success in males: aggressive competition and sneaking. Even in species without jacks, males vary more in age at maturity than females. For example, Beacham and Starr (1982) sampled Fraser River chum salmon and found that males were more common at ages 3 (54%) and 5 (57%) whereas 56% of the fish aged 4 were females. There seems to be a single female strategy, dedicated to egg production and nest defense. There is little payoff for a female to mature at a very small size because she could not produce enough eggs to match the lifetime mortality of her progeny. Thus we regularly get jacks but almost never jills. The longer a female stays at sea, the larger she grows, and so the greater her potential reproduction. However, the benefits of increased size at maturity must be balanced against the risk of mortality with every passing year at sea. Healey (1986) modeled the benefits of longevity in terms of growth and fecundity against the risk of mortality and showed that each set of conditions will produce an optimum age at maturity for females. Interestingly, the calculated optimal ages at maturity tended to be lower than the observed ages, indicating that there may be benefits of large size for females in addition to fecundity (such as egg size, nest depth, etc.).

In addition to the variation in age at maturity between males and females, there is also a well-known (but seldom studied) tendency for males to be larger at a given age than females. In Bristol Bay, Alaska, sockeye, for example (table 17-3), males were consistently larger than females, for a given age. Not surprisingly, the fish that spent 3 years at sea were larger than those spending 2 years, and an additional year in freshwater provided a small but consistent size advantage later in life as well. From an evolutionary

TABLE 17-3. Length (mideye to fork of tail, in mm) of mature sockeye salmon in Bristol Bay, Alaska, averaged from 1958 through 2002 annual averages (Donald Rogers, University of Washington, unpublished summary of Alaska Department of Fish and Game data).

	Age designation			
	1.2	*2.2*	*1.3*	*2.3*
Males	512.9	527.1	582.1	586.3
Females	497.5	511.4	558.5	567.8

standpoint we might explain the larger size of males by inferring a greater benefit for reproductive success from large size in males than in females. This is plausible, and there are studies linking various attributes of reproductive success to size in males and females (see chapter 6). However, the few studies so far that have determined the realized reproductive success of individual fish spawning naturally (using DNA parentage analysis) have not found strong correlations between the body size of the parent and the number of surviving offspring (e.g., Dickerson 2003, McLean et al. 2003; Seamons et al. 2004). Thus there is still much to be learned about the selection on body size within populations. From a mechanistic standpoint we might ask how males actually accomplish their more rapid growth. Do they have a faster metabolism and a higher intrinsic growth rate, or do they eat more, or show other behavior patterns that affect growth? It would seem possible to do experiments or field sampling to address these questions, but this subject has not been extensively investigated.

The generalization that males are larger than females, for a given age, has a fascinating exception. Holtby and Healey (1990) pointed out that in many coho salmon populations, females were larger than males of the same age (1 full year at sea). In these populations, the sex ratio was skewed, with more males than females. Holtby and Healey hypothesized that relative growth and survival are linked. In some populations, both sexes enjoy similar survival rates and males grow larger than females. However, in other populations, for unknown reasons, large size is especially favored in females but their growth requires more aggressive feeding, which comes at the cost of increased mortality rates. Female coho salmon were caught in troll fisheries (i.e., baited hooks or lures) more often than males, and the authors interpreted this as evidence that the females were feeding more aggressively.

One might ask whether the sexes experience skewed survival and growth rates throughout life or just at sea. Coho from Big Beef Creek, Washington, show a preponderance of males and a tendency for females to be larger as adults (David Seiler, Washington Department of Fish and Wildlife, personal communication). Over a 16-year period, females averaged 58.2 cm versus 53.5 cm for males, not including jacks, but there were 10,141 adult males and only 8562 females. Genetic examination of smolts revealed an even sex ratio (Spidle et al. 1998), so the bias in survival occurs at sea. At Big Beef Creek, counts of upstream migrants in 12 of the years included jacks, and in these years only 34.8% of the coho salmon were females, with 42.9% adult males and 22.3% jacks. The fishery, of course, takes almost exclusively the older males and the females, so there are more jacks in the creek than the actual "production" of this life-history type.

Among the more extensive data sets I have seen regarding sex-specific growth in freshwater and at sea are the records of coho salmon in the size and time of release studies led by Bilton (Bilton, Morley et al. 1984; Morley et al. 1988). Male smolts were slightly but consistently larger, not smaller, than females; in thirty out of thirty-six release groups the males were larger, on average, than the females. In contrast, the females were much larger than the males as adults. These patterns are intriguing, and certainly merit further investigation.

Besides the differences in age composition and size between sexes, there are well-documented patterns among populations at both regional and local scales. At the regional level, figure 16-7 showed the general decrease in size of pink and chum salmon at higher latitudes. In the chum salmon this tendency is matched by a tendency to mature at an older age farther north, so the absolute size at maturity is compensated (Salo 1991). Sockeye salmon from the southern part of the range tend to spend only 2 years at sea whereas those from more northerly populations more often spend 3 years at sea (Burgner 1991), and the modal age at maturity for chinook salmon shifts from mostly 3 years in California, to 4 in Oregon and Washington, to 4 and 5 in British Columbia, and to mostly 5 in Alaska (J. M. Myers et al. 1998). Some of this shift in total age results from the increased proportion of stream-type chinook in the north, but the marine age also varies. Such regional variation is presumably a consequence of general patterns in productivity of the ocean, though rivers in different regions might also have some common traits that would favor fish of one size or another.

There is also a great deal of variation in size at much more local scales, and this cannot be easily explained by differences in growing conditions. For example, Rogers (1987) found that sockeye salmon spawning in larger tributaries of the Wood River lakes of Bristol Bay tended to spend 3 years at sea, whereas those in smaller streams tended to mature after 2 years (fig. 17-2). At still finer scales, there is variation in length at a given age and weight at a given length for salmon spawning in discrete habitats. Hansen Creek sockeye salmon are younger on average, shorter for a given age, and lighter for a given length than those from Bear Creek, located just a few kilometers away in Lake Aleknagik, within the Wood River system (table 17-4). The juveniles rear in the same lake, migrate

FIGURE 17-2. Percentage of female sockeye salmon that spent 3 years at sea (rather than 2), averaged over about 4 decades of spawning-ground samples from 17 rivers of different sizes in the Wood River system, Bristol Bay, Alaska (Rogers 1987; Quinn, Wetzel et al. 2001, and unpublished data).

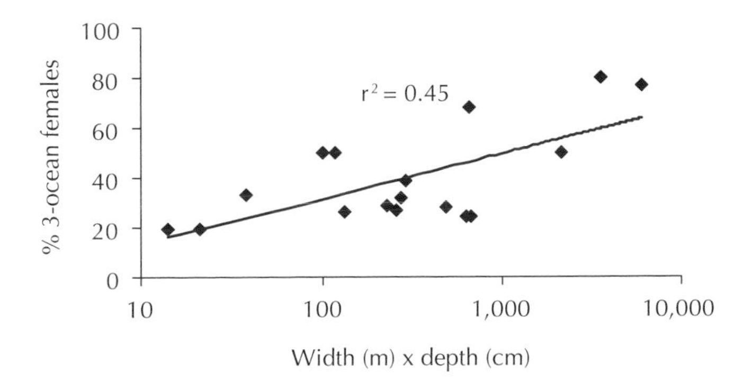

TABLE 17-4. Percentage of the sockeye salmon on the spawning grounds aged 1.2 (rather than 1.3), the average length (mideye to hypural plate, in mm) of fish of these two age groups, and the average weight, adjusted to a standard length of 450 mm for populations spawning in two tributaries of Lake Aleknagik, Alaska (based on 100 fish of each sex and population, in each year 1998–2003; Quinn, unpublished data).

Sex	*Creek*	*% 1.2*	*1.2 length*	*1.3 length*	*Weight (g) at 450 mm*
Females	Bear	46.8	420.6	485.4	2201
	Hansen	73.2	414.4	475.7	1995
Males	Bear	42.3	445.6	507.5	2531
	Hansen	70.3	424.6	484.4	2144

to sea at similar dates (almost exclusively after 1 year in freshwater), presumably have access to similar areas at sea, and return and spawn at the same time. So, the variation in size and age seems to be caused by natural selection related to breeding habitat, discussed in more detail in the next chapter.

Causes of variation in age at maturity

As with so many things in life, the age at which salmonids mature is controlled by a combination of nature and nurture: genetic and environmental factors. Genetic control over age at maturity might be inferred from the patterns of local variation among populations just described, and a number of experiments support this inference. Hankin et al. (1993) took adult chinook salmon from the Elk River, Oregon, and mated age-2 males with females aged 4 and over (4+) and also mated age-4+ males with age-4+ females. In this experiment the jacks produced more jacks and fewer age-4 males than did the older males (fig. 17-3). Further evidence for a genetic control over age at maturity comes from other experimental breeding studies. Iwamoto et al. (1984) reported that University of Washington

FIGURE 17-3. Age at maturity of male Elk River, Oregon, chinook salmon produced by mating females (aged 4 or greater) with jacks (age 2: open bars, designated 2 x 4+) or by females with older males (hatched bars, designated 4+ x 4+). The data from brood years 1979 and 1980 were averaged (from Hankin et al. 1993).

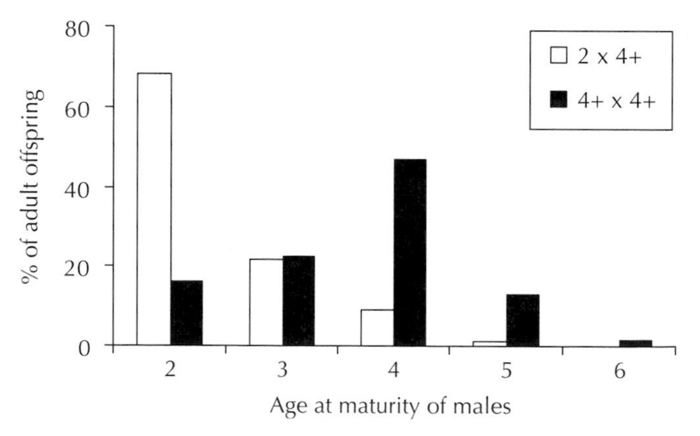

hatchery coho salmon sired by jacks had 4.6 times as many jacks as those sired by older males. Heath et al. (1994) reported a smaller but still significant effect in chinook salmon; under controlled rearing conditions, 44.8% of the males sired by jacks became jacks compared to 26.9% of the males with nonjack fathers. Tipping (1991) produced Cowlitz River steelhead from parents that had spent 2 or 3 years at sea and the progeny tended to follow the parental age at maturity. The age-2 parents produced 57.0% age-2 offspring and 43.0% age-3, whereas the age-3 parents produced 20.6% age-2 offspring and 79.4% age-3. Finally, Hard et al. (1985) reared chinook salmon from the Chickamin and Unuk rivers, Alaska, and released them from a common hatchery. The rates of early maturity (age 4 or younger) were much lower in the Chickamin River males compared to Unuk River males (5.4–16.7% vs. 70.6–96.0%), reflecting the patterns in the native populations.

The abundant correlative and experimental evidence for genetic control over age at maturity must be balanced by equally abundant evidence of environmental control. The environmental control operates in a manner similar to that controlling the age of smolt transformation; rapid growth is associated with early transition. It has been known for some time that the fastest growing salmon of a given cohort (population and year class) tend to mature early and that the slower growing individuals stay at sea. Thus the fish that are large at maturity are actually the slowest growing members of the population. For example, Parker and Larkin (1959) developed the relationship between the size of chinook salmon and their scales, and they estimated the growth history of salmon that matured at different ages. The older salmon were larger on an absolute basis than those maturing earlier in life, but the oldest fish were consistently the slowest growing representatives of the group still at sea (table 17-5). Another way of showing the linkage between growth and maturation is to examine the average sizes of salmon caught in a given area, separated by number of years spent at sea and by state of maturation. LaLanne (1971) provided such data from chum salmon; figure 17-4 reveals that almost all the year's growth seemed to occur from the beginning of June through the end of August and that the maturing fish of a given age were much larger than the immature fish.

Because maturity is linked to size or growth rate earlier in life, we may expect both smolt size and growth rate at sea to affect maturation. The kinds of studies conducted to determine the optimal combination of smolt size and release date for survival described in chapter 15 also revealed that smolts released at a large size tended to mature as jacks. For example, almost all male chinook salmon from the University of Washington

TABLE 17-5. Back-calculated lengths of chinook salmon (in mm) at different ages, as a function of their final age at maturity (data from Parker and Larkin 1959).

		Age				
Age at maturity	*Sample*	*1*	*2*	*3*	*4*	*5*
0.1	4	266				
0.2	27	220	508			
0.3	150	183	466	676		
0.4	60	171	425	622	813	
0.5	8	122	368	549	730	917

FIGURE 17-4. Lengths of Asian chum salmon sampled at sea west of 180° longitude that were maturing or immature, at four successive ages. Note that the maturing fish were consistently larger than the immature fish of their age and also that most growth occurred in only a few months of the year (from LaLanne 1971).

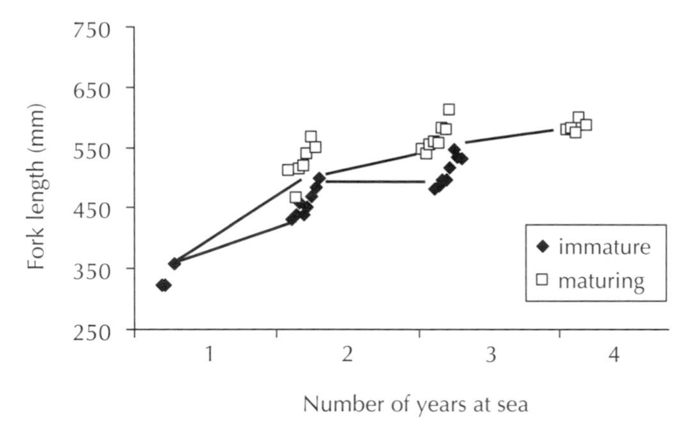

hatchery matured at ages 1–4 (ages 1 and 2 being jacks), and females matured at ages 3 and 4. The proportion of jacks increased with smolt size, and age-1 jacks were rare when smolts were released weighing less than about 12 g (fig. 17-5) . Likewise, releases of larger smolts increased the proportion of females maturing at age 3 rather than staying at sea for another year.

The effects of smolt size on age at maturity can be seen in wild populations as well. For example, data on ages of 22,468 adult sockeye returning to Iliamna Lake, Alaska (Fisheries Research Institute, University of Washington, unpublished data) revealed that there were more than five times as many jacks from age-2 smolts than from age-1 smolts (1.79 vs 0.33% of all adults). The age-2 smolts also produced fewer adults that spent 3 rather than 2 years at sea compared to age-1 smolts (25.0 vs 32.6%). Growth rates at sea also affect age at maturity, as shown by extensive data on Japanese chum salmon. Hatcheries

FIGURE 17-5. Relationship between the annual average size of chinook salmon smolts and the age composition of returning adults at the University of Washington hatchery. The left panel shows the percent of all males from that brood year that matured as jacks (ages 1 and 2, and the line) and "mini-jacks" (age 1). The right panel shows the percentage of females that matured at age 3 rather than age 4. Data are for brood years 1955 to 1996 (Quinn, Vøllestad et al. 2004; Vøllestad et al. 2004).

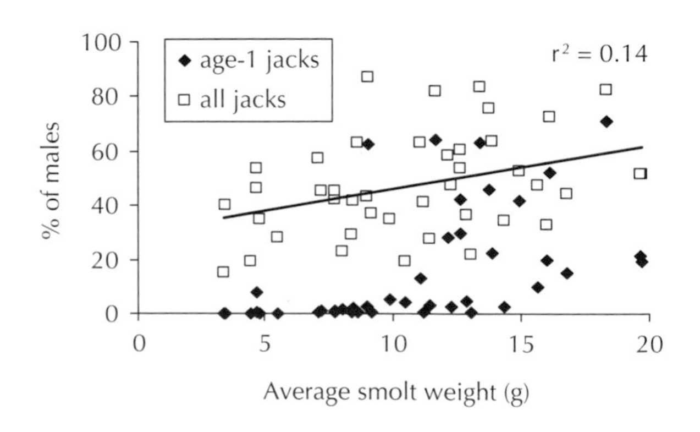

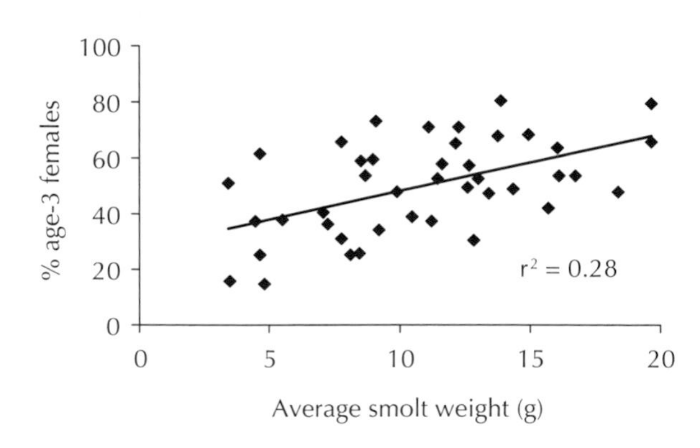

FIGURE 17-6. Relationship between the number of adult chum salmon returning to Hokkaido, Japan, hatcheries and the average age at maturity for brood years 1963–1993 (from Kaeriyama 1998).

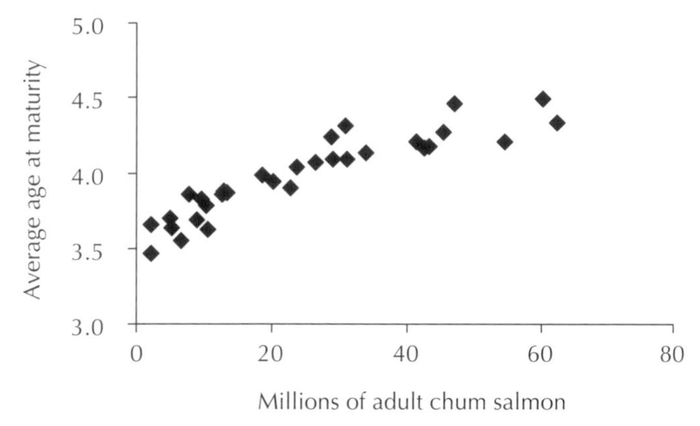

have been releasing more juveniles, leading to a reduction in size at a given age and an increase in average age (fig. 17-6).

Pacific salmon mature only once and die after the breeding season. However, age and size at maturity is much more complicated in the iteroparous species, and comparisons between anadromous and nonanadromous populations are also tricky. First, salmonids often grow faster at sea than they do in freshwater. So, for example, a sockeye is typically much larger after 1 year in a lake and 2 years at sea than a kokanee is after 3 years in the lake. Trout residing in small streams also grow slowly; Fuss (1984) reported that age-5 cutthroat in the Clearwater River system of coastal Washington were only 234 mm long. There are many exceptions, especially when salmonids rearing in lakes get large enough to eat fish, as this can accelerate their growth considerably. Such nonanadromous trout may be larger than some anadromous ones. For example, the age-6 anadromous cutthroat at Eva Lake, Alaska, were only 306 mm (Armstrong 1971) and those from Sand Creek, Oregon, were 366 mm (Sumner 1962), but the adfluvial ones in Lake Washington averaged 491 mm (Nowak et al. 2004).

The cutthroat trout from Pyramid Lake, Nevada, were as large as chinook salmon, and rainbow trout in some lakes also grow very large (photo 17-3). Iteroparity and longevity are the rules in these cases. For example, Iliamna Lake is known for large rainbow trout. Seidelman et al. (1973) trapped rainbows in two of their major spawning rivers (both of which also support large sockeye populations, on whose eggs the trout feed). The trout matured late in life, and grew to a large size, and those in the slower-growing population (the Copper River) matured later than those in Lower Talarik Creek (table 17-6). In addition to the differences between anadromous and nonanadromous populations, and the rapid growth of some trout in lakes, there are also differences in life history between stream-resident populations and fluvial populations that migrate within river systems. Meyer et al. (2003) sampled Yellowstone cutthroat trout in eleven Idaho rivers and creeks and found that trout in the migratory populations were larger at maturity and matured later, as indicated by the maximum size and age of immature fish and the minimum size and age of mature fish. Tallman et al. (1996) reported similar linkages between life history and migration in Arctic char. Char in the anadromous

17-3 The Gerrard strain of adfluvial rainbow trout is known for its large size, often over 10 kg. Fish such as this one rear in Kootenay Lake, British Columbia, and migrate up the Lardeau River to spawn near the outlet of Trout Lake. Photograph by Ernest Keeley, Idaho State University.

populations were larger than nonanadromous char at age 5 (median: 270 vs. 208 mm), matured later in life (age 8.0 vs. 4.5), lived longer (13.5 vs. 10.0 years), and had much higher fecundity (2648 vs. 264 eggs).

Downward trends in age and size of Pacific salmon

For more than 20 years, researchers have noticed declines in the average age, size, or size at a given age in many (but by no means all) salmon populations (e.g., Ricker 1981, 1995; Bigler et al. 1996; Helle and Hoffman 1998). Ricker (1980) reviewed the data and proposed several hypotheses to explain such declines. He was writing specifically about chinook salmon so some hypotheses do not apply to all salmon species. The hypotheses are also not mutually exclusive; several may operate on some populations. Taken together, however, they are an excellent review of the subject. I have reordered his hypotheses into themes, rephrased them, and added some hypotheses but tried to preserve the essence of his ideas.

Theme 1: Biased data

It is possible that the declines in size, or at least some part of them, do not result from changes in salmon themselves but from statistical artifacts. First, if there is a shift from catching mature fish near the river mouth to fishing earlier in the season and farther away, the average size of fish in the catch will be smaller just because they have not finished growing. These and other shifts in the gear, timing, and location of fishing can affect the average size in the catch, and they complicate analysis of trends in fishery statistics. Second, increased fishing pressure might cause more fish to get caught in the year before they would have matured, so the average size in the catch will go down. This

TABLE 17-6. Growth and maturity schedules of rainbow trout in two tributaries of Iliamna Lake, Alaska, sampled in summer 1972 (Seidelman et al. 1973). Age refers to the number of full years (i.e., winters) of life, percentage of spawners refers to the percentage of all spawning fish of each age, and length is in mm. A single age 11 fish 808 mm long from Lower Talarik Creek was omitted from the table.

	Lower Talarik Creek		*Copper River*	
Age	*% spawners*	*Length*	*% spawners*	*Length*
0	0	no data	0	40
1	0	no data	0	71
2	0	143	0	85
3	0	179	0	154
4	0	216	0	212
5	2.3	270	0	263
6	9.3	353	0.9	351
7	38.4	496	12.8	416
8	34.9	584	48.7	467
9	9.3	661	26.5	527
10	4.7	688	11.1	598

hypothesis is particularly relevant to chinook salmon, and to the period in the latter half of the twentieth century when there was significant fishing on the high seas (e.g., Harris 1987), but is less applicable now to the other species because almost all are caught in the year they mature.

Theme 2: Genetic changes

Three types of genetic changes might contribute to declines in size or age at maturity. First, there is variation in the fish size among populations, and if the populations producing very large fish have been selectively reduced or eliminated, then the overall average would go down. For example, if the chinook salmon produced in the upper Columbia River were especially large, their elimination by Grand Coulee Dam would reduce the overall average. Second, a long-term tendency of fisheries (e.g., trolling for chinook or high-seas fisheries) to catch immature salmon would be a form of genetic selection against delayed maturity because fish maturing at a younger age would be less likely to get caught. Third, fishing gear (especially gillnets) can selectively catch larger fish (either older fish, e.g., table 17-2, or larger fish for a given age), changing the genetic makeup of the population by favoring younger, smaller fish. Growth rate and age composition are heritable, as we have seen, so this is plausible in some cases.

Theme 3: Environmental changes

Physical changes in the ocean environment such as temperature, salinity, upwelling, and competition (salmon density) could affect growth and thus size at a given age or age structure. As we have seen, Ricker was prescient in this regard, as the 2 decades that followed his paper have seen abundant evidence of these kinds of effects over broad geographic areas (e.g., Pyper et al. 1999). However, there is a great deal of local variation and even inverse correlations among areas, so it is not clear that this is a sufficient overall explanation, though it is obviously an important part.

Theme 4: Effects related to hatchery propagation

Hatchery propagation of salmon can affect size and age at maturity in two ways. First, hatchery smolts tend to be larger than local wild fish, as a consequence of heavy feeding and also elevated temperatures in some cases, and smolt size affects age at maturity. For example, Norris et al. (2000) reported that wild chinook salmon from a population known as "upriver brights" in the Columbia River system were significantly older than representatives of their population produced in the nearby Priest Rapids Hatchery. Of the wild fish, 33% were age 4 and 46% were age 5, compared to 70% age 4 and only 30% age 5 among the hatchery fish. The second possible hatchery effect is a genetic one. Staff in many hatcheries have favored large fish and have often avoided using jacks for spawning. This has probably stemmed from a sense that large fish would produce large fish. Old fish may beget old fish but such mating schemes also select for slow growth, so the size at a given age might go down.

All of these hypotheses have merit, though some are more broadly relevant than others. Taken together, they give a good picture of the kinds of factors that can affect age and size of mature salmon. It is important to bear in mind that not all trends are downward. Indeed, Ricker (1995) reported several areas where chinook size declined until the late 1970s (when his earlier paper was written) but then increased. Moreover, many populations are exposed to a combination of these effects, depending on the nature of the data being examined, whether the salmon are wild or hatchery produced, the gear and regulations of the fisheries exploiting them, and the region of the ocean where they feed.

Summary

Information on the age of salmon is critical for understanding their basic biology and population dynamics. Age is determined by examination of annual rings, visible on scales, or otoliths, and from tags or marks left by humans. The age-designation systems are designed to reveal not only the total age of the fish but the number of years spent in freshwater and at sea. Salmon species have typical periods of time spent at sea (e.g., 1 year plus a summer for coho and pink salmon, 2 or 3 years for sockeye, 2 to 4 years for chum and chinook), but populations in the northern part of the range tend to be older, spending more time in both freshwater and at sea before maturing. In addition to these latitudinal patterns, there is local variation in age, apparently related to site-specific regimes of selection and growing conditions. In general, rapid growth leads to earlier maturity, but age at maturity is also under genetic influence. Besides variation in growth and maturity schedule among species and populations, males tend to grow faster at sea than females (with coho being a noteworthy and puzzling exception) and vary more in age at maturity. Finally, declines in age or size have been observed in many populations of Pacific salmon over the past decades. Several hypotheses, related to natural and human-related processes, have been proposed to explain these declines and all have merit for at least some species and populations.

18

The Evolution and Structure of Salmon Populations

By now, the salmon are essentially back where they were at the beginning of the book: starting the process of sexual maturation and homeward migration from the open ocean to their natal river. The book's final two chapters bring together themes from throughout their life history by way of conclusion. This chapter considers the evolutionary processes underlying the population structure of salmon, and the last one will consider patterns of abundance. In so doing, I will connect the basic behavior and ecology of salmon to some pressing conservation issues.

In 1938, the American Association for the Advancement of Science organized a remarkable symposium called "The Migration and Conservation of Salmon" (Moulton 1939). The papers and published discussions from this gathering make it clear that these scientists fully understood the fact that homing behavior leads to the isolation of populations and that salmon must be managed on a population-by-population basis. Rich (1939, 45–46) wrote, "It is apparent then that one of the first requirements of a sound conservation program must be the determination of the extent to which the species to be conserved is broken up into local populations." He pointed to three lines of evidence that Pacific salmon populations are isolated: (1) differences in phenotypic traits such as body size, egg size, fat content when leaving the ocean, chronology of migration and spawning, and counts of fin rays, vertebrae, and other body parts among salmon from different rivers; (2) persistent differences in age composition and patterns of abundance (both average density and tendency to cycle) of salmon from different rivers; and (3) marking experiments showing that juveniles leaving one river tend to return there rather than to other nearby rivers.

These early scientists did not freely use the terms "evolution" and "natural selection" but we now integrate those concepts into the fundamental equation for salmon populations: homing + local patterns of selection + heritability = evolution of populations. The first element, homing, isolates the fish from different rivers at the time of reproduction. Thus salmon that originated from one stream are more likely to breed with other salmon from that stream than they are to breed with salmon from other streams. As discussed in chapter 5, the extent to which populations are isolated depends partly on distance. So, for example, sockeye salmon spawning in two tributaries of a lake may occasionally stray between creeks, but may less often stray to creeks flowing into a different lake in the same drainage basin, and still less often stray to creeks in entirely different basins. Straying among basins becomes progressively less common with distance, so there is essentially no interbreeding between members of a given species at the extremes of the geographical range. Straying and genetic isolation are probably also influenced by habitat features, as salmon may be more likely to stray to similar rather than different habitats (for example, among tributaries of a lake rather than between inlet and outlet populations). Finally, the degree of overlap in breeding dates will strongly affect opportunities for interbreeding between populations.

Genetic isolation

This genetic isolation is indicated by the kinds of evidence reported by Rich (1939), but it is revealed more elegantly by biochemical and molecular techniques. As briefly described in the introductory chapter on the evolution of salmon and trout, there are proteins that occur in more than one form with no apparent difference in function. We can designate these forms "*A*" and "*a*" for simplicity, and assume that they exist in equal proportions (i.e., 50% *A* and 50% *a*) in the salmon in two different rivers. After a breeding season, we would expect the proportions to remain the same, if each population was large and the fish freely interbred. However, this is like flipping a coin. If we flipped it a vast number of times we would expect to get heads 50% of the time, but if we flipped it only fifty times we might not be surprised to get 60% heads, or 40%. If the salmon populations were small, after a breeding season, type *A* might constitute 60% of the fish in one population and only 40% in the other. In the next generation, if we combined all the genes we would again expect a 50:50 ratio. However, if the populations were isolated, they would tend to retain the difference that had arisen by chance in the previous generation, rather than return to the original ratio. Persistent differences in gene frequencies, resulting from such "genetic drift," indicate that the populations are not freely interbreeding.

Pioneering work (e.g., Utter et al. 1973) led to numerous studies determining the relative similarity among salmonid populations in the alternate forms of their proteins, known as allozymes (reviewed by Utter 1991). More recently, molecular markers (mitochondrial and nuclear DNA) have been developed for similar purposes (Carvalho and Hauser 1994). The general conclusion of these studies is that salmon populations exist in genetic isolation from each other, along a continuum as described in the paragraph above. In general, populations are more genetically similar (in these selectively neutral traits) to nearby populations than they are to ones farther away, and this is termed "isolation by distance." However, these patterns vary among species (Hendry, Castric et al. 2004). For a given geographical distance, pink and chum salmon populations tend to be

FIGURE 18-1. The relationship between the geographical distance among populations and the degree of genetic difference (FST). The lines are based on relationships for specific populations (Hendry, Castric et al. 2004).

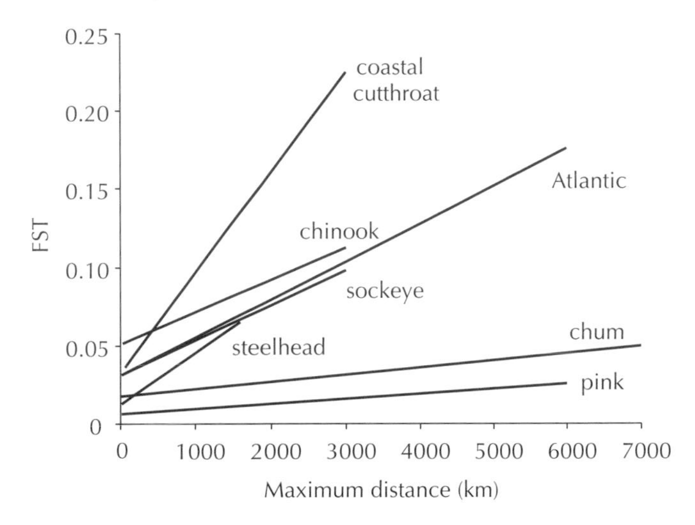

more similar to each other than to populations of the other anadromous species (fig. 18-1). Chinook and sockeye salmon, steelhead, and Atlantic salmon show strong genetic differentiation with distance, and these patterns are consistent with predictions about variation in straying from the qualitative models presented in chapter 5. The few studies on anadromous cutthroat trout showed particularly high levels of genetic differentiation, perhaps resulting from their limited marine migrations rather than their superior homing abilities. Consistent with theory, nonanadromous cutthroat trout populations isolated by barrier waterfalls are very distinct from each other (Latterell 2001).

Gene flow in salmonid populations is controlled by two opposing forces: homing (leading to isolation) and straying (leading to homogenization). However, these contemporary forces are only partially responsible for the genetic structure of salmonid populations. Much of the present range of salmon was glaciated quite recently, and the attributes of today's populations are affected by whether they descended from lineages that spread southward from the Bering Refuge or northward from the Cascade Refuge. Thus the levels of genetic differentiation among populations reflect both current rates of straying and gene flow and postglacial colonization.

The genetic diversity of North American sockeye salmon illustrates the importance of both past and present processes (Wood 1995). "River-type" sockeye rear in rivers for a year before migrating to the ocean, and "ocean-type" sockeye migrate to sea in their first year of life. Both these forms are often associated with glacial rivers (Gustafson and Winans 1999). Sockeye, coho, and Dolly Varden were the first salmonids to colonize new streams in Glacier Bay, Alaska (Milner and Bailey 1989). Populations of river- and ocean-type sockeye are less genetically differentiated, for a given geographical distance, than the more typical lake-type populations (Gustafson and Winans 1999). Wood (1995) hypothesized that the sea and river types are the colonizing forms of sockeye, and the lake type is the derived (though presently most common) form. The evolution of sockeye populations may proceed from postglacial colonization by ocean-type fish, to lake-type populations if a suitable lake is present, and then to kokanee if there is some combination of good growing conditions and an arduous migration.

The existence of stream-type and ocean-type chinook salmon is another case in which we can view present conditions from different perspectives. There are strong environmental correlates of these life-history patterns (ocean-type prevail in areas facilitating growth; E. B. Taylor 1990a). We might imagine that each river was colonized by some generalized chinook salmon, which then gave rise to stream-type fish in the headwaters (with such traits as upstream orientation, territorial behavior, delayed migration, and slow growth; Taylor 1990b) and ocean-type in the lower reaches (Brannon et al. 2002). This is consistent with the rapid establishment of both life-history types among the newly evolved populations in New Zealand (Unwin et al. 2000). On the other hand, Healey (1983, 1991) proposed that chinook salmon exist as two distinct "races" (stream-type and ocean-type) that are more similar to distant populations of their race than to nearby populations of the other race. Healey hypothesized that the stream-type race might have colonized southward from the Bering Refuge and that the ocean-type race colonized northward from the Cascade Refuge after the last glacial period. Thus, populations that currently share the same river might be very distinct.

Recent synthesis of phenotypic and genetic data on chinook salmon from California to British Columbia suggests that the current population structure in general reflects the parallel evolution of spring and fall patterns of adult return migration timing after post-glacial colonization in many different rivers (Waples, Teel et al., 2004). However, the chinook salmon populations in the interior of the Columbia River basin seem to exist as two divergent lineages, corresponding to spring and fall migration timing, more consistent with Healey's model. Thus, in most rivers, fall and spring chinook may represent recent divergence, but the interior of the Columbia basin seems to have been colonized by two separate lineages that remain very distinct despite their current proximity to each other (Waples, Teel et al., 2004). Ecological and genetic data provide complementary sources of information on population structure; Utter et al. (1993) concluded that the biochemical and molecular markers clarify ancestral relationships whereas the adaptive traits (see "Heritable variation in phenotypic traits," below) better indicate current ecological pressures, and together these two categories of traits "constitute a hierarchy of genetic variation that is useful in defining distinct population segments" (69).

This hierarchy results from the fact that studies of selectively neutral traits can demonstrate the patterns of ancestry and present level of gene flow (or lack thereof) between populations, but such studies provide no information on the functional differences between the populations. That is, the variant forms of proteins and DNA fragments are genetically controlled but have little or no effect on fitness. The level of gene flow between two populations results in part from the balance between homing and straying, and this is indicated by the first part of my conceptual equation (remember, homing + local patterns of selection + heritability = evolution of populations). However, gene flow is not the same thing as straying; the reproductive success of the strays determines the gene flow between populations. Tallman and Healey (1994) reported that the level of straying by chum salmon between two small streams on

Vancouver Island (demonstrated by marking juveniles and recovering them as adults) was too high to be compatible with the observed level of genetic differentiation between populations. The authors hypothesized that the strays were less fit and so produced fewer offspring, per capita, than the local fish. Similarly, Hendry, Wenburg et al. (2000) studied populations of sockeye salmon in the Lake Washington basin (in Washington State), including a large population in the Cedar River and a very small one on Pleasure Point beach. The populations differed in phenotypic traits, despite some straying by Cedar River fish into the beach population. The existence of significant genetic differences indicated that the strays were reproducing at a much lower rate than the local fish, so the phenotypic differences seemed to affect fitness.

Heritable variation in phenotypic traits

When considering the differences in fitness among populations, we must acknowledge Ricker's (1972) landmark work. He reviewed numerous studies, presenting evidence that Pacific salmon populations home and that they differ in various phenotypic traits such as body size, age at maturity, flesh color, timing of migration, and so on. He also summarized evidence that populations transplanted from one river to another retained the ancestral traits (such as timing of migration and spawning), as well as evidence that transplanted populations did not survive as well as local ones. This monumental document solidified "the stock concept" in salmon management and conservation: the idea that rivers produce salmon that are numerically and genetically distinct from those produced in other rivers and so must be conserved as separate units. This concept was further elaborated at a symposium, published as a special issue of the *Canadian Journal of Fisheries and Aquatic Sciences* in 1981, including many papers on salmon and trout (e.g., McDonald 1981). Another symposium, published by the American Fisheries Society, developed many of these concepts further (Nielsen 1995).

The 1970s and 1980s saw a tremendous increase in the quantity and variety of work on differences in traits among salmon populations. Some have been described in this book so far, but it useful to consider such traits in the context of the entire life cycle. Populations differ in the timing of migration from the ocean to coastal areas, fat content in their bodies upon entering freshwater, timing of reproduction, size and number of eggs produced by females of a given body size, morphology of sexually mature males and females, developmental rate and temperature tolerance of embryos, responses of newly emerged fry to the direction of water currents, growth rate, agonistic behavior and duration of freshwater residence of juveniles, disease resistance, timing of seaward migration (or tendency to remain in freshwater), regions of the ocean that they occupy, growth rates at sea, and patterns of age and size at maturity.

The working hypothesis is that these population-specific traits reflect the operation of natural selection on heritable traits, with some degree of environmental influence as well, and they are often referred to as "local adaptations." Taylor (1991) reviewed many such examples in Pacific and Atlantic salmon and concluded that a rigorous proof of local adaptation involves not merely showing variation in traits among populations, but three further criteria: there must be a genetic basis for the trait, differential expression of the trait must affect survival or reproduction, and a mechanism of selection responsible for maintaining the trait must be identified. It is easy to find a measurable trait that varies among populations, but demonstration of a genetic basis for the trait requires a controlled breeding experiment and either captive rearing of the fish or a mark-release-recapture design. Such experiments can be expensive and time-consuming, depending on the trait of interest, but they have been done for many traits. The second and third criteria (demonstration of fitness and a selection mechanism) are especially important because they rule out traits with no adaptive benefit and traits that are universally beneficial in all populations. However, it is difficult to obtain such a full understanding of the fitness consequences of traits and the regimes of selection.

There are a few traits for which Taylor's rigorous criteria are clearly met, some traits for which a plausible case can be made but for which there is no hard, experimental evidence of fitness benefits, and some traits for which benefits may exist but are not obvious. Resistance to pathogens provides some of the most convincing examples of local adaptations. A myxosporean protozoan, *Ceratomyxa shasta*, occurs in the Columbia River drainage system but is generally absent from Oregon coastal streams. Buchanan et al. (1983) reported almost no mortality attributable to the disease in steelhead from three Columbia River basin populations (0–4% mortality in four different experiments for Skamania, Clearwater, and Deschutes fish), but 89–98% mortality in steelhead from the Siletz River, on the Oregon coast. Hemmingsen et al. (1986) demonstrated genetic control over resistance when they observed differences in mortality among coho salmon from Big Creek, Oregon (a resistant population), vulnerable populations from the Smith River, Oregon, and from the Soleduck River, Washington, and hybrids between populations.

Bower et al. (1995) provided another example of resistance to a prevalent pathogen as an example of local adaptation. *Cryptobia salmositica*, a hemoflagellate parasite, and *Piscicola salmositica*, its leech vector, were noted in coho salmon from the Big Qualicum River on Vancouver Island, but not in the Kitimat River on the British Columbia coast; and the organisms occurred in sockeye salmon from the lower Fraser River system, but not from the Skeena River. Bower and colleagues obtained adult salmon from these systems, produced pure and hybrid progeny in the lab, and exposed them to the parasite. The populations that evolved with the parasite were resistant, the naïve populations were susceptible, and the hybrids showed intermediate susceptibility (table 18-1).

TABLE 18-1. Survival of coho salmon from the Kitimat and Big Qualicum (BQ) rivers and their reciprocal hybrids, and of sockeye salmon from Fulton River, Weaver Creek, and their hybrids after experimental exposure to *Cryptobia salmositica* (Bower et al. 1995).

	Experimental populations			
Species	*vulnerable*	*hybrids*		*resistant*
coho	Kitimat x Kitimat	BQ x Kitimat	Kitimat x BQ	BQ x BQ
% survival	3	22	38	95
sockeye	Fulton x Fulton	Fulton x Weaver	Weaver x Fulton	Weaver x Weaver
% survival	43	66	74	88

FIGURE 18-2. Relationship between the median spawning date and latitude of the site for sockeye salmon populations throughout their North American range (from Hodgson and Quinn 2002 and additional records).

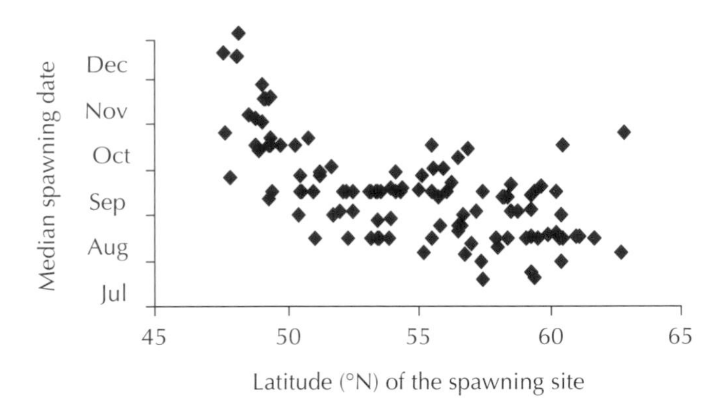

In these cases, the trait (disease resistance) confers obvious fitness benefits, and salmon from susceptible populations that strayed into rivers having the pathogen would fare very poorly. There are other traits that differ among populations, have a genetic basis, and probably influence survival but for which no specific experiment to demonstrate the fitness benefit has been conducted. For example, Hodgson and Quinn (2002) summarized the spawning dates of sockeye salmon populations throughout the North American range and showed a clear pattern of earlier spawning at higher latitudes, consistent with colder temperatures and longer incubation periods (fig. 18-2; see also fig. 8-1). Brannon (1972) and others showed that sockeye salmon fry tended to orient to flowing water in the direction that would facilitate migration to their nursery lake and that this response was genetically determined. It seems likely that fry entering the lake would have a higher probability of surviving than fry migrating downstream, but this was not specifically demonstrated. Similarly, there are correlations between body size and shape among sockeye salmon populations and the depth of the water where they spawn (fig. 18-3), and between stream size (width and depth) and proportion of salmon being killed by bears (Quinn, Wetzel et al. 2001; fig. 7-1). These relationships suggest higher fitness for small, slim salmon in small streams, and higher fitness for large, deep-bodied salmon in large rivers and lake beaches, but the final proof has not been obtained. Sockeye salmon spawning in large gravel have larger eggs than those spawning nearby in very fine sand (Quinn, Hendry, and Wetzel 1995; fig. 18-4). Given the relationship between gravel size and flow of water to the embryos, these differences in egg size probably affect fitness, but experiments have not been done to prove it.

We generally envision the evolution of populations that are isolated, but it is important to point out that some "sympatric" divergence can occur as well. Some of the most elegant examples of this are the divergence of Arctic char into multiple forms within a single lake (Jonsson and Jonsson 2001). These forms differ in habitat (e.g., pelagic vs. benthic), food habitats (zooplankton, fish, benthos), morphology (size, position, and shape of mouth; shape and number of gillrakers), color (silvery in the

FIGURE 18-3. Relationship between the depth of the spawning site and the dorso-ventral body depth of male sockeye salmon (standardized to a body length of 450 mm from mideye to hypural plate) from the Iliamna and Wood River systems, Bristol Bay (from Quinn, Wetzel et al. 2001 and additional unpublished information).

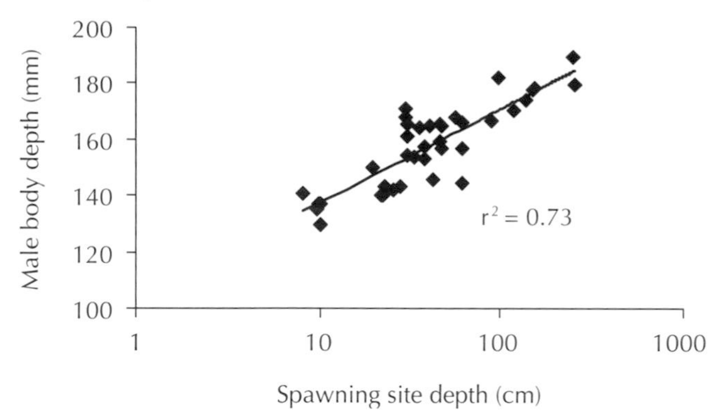

pelagic form or dark in the benthic form), and size at maturity. Reproductive isolation can occur if the forms breed in different habitats and mate assortatively, reinforcing the differences. Kurenkov (1977) presented similar evidence for the divergence of kokanee populations into benthic and pelagic forms within one lake on Kamchatka Peninsula. However, not all cases of multiple forms of a species in a single lake result from sympatric divergence. Lough Melvin, Ireland, has three types of brown trout, locally known as gillaroo, sonaghen and ferox (Ferguson, forthcoming). These three types differ in morphology and ecology and are genetically distinct, due to independent ancestry. They presently home to separate spawning areas, minimizing hybridization since they became sympatric.

FIGURE 18-4. Relationship between the size of gravel used for spawning (based on surface substrate near redds) and the size of eggs, adjusted to a common body length of 450 mm (mideye to hypural plate) for twenty-one sockeye salmon populations in the Iliamna and Wood River systems, Bristol Bay (Quinn et al. 1995 and unpublished data).

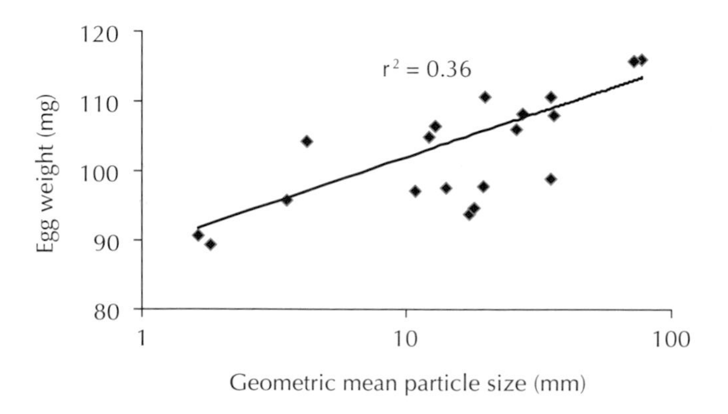

Defining population diversity

At present, salmonid species exist as semi-isolated populations, over large geographical ranges. How do we define and conserve these populations? The answer to the commonly posed question, "What constitutes a population?" depends in part on the management context. For example, early research determined the continent of origin of sockeye salmon caught on the open ocean. In that context, "North America" was a population, as distinct from Asia. The nature of geography and fisheries makes Bristol Bay, Alaska, a meaningful level of population structure within North America. Sockeye from lakes draining into the bay seem to share a common ocean distribution, life history, timing of adult migration, and dynamics, so this is a "stock complex." The Bristol Bay commercial fisheries operate in five major districts, each of which includes one or more large lakes and associated rivers. For example, fisheries in the Naknek-Kvichak district catch sockeye returning to three separate river systems: Naknek, Kvichak, and Alagnak. Only with great management effort can the sockeye returning to each river be fished at different rates. Thus, for example, the Iliamna Lake system, draining into the Kvichak River, is the finest level of population structure that can be accommodated by fishery management. Its spawning areas can be categorized into four primary types of habitats: small creeks, rivers, mainland beaches, and island beaches. So, are there 4 populations or about 100, the number of identified spawning areas, or something in between? There are many islands; is each one a population, or is each discrete spawning beach on each island a population? At what point does homing give way to habitat selection? How much straying between areas makes them functionally a single population?

Such questions about population structure inevitably lead to the question, what level of conservation and management do the salmon using major habitats and discrete sites require? Within the Bristol Bay system, the demographic stability of the stock complex as a whole depends greatly on the persistence of sockeye using different lakes and habitats within them. During the past century the abundance and productivity of discrete populations has fluctuated markedly, perhaps because they respond differently to climate changes (Hilborn et al. 2003). Similar questions about population structure and management could be posed for other species of salmon in large river systems such as the Yukon, Columbia, Fraser, and smaller systems as well. Like most matters of taxonomy, the definition of a "population" is drawn more in shades of gray rather than black and white. We can quantify similarities and differences between populations in life-history traits, reflecting current selection, and differences in selectively neutral traits, reflecting ancestral lineages, but how do we integrate the two kinds of information?

Practical application of the population concept is driven by the fact that differences in both carrying capacity and productivity among populations require that fisheries exploiting them be regulated separately, and these differences in population dynamics depend on life-history traits and local variation in habitat quantity and quality (Thompson 1962). In the United States, the trend has been toward increasingly fine-scale management of populations over the past 30 years, spurred in part by three major events: the allocation of salmon fishing rights between Native American and nonnative groups, the

TABLE 18-2. Diversity, in terms of ecology, life history, and genetics, of salmonids from Washington, Oregon, Idaho, and California, and number of recognized Evolutionarily Significant Units (ESUs) (from Waples et al. 2001).

Diversity	*pink*	*sockeye*	*chum*	*coho*	*chinook*	*steelhead*	*cutthroat*
Ecology	2	4	4	6	11	11	6
Life history	1	6	1	1	7	7	2
Genetics	2	9	2	2	10	7	3
ESUs	2	7	4	7	17	15	6

U.S.–Canada treaty governing interceptions of salmon, and the application of the U.S. Endangered Species Act (ESA) to salmon populations (see also chapter 19). In response to petitions to list salmon populations under the ESA, the National Marine Fisheries Service conducted "status reviews" of the species, and they have been cited throughout this book as valuable sources of information.

Faced with the need to conserve salmonid populations in a comprehensive yet practical manner, Waples (1991, 11) applied the Evolutionarily Significant Unit (ESU) concept, stating that for a population or complex of populations to constitute an ESU it "must be substantially reproductively isolated from other conspecific population units, and it must represent an important component in the evolutionary legacy of the species" (see Waples 1995 for a review of this concept and its application to salmon). In some cases, the ESU was effectively one population (e.g., the sockeye salmon in Redfish Lake, Idaho), but other ESUs include salmon spawning in several distinct rivers that share many life-history traits and ecological conditions and that share a genetic lineage. As summarized in table 18-2, the greatest diversity within the U.S. Pacific Northwest region is in chinook and steelhead; intermediate diversity is found in coho, cutthroat, and sockeye; and least diversity occurs in chum and pink. These assessments reflect overall levels of diversity and also the fact that there are fewer pink, chum, and sockeye populations within the region.

Transplants within the species' range

Some of the most powerful evidence for local adaptation of salmon populations and the greatest insights into the evolution of populations comes from efforts to transplant salmon and trout to rivers within and outside their native range. Transplanting salmon to rivers with existing runs of the same species was a common practice in the early and mid-1900s. However, these efforts were seldom accompanied by proper marking and there is virtually no evidence that existing runs were significantly augmented. There were also many attempts to establish runs of salmon in rivers where they were absent, and the failure of these efforts is truly remarkable. A few examples illustrate the point, taken from the excellent reviews by Withler (1982) and Fedorenko and Shepherd (1986).

Puget Sound has many rivers with adult pink salmon in odd-numbered years, but only one with even-year pinks, and this was regarded as a waste of potential production.

To rectify this situation, 85 million fry were released into Puget Sound streams in nine of ten odd-numbered years between 1915 and 1933. These fry were produced from even-year runs in Afognak, Cordova, and Yes Bay streams, Alaska. Attempts ceased after returns were poor. From 1949 to 1959, about 1.7 million fingerlings from Alaska and from the Lakelse River, British Columbia, were reared and released from saltwater ponds adjacent to Puget Sound streams in an effort to avoid size-selective mortality. The returns were again poor and no even-year stocks were established. Likewise, 32 million eggs planted in a spawning channel on Robertson Creek, British Columbia, from 1959 to 1964 failed to establish a run, despite good production of fry. There were also extensive efforts to augment a weak pink salmon run in the Big Qualicum River, a Vancouver Island river that supports other salmon runs. In 1963 and 1964, 12.6 million eggs were planted. Survival to emergence was poor, probably owing to siltation and dig-up by chum salmon. However, in 1963, 5.8 million eggs produced 1.55 million fry, and only about 100 adults in 1965. In 1964, 6.85 million eggs produced 2.97 million fry and 11,940 adults returned in 1966. They spawned naturally and produced about 3,000 adults in 1968, which produced about 300 adults in 1970, and the run dwindled further.

A particularly innovative effort was made to develop the odd-year pink salmon run to the Bear River, flowing in Johnstone Strait, British Columbia. This river had an even-year run of up to 100,000 but very few odd-year pinks, so scientists sought to mix local male genes with eggs from a donor source, the Glendale River, British Columbia. Three experimental groups were produced: (1) Glendale River eggs and cryopreserved sperm from even-year Bear River males; (2) Glendale eggs and sperm from Bear River males from the even-year line that had been reared in captivity on an accelerated schedule so that they matured at age 1; and (3) Glendale eggs and sperm. In total, 1.6 million fry were released in 1976. An intensive survey of commercial catches and canneries revealed very few fish, and no marked fish returned to the Bear River. These and other efforts led Withler (1982, 22) to conclude that "the record for success in establishing natural self-sustaining runs in barren waters within the salmon's native range is dismal." With the exception of the transplant of sockeye to Lake Washington (see below), Withler found that "no record of undisputed successful transplantation exists in situations where no obvious physical barrier was apparent" (22). It would seem that salmon have colonized all habitat that is presently suitable for them, and if they are not present in a river there is a good (though not always obvious) reason why not.

The failures of these transplant efforts cannot be readily ascribed to a mismatch between specific local conditions and specific traits in the salmon. However, transplant failures and poor performance of nonlocal populations are not restricted to pink salmon. Brannon and Hershberger (1984) released chinook salmon from the University of Washington (UW) hatchery that included the local population, fish from the Elwha River, and hybrids. Tag recoveries not only revealed differences in spatial distribution (see table 14-6), but they showed higher survival (catch plus return to the hatchery) of UW fish (11.9%) compared to Elwha fish (3.1%), and hybrids showed intermediate performance (6.2%). Some of this might have been related to the difference in age composition between the populations (Elwha, being older, might be expected to have fewer returns), but it is also consistent with poorer performance of the transplanted population.

TABLE 18-3. Survival rate of anadromous masu salmon originating from two rivers in Hokkaido, Japan, reared and released from their own and from the other river in brood years 1984 and 1985 (Mayama et al. 1989). Bold numerals indicate local salmon.

	Release site			
	Shiribetsu		*Shari*	
Origin	*1984*	*1985*	*1984*	*1985*
Shiribetsu	**0.592**	**0.222**	0.106	0.07
Shari	0.046	0.024	**0.969**	**0.91**

One of the most elegant transplant experiments was conducted on masu salmon from rivers on opposite sides of Hokkaido, Japan. Mayama (1989) reported large and consistent differences in phenotypic traits between the Shiribetsu and Shari populations. A reciprocal transplant experiment was carried out; each population was reared and released from its own river and the other river. Each population survived at a higher rate when released from its local site than it did when released from the other site, and at a higher rate than the nonlocal population (table 18-3).

The review of transplants by Fedorenko and Shepherd (1986) identified some cases in which the transplanted population performed as well as the local one, so failure is not inevitable. However, Reisenbichler's (1988) summary of data on thirteen paired releases of local and nonlocal coho salmon from hatcheries is instructive (fig. 18-5). In all cases, the nonlocal fish were brought as milt and unfertilized eggs, or as eyed eggs, and were reared and released in common with the local fish. Coho from within a drainage showed comparable survival to the local fish, but those from farther away showed an apparently exponential decrease in relative survival with distance. The only exceptions to the pattern were the Umpqua and Soleduck coho released from Big Creek Hatchery. The local fish were resistant to the *Ceratomyxa shasta* whereas the transplanted fish were vulnerable, and this may explain why their survival rate was lower than might otherwise be expected for populations so near to each other.

One final example will suffice to show the importance of local adaptation to fitness. The very large population of sockeye salmon spawning in the Upper Adams River, British Columbia, was entirely wiped out by the middle of the twentieth century from a combination of overfishing, local splash dams, and the blockage at Hell's Gate (Williams 1987). These sockeye spawn in a tributary to Adams Lake and are not to be confused with the strong population spawning in Lower Adams River, tributary to Shuswap Lake. As reviewed by Williams (1987), prolonged efforts to reestablish a population in the Upper Adams River met with very limited success, long after the problems causing the extinction had been rectified.

Even the exceptions to Withler's (1982) conclusion that transplants have not created new salmon populations are instructive. In the two best-known cases, the river system itself was modified prior to the transplant effort. Sockeye salmon were successfully established in Frazer Lake, on Kodiak Island, Alaska, but only after a ladder was built allowing them to circumvent an otherwise impassable waterfall. Genetic analysis by Burger et al. (2000) indicated that the habitat types of the donors (e.g., tributary or

FIGURE 18-5. Relative recovery rates of local and nonlocal coho salmon, as a function of the distance between the release site and the river where the nonlocal population originated. Releases from Big Creek (open squares) were not included in the relationship because vulnerability to disease greatly reduced their fitness when transplanted (Reisenbichler 1988).

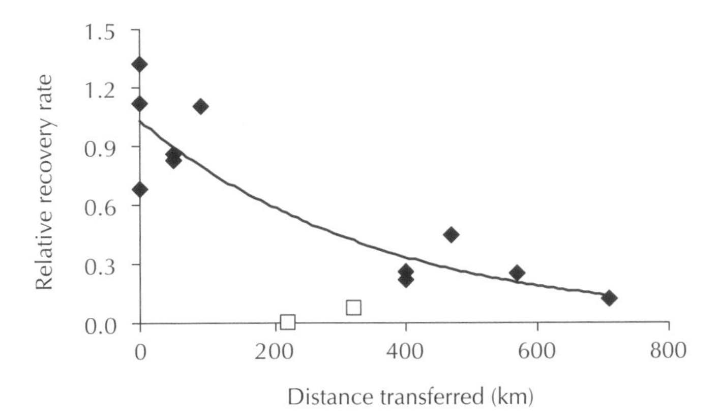

beach spawners) strongly influenced the habitats in the lake where transplants became established, but that some exchange among types also occurred. The other famous case is the transplant of sockeye salmon (photo 18-1) into Lake Washington, starting in the mid-1930s. The basin had kokanee populations prior to the transplant but the presence and distribution of sea-run sockeye is uncertain. There are now self-sustaining anadromous populations in several rivers and some of them (the Cedar River and Issaquah Creek) are almost certainly not native (reviewed by Hendry 2001). However, the Cedar River did not originally flow into the lake, but was diverted into it some 20 years prior to the transplants, and the limnology of lake was also greatly altered in the years following the transplants. It is thus unclear whether the transplants would have succeeded in the lake's original condition.

18-1 Adult male sockeye salmon. Photograph by Thomas Quinn, University of Washington.

Transplants outside the species' range

The transplants of salmonids confront us with a paradox. Most of the local adaptations that we find seem to reflect selection on traits in the freshwater environment (adult size and morphology, timing of migration and spawning, egg size, developmental rate, fry orientation and agonistic behavior, disease resistance, etc.). This is not surprising because streams and lakes differ in so many obvious physical and biological attributes. In contrast, the ocean is available as a pasture for all and is structured on a much broader scale. These facts would lead us to expect that transplants of nonanadromous populations might fail whereas anadromous ones would be no more difficult to establish, and perhaps even less so. However, more than a century of experience has shown us the opposite pattern. Nonanadromous populations of salmonids have been very widely established throughout the world (see reviews by MacCrimmon 1971 for rainbow trout, MacCrimmon and Campbell 1969 for brook trout, MacCrimmon and Gots 1979 for Atlantic salmon, and MacCrimmon and Marshall 1968 for brown trout). For example, within the Northern Hemisphere, rainbow trout were transplanted from the west to the Great Lakes and eastern North America, brown trout to North America from Europe, brook trout from eastern to western North America, and kokanee to numerous lakes within and outside the species' range. Transplants also established self-sustaining populations of salmonids in the Southern Hemisphere, including Australia, New Zealand, South Africa, and Argentina. For example, MacCrimmon (1971) reported naturalized populations of rainbow trout in central and eastern North America, parts of Mexico, much of the west coast of South America, several countries in Europe, parts of India, Sri Lanka and Malaysia, several countries in southeastern Africa, Australia, and in New Zealand. In only a few cases did these transplants result in anadromous populations, whereas self-sustaining nonanadromous populations are common. Thus, salmonids are often capable of establishing self-sustaining populations but anadromy is difficult to transplant. Three examples illustrate this point.

Considerable planning went into an effort to establish pink salmon in Newfoundland, Canada (Lear 1980). Pinks were selected because they were judged least likely to compete with local species and because their lack of specializations for freshwater rearing made them promising. William E. Ricker advised planting large numbers to allow for losses, and he also recommended that care be taken to match the donor population with local features. Newfoundland's marine waters are very cold, so there was something to be said for a northern donor, but the latitude is not high (47–50°). The position of Newfoundland in the western Atlantic Ocean suggested that a Russian donor population might fare better than one from western North America. Incubation rates, temperatures, and the timing of food production in spring were all considered. In the end, eyed eggs from three British Columbia populations were transported and placed in spawning channels in two rivers. In 1959, 1962, 1964, 1965, and 1966, a total of 15.35 million embryos were transplanted. Egg-to-fry survival was poor (38%) in the first year, but it averaged a very acceptable 77.5% thereafter. However, the fry-to-adult (marine) survival was very poor, averaging only 0.09%, with no year even as high as 0.3%. These low survival rates, combined with considerable straying, destined the effort to failure. Only the runs in 1968 and 1973 gave rise to adult generations more numerous than themselves, so the populations dwindled and the transplant failed.

The Great Lakes of North America have seen numerous transplants of Atlantic and Pacific salmonids, dating back into the nineteenth century, and Crawford (2001) provided an excellent historical review. Efforts to establish populations of Atlantic salmon, brown and rainbow/steelhead trout, and coho and chinook salmon have met with varying degrees of success. The Pacific species were stocked heavily and widely for decades and came to support extensive recreational fisheries. Chinook and coho salmon grew rapidly on the abundant forage fish, especially alewife (*Alosa pseudoharengus*) and rainbow smelt (*Osmerus mordax*), and exhibited traits reminiscent of anadromous populations, such as silvery color and epipelagic distribution in the lakes. There has certainly been natural reproduction but it is not clear to what extent the populations would continue, much less support the current levels of fishing, without continued stocking from hatcheries.

In contrast to the moderate results with these cherished species, the notable success story in the Great Lakes was an accident (Kwain 1987; Crawford 2001). The plan was to establish pink salmon populations in Goose Creek, a tributary to Hudson Bay. In January 1956, 513,000 pink salmon embryos and alevins from Lakelse River, British Columbia, were brought to a hatchery in Port Arthur, Ontario, and then planted in Goose Creek. In 1957, 224,112 more juveniles were planted. In 1957 and 1958, no spawners were found in Goose Creek, and this was deemed one more in a long line of failed transplants. However, in 1959, two adult pink salmon were caught by anglers in the mouth of Cross Creek, on the Minnesota side of Lake Superior. Inquiries revealed that about 21,000 "surplus" fry were released down a drain from the hatchery after the rest had been taken to Goose Creek. The drain led into the Current River, a tributary to Lake Superior in Ontario. Presumably, some of those fry survived, reproduced at some unknown location in the fall of 1957, and two of their grandchildren strayed to Minnesota where they were caught in 1959. In 1969 the first pink salmon was found in Lake Huron and in 1973 one was found in Lake Michigan. In 1979 pinks were found in Lakes Erie and Ontario, completing their colonization of the five Great Lakes, with no assistance from humans after the first release.

There are several remarkable facts concerning the establishment of pink salmon in the Great Lakes. First, they would have been considered a very unlikely species to survive there, as they migrate to saltwater immediately after emergence, and freshwater populations are unknown in their native range. Nevertheless, they thrived, even though the number of fry that entered the lake was a tiny fraction of the number planted deliberately in other places on many occasions without success. The number of adults that survived in the first few generations must have been very small. Second, they strayed extensively in the years and decades following the initial invasion. It is unclear if this is because pink salmon stray a lot in general, or if there is a greater tendency for salmon to stray in the early stages of colonization. Third, the apparently fixed, 2-year life cycle so characteristic of the species changed. In 1976, the first even-year spawners were found in two rivers, and it seems that in Great Lakes pink salmon mature at 1 and 3 years of age as well as the more typical 2 years. This indicates that it is not the age at maturity per se that is fixed but some linkage between growth rate and maturity. In their normal range it seems that all pink salmon grow at a rate that triggers maturity at age 2, whereas in the Great Lakes the trophic conditions are so different that growth rates can delay or even accelerate the maturation process. Having said this, I must admit to being puzzled. Pink salmon vary greatly in size within their native range (up to an eightfold range in weights of males in one tiny creek in southeast Alaska: Dickerson 2003). It seems odd that age-3 adults would be so extraordinarily rare in their native range (Heard 1991) but would occur soon after a transplant. One final curiosity may be mentioned in the context of the Great Lakes pink salmon. The founder population, from the Lakelse River, not only failed to establish a run in Hudson Bay, but it had been among the populations that failed in Newfoundland and also failed in Puget Sound. Presumably, they thrived in the Great Lakes from some combination of luck and match between genotype and local conditions.

The most well-documented, successful transplant of anadromous salmon outside their native range is the chinook salmon in New Zealand. Efforts with Atlantic salmon failed to establish anadromous populations, and transplants of spring chinook in the 1870s failed outright (McDowall 1990). However, between 1901 and 1907, 2.1 million fertilized eggs, probably ocean-type chinook from Battle Creek, a tributary of the Sacramento River, were released into the Hakataramea River, a tributary of the Waitaki River in the central part of the South Island's east coast. These salmon survived and quickly colonized the major rivers of the coast, forming populations that not only sustain themselves but support significant fishing as well. The rivers occupied by chinook salmon in New Zealand differ from the Sacramento in physical attributes, especially their steep gradients and absence of significant estuaries, and of course the suite of predators, prey, and competitors are all different as well. It is therefore unclear what aspects of this environment were responsible for the outstanding success of this transplant against the backdrop of so many failures elsewhere.

Regardless of the reasons for their success, the New Zealand chinook salmon provided us with a remarkable opportunity to examine the processes of evolution that must have occurred during the postglacial colonization of rivers now occupied by salmon. M. Unwin, M. Kinnison, and I took advantage of this opportunity and conducted a large-scale study of the environmental and genetic basis for life-history traits (reviewed by Quinn, Kinnison, and Unwin 2001). Our first finding was that salmon from different rivers varied in the kinds of traits that are commonly used to characterize distinct populations in their native range: age at maturity, length at a given age, weight at a given length, fecundity, and the timing of adult migration and spawning. We then bred salmon from families representing two populations, marked them, and reared them under common conditions either to adulthood or until we released them to migrate to the sea from two different hatcheries. These experiments showed a very strong genetic control over some traits (especially the timing of migration and spawning), a more balanced blend of environmental and genetic control over other traits (e.g., growth rate, age at maturity, and egg production), and less genetic variation in other traits such as developmental rate and seawater tolerance.

In less than a century, the New Zealand chinook salmon colonized rivers and evolved significant differences in heritable traits as a result of the isolation of populations by homing and the action of local regimes of natural selection. We believe that the genetic control over timing of adult migration and reproduction is especially important for two reasons. First, selection will act strongly on successful return to the spawning grounds by adults and on emergence of fry at a locally appropriate time in the spring to feed and grow. Thus, adults spawning at an inappropriate date will be quickly culled from the population, moving the mean spawning date. Deliberate selection in hatcheries can rapidly change the spawning date, so such evolution in nature is not unexpected. Second, the spatial isolation that new populations start to experience from homing to natal sites will be compounded by temporal isolation as their spawning dates come to differ. Thus spawning date is both an important, fitness-related trait, and its evolution accelerates the divergence of populations in other traits.

There is also an interaction between environmental effects and genetic change as populations adapt to new conditions. Growth of juvenile chinook salmon varies among rivers, and this is a primary initial control over their life history: ocean-type in high-growth and stream-type in low-growth environments. In New Zealand, we not only saw such phenotypic differences, but a population that produces stream-type fish in nature grew much slower than a predominantly ocean-type population, under controlled conditions of diet and temperature (Unwin et al. 2000). It was not possible to meet Taylor's (1991) rigorous demands for demonstration of the specific links between survival and the specific traits that differ among the populations (see "Heritable variation in phenotypic traits," above). However, we found significant variation in survival among families, indicating a genetic control over traits affecting survival. Moreover, the survival rates of the populations depended on the release site. Fish from one population performed better in their home site than fish from another population (Unwin et al. 2003), but the two populations performed similarly when released from a site not native to either population. This suggests that the kind of fitness-related differences showed by Mayama et al. (1989) can occur over less than a century.

My focus with respect to introduced populations has been on the insights that they can give us into the evolutionary process. However, there are also many questions about the ecological interactions between the introduced species and the local fauna, including native salmonids in the Northern Hemisphere and non-salmonids in all areas. This is too large a subject to tackle, but readers might consult McDowall (2003) for a review of interactions between salmonids and galaxiids in New Zealand streams, Crawford (2001) for a review of the Great Lakes, and Fausch (1988) for studies on salmonids. In general, it is easy to conduct laboratory studies showing predatory or competitive interactions, and in the field one can find evidence of predation or diet overlap, but determining the role of the introduced species in the status of the native species is often complicated by other contemporary events such as habitat degradation, fishing, interactions with other introduced species, and so on. Such introductions are now viewed very critically and less often permitted than in the past. Despite our reservations about the wisdom of such introductions, they can provide special opportunities to learn about salmonid life history and ecology, and we should learn as much as possible from these cases.

Summary

The present genetic structure of salmonid species reflects past and present processes. The postglacial colonization process has three main phases. First, there must be straying into the unoccupied habitat, and work in the newly formed streams in Glacier Bay, Alaska, gives insights into this phase (A. M. Milner et al. 2000). Homing then prevails, and the second phase is marked by relatively rapid evolution of adaptive traits, as the individuals who are less fit for the new environment are culled at one life-history stage or another. Some traits evolve faster than others, and some are more susceptible to environmental effects. The third phase of divergence from the source population, and from other new populations, is long, slow, and dominated by drift in selectively neutral traits. These processes—first straying, then homing; first strong selection, then drift—were essential in establishing the present distribution and diversity of salmonids. Salmonid populations now exist in a balance between the processes of isolation (from homing, leading to divergence in selected and neutral traits) and communication with other populations (from straying, leading to gene flow and similarity). Like all evolutionary processes, they are dynamic and ongoing.

Despite some straying (generally between nearby streams), populations vary in a wide variety of traits under at least partial genetic control. In some cases these traits obviously enhance the fitness of the local population (e.g., resistance to endemic parasites or bacteria), but in other cases the benefits are less obvious. Nevertheless, the overall benefits are evident from numerous attempts at transplanting populations within the native range of the species; local fish outperform transplants, and new populations are only very rarely established. However, there have been many successful introductions of salmonid populations (especially nonanadromous ones) outside the range of the species, and these have revealed the capacity for rapid evolution in adaptive traits in salmon (photo 18-2).

18-1 The author with a chinook salmon that had returned to New Zealand as part of a study of the evolution of salmon introduced there in the early 1900s. Photograph by Martin Unwin, National Institute of Water and Atmospheric Research.

19

The Abundance and Diversity of Pacific Salmon: Past, Present, and Future

It is clear from the work cited in previous chapters that salmon populations have been affected by broad-scale, climate-driven forces during the last century, and they have been affected by a variety of human factors as well. Estimates of salmon abundance prior to about 1900 rely on many assumptions (e.g., Chapman 1986 for the Columbia River, and Ricker 1950 and 1987 for Fraser River sockeye), but indirect techniques allow us to estimate abundance in past decades, centuries, and millennia. Drake et al. (2002) found that trees near rivers in southeast Alaska grew faster (as indicated by the spacing of their rings) in years when salmon were more abundant. Using this relationship, developed for years when salmon were counted, they estimated how many salmon the rivers might have supported back to 1820. The tree cores indicated high salmon abundance in that area at the turn of the twentieth century but significant variation during the nineteenth century, including apparently low density in the early 1800s.

Tree rings, although informative, can only provide insights over the period when the tree was alive. Looking farther back in time, Finney et al. (2000) took advantage of the difference in ratio of nitrogen isotopes between salt- and freshwater ecosystems, described in chapter 7. They developed a relationship between the density of salmon in Karluk Lake on Kodiak Island and in Ugashik and Becharof lakes in Bristol Bay (Alaska) and the proportion of the nitrogen of marine origin in each lake's sediments. Cores dating back about 300 years indicated that salmon populations had been quite abundant about the time when commercial fishing began in the early 1900s, and that salmon had also been very abundant in the mid-1700s, but that the early 1800s had seen weak populations, presumably for natural reasons. Interestingly, these data are consistent with the entirely independent tree-ring data reported by Drake et al. (2002) for a different region of Alaska. Finney et al. (2002) then extended their analyses of Karluk Lake cores back 2200 years. This longer record indicated high levels of salmon abundance during the first 200 years B.C., followed by an abrupt decline in abundance. Salmon populations reached a nadir around 100 A.D. and gradually increased until about 1200 A.D. From this point on, comparatively smaller changes in abundance seem to have taken place, and the sediment cores showed the drop in marine nutrients resulting from the heavy commercial fishing during the twentieth century.

There are three important points to make on the basis of these glimpses into the past. First, the evidence (albeit indirect) indicates that salmon population densities fluctuated greatly. There was certainly fishing by native peoples during these years, but it seems implausible that they caught enough salmon to explain the variation. It is more likely that these changes reflect natural (probably climatic) influences. Second, the brief experience of Euro-Americans and their written records does not necessarily represent the long-term pattern. Indeed, the large-scale commercial fisheries in Alaska developed at a time of exceptionally high salmon densities. Viewed from this perspective, "catastrophic declines" of Alaskan salmon in the middle of the twentieth century do not seem so extraordinary. Third, the levels of nutrients returning to these lakes in the years after the development of heavy commercial fishing have been very low, though not unprecedented. The effects of density-dependent processes, such as competition for breeding space by adults and food for juveniles, and climate-driven processes on salmon abundance certainly need further investigation.

Chatters et al. (1995) sought an even longer perspective on salmon abundance in the Columbia River basin. First, they estimated the likely flow regime of the river based on geomorphology, reconstructed plant communities from fossil pollen samples, and on layered depositions of freshwater mussel shells. The mussels were especially useful because one species favors faster, gravel-rich streams whereas the other favors streams with more sand and silt. These analyses suggested that climate conditions have been favorable for salmon during most of the last 4000 years. However, earlier records suggested unfavorable conditions from about 6000 to 8000 years ago, followed by gradually improving conditions to the optimum period, from around 2000 to 4000 years ago. These analyses are much less precise than those from the other techniques, but they gain some support from the presence of salmon bones left by humans in archeological sites. The proportion of salmon bones was low in the earliest sites sampled (3800 years ago and older) but much more abundant starting around 3000 to 3300 years ago.

If we take an even longer look back in time, we realize Pacific salmon were extirpated from much of their present range for thousands of years during the last glacial period. Pielou (1991) provided fascinating insights into the research methods and findings of scientists studying the changing climate, flora, and fauna during the retreat of the glaciers. Indeed, the term glacial "retreat" is a bit misleading, as it implies a steady and gradual process. Instead, conditions were probably quite variable, especially for fishes, as huge lakes were formed by melting water behind ice dams, followed by cataclysmic floods when the dams broke. A series of such floods scoured the coulees in eastern Washington, and the effects on fishes in the Columbia River must have been devastating. As deglaciated habitat became suitable, straying salmon would have founded populations. Thus, salmon are millions of years old as species, but many populations are 10,000 years old or younger.

How many salmon are there now?

The purpose of this chapter is to consider the past and present abundance of salmon, with comments on prospects for the future. We therefore need to address the simplest question, how many salmon are there? This is actually a very difficult estimate to make because there are so many governmental jurisdictions responsible for collecting data and no centralized process for reporting it. However, Rogers (2001) estimated adult salmon abundance for each year from 1951 through 2000 based on a variety of governmental sources, and I took the last 20 years to represent present conditions (table 19-1). Pink salmon are the most abundant species, especially in Russia and central and southeast Alaska. Sockeye (essentially all wild) are especially abundant in western Alaska. Chum are naturally abundant but there is also tremendous production from hatcheries in Japan. These three species have freed themselves, figuratively speaking, from the typical salmonid pattern of rearing in the natal stream. As a consequence, they are by far the most abundant species in this family, and they spawn at the highest densities. Coho and chinook salmon are markedly less abundant than the other three species, and almost certainly always were, for the basic ecological reason that streams can support a higher density of incubating embryos than hungry fry. Rogers did not make estimates for the anadromous trout but they would be less abundant than the salmon, as would masu.

As described in chapter 15, there have been substantial changes in catch, abundance, and survival of salmon during the past century. These changes have affected all species, regardless of life-history type or rearing habitat. Figure 19-1 shows the catch trends in Alaska. The absolute scales differ among species, but the timing of the changes is strikingly similar. The low catches in the early years are attributable to low fishing effort rather than low abundance, but after that there is probably a reasonably good correspondence between catch and abundance. Needless to say, at finer spatial scales abundance and catch are affected by many factors, including natural variation in local habitat conditions, consequences of development by humans, fishing regulations, and so on. The broad patterns are thus like a mosaic, being clearest when seen from a distance. Marine conditions are playing a large role in the changes in abundance, but climatic changes would affect freshwater life-history phases as well through patterns of temperature and flow regime. One of the challenges is to understand the ways in which different

TABLE 19-1. Estimated average annual abundance of adult Pacific salmon (catch and escapement, wild and hatchery) from 1981 through 2000 by geographical area, expressed as percentages of the total (in millions of salmon; from Rogers 2001).

species	*Japan*	*Russia*	*Western Alaska*	*Central Alaska*	*SE Alaska to California*	*Total*	*% of total*
pink	5.1	46.9	0.4	18.9	28.7	327.6	61.4
chum	53.3	16.5	7.5	6.2	16.5	102.0	19.1
sockeye	0.0	10.0	50.4	20.3	19.3	83.0	15.6
coho	0.0	12.8	11.7	15.4	60.1	16.5	3.1
chinook	0.0	6.2	14.3	6.0	73.5	4.7	0.9

FIGURE 19-1. Commercial catches of salmon in Alaska (Rigby et al. 1991 and unpublished data from the Alaska Department of Fish and Game).

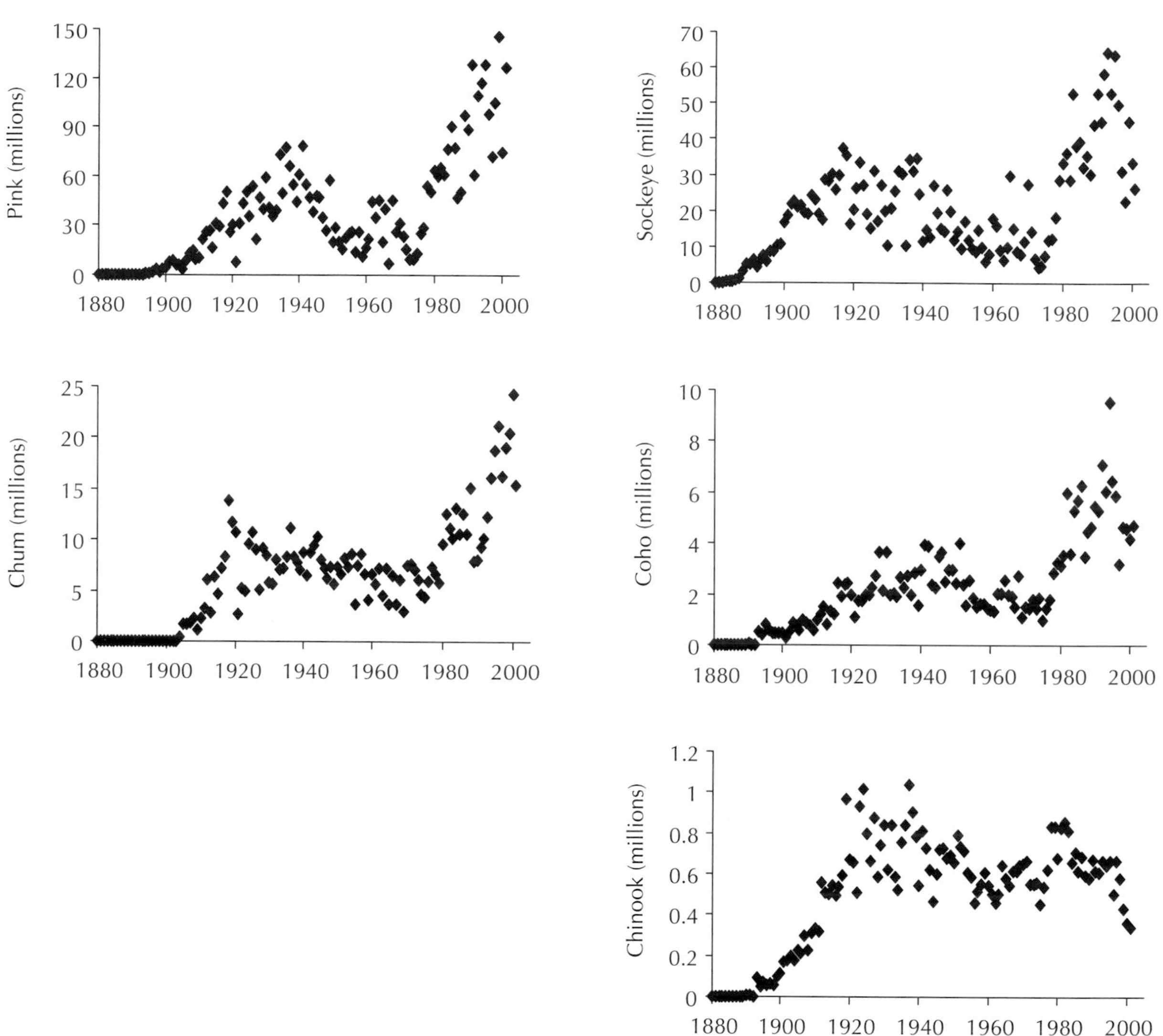

life-history stages are affected by climate conditions and whether the effects at early stages are magnified or negated later in life.

Considering only North America, we can include an estimate by Burgner et al. (1992) for steelhead with Rogers's estimates of salmon abundance and combine them in table 19-2 with estimates of the number of discrete populations by Waples, Gustafson et al. (2001). For this purpose the longer period of record, 1951–2001, was used, as it may better represent the general levels of abundance than the recent period of large runs. Pink salmon are by far the most numerous, in terms of species and populations. There are many coho populations but few fish, so most populations are small. Sockeye, chinook, and steelhead have similar numbers of populations, but the sockeye populations tend to be large whereas those of the other species, especially steelhead, tend to be small.

TABLE 19-2. Estimated abundance (catch plus escapement) of North American salmon from 1951 through 2001 (Rogers 2001) and steelhead (Burgner et al. 1992; period of record not specified), and number of populations (Waples et al. 2001). The number of fish per population was calculated from the first two rows (before rounding) and so is not an average of actual population sizes.

Abundance	*pink*	*sockeye*	*chum*	*coho*	*chinook*	*steelhead*
Millions of adults	105.5	51.4	23.7	13.5	5.1	1.6
Number of populations	8165	1400	3600	5400	1400	1500
Fish per population	12,918	36,681	6571	2501	3627	1067

The present status of salmon

As should be clear from the figures and tables presented above, the last two decades have been marked by high overall abundance of salmon. However, the abundance in North America largely reflects strong runs in Alaska, where salmon have always been most numerous, and the recovery of Fraser River sockeye and pink salmon after the construction of fishways at Hell's Gate. These successes mask declines elsewhere, especially to the south, and we must acknowledge this fact.

Since the late nineteenth century, diverse human activities have depleted salmon populations, affecting them at all life-history stages. Some activities (including mining, logging, agricultural practices, and urban development) affect the survival of embryos and the quality of freshwater habitats. Construction of dams for water storage, diversion, and hydroelectric power generation blocked all access to some habitats and changed the flow regimes, temperature, habitat, and community dynamics in many large rivers. Modification of estuaries for agriculture, commerce, and urban development also contributed to declines in salmon abundance (NRC 1996), and of course fishing has played a large role too, at least in some cases. Some of the earliest fisheries scientists were aware of the effects of these activities on salmon, and a number of them wrote eloquently and repeatedly on the subject (see histories by Dunn 1996 and Lichatowich 1999). Artificial propagation of salmon in hatcheries was seen by many as a way to counter the effects of overfishing and return the former levels of abundance (reviewed by Bottom 1996), but such efforts masked the decline of wild populations and often accelerated it by encouraging fishing levels that only hatchery runs could sustain.

Bruce Brown's 1982 book, *Mountain in the Clouds*, presented the public with historical factors leading to the decline of salmon in a passionate, highly critical (though perhaps less than objective) manner, and I was greatly moved by it. In the 1970s and 1980s, many fisheries scientists documented the effects of logging, dams, fishing, and other activities on salmon. I will not try to summarize this vast literature, but readers will find many excellent reviews in the books by Salo and Cundy (1987), Meehan (1991), the National Research Council (NRC 1996), Stouder et al. (1996), and Knudsen et al. (2000). However, there was little study or discussion of the overall status of salmon until the 1990s. Perhaps this is because so much research was local in focus, conducted by people working in the many agencies with specific jurisdictions defined by city, county, state, provincial, and national borders. Broad-scale data were not readily available (though perhaps we did not look hard enough for them), and in any case trends were difficult to detect because individual populations fluctuated so much and because declines in one area were often matched by increases elsewhere. Finally, the effects of climate change and ocean conditions on survival and abundance were not fully appreciated.

In 1991, Willa Nehlsen, Jack Williams, and James Lichatowich published a paper that was a clarion call, uniting many fisheries biologists who had independently become concerned about the status of salmon. Nehlsen et al. (1991) surveyed biologists in Washington, Oregon, Idaho, and California to estimate the number of populations of anadromous salmon and trout at risk of extinction. Stocks were categorized as being already extinct, at high or moderate risk of extinction, or of special concern for various reasons (they did not catalog healthy populations). There were several shocking findings, notably the high number (107) of populations that had apparently gone extinct and the number in some form of jeopardy (214). Perhaps just as shocking as the number of populations in jeopardy was the dismal state of information on the status of salmon. In many cases there was no quantitative assessment, and only the opinion of the local biologist was available as a basis for categorizing the status of the population. A corollary of this lack of good information was the gross manner in which many stocks were lumped together. For example, the report listed only two "stocks" of sea-run cutthroat trout in all of Washington outside the Columbia River. One stock combined all the populations in Hood Canal and Grays Harbor (two areas that are not even close to each other, much less adjacent), and the other stock included everything else! Had the populations been defined more sharply, more would probably have been found at risk.

Nehlsen et al. (1991) reported that habitat loss and degradation contributed to the risk of extinction in 91% of the cases, biotic factors (hybridization, competition or predation by nonnative species, and interactions with hatchery-produced conspecifics) contributed in 53% of the cases, and overfishing contributed in 50%. These were subjective assessments but they do not seem out of line with our general understanding of human impacts on salmon. Even this assessment, however, serves to remind us of the natural as well as anthropogenic influences on salmon. A map of the Columbia River basin reveals that about one-third of the area was not historically accessible to anadromous salmon (because of natural barriers), about a third was made inaccessible by dams without fish ladders, and about a third is currently accessible (fig. 19-2). In terms of stream length, the Northwest Power Planning Council estimated that "salmon and steelhead habitat in the entire basin has decreased from about 14,666 miles of stream before 1850 to 10,073 miles presently, a 31 percent loss all due to water development." (NWPPC 1986, 4).

Despite (or perhaps because of) its limitations, the study by Nehlsen and her colleagues had a galvanic effect. A number of agencies within the region covered by the report soon produced their own reports (e.g., Nickelson et al. 1992; WDF et al. 1993) and contributed to books (Stouder et al. 1996; Knudsen et al. 2000) and journal articles (Slaney et al. 1996; Baker et al. 1996). A panel of the National Research Council was convened to study the status of anadromous salmonids in the Pacific Northwest (NRC 1996), and countless workshops, symposia, and other meetings have taken place to discuss salmon abundance. Table 19-3 summarizes the data from four of these regions, and several patterns are obvious. First, there are far fewer populations in coastal Oregon and Washington than in British Columbia, and southeast Alaska alone has as many populations as

FIGURE 19-2. Map of the Columbia River basin, showing the area not historically accessible to anadromous salmonids, the area formerly accessible that was cut off by impassible dams, and the area still accessible (modified from a map by Blake Feist, National Marine Fisheries Service).

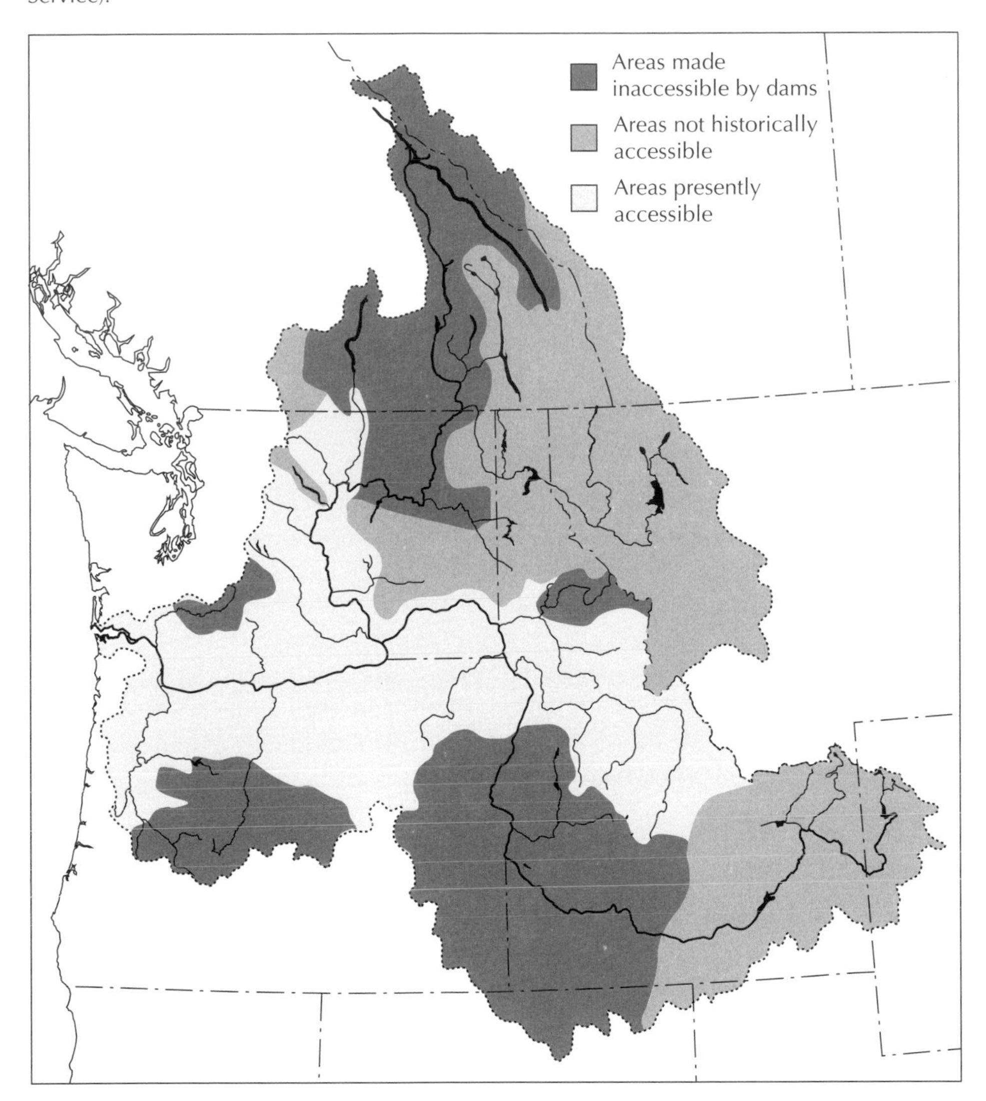

British Columbia. Second, the proportion of extinctions is greatest in the south, as is the proportion of populations in some degree of jeopardy. I have not seen a comparable review for anadromous salmonids in California, but work on coho (L. R. Brown et al. 1994) and Central Valley chinook (Yoshiyama et al. 1998) indicated that many populations are in serious jeopardy there too.

The third important message from these reviews is the large proportion of populations of unknown status, especially toward the north. Given the thousands of populations in the vast, often remote areas of British Columbia and Alaska, this is not surprising. Consistent with these assessments, Knudsen (2000) found that the methods for estimating salmon

TABLE 19-3. Status of Pacific salmon and steelhead populations in southeast Alaska (Baker et al. 1996), British Columbia and the Yukon (Slaney et al. 1996), Washington (including Columbia River populations: WDF et al. 1993), and coastal Oregon (Nickelson et al. 1992). Each study listed populations as "healthy" and in "unknown status." My category, "in jeopardy," includes different categories indicating more or less danger of extinction (Alaska and British Columbia: special concern, moderate, and high risk; Washington: depressed and critical; coastal Oregon: special concern and depressed). Extinctions in Washington and coastal Oregon are from Nehlsen et al. (1991). Data are expressed as the percentage of the total number of populations in the different status categories.

Region	*Healthy*	*In jeopardy*	*Extinct*	*Unknown*	*Total*
Southeast Alaska	10.0	0.1	< 0.1	89.9	9228
British Columbia	48.3	9.7	1.3	40.7	9038
Washington	37.5	22.2	16.1	24.2	248
Coastal Oregon	32.6	49.7	6.4	11.3	141

escapements are often weak. Only 2.6% of the 9430 populations that he tallied in the United States had estimation methods (e.g., traps, towers, or mark-recapture methods) deemed excellent, good, or fair. More than 30% of the populations had methods deemed poor (mostly aerial surveys), and more than 60% of the populations had no method at all, or no information on the method was available.

Recognition of the declines in salmon abundance reached the highest possible levels in the United States when the full power of the federal Endangered Species Act (ESA) was applied to discrete population complexes (termed "evolutionarily significant units" or ESUs) of salmon and to species or subspecies of trout. Under this law, a species is *endangered* if it "is in danger of extinction throughout all or a significant portion of its range," and a species is *threatened* if it is "likely to become an endangered species within the foreseeable future throughout all or a significant portion of its range." An ESU is defined by two criteria: it "(1) is substantially reproductively isolated from other conspecific population units, and (2) represents an important component in the evolutionary legacy of the species" (Waples 1991, 127).

The anadromous species in greatest jeopardy are chinook salmon, with nine listed ESUs (two endangered and seven threatened as of spring 2003), and steelhead with ten listed ESUs (two endangered and eight threatened). There are also three coho ESUs listed as threatened, two of chum listed as threatened, and two of sockeye, one endangered and one threatened. In addition to these listings of anadromous ESUs, seven taxa of nonanadromous salmonids are also listed under the ESA: Apache trout, bull trout, greenback cutthroat trout, Lahontan cutthroat trout, Little Kern golden trout, and Paiute trout as threatened, and Gila trout as endangered.

Collectively, the ranges of these salmonid taxa listed under the ESA and those already extinct comprise the majority of the area south of the Canadian border that historically supported salmonids. Thus we have the paradox of great abundance of salmon as a commodity while in large parts of their range they are in dire straits. Hyatt and Riddell (2000) commented that we have done a fair job managing salmon for biomass but a poor job managing for biodiversity, and Scudder (1989) and Riddell (1993) pointed out the importance of small populations in the overall genetic and ecological health of the species. The depletions of salmon have been most severe toward the southern end of their distribution. There has been little numerical loss of pink salmon, whose distribution largely ends at Puget Sound. Summer runs of chum salmon in Washington and runs in the Columbia River are in jeopardy but, like pinks, the historical center of distribution of the species is farther north. Sockeye are most abundant in North America from the Fraser River to Bristol Bay. Some moderately large populations in the Columbia River have gone extinct or have been drastically reduced (Chapman 1986), but this system has fewer large lakes, for its size, than the Fraser. Coho salmon exist as many small populations, mostly along the coast, and habitat degradation has affected them. The extensive damming and degradation of large river systems (notably the Sacramento and Columbia) particularly affected chinook salmon and steelhead, as these systems supported large and diverse assemblages of those species. Thus it is a bit deceptive to state that salmon are as abundant now as they were at any time during the last century. Coho, chinook, and steelhead have the most southerly distributions and have been most affected by human activities, but were always the least abundant, so their proportionally larger losses have comparatively little influence on the total count of salmon.

Having driven these salmonid taxa to (or over) the brink of extinction, we are now spending great sums of money trying to restore them. I will not pass judgment on these efforts, other than to make two points. First, it is critical that any restoration projects be firmly grounded in the basic biology of the species in question. The scope for gain needs to be realistically assessed in terms of the quantity and quality of habitat and the factors limiting the density or productivity of the species. For examples, figures 11-1 and 11-2 indicate that the quantity of coho salmon smolts is fundamentally constrained by the size of the river, and secondarily by the quality of the habitat, regardless of how many females spawn there. In addition, we must remember that there are ecological relationships among salmonids, and that changes in a river may not benefit all species equally. Interspecific interactions within the salmonid community and with non-salmonids need to be carefully considered.

The second point to consider is the difficulty in detecting the effects of human activities, either harmful or beneficial. Lichatowich and Cramer (1979) examined the levels of variation in salmonid abundance and concluded that the count of adults is an extremely insensitive measure of the status of the population. This is frustrating because we are most interested in changes in adult abundance. Lichatowich and Cramer pointed out that natural variation in freshwater and marine survival makes it difficult to detect, with accepted levels of statistical confidence, even large underlying changes in abundance: "Studies of survival and abundance may require 20–30 years to produce an 80% chance of detecting a 50% change" (1). The effects of habitat degradation on the population might be masked by high marine survival rates, or the effects of habitat restoration might be negated by poor conditions at sea. Similar analysis by Bisson et al. (1996) indicated that the coefficient of variation (standard deviation divided by the mean) is about 50% in anadromous populations and 25% in nonanadromous ones. It is therefore somewhat easier to detect changes in nonanadromous populations because their responses to events in freshwater are not complicated by marine survival. However, even in these species the "background" level of variation makes it difficult to quantify the magnitude

FIGURE 19-3. The number of years of data (monitoring period) needed to detect changes of a given magnitude in populations of anadromous and non-anadromous salmonids due to some treatment (beneficial or deleterious), based on observed variation in abundance (from Bisson et al. 1996). For example, detection of a 30% change might require about 20 years for non-anadromous and 70 years for anadromous salmonids.

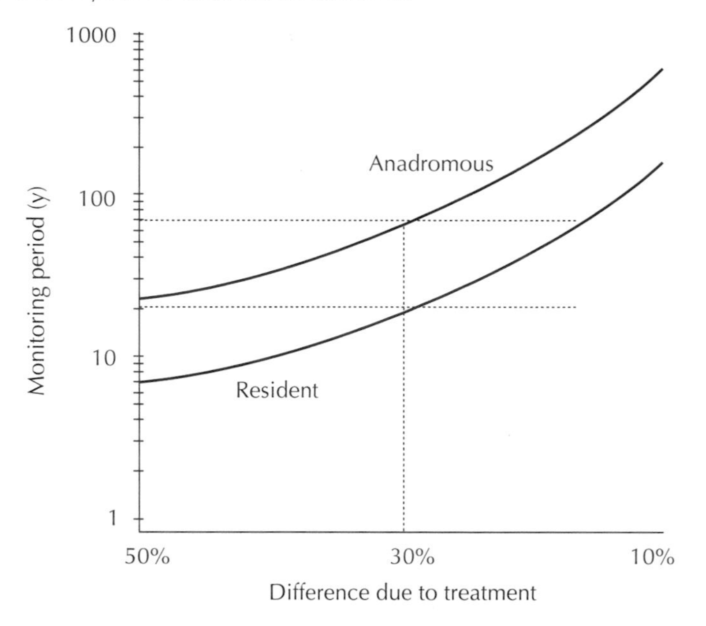

of loss in fish abundance resulting from a particular action or to ascertain whether the population has recovered after restoration efforts (fig. 19-3).

The idea that salmon conservation must be grounded in good science is hardly new. Thompson's (1962) review of the salmon research program set up by the University of Washington's Fisheries Research Institute in western Alaska showed a full appreciation of not only the key scientific issues but also the institutional ones. He emphasized the "value (1) of providing stabilized conditions for specialization of young [researchers], (2) of shielding them from the effects of hampering administrative and political conditions, and (3) of long, continued effort to secure adequate basic data in any project undertaken" (35). Many of our agencies might benefit from his advice.

The capacity for recovery of Pacific salmon

Notwithstanding the strength of salmon in the northern part of their range, we must be concerned for the populations at risk in the south. Salmon have sometimes been characterized as the "canary in the mineshaft," using as a metaphor the birds whose death signaled a drop in air quality that would soon threaten coal miners. I believe that this image of salmon, as the most fragile element of their ecosystem, is incorrect and inappropriate. The truth is that salmon populations are highly productive and mobile. Withler (1982) reviewed historical reports indicating that millions of pink salmon spawned in tributaries of the Fraser and Thompson rivers above Hell's Gate prior to the rockslide in 1913 but that none successfully negotiated the rapids that year. For five generations no pinks were seen above the canyon, but they eventually reappeared. When the fishways were completed, the annual escapements increased dramatically, from under 2000 fish in 1947 and 1949, to tens of thousands in 1951 and 1953, to hundreds of thousands through 1975, and then into the millions. The sockeye salmon populations were not exterminated by the rockslide but were greatly reduced, and they too recovered.

Given the high fishing rates, habitat loss and degradation, careless transfers of fish among basins, overzealous hatchery propagation, and other stressors, the remarkable thing is not that salmon are in danger but that they still persist at all. It is my view that their chances of recovery are good if we would only take our collective foot off their neck. I am not so naïve as to believe that recovery is possible in all areas. However, the southern part of Puget Sound, hardly pristine habitat by any definition, has had truly phenomenal runs of chum salmon during the past decade. Whatever the reason (favorable ocean conditions, appropriate fishery management, renewed habitat protection, or something else), success stories such as this should encourage us not to give up on salmon and trout.

I realize that many knowledgeable, experienced scientists do not share my optimism. Hartman et al. (2000) expressed concern that the combination of global climate change, increasing human population, and the complex changes in land use that will inevitably follow make the future of salmon very dim. These researchers also cautioned that scientists should not give the public false hopes by promising "sustainable" salmon runs when the real problems are beyond our control. Lackey (2001) built on this theme, predicting that in the Pacific Northwest by the year 2100 "many stocks or populations will have disappeared, and those that remain will have small runs incapable of supporting appreciable fishing" (26). He criticized fisheries scientists and policy makers for providing the kind of "narrow, reductionist scientific information and assessments . . . [that] often mislead the public into endorsing false expectations of the likelihood of the recovery of wild salmon" (27). He asserted that when scientists do not give the public realistic information on the probability of restoring salmon runs, "we simply squander our professional credibility to become acolytes of delusion" (27).

Final thoughts

Pacific salmon have existed as species for millions of years, yet our perspective on their abundance is largely limited to the twentieth century. However, new techniques allowing estimates of relative abundance in earlier centuries and millennia indicate that there have been large, natural shifts in the numbers of salmon. The latter part of the twentieth century has been a period of high abundance, especially for the northern part of the range of salmon. In addition to these natural changes in abundance, the twentieth century has been marked by human-induced degradation and loss of salmon habitat, overfishing, and unwise reliance on hatcheries to make up for the shortfalls. These losses of wild salmon have been particularly acute at the southern end of the distribution, where the populations tended to be sparse to begin with. However, the present abundance of salmon, the magnitude of declines, and the loss of discrete populations are shrouded by poor data from the field.

The declines in salmon abundance and shrinkage of their range have caused some to question whether they will remain a part of our ecosystems into the future. The seemingly relentless pressures from expanding human populations have made some salmon experts despair for them. However, as this book has revealed, wild salmon are very resilient in the face of short-term changes (e.g., habitat modification and decadal climate shifts) and long-term (century-scale) processes as well. There is reason for optimism; they persist in many areas of California, Oregon, Washington, Idaho, and southern British Columbia despite many insults.

Still, we should not be naïve. A great deal of habitat from southern British Columbia to California is no longer accessible to salmon or has been altered to their detriment. I have no desire to end my book by deluding readers with rosy pictures of the future of salmon populations. However, if we write salmon off as incapable of recovery and so justify actions and inactions that harm them, we will turn our pessimism into their reality, and that would be unforgivable.

By way of conclusion, let us return to the themes that began the book. Salmon are important to so many of us, in so many ways. They are our food, our recreation, our symbol and inspiration, and a critical component in the ecosystems that we value and depend on. If we dedicate ourselves to ensuring that they continue to play all these roles, I believe the salmon will do the rest. If we preserve habitat they will use it, and if we restore habitat and make it accessible they will find it. We must be patient, and we must ground all conservation efforts in a thorough knowledge of salmon behavior and ecology. Conserving salmon populations will depend on firm political resolve, some economic sacrifices, and continued investment in both fundamental research and the monitoring needed for good management of fisheries and habitat protection. I hope my book serves this goal in some small way by providing information about salmon. It is up to us all to put the knowledge to use and to give wild salmon the chance they need to not only persist but to thrive and prosper. Salmon will respond, so the choice is ours.

References Cited

Abbott, J. C., R. L. Dunbrack, and C. D. Orr. 1985. The interaction of size and experience in dominance relationships of juvenile steelhead trout (*Salmo gairdneri*). *Behaviour* 92:241–253.

Alderdice, D. F., W. P. Wickett, and J. R. Brett. 1958. Some effects of temporary exposure to low dissolved oxygen levels on Pacific salmon eggs. *J. Fish. Res. Board Can.* 15:229–249.

Allen, G. H. 1959. Behavior of chinook and silver salmon. *Ecology* 40:108–113.

Anderson, A. D., and T. D. Beacham. 1983. The migration and exploitation of chum salmon stocks of the Johnstone Strait–Fraser River study area, 1962–1970. *Can. Tech. Rept. Fish. Aquat. Sci.* 1166. Nanaimo, BC. 125 p.

Ankenbrandt, L. G. 1988. The phylogenetic relationships of the Pacific fishes contained in the teleost genera *Oncorhynchus* and *Salmo* based on restriction fragment analysis of mitochondrial DNA. M.S. thesis, University of Washington, Seattle.

Appleby, A. E., J. M. Tipping, and P. R. Seidel. 2003. The effect of using two-year-old male coho salmon in hatchery broodstock on adult returns. *N. Am. J. Aquacult.* 65:60–62.

Armstrong, R. H. 1965. *Some migratory habits of the anadromous Dolly Varden* Salvelinus malma *(Walbaum) in southeastern Alaska.* Alaska Dept. Fish Game, Res. Rept. 3. Juneau. 36 p.

Armstrong, R. H. 1971. Age, food, and migration of sea-run cutthroat trout, *Salmo clarki*, at Eva Lake, Southeastern Alaska. *Trans. Am. Fish. Soc.* 100:302–306.

———. 1984. Migration of anadromous Dolly Varden charr in southeastern Alaska—a manager's nightmare. Pages 559–570 in L. Johnson and B. Burns, eds., *Biology of the Arctic charr: Proceedings of the international symposium on Arctic charr.* University of Manitoba Press, Winnipeg, Manitoba.

Armstrong, R. H., and J. E. Morrow. 1980. The Dolly Varden charr, *Salvelinus malma*. Pages 99–140 in E. K. Balon, ed., *Charrs: Salmonid fishes of the genus* Salvelinus. Dr. W. Junk, Publishers, The Hague.

Azuma, T. 1992. Diel feeding habits of sockeye and chum salmon in the Bering Sea during the summer. *Nippon Suisan Gakkaishi* 58:2019–2025.

Baker, T. T., A. C. Wertheimer, R. D. Burkett, R. Dunlap, D. M. Eggers, E. I. Fritts, A. J. Gharrett, R. A. Holmes, and R. L. Wilmot. 1996. Status of Pacific salmon and steelhead escapements in southeastern Alaska. *Fisheries* 21(10):6–18.

Ballantyne, A. P., M. T. Brett, and D. E. Schindler. 2003. The importance of dietary phosphorus and highly unsaturated fatty acids for sockeye (*Oncorhynchus nerka*) growth in Lake Washington—a bioenergetics approach. *Can. J. Fish. Aquat. Sci.* 60:12–22.

Bams, R. A. 1967. Differences in performance of naturally and artificially propagated sockeye salmon migrant fry, as measured with swimming and predation tests. *J. Fish. Res. Board Can.* 24:1117–1153.

———. 1969. Adaptations of sockeye salmon associated with incubation in stream gravels. Pages 71–87 in T. G. Northcote, ed., *Symposium on salmon and trout in streams*. H. R. MacMillan Lectures in Fisheries, University of British Columbia, Vancouver.

———. 1976. Survival and propensity for homing as affected by presence or absence of locally adapted paternal genes in two transplanted populations of pink salmon *(Oncorhynchus gorbuscha)*. *J. Fish. Res. Board Can.* 33:2716–2725.

Barlaup, B. T., H. Lura, H. Sægrov, and R. C. Sundt. 1994. Inter- and intra-specific variability in female salmonid spawning behaviour. *Can. J. Zool.* 72:636–642.

Bax, N. J. 1983. Early marine mortality of marked juvenile chum salmon (*Oncorhynchus keta*) released into Hood Canal, Puget Sound, Washington, in 1980. *Can. J. Fish. Aquat. Sci.* 40:426–435.

Beacham, T. D. 1986. Type, quantity, and size of food in Pacific salmon (*Oncorhynchus*) in the Strait of Juan de Fuca, British Columbia. *Fish. Bull.* 84:77–89.

Beacham, T. D., and C. B. Murray. 1985. Variation in length and body depth of pink salmon (*Oncorhynchus gorbuscha*) and chum salmon (*O. keta*) in southern British Columbia. *Can. J. Fish. Aquat. Sci.* 42:312–319.

———. 1989. Variation in developmental biology of sockeye salmon (*Oncorhynchus nerka*) and chinook salmon (*O. tshawytscha*) in British Columbia. *Can. J. Zool.* 67:2081–2089.

———. 1990. Temperature, egg size, and development of embryos and alevins of five species of Pacific salmon: A comparative analysis. *Trans. Am. Fish. Soc.* 119:927–945.

———. 1993. Fecundity and egg size variation in North American Pacific salmon (*Oncorhynchus*). *J. Fish Biol.* 42:485–508.

Beacham, T. D., and P. Starr. 1982. Population biology of chum salmon, *Oncorhynchus keta*, from the Fraser River, British Columbia. *Fish. Bull.* 80:813–825.

Beamesderfer, R. C. P., D. L. Ward, and A. A. Nigro. 1996. Evaluation of the biological basis for a predator control program on northern squawfish (*Ptychocheilus oregonensis*) in the Columbia and Snake rivers. *Can. J. Fish. Aquat. Sci.* 53:2898–2908.

Beamish, R. J., and D. R. Bouillon. 1993. Pacific salmon production trends in relation to climate. *Can. J. Fish. Aquat. Sci.* 50:1002–1016.

Beamish. R. J., C. Mahnken, and C. M. Neville. 2004. Evidence that reduced early marine growth is associated with lower marine survival of coho. *Trans. Am. Fish. Soc.* 133:26–33.

Beamish, R. J., D. McCaughran, J. R. King, R. M. Sweeting, and G. A. MacFarlane. 2000. Estimating the abundance of juvenile coho salmon in the Strait of Georgia by means of surface trawls. *N. Am. J. Fish. Mgmt.* 20:369–375.

Beamish, R. J., and C. M. Neville. 1995. Pacific salmon and the Pacific herring mortalities in the Fraser River plume caused by river lamprey (*Lampetra ayresi*). *Can. J. Fish. Aquat. Sci.* 52:644–650.

Beamish, R. J., I. A. Pearsall, and M. C. Healey. 2003. A history of the research on the early marine life of Pacific salmon off Canada's Pacific coast. *N. Pac. Anad. Fish Comm. Bull.* 3:1–40.

Beamish, R. J., B. L. Thomson, and G. A. McFarlane. 1992. Spiny dogfish predation on chinook and coho salmon and the potential effects on hatchery-produced salmon. *Trans. Am. Fish. Soc.* 121:444–455.

Beauchamp, D. A., C. M. Baldwin, J. L. Vogel, and C. P. Gubala. 1999. Estimating diel, depth-specific foraging opportunities with a visual encounter rate model for pelagic piscivores. *Can. J. Fish. Aquat. Sci.* 56:128–139.

Beauchamp, D. A., M. G. LaRiviere, and G. L. Thomas. 1995. Evaluation of competition and predation as limits to juvenile kokanee and sockeye production in Lake Ozette, Washington. *N. Am. J. Fish. Mgmt.* 15:193–207.

Becker, C. D. 1962. Estimating red salmon escapements by sample counts from observation towers. *Fish. Bull.* 61:355–369.

Beckman, B. R., and W. W. Dickhoff. 1998. Plasticity of smolting in spring chinook salmon: relation to growth and insulin-like growth factor-I. *J. Fish Biol.* 53:808–826.

Beckman, B. R., D. A. Larsen, B. Lee-Pawlak, and W. W. Dickhoff. 1998. Relation of fish size and growth rate to migration of spring chinook salmon smolts. *N. Am. J. Fish. Mgmt.* 18:537–546.

Behnke, R. J. 1992. *Native trout of western North America.* American Fisheries Society Monograph 6. Bethesda, Maryland.

Behnke, R. J. 2002. *Trout and salmon of North America.* The Free Press, New York.

Bell, E., W. G. Duffy, and T. D. Roelofs. 2001. Fidelity and survival of juvenile coho salmon in response to a flood. *Trans. Am. Fish. Soc.* 130:450–458.

Ben-David, M., T. A. Hanley, and D. M. Schell. 1998. Fertilization of terrestrial vegetation by spawning Pacific salmon: The role of flooding and predator activity. *Oikos* 83:47–55.

Berejikian, B. A., E. P. Tezak, and A. L. LaRae. 2000. Female mate choice and spawning behaviour of chinook salmon under experimental conditions. *J. Fish Biol.* 57:647–661.

Berggren, T. J., and M. J. Filardo. 1993. An analysis of variables influencing the migration of juvenile salmonids in the Columbia River basin. *N. Am. J. Fish. Mgmt.* 13:48–63.

Berman, C. H., and T. P. Quinn. 1991. Behavioural thermoregulation and homing by spring chinook salmon, *Oncorhynchus tshawytscha* (Walbaum), in the Yakima River. *J. Fish Biol.* 39:301–312.

Bernard, D. R., K. R. Helper, J. D. Jones, M. E. Whalen, and D. N. McBride. 1995. Some tests of the "migration hypothesis" for anadromous Dolly Varden (southern form). *Trans. Am. Fish. Soc.* 124:297–307.

Bernard, R. L., and K. W. Myers. 1997. The performance of quantitative scale pattern analysis in the identification of hatchery and wild steelhead (*Oncorhynchus mykiss*). *Can. J. Fish. Aquat. Sci.* 53: 1727–1735.

Beschta, R. L., R. E. Bilby, G. W. Brown, L. B. Holtby, and T. Hofstra. 1987. Stream temperature and aquatic habitat: fisheries and forestry interactions. Pages 191–232 in E.O. Salo and T. W. Cundy, eds., *Streamside management: forestry and fishery interactions.* University of Washington College of Forest Resources, Seattle.

Bevelhimer, M. S., and S. M. Adams. 1993. A bioenergetics analysis of diel vertical migration by kokanee salmon, *Oncorhynchus nerka*. *Can. J. Fish. Aquat. Sci.* 50:2336–2349.

Beverton, R. J. H., and S. J. Holt. 1957. *On the dynamics of exploited fish populations.* H. M. Stationary Office, London.

Bickford, S. A., and J. R. Skalski. 2000. Reanalysis and interpretation of 25 years of Snake–Columbia River juvenile salmonid survival studies. *N. Am. J. Fish. Mgmt.* 20:53–68.

Biette, R. M., and G. H. Geen. 1980. Growth of underyearling sockeye salmon (*Oncorhynchus nerka*) under constant and cyclic temperatures in relation to live zooplankton ration size. *Can. J. Fish. Aquat. Sci.* 37:203–210.

Bigler, B. S., D. W. Welch, and J. H. Helle. 1996. A review of size trends among North Pacific salmon (*Oncorhynchus* spp.). *Can. J. Fish. Aquat. Sci.* 53:455–465.

Bilby, R. E., B. R. Fransen, and P. A. Bisson. 1996. Incorporation of nitrogen and carbon from spawning coho salmon into the trophic system of small streams: evidence from stable isotopes. *Can. J. Fish. Aquat. Sci.* 53:164–173.

Bilby, R. E., B. R. Fransen, P. A. Bisson, and J. K. Walter. 1998. Responses of juvenile coho salmon (*Oncorhynchus kisutch*) and steelhead (*Oncorhynchus mykiss*) to the addition of salmon carcasses to two streams in southwestern Washington, U.S.A. *Can. J. Fish. Aquat. Sci.* 55:1909–1918.

Bilby, R. E., B. R. Fransen, J. K. Walter, C. J. Cederholm, and W. J. Scarlett. 2001. Preliminary evaluation of the use of nitrogen stable isotope ratios to establish escapement levels for Pacific salmon. *Fisheries* 26 (1):6–14.

Bilby, R. E., and J. W. Ward. 1989. Changes in characteristics and function of woody debris with increasing size of streams in western Washington. *Trans. Am. Fish. Soc.* 118:368–378.

Bilton, H. T., D. F. Alderdice, and J. T. Schnute. 1982. Influence of time at release of juvenile coho salmon (*Oncorhynchus kisutch*) on returns at maturity. *J. Fish. Res. Board Can.* 39:426–447.

Bilton, H. T., R. B. Morley, A. S. Coburn, and J. van Tyne. 1984. The influence of time and size at release of juvenile coho salmon (*Oncorhynchus kisutch*) on returns at maturity; results from releases from Quinsam River Hatchery, B.C., in 1980. *Can. Tech. Rept. Fish. Aquat. Sci.* 1306. Nanaimo, BC. 98 p.

Birtwell, I. K., M. D. Nassichak, and H. Beune. 1987. Underyearling sockeye salmon (*Oncorhynchus nerka*) in the estuary of the Fraser River. *Can. Spec. Publ. Fish. Aquat. Sci.* 96:25–35.

Bisson, P. A., and R. E. Bilby. 1998. Organic matter and trophic dynamics. Pages 373–398 in R. J. Naiman and R. E. Bilby, eds., *River ecology and management.* Springer-Verlag, New York.

Bisson, P. A., and D. R. Montgomery. 1996. Valley segments, stream reaches, and channel units. Pages 23–52 in F. R. Hauer and G. A. Lamberti, eds., *Methods in stream ecology*. Academic Press, New York.

Bisson, P. A., G. H. Reeves, R. E. Bilby and R. J. Naiman. 1996. Watershed management and Pacific salmon: Desired future conditions. Pages 447–474 in D. J. Stouder, P. A. Bisson, and R. J. Naiman, eds., *Pacific salmon and their ecosystems: Status and future options.* Chapman and Hall, New York.

Bisson, P. A., K. Sullivan, and J. L. Nielsen. 1988. Channel hydraulics, habitat use, and body form of juvenile coho salmon, steelhead, and cutthroat trout in streams. *Trans. Am. Fish. Soc.* 117:262–273.

Bjornn, T. C., and D.W. Reiser. 1991. Habitat requirements of salmonids in streams. In W. R. Meehan, ed., Influences of forest and rangeland management on salmonid fishes and their habitats. Pages 83–138 in Am. Fish. Soc. Spec. Publ. 19. Bethesda, MD.

Blackbourn, D. J. 1987. Sea surface temperature and pre-season prediction of return timing in Fraser River sockeye salmon (*Oncorhynchus nerka*). *Can. Spec. Publ. Fish. Aquat. Sci.* 96:296–306.

Bodznick, D. 1975. The relationship of the olfactory EEG evoked by naturally-occurring stream waters to the homing behavior of sockeye salmon (*Oncorhynchus nerka*, Walbaum). *Comp. Biochem. Physiol.* 52A:487–495.

———. 1978. Water source preference and lakeward migration of sockeye salmon fry *(Oncorhynchus nerka). J. Comp. Physiol.* 127:139–146.

Boss, S. M., and J. S. Richardson. 2002. Effects of food and cover on the growth, survival, and movement of cutthroat trout (*Oncorhynchus clarki*) in coastal streams. *Can. J. Fish. Aquat. Sci.* 59:1044–1053.

Bottom, D. L. 1996. To till the water: A history of ideas in fisheries conservation. Pages 569–597 in D. J. Stouder, P. A. Bisson, and R. J. Naiman, eds., *Pacific salmon and their ecosystems: Status and future options.* Chapman and Hall, New York.

Bottom, D. L., and K. K. Jones. 1990. Species composition, distribution, and invertebrate prey of fish assemblages in the Columbia River estuary. *Prog. Oceanogr.* 25:243–270.

Bottom, D. L., C. A. Simenstad, A. M. Baptista, D. A. Jay, J. Burke, K. K. Jones, E. Casillas, and M. H. Schiewe. 2001. *Salmon at river's end: The role of the estuary in the decline and recovery of Columbia River salmon.* Rept. to the Bonneville Power Administration. Portland, Oregon. 271 p.

Bower, S. M., R. E. Withler, and B. E. Riddell. 1995. Genetic variation in resistance to the hemoflagellate *Cryptobia salmositica* in coho and sockeye salmon. *J. Aquat. Anim. Health* 7:185–194.

Bowler, B. 1975. Factors influencing genetic control in lakeward migrations of cutthroat trout fry. *Trans. Am. Fish.* Soc. 104:474–482.

Boyce, N. P. J. 1974. Biology of *Eubothrium salvelini* (Cestoda: Pseudophyllidea), a parasite of juvenile sockeye salmon (*Oncorhynchus nerka*) of Babine Lake, British Columbia. *J. Fish. Res. Board Can.* 31:1935–1742.

Bradford, M. J. 1995. Comparative review of Pacific salmon survival rates. *Can. J. Fish. Aquat. Sci.* 52:1327–1338.

Bradford, M. J., G. C. Taylor, and J. A. Allan. 1997. Empirical review of coho salmon smolt abundance and the prediction of smolt production at the regional level. *Trans. Am. Fish. Soc.* 126:49–64.

Brännäs, E. 1995. First access to territorial space and exposure to strong predation pressure: a conflict in early emerging Atlantic salmon (*Salmo salar* L.) fry. *Evol. Ecol.* 9:411–420.

Brannon, E. L. 1972. Mechanisms controlling migration of sockeye salmon fry. *Int. Pac. Salmon Fish. Comm. Bull.* 21. New Westminster, BC. 86 p.

———. 1987. Mechanisms stabilizing salmonid fry emergence timing. *Can. Spec. Publ. Fish. Aquat. Sci.* 96:120–124.

Brannon, E. L., and W. K. Hershberger. 1984. Elwha River fall chinook salmon. Pages 169–172 in J. M. Walton and D. B. Houston, eds., *Proceedings of the Olympic wild fish conference.* Peninsula College, Port Angeles, Washington.

Brannon, E. L., M. Powell, T. Quinn, and A. Talbot. 2002. *Population structure of Columbia River basin chinook salmon and steelhead trout.* Rept. to the Bonneville Power Administration, Portland, Oregon. 178 p.

Brannon, E. L., and T. P. Quinn. 1990. A field test of the pheromone hypothesis for homing by Pacific salmon. *J. Chem. Ecol.* 16:603–609.

Brannon, E. L., and A. Setter. 1989. Marine distribution of a hatchery fall chinook salmon population. Pages 63–69 in E. Brannon and B. Jonsson, eds., *Proceedings of the salmonid migration and distribution symposium.* University of Washington School of Fisheries, Seattle.

Brett, J. R. 1971. Energetic response of salmon to temperature. A study of some thermal relations in the physiology and freshwater ecology of sockeye salmon (*Oncorhynchus nerka*). *Am. Zool.* 11: 99–113.

———. 1983. Life energetics of sockeye salmon, *Oncorhynchus nerka*. Pages 29–63 in W. P. Aspey and S. I. Lustick, eds., *Behavioral energetics: The cost of survival in vertebrates.* Ohio State University Press, Columbus, Ohio.

———. 1995. Energetics. Pages 1–68 in C. Groot, L. Margolis, and W. C. Clarke, eds.), *Physiological ecology of Pacific salmon.* University of British Columbia Press, Vancouver.

Brett, J. R., J. E. Shelbourn, and C. T. Shoop. 1969. Growth rate and body composition of fingerling sockeye salmon, *Oncorhynchus nerka*, in relation to temperature and ration size. *J. Fish. Res. Board Can.* 26:2363–2394.

Brodeur, R. D. 1989. Neustonic feeding by juvenile salmonids in coastal waters of the Northeast Pacific. *Can. J. Zool.* 67:1995–2007.

Brodeur, R. D. 1990. *A synthesis of the food habits and feeding ecology of salmonids in marine waters of the North Pacific.* International North Pacific Fisheries Commission document. University of Washington School of Fisheries, Fish. Res. Inst., FRI-UW-9016. Seattle. 38 p.

Brodeur, R. D., B. W. Frost, S. R. Hare, R. C. Francis, and W. J. J. Ingraham. 1996. Interannual variations in zooplankton biomass in the Gulf of Alaska, and covariation with California Current zooplankton biomass. *Cal. Coop. Fish. Invest. Rept.* 37:80–99.

Brodeur, R. D., K. W. Myers, and J. H. Helle. 2003. Research conducted by the United States on the early ocean life history of Pacific salmon. *N. Pac. Anad. Fish Comm. Bull.* 3:89–131.

Brown, B. 1982. *Mountain in the clouds: A search for wild salmon.* Simon and Schuster, New York.

Brown, G. E., and J. A. Brown. 1996. Kin discrimination in salmonids. *Rev. Fish Biol. Fish.* 6:201–219.

Brown, L. R., P. B. Moyle, and R. M. Yoshiyama. 1994. Historical declines and current status of coho salmon in California. *N. Am. J. Fish. Mgmt.* 14:237–261.

Brown, R. S., and W. C. Mackay. 1995a. Spawning ecology of cutthroat trout (*Oncorhynchus clarki*) in the Ram River, Alberta. *Can. J. Fish. Aquat. Sci.* 52:983–992.

———. 1995b. Fall and winter movements of and habitat use by cutthroat trout in the Ram River, Alberta. *Trans. Am. Fish. Soc.* 124:873–885.

Buchanan, D. V., J. E. Sanders, J. L. Zinn, and J. L. Fryer. 1983. Relative susceptibility of four strains of summer steelhead to infection by *Ceratomyxa shasta. Trans. Am. Fish. Soc.* 112:541–543.

Burger, C. V., K. T. Scribner, W. J. Spearman, C. O. Swanton, and D. E. Campton. 2000. Genetic contribution of three introduced life history forms of sockeye salmon to colonization of Frazer Lake, Alaska. *Can. J. Fish. Aquat. Sci.* 57:2096–2111.

Burgner, R. L. 1962. Studies of red salmon smolts from the Wood River lakes, Alaska. Pages 247–314 in T. S. Y. Koo, ed., *Studies of Alaska red salmon.* University of Washington Press, Seattle.

———. 1980. Some features of ocean migrations and timing of Pacific salmon. Pages 153–164 in W. J. McNeil and D. C. Himsworth, eds., *Salmonid ecosystems of the North Pacific.* Oregon State University Press, Corvallis.

———. 1991. Life history of sockeye salmon (*Oncorhynchus nerka*). Pages 1–117 in C. Groot and L. Margolis, eds., *Pacific salmon life histories.* University of British Columbia Press, Vancouver.

Burgner, R. L., J. T. Light, L. Margolis, T. Okazaki, A. Tautz, and S. Ito. 1992. Distribution and origins of steelhead trout (*Oncorhynchus mykiss*) in offshore waters of the North Pacific Ocean. *Int. N. Pac. Fish. Comm. Bull.* 51. Vancouver, BC. 92 p.

Burner, C. J. 1951. Characteristics of spawning nests of Columbia River salmon. *Fish. Bull.* 52:97–110.

Busby, P. J., T. C. Wainwright, G. J. Bryant, L. J. Lierheimer, R. S. Waples, F. W. Waknitz, and I. V. Lagomarsino. 1996. *Status review of west coast steelhead from Washington, Idaho, Oregon and California.* National Marine Fisheries Service, NOAA Tech. Memo. NMFS-NWFSC-27. Seattle. 261 p.

Butler, J. A., and R. E. Millemann. 1971. Effect of the "salmon poisoning" trematode, *Nanophyetus salmincola*, on the swimming ability of juvenile salmonid fishes. *J. Parasit.* 57:860–865.

Candy, J. R., and T. D. Beacham. 2000. Patterns of homing and straying in southern British Columbia coded-wire tagged chinook salmon (*Oncorhynchus tshawytscha*) populations. *Fish. Res.* 47:41–56.

Candy, J. R., and T. P. Quinn. 1999. Behavior of adult chinook salmon (*Oncorhynchus tshawytscha*) in British Columbia coastal waters determined from ultrasonic telemetry. *Can. J. Zool.* 77:1161–1169.

Carey, W. E., and D. L. G. Noakes. 1981. Development of photobehavioural responses in young rainbow trout, *Salmo gairdneri* Richardson. *J. Fish Biol.* 19:285–296.

Carl, G. C., W. A. Clemens, and C. C. Lindsey. 1977. *The freshwater fishes of British Columbia.* British Columbia Provincial Museum, Victoria.

Carl, L. M., and M. C. Healey. 1984. Differences in enzyme frequency and body morphology among three juvenile life history types of chinook salmon (*Oncorhynchus tshawytscha*) in the Nanaimo River, British Columbia. *Can. J. Fish. Aquat. Sci.* 41:1070–1077.

Carvalho, G. R., and L. Hauser. 1994. Molecular genetics and the stock concept in fisheries. *Rev. Fish Biol Fish.* 4:326–350.

Cederholm, C. J., R. E. Bilby, P. A. Bisson, T. W. Bumstead, B. R. Fransen, W. J. Scarlett, and J. W. Ward. 1997. Response of juvenile coho salmon and steelhead to placement of large woody debris in a coastal Washington stream. *N. Am. J. Fish. Mgmt.* 17:947–963.

Cederholm, C. J., D. B. Houston, D. L. Cole, and W. J. Scarlett. 1989. Fate of coho salmon (*Oncorhynchus kisutch*) carcasses in spawning streams. *Can. J. Fish. Aquat. Sci.* 46:1347–1355.

Cederholm, C. J., M. D. Kunze, T. Murota, and A. Sibatani. 1999. Pacific salmon carcasses: essential contributions of nutrients and energy for aquatic and terrestrial ecosystems. *Fisheries* 24(10):6–15.

Cederholm, C. J., and L. M. Reid. 1987. Impact of forest management on coho salmon (*Oncorhynchus kisutch*) populations of the Clearwater River, Washington: A project summary. Pages 373–398 in E. O. Salo and T. W. Cundy, eds., *Streamside management: Forestry and fishery interactions*. University of Washington College of Forest Resources, Seattle.

Chapman, D. W. 1966. Food and space as regulators of salmonid populations in streams. *Am. Nat.* 100:345–357.

———. 1986. Salmon and steelhead abundance in the Columbia River in the nineteenth century. *Trans. Am. Fish. Soc.* 115:662–670.

———. 1988. Critical review of variables used to define effects of fines in redds of large salmonids. *Trans. Am. Fish. Soc.* 117:1–21.

Chatters, J. C., V. L. Butler, M. J. Scott, D. M. Anderson, and D. A. Neitzel. 1995. A paleoscience approach to estimating the effects of climate warming on salmonid fisheries of the Columbia River basin. *Can. Spec. Publ. Fish. Aquat. Sci.* 121:489–496.

Clark, C. W., and D. A. Levy. 1988. Diel vertical migrations by juvenile sockeye salmon and the antipredation window. *Am. Nat.* 131:271–290.

Clark, W. K. 1959. Kodiak bear–red salmon relationships at Karluk Lake, Alaska. *Trans. N. Am. Wildl. Conf.* 24:337–345.

Clarke, W. C., and T. Hirano. 1995. Osmoregulation. Pages 317–377 in C. Groot, L. Margolis, and W. C. Clarke, eds., *Physiological ecology of Pacific salmon*. University of British Columbia Press, Vancouver.

Clarke, W. C., and H. D. Smith. 1972. Observations on the migration of sockeye salmon fry (*Oncorhynchus nerka*) in the lower Babine River. *J. Fish. Res. Board Can.* 29:151–159.

Clemens, W. A., R. E. Foerster, and A. L. Pritchard. 1939. The migration of Pacific salmon in British Columbia waters. Pages 51–59 in Am. Assoc. Advanc. Sci. Publ. 8. Lancaster, PA.

Coble, D. W. 1961. Influence of water exchange and dissolved oxygen in redds on survival of steelhead trout embryos. *Trans. Am. Fish. Soc.* 90:469–474.

Connolly, P. J., and J. H. Petersen. 2003. Bigger is not always better for overwintering young-of-year steelhead. *Trans. Am. Fish. Soc.* 132:262–274.

Cooke, K., R. Hungar, C. Groot, and G. Tolson. 1987. Data record of adult sockeye salmon and other fish species captured by purse seine in Queen Charlotte Strait, Johnstone Strait and the Strait of Georgia in 1985 and 1986. *Can. Data Rept. Fish. Aquat. Sci.* 680. Nanaimo, BC. 127 p.

Cooke, S. J., S. G. Hinch, A. P. Farrell, M. Lapointe, S. R. M. Jones, J. S. Macdonald, D. Patterson, M. C. Healey, and G. Van Der Kraak. 2004. Abnormal migration timing and high enroute mortality of sockeye salmon in the Fraser River, British Columbia. *Fisheries* 29(2):22–33.

Cooney, R. T., J. R. Allen, M. A. Bishop, D. L. Eslinger, T. Kline, B. L. Norcross, C. P. McRoy, J. Milton, J. Olsen, V. Patrick, A. J. Paul, D. Salmon, D. Scheel, G. L. Thomas, S. L. Vaughan, and T. M. Willette. 2001. Ecosystem controls of juvenile pink salmon (*Oncorhynchus gorbuscha*) and Pacific herring (*Clupea pallasi*) populations in Prince William Sound, Alaska. *Fish. Oceanogr.* 10 (Suppl. 1):1–13.

Cooney, R. T., T. M. Willette, S. Sharr, D. Sharp, and J. Olsen. 1995. The effect of climate on North Pacific pink salmon (*Oncorhynchus gorbuscha*) production: examining some details of a natural experiment. *Can. Spec. Publ. Fish. Aquat. Sci.* 121:475–482.

Courtenay, S. C., T. P. Quinn, H. M. C. Dupuis, C. Groot, and P. A. Larkin. 1997. Factors affecting the recognition of population-specific odours by juvenile coho salmon (*Oncorhynchus kisutch*). *J. Fish Biol.* 50:1042–1060.

Crawford, D. L., J. D. Woolington, and B. A. Cross. 1992. *Bristol Bay sockeye salmon smolt studies for 1990*. Alaska Dept. Fish Game, Tech. Rept. 92–19. Juneau. 74 p.

Crawford, S. S. 2001. Salmonine introductions to the Laurentian Great Lakes: An historical review and evaluation of ecological effects. *Can. Spec. Publ. Fish. Aquat. Sci.* 132. Ottawa. 205 p.

Crespi, B. J., and R. Teo. 2002. Comparative phylogenetic analysis of the evolution of semelparity and life history in salmonid fishes. *Evolution* 56:1008–1020.

Crisp, D. T. 1988. Prediction, from temperature, of eyeing, hatching and "swim-up" times for salmonid embryos. *Freshw. Biol.* 19:41–48.

Crone, R. A., and C. E. Bond. 1976. Life history of coho salmon, *Oncorhynchus kisutch*, in Sashin Creek, southeastern Alaska. *Fish. Bull.* 74:897–923.

Cunjak, R. A. 1996. Winter habitat of selected stream fishes and potential impacts from land-use activity. *Can. J. Fish. Aquat. Sci.* 53 (Suppl. 1):267–282.

Dahlberg, M. L. 1968. Analysis of the dynamics of sockeye salmon returns to the Chignik Lakes, Alaska. Ph.D. dissertation, University of Washington, Seattle.

Dalton, T. J. 1991. Variation in prevalence of *Nanophyetus salmincola*, a parasite tag indicating U.S. northwest origin, in steelhead trout (*Oncorhynchus mykiss*) caught in central north Pacific Ocean. *Can. J. Fish. Aquat. Sci.* 48:1104–1108.

Dangel, J. R., and J. D. Jones. 1988. Southeast Alaska pink salmon total escapement and stream life studies, 1987. Alaska Dept. Fish Game, Division of Commercial Fisheries, Reg. Info. Rept, 1J88-24. Juneau. 56 p.

Dat, C. G., P. H. LeBlond, K. A. Thompson, and W. J. Ingraham, Jr. 1995. Computer simulations of homeward-migrating Fraser River sockeye salmon: Is compass orientation a sufficient direction-finding mechanism in the north-east Pacific Ocean? *Fish. Oceanogr.* 4:209–219.

Daum, D. W., and B. M. Osborne. 1998. Use of fixed-location, split-beam sonar to describe temporal and spatial patterns of adult fall chum salmon migration in the Chandalar River, Alaska. *N. Am. J. Fish. Mgmt.* 18:477–486.

Davidson, F. A., E. Vaughan, and S. J. Hutchinson. 1943. Factors influencing the upstream migration of the pink salmon (*Oncorhynchus gorbuscha*). *Ecology* 24:149–168.

Davis, J. C. 1975. Minimal dissolved oxygen requirements of aquatic life with emphasis on Canadian species: A review. *J. Fish. Res. Board Can.* 32:2295–2332.

Davis, N. D., K. W. Myers, R. V. Walker, and C. K. Harris. 1990. The Fisheries Research Institute's high seas salmonid tagging program and methodology for scale pattern analysis. Pages 863–879 in Am. Fish. Soc. Symp. 7. Bethesda, MD.

Dawley, E. M., R. D. Ledgerwood, T. H. Blahm, C. W. Sims, J. T. Durkin, R. A. Kirn, A. E. Rankis, G. E. Monan, and F. J. Ossiander. 1986. *Migrational characteristics, biological observations, and relative survival of juvenile salmonids entering the Columbia River estuary, 1966–1983.* Final report to the Bonneville Power Administration, Project 81-102. Portland, Oregon. 256 p.

DeCicco, A. L. 1992. Long-distance movements of anadromous Dolly Varden between Alaska and the U.S.S.R. *Arctic* 45:120–123.

Dickerson, B. R. 2003. Reproductive success in wild pink salmon, *Oncorhynchus gorbuscha*. Ph.D. dissertation, University of Washington, Seattle.

Dickerson, B. R., M. F. Willson, P. Bentzen, and T. P. Quinn. Forthcoming. Size-assortative mating in salmonids: Negative evidence for pink salmon in natural conditions. *Anim. Behav.*

Diebel, C. E., R. Proksch, C. R. Green, P. Nielson, and M. M. Walker. 2000. Magnetite defines a vertebrate magnetoreceptor. *Nature* 406:299–302.

Dickhoff, W. W., B. R. Beckman, D. A. Larsen, C. Duan, and S. Moriyama. 1997. The role of growth in endocrine regulation of salmon smoltification. *Fish Physiol. Biochem.* 17:231–236.

Dickhoff, W. W., D. S. Darling, and A. Gorbman. 1982. Thryoid function during smoltification of salmonid fish. *Gunma Symp. Endocrinol.* (Tokyo) 19:45–61.

Dickhoff, W. W., and C. V. Sullivan. 1987. Involvement of the thyroid gland in smoltification, with special reference to metabolic and developmental processes. Pages 197–210 in Am. Fish. Soc. Symp. 1. Bethesda, MD.

Dill, L. M., and T. G. Northcote. 1970. Effects of gravel size, egg depth, and egg density on intragravel movement and emergence of coho salmon (*Oncorhynchus kisutch*) alevins. *J. Fish. Res. Board Can.* 27:1191–1199.

Dill, L. M., R. C. Ydenberg, and A. H. G. Fraser. 1981. Food abundance and territory size in juvenile coho salmon (*Oncorhynchus kisutch*). *Can. J. Zool.* 59:1801–1809.

Dittman, A. H., and T. P. Quinn. 1996. Homing in Pacific salmon: Mechanisms and ecological basis. *J. Exp. Biol.* 199:83–91.

Dittman, A. H., T. P. Quinn, and G. A. Nevitt. 1996. Timing of imprinting to natural and artificial odors by coho salmon (*Oncorhynchus kisutch*). *Can. J. Fish. Aquat. Sci.* 53:434–442.

Dittman, A. H., T. P. Quinn, G. A. Nevitt, B. Hacker, and D. R. Storm. 1997. Sensitization of olfactory guanylyl cyclase to a specific imprinted odorant in coho salmon. *Neuron* 19:381–389.

Dolloff, C. A. 1993. Predation by river otters (*Lutra canadensis*) on juvenile coho salmon (*Oncorhynchus kisutch*) and Dolly Varden (*Salvelinus malma*) in southeast Alaska. *Can. J. Fish. Aquat. Sci.* 50:312–315.

Dolloff, C. A., and G. H. Reeves. 1990. Microhabitat partitioning among stream-dwelling juvenile coho salmon, *Oncorhynchus kisutch*, and Dolly Varden, *Salvelinus malma*. *Can. J. Fish. Aquat. Sci.* 47: 2297–2306.

Donaldson, J. R. 1967. The phosphorous budget of Iliamna Lake, Alaska as related to the cyclic abundance of sockeye salmon. Ph.D. dissertation, University of Washington, Seattle.

Donaldson, L. R., and G. H. Allen. 1957. Return of silver salmon, *Oncorhynchus kisutch* (Walbaum) to point of release. *Trans. Am. Fish. Soc.* 87:13–22.

Døving, K. B., H. Westerberg, and P. B. Johnsen. 1985. Role of olfaction in the behavioral neuronal responses of Atlantic salmon, *Salmo salar*, to hydrographic stratification. *Can. J. Fish. Aquat. Sci.* 42:1658–1667.

Drake, D. C., R. J. Naiman, and J. M. Helfield. 2002. Reconstructing salmon abundance in rivers: An initial dendrochronological evaluation. *Ecology* 83:2971–2977.

Dunn, J. R. 1969. Direction of movement of salmon in the North Pacific Ocean, Bering Sea, and Gulf of Alaska as indicated by surface gillnet catches. *Int. N. Pac. Fish. Comm. Bull.* 26:27–55.

———. 1996. Charles Henry Gilbert (1859–1928): An early fishery biologist and his contributions to knowledge of Pacific salmon (*Oncorhynchus* spp.). *Rev. Fish. Sci.* 4:133–184.

Eastman, D. E. 1996. Response of freshwater fish communities to spawning sockeye salmon (*Oncorhynchus nerka*). M.S. thesis, University of Washington, Seattle.

Ebel, W. J. 1980. Transportation of chinook salmon, *Oncorhynchus tshawytscha*, and steelhead, *Salmo gairdneri*, smolts in the Columbia River and effects on adult returns. *Fish. Bull.* 78:491–505.

Ebel, W. J., D. L. Park, and R. C. Johnsen. 1973. Effects of transportation on survival and homing of Snake River chinook salmon and steelhead trout. *Fish. Bull.* 71:549–563.

Edmundson, J. A., and A. Mazumder. 2001. Linking growth of juvenile sockeye salmon to habitat temperature in Alaskan lakes. *Trans. Am. Fish. Soc.* 130:644–662.

Eggers, D. M. 1978. Limnetic feeding behavior of juvenile sockeye salmon in Lake Washington and predator avoidance. *Limnol. Oceanogr.* 23:1114–1125.

———. 1982. Planktivore preference by prey size. *Ecology* 63:381–390.

Einum, S., and I. A. Fleming. 1999. Maternal effects of egg size in brown trout (*Salmo trutta*): norms of reaction to environmental quality. *Proc. Royal Soc. Lond. B* 266:2095–2100.

Emanuel, M. E., and J. J. Dodson. 1979. Modification of the rheotropic behavior of male rainbow trout (*Salmo gairdneri*) by ovarian fluid. *J. Fish. Board Can.* 36:63–68.

Erickson, D. L., and E. K. Pikitch. 1994. Incidental catch of chinook salmon in commercial bottom trawls off the U.S. west coast. *N. Am. J. Fish. Mgmt.* 14:550–563.

Essington, T. E., T. P. Quinn, and V. E. Ewert. 2000. Intra- and interspecific competition and the reproductive success of sympatric Pacific salmon. *Can. J. Fish. Aquat. Sci.* 57:205–213.

Essington, T. E., P. W. Sorensen, and D. G. Paron. 1998. High rate of redd superimposition by brook trout (*Salvelinus fontinalis*) and brown trout (*Salmo trutta*) in a Minnesota stream cannot be explained by habitat availability alone. *Can. J. Fish. Aquat. Sci.* 55:2310–2316.

Everest, F. H., and D. W. Chapman. 1972. Habitat selection and spatial interaction by juvenile chinook salmon and steelhead trout in two Idaho streams. *J. Fish. Res. Board Can.* 29:91–100.

Fast, D. E. 1987. The behavior of salmonid alevins in response to changes in dissolved oxygen, velocity and light during incubation. Ph.D. dissertation, University of Washington, Seattle.

Fausch, K. D. 1988. Test of competition between native and introduced salmonids in streams: What have we learned? *Can. J. Fish. Aquat. Sci.* 45:2238–2246.

Favorite, F., A. J. Dodimead, and K. Nasu. 1976. Oceanography of the subarctic Pacific Region, 1960–71. *Int. N. Pac. Fish. Comm. Bull.* 33. Vancouver, BC. 187 p.

Fedorenko, A. Y., and B. G. Shepherd. 1986. Review of salmon transplant procedures and suggested transplant guidelines. *Can. Tech. Rept. Fish. Aquat. Sci.* 1479.Vancouver, BC. 144 p.

Ferguson, A. Forthcoming. "The importance of identifying conservation units: Brown trout and pollan diversity in Ireland." *Biology and environment: Proceedings of the Royal Irish Academy*.

Finney, B. P., I. Gregory-Eaves, M. S. V. Douglas, and J. P. Smol. 2002. Fisheries productivity in the northeastern Pacific Ocean over the past 2,200 years. *Nature* 416:729–733.

Finney, B. P., I. Gregory-Eaves, J. Sweetman, M. S. V. Douglas, and J. P. Smol. 2000. Impacts of climate change and fishing on Pacific salmon abundance over the past 300 years. *Science* 290:795–799.

Fisher, F. W. 1994. Past and present status of Central Valley chinook salmon. *Cons. Biol.* 8:870–873.

Fisher, J. P., and W. G. Pearcy. 1988. Growth of juvenile coho salmon (*Oncorhynchus kisutch*) in the ocean off Oregon and Washington, USA, in years of differing coastal upwelling. *Can. J. Fish. Aquat. Sci.* 45:1036–1044.

Fleming, I. A. 1998. Pattern and variability in the breeding system of Atlantic salmon (*Salmo salar*), with comparisons to other salmonids. *Can. J. Fish. Aquat. Sci.* 55 (Suppl. 1):59–76.

Fleming, I. A., and M. R. Gross. 1994. Breeding competition in a Pacific salmon (coho: *Oncorhynchus kisutch*): Measures of natural and sexual selection. *Evolution* 48:637–657.

Flynn, L., R. Hilborn, and A. E. Punt. 2003. Identifying the spatial distribution of stocks of migrating adult sockeye salmon using age composition data. *Alaska Fish. Res. Bull.* 10:50–60.

Foerster, R. E. 1936. The return from the sea of sockeye salmon (*Oncorhynchus nerka*) with special reference to percentage survival, sex proportions and progress of migration. *J. Biol. Board Can.* 3:26–42.

———. 1968. The sockeye salmon. *Fish. Res. Board Can. Bull.* 162. Ottawa. 422 p.

Foerster, R. E., and W. E. Ricker. 1953. The coho salmon of Cultus Lake and Sweltzer Creek. *J. Fish. Res. Board Can.* 10:293–319.

Foote, C. J. 1988. Male mate choice dependent on male size in salmon. *Behaviour* 106:63–80.

———. 1989. Female mate preference in Pacific salmon. *Anim. Behav.* 38:721–722.

———. 1990. An experimental comparison of male and female spawning territoriality in a Pacific salmon. *Behaviour* 115:283–314.

Foote, C. J., and G. S. Brown. 1998. Ecological relationship between freshwater sculpins (genus *Cottus*) and beach-spawning sockeye salmon (*Oncorhynchus nerka*) in Iliamna Lake, Alaska. *Can. J. Fish. Aquat. Sci.* 55:1524–1533.

Foote, C. J., G. S. Brown, and C. C. Wood. 1997. Spawning success of males using alternative mating tactics in sockeye salmon, *Oncorhynchus nerka. Can. J. Fish. Aquat. Sci.* 54:1785–1795.

Foote, C. J., and P. A. Larkin. 1988. The role of male choice in the assortative mating of anadromous and non-anadromous sockeye salmon (*Oncorhynchus nerka*). *Behaviour* 106:43–62.

Foote, C. J., C. C. Wood, and R. E. Withler. 1989. Biochemical genetic comparison of sockeye salmon and kokanee, the anadromous and nonanadromous forms of *Oncorhynchus nerka. Can. J. Fish. Aquat. Sci.* 46:149–158.

Ford, J. K. B., G. M. Ellis, L. G. Barrett-Lennard, A. B. Morton, R. S. Palm, and K. C. I. Balcomb. 1998. Dietary specialization in two sympatric populations of killer whales (*Orcinus orca*) in coastal British Columbia and adjacent waters. *Can. J. Zool.* 76:1456–1471.

Foster, N. R. 1985. Lake trout reproductive behavior: Influence of chemosensory cues from young-of-the-year by-products. *Trans. Am. Fish. Soc.* 114:794–803.

Fraley, J. J., and B. B. Shepard. 1989. Life history, ecology and population status of migratory bull trout (*Salvelinus confluentus*) in the Flathead Lake and river system, Montana. *Northwest Sci.* 63:133–143.

Francis, R. C., and S. R. Hare. 1994. Decadal-scale regime shifts in the large marine ecosystems of the North-east Pacific: A case for historical science. *Fish. Oceanogr.* 3:279–291.

Francis, R. C., and T. H. Sibley. 1991. Climate change and fisheries: What are the real issues? *Northwest Envir. J.* 7:295–307.

Fransen, B. R., P. A. Bisson, J. W. Ward, and R. E. Bilby. 1993. Physical and biological constraints on summer rearing of juvenile coho salmon (*Oncorhynchus kisutch*) in small western Washington streams. Pages 271–288 in L. Berg and P. W. Delaney, eds., *Coho salmon workshop*. Dept. of Fisheries and Oceans, Canada, Nanaimo, BC.

French, R. R., H. Bilton, M. Osako, and A. Hartt. 1976. Distribution and origin of sockeye salmon (*Oncorhynchus nerka*) in offshore waters of the North Pacific Ocean. *Int. N. Pac. Fish. Comm. Bull.* 34. Vancouver, BC. 113 p.

French, R. R., and W. B. McAlister. 1970. Winter distribution of salmon in relation to currents and water masses in the northeastern Pacific Ocean and migrations of sockeye salmon. *Trans. Am. Fish. Soc.* 99:649–663.

Friedland, K. D., R. V. Walker, N. D. Davis, K. W. Myers, G. W. Boehlert, S. Urawa, and Y. Ueno. 2001. Open-ocean orientation and return migration routes of chum salmon based on temperature data from data storage tags. *Mar. Ecol. Progr. Ser.* 216:235–252.

Fukuwaka, M., and T. Suzuki. 2002. Early sea mortality of mark-recaptured juvenile chum salmon in open coastal waters. *J. Fish Biol.* 60:3–12.

Fuss, H. J. 1984. Age, growth and instream movement of Olympic Peninsula coastal cutthroat trout (*Salmo clarki clarki*). Pages 125–134 in J. M. Walton and D. B. Houston, eds., *Proceedings of the Olympic wild fish conference*. Peninsula College, Port Angeles, Washington.

García de Leániz, C., N. Fraser, and F. A. Huntingford. 2000. Variability in performance in wild Atlantic salmon, *Salmo salar* L., fry from a single redd. *Fish. Mgmt. Ecol.* 7:489–502.

Gaudet, D. M. 1990. Enumeration of migrating salmon populations using fixed-location sonar counters. *Procès-verbal de la Réunion, Conseil International pour l'Exploration de la Mer* 189:197–209.

Geiger, H. J., W. W. Smoker, L. A. Zhivotovsky, and A. J. Gharrett. 1997. Variability of family size and marine survival in pink salmon (*Oncorhynchus gorbuscha*) has implications for conservation biology and human use. *Can. J. Fish. Aquat. Sci.* 54:2684–2690.

Gende, S. M., R. T. Edwards, M. F. Willson, and M. S. Wipfli. 2002. Pacific salmon in aquatic and terrestrial ecosystems. *BioScience* 52:917–928.

Gende, S. M., and T. P. Quinn. 2004. The relative importance of prey density and social dominance in determining energy intake by bears feeding on Pacific salmon. *Can. J. Zool.* 82:75–85.

Gende, S. M., T. P. Quinn, and M. F. Willson. 2001. Consumption choice by bears feeding on salmon. *Oecologia* 127:372–382.

Gende, S. M., T. P. Quinn, M. F. Willson, R. Heintz, and T. M. Scott. 2004. Magnitude and fate of salmon-derived nutrients and energy in a coastal stream ecosystem. *J. Freshw. Ecol.* 19:149–160.

Gende, S. M., T. P. Quinn, R. Hilborn, A. P. Hendry, and B. Dickerson. 2004. Brown bears selectively kill salmon with higher energy content but only in habitats that facilitate choice. *Oikos* 104:518–528.

Gende, S. M., and M. F. Willson. 2001. Passerine densities in riparian forests of southeast Alaska: potential effects of anadromous spawning salmon. *Condor* 103:624–629.

Gharrett, A. J., and S. M. Shirley. 1985. A genetic examination of spawning methodology in a salmon hatchery. *Aquaculture* 47:245–256.

Gilbert, C. H. 1922. The salmon of the Yukon River. *Bull. U.S. Bureau Fish.* 38:317–332.

Gilhousen, P. 1980. Energy sources and expenditures in Fraser River sockeye salmon during their spawning migration. *Int. Pac. Salmon Fish. Comm. Bull.* 22. New Westminster, BC. 51 p.

———. 1990. Prespawning mortalities of sockeye salmon in the Fraser River system and possible causes. *Int. Pac. Salmon Fish. Comm. Bull.* 26. New Westminster, BC. 58 p.

Giorgi, A. E., T. W. Hillman, J. R. Stevenson, S. G. Hays, and C. M. Peven. 1997. Factors that influence the downstream migration rates of juvenile salmon and steelhead through the hydroelectric system in the mid–Columbia River basin. *N. Am. J. Fish. Mgmt.* 17:268–282.

Giorgi, A. E., G. A. Swan, W. S. Zaugg, T. Coley, and T. Y. Barila. 1988. Susceptibility of chinook salmon smolts to bypass systems at hydroelectric dams. *N. Am. J. Fish. Mgmt.* 8:25–29.

Glova, G. J. 1987. Comparison of allopatric cutthroat trout stocks with those sympatric with coho salmon and sculpins in small streams. *Env. Biol. Fish.* 20:275–284.

Godfrey, H., K. A. Henry, and S. Machidori. 1975. Distribution and abundance of coho salmon in offshore waters of the North Pacific Ocean. *Int. N. Pac. Fish. Comm. Bull.* 31. Vancouver, BC. 80 p.

Godfrey, H., W. R. Hourston, J. W. Stokes, and F. C. Withler. 1954. Effects of a rock slide on Babine River salmon. *Fish. Res. Board Can. Bull.* 101. Ottawa. 100 p.

Godfrey, H., W. R. Hourston, and F. C. Withler. 1956. Babine River salmon after removal of the rock slide. *Fish. Res. Board Can. Bull.* 106. Ottawa. 41 p.

Godin, J.-G. J. 1980. Temporal aspects of juvenile pink salmon (*Oncorhynchus gorbuscha* Walbaum) emergence from a stimulated gravel redd. *Can. J. Zool.* 58:735–744.

———. 1982. Migrations of salmonid fishes during early life history phases: daily and annual timing.Pages 22–50 in E. L. Brannon and E.O. Salo, eds., *Proceedings of the salmon and trout migratory behavior symposium*. University of Washington School of Fisheries, Seattle.

Grachev, L. Y. 1971. Alteration in the number of oocytes in the pink salmon (*Oncorhynchus gorbuscha* [Walbaum]) in the marine period of life. *J. Ichthyol.* 11:199–206.

Grant, J. W. A., S. Ó. Steingrímsson, E. R. Keeley, and R. A. Cunjak. 1998. Implications of territory size for the measurement and prediction of salmonid abundance in streams. *Can. J. Fish. Aquat. Sci.* 55 (Suppl. 1):181–190.

Gray, A., C. A. Simenstad, D. L. Bottom, and T. J. Cornwell. 2002. Contrasting functional performance of juvenile salmon habitat in recovering wetlands of the Salmon River estuary, Oregon, U.S.A. *Rest. Ecol.* 10:514–526.

Gresh, T., J. A. Lichatowich, and P. Schoonmaker. 2000. An estimation of historic and current levels of salmon production in the Northeast Pacific ecosystem: evidence of a nutrient deficit in the freshwater systems of the Pacific Northwest. *Fisheries* 25(1):15–21.

Groot, C. 1965. On the orientation of young sockeye salmon (*Oncorhynchus nerka*) during their seaward migration out of lakes. *Behaviour* 14 (Suppl.):1–198.

———. 1972. Migration of yearling sockeye salmon (*Oncorhynchus nerka*) as determined by time-lapse photography of sonar observations. *J. Fish. Res. Board Can.* 29:1431–1444.

Groot, C., R. E. Bailey, L. Margolis, and K. Cooke. 1989. Migratory patterns of sockeye salmon (*Oncorhynchus nerka*) smolts in the Strait of Georgia, British Columbia, as determined by analysis of parasite assemblages. *Can. J. Zool.* 67:1670–1678.

Groot, C., and K. D. Cooke. 1987. Are the migrations of juvenile and adult Fraser River sockeye salmon (*Oncorhynchus nerka*) in near-shore waters related? *Can. Spec. Publ. Fish. Aquat. Sci.* 96:53–60.

Groot, C., and L. Margolis, eds. 1991. *Pacific salmon life histories*. University of British Columbia Press, Vancouver.

Groot, C., and T. P. Quinn. 1987. The homing migration of sockeye salmon to the Fraser River. *Fish. Bull.* 85:455–469.

Groot, C., K. Simpson, I. Todd, P. D. Murray, and G. A. Buxton. 1975. Movements of sockeye salmon (*Oncorhynchus nerka*) in the Skeena River estuary as revealed by ultrasonic tracking. *J. Fish. Res. Board Can.* 32:233–242.

Gross, M. R. 1985. Disruptive selection for alternative life history strategies in salmon. *Nature* 313:47–48.

———. 1987. Evolution of diadromy in fishes. Pages 14–25 in Am. Fish. Soc. Symp. 1. Bethesda, MD.

Gross, M. R., R. M. Coleman, and R. M. McDowall. 1988. Aquatic productivity and the evolution of diadromous fish migration. *Science* 239:1291–1293.

Gross, M. R., and R. C. Sargent. 1985. The evolution of male and female parental care in fishes. *Am. Zool.* 25:807–822.

Groves, P. A., and J. A. Chandler. 1999. Spawning habitat used by fall chinook salmon in the Snake River. *N. Am. J. Fish. Mgmt.* 19:912–922.

Grunbaum, J. B. 1996. Geographical and seasonal variation in diel habitat use by juvenile (age 1+) steelhead trout (*Oncorhynchus mykiss*) in Oregon coastal and interior streams. M.S. thesis, Oregon State University, Corvallis.

Gunnes, K. 1979. Survival and development of Atlantic salmon eggs and fry at three different temperatures. *Aquaculture* 16:211–218.

Gunstrom, G. K. 1968. Gonad weight/body weight ratio of mature chinook salmon males as a measure of gonad size. *Prog. Fish-Cult.* 30:23–25.

Gustafson, R. G., T. C. Wainwright, G. A. Winans, F. W. Waknitz, L. T. Parker, and R. S. Waples. 1997. *Status review of sockeye salmon from Washington and Oregon*. National Marine Fisheries Service, NOAA Tech. Memo. NMFS-NWFSC-33. Seattle. 282 p.

Gustafson, R. G., and G. A. Winans. 1999. Distribution and population genetic structure of river- and sea-type sockeye salmon in western North America. *Ecol. Freshw. Fish.* 8:181–193.

Gyllensten, U. 1985. The genetic structure of fish: Differences in the intraspecific distribution of biochemical genetic variation between marine, anadromous, and freshwater species. *J. Fish Biol.* 26:691–699.

Haas, G. R., and J. D. McPhail. 1991. Systematics and distributions of Dolly Varden (*Salvelinus malma*) and bull trout (*Salvelinus confluentus*) in North America. *Can. J. Fish. Aquat. Sci.* 48:2191–2211.

Hager, R. C., and C. W. J. Hopley. 1981. *A comparison of the effect of adult return timing of Cowlitz and Toutle hatchery coho on catch and escapement.* Washington Dept. Fisheries, Tech. Rept. 58. Olympia. 41 p.

Hallock, R. J., and D. H. Fry, Jr. 1967. Five species of salmon, *Oncorhynchus*, in the Sacramento River, California. *Cal. Fish Game* 53:5–22.

Hankin, D. G., J. W. Nicholas, and T. W. Downey. 1993. Evidence for inheritance of age at maturity in chinook salmon (*Oncorhynchus tshawytscha*). *Can. J. Fish. Aquat. Sci.* 50:347–358.

Hanson, A. J., and H. D. Smith. 1967. Mate selection in a population of sockeye salmon (*Oncorhynchus nerka*) of mixed age-groups. *J. Fish. Res. Board Can.* 24:1955–1977.

Hara, T. J., K. Ueda, and A. Gorbman. 1965. Electroencephalographic studies of homing salmon. *Science* 149:884–885.

Hard, J. J., and W. R. Heard. 1999. Analysis of straying variation in Alaskan hatchery chinook salmon (*Oncorhynchus tshawytscha*) following transplantation. *Can. J. Fish. Aquat. Sci.* 56:578–589.

Hard, J. J., R. G. Kope, W. S. Grant, F. W. Waknitz, L. T. Parker, and R. S. Waples. 1996. Status review of pink salmon from Washington, Oregon, and California. National Marine Fisheries Service, NOAA Tech. Memo. NMFS-NWFSC-25. Seattle. 131 p.

Hard, J. J., A. C. Wertheimer, W. R. Heard, and R. M. Martin. 1985. Early male maturity in two stocks of chinook salmon (*Oncorhynchus tshawytscha*) transplanted to an experimental hatchery in southeastern Alaska. *Aquaculture* 48:351–359.

Harden Jones, F. R. 1968. *Fish migration*. Arnold, London.

Hare, S. R., N. J. Mantua, and R. C. Francis. 1999. Inverse production regimes: Alaska and west coast Pacific salmon. *Fisheries* 24(1):6–14.

Hargreaves, N. B., and R. J. LeBrasseur. 1985. Species selective predation on juvenile pink (*Oncorhynchus gorbuscha*) and chum salmon (*O. keta*) by coho salmon (*O. kisutch*). *Can. J. Fish. Aquat. Sci.* 42:659–668.

———. 1986. Size selectivity of coho (*Oncorhynchus kisutch*) preying on juvenile chum salmon (*O. keta*). *Can. J. Fish. Aquat. Sci.* 43:581–586.

Harris, C. K. 1987. Catches of North American sockeye salmon (*Oncorhynchus nerka*) by the Japanese high seas salmon fisheries, 1972–84. *Can. Spec. Publ. Fish. Aquat. Sci.* 96:458–479.

Hart, J. L. 1973. Pacific fishes of Canada. *Fish. Res. Board Can. Bull* 180. Ottawa. 740 p.

Hartman, G. F., B. C. Anderson, and J. C. Scrivener. 1982. Seaward migration of coho salmon (*Oncorhynchus kisutch*) fry in Carnation Creek, an unstable coastal stream in British Columbia. *Can. J. Fish. Aquat. Sci.* 39:588–597.

Hartman, G. F., C. Groot, and T. G. Northcote. 2000. Science and management in sustainable salmonid fisheries: the ball is not in our court. Pages 31–50 in E. E. Knudsen, C. R. Steward, D. D. MacDonald, J. E. Williams, and D. W. Reiser, eds., *Sustainable fisheries management: Pacific salmon*. Lewis Publishers, Boca Raton, FL.
Hartman, G. F., J. C. Scrivener, L. B. Holtby, and L. Powell. 1987. Some effects of different streamside treatments on physical conditions and fish population processes in Carnation Creek, a coastal rain forest stream in British Columbia. Pages 330–372 in E. O. Salo and T. W. Cundy, eds., *Streamside management: Forestry and fishery interactions*. University of Washington College of Forest Resources, Seattle.
Hartt, A. C., and M. B. Dell. 1986. Early oceanic migrations and growth of juvenile Pacific salmon and steelhead trout. *Int. N. Pac. Fish. Comm. Bull.* 46:1–105.
Hasler, A. D. 1966. *Underwater guideposts: Homing in salmon*. University of Wisconsin Press, Madison.
Hasler, A. D., and A. T. Scholz. 1983. *Olfactory imprinting and homing in salmon*. Springer-Verlag, Berlin, New York.
Hasler, A. D., and W. J. Wisby. 1951. Discrimination of stream odors by fishes and its relation to parent stream behavior. *Am. Nat.* 85:223–238.
Hawke, S. P. 1978. Stranded redds of quinnat salmon in the Mathias River, South Island, New Zealand. *N. Z. J. Mar. Freshw. Res.* 12:167–171.
Healey, M. C. 1979. Detritus and juvenile salmon production in the Nanaimo estuary: I. Production and feeding rates of juvenile chum salmon (*Oncorhynchus keta*). *J. Fish. Res. Board Can.* 36:488–496.
———. 1980. Utilization of the Nanaimo River estuary by juvenile chinook salmon, *Oncorhynchus tshawytscha*. *Fish. Bull.* 77:653–668.
———. 1982a. Juvenile Pacific salmon in estuaries: the life support system. Pages 315–341 in V. S. Kennedy, ed., *Estuarine comparisons*. Academic Press, New York.
———. 1982b. Timing and relative intensity of size-selective mortality of juvenile chum salmon (*Oncorhynchus keta*) during early sea life. *Can. J. Fish. Aquat. Sci.* 39:952–957.
———. 1983. Coastwide distribution and ocean migration patterns of stream- and ocean-type chinook salmon, *Oncorhynchus tshawytscha*. *Can. Field-Nat.* 97:427–433.
———. 1986. Optimum size and age at maturity in Pacific salmon and effects of size-selective fisheries. *Can. Spec. Publ. Fish. Aquat. Sci.* 89:39–52.
———. 1991a. Diets and feeding rates of juvenile pink, chum, and sockeye salmon in Hecate Strait, British Columbia. *Trans. Am. Fish. Soc.* 120:303–318.
———. 1991b. Life history of chinook salmon (*Oncorhynchus tshawytscha*). Pages 311–393 in C. Groot and L. Margolis, eds., *Pacific salmon life histories*. University of British Columbia Press, Vancouver.
Healey, M. C., and C. Groot. 1987. Marine migration and orientation of ocean-type chinook and sockeye salmon. Pages 298–312 in Am. Fish. Soc. Symp. 1. Bethesda, MD.
Healey, M. C., M. A. Henderson, and I. Burgetz. 2000. Precocial maturation of male sockeye salmon in the Fraser River, British Columbia, and its relationship to growth and year-class strength. *Can. J. Fish. Aquat. Sci.* 57:2248–2257.
Healey, M. C., and A. Prince. 1998. Alternative tactics in the breeding behaviour of male coho salmon. *Behaviour* 135:1099–1124.
Healy, B. D., and D. G. Lonzarich. 2000. Microhabitat use and behavior of overwintering juvenile coho salmon in a Lake Superior tributary. *Trans. Am. Fish. Soc.* 129:866–872.
Heard, W. R. 1964. Phototactic behavior of emerging sockeye salmon fry. *Anim. Behav.* 12:382–388.
———. 1991. Life history of pink salmon (*Oncorhynchus gorbuscha*). Pages 119–230 in C. Groot and L. Margolis, eds., *Pacific salmon life histories*. University of British Columbia Press, Vancouver.
Heath, D. D., R. H. Devlin, J. W. Heath, and G. K. Iwama. 1994. Genetic, environmental and interaction effects on the incidence of jacking in *Oncorhynchus tshawytscha* (chinook salmon). *Heredity* 72: 146–154.
Heath, D. D., C. W. Fox, and J. W. Heath. 1999. Maternal effects on offspring size: variation through early development of chinook salmon. *Evolution* 53:1605–1611.
Helfield, J. M. 2001. Interactions of salmon, bear, and riparian vegetation in Alaska. Ph.D. dissertation, University of Washington, Seattle.
Helfield, J. M., and R. J. Naiman. 2001. Effects of salmon-derived nitrogen on riparian forest growth and implications for stream habitat. *Ecology* 82:2403–2409.
Helle, J. H. 1989. Relation between size-at-maturity and survival of progeny in chum salmon, *Oncorhynchus keta* (Walbaum). *J. Fish Biol.* 35 (Suppl. A):99–107.

Helle, J. H., and M. S. Hoffman. 1998. Changes in size and age at maturity of two North American stocks of chum salmon (*Oncorhynchus keta*) before and after a major regime shift in the North Pacific Ocean. *N. Pacific Anad. Fish Comm. Bull.* 1:81–89.

Heming, T. A. 1982. Effects of temperature on utilization of yolk by chinook salmon (*Oncorhynchus tshawytscha*) eggs and alevins. *Can. J. Fish. Aquat. Sci.* 39:184–190.

Hemmingsen, A. R., R. A. Holt, R. D. Ewing, and J. D. McIntyre. 1986. Susceptibility of progeny from crosses among three stocks of coho salmon to infection by *Ceratomyxa shasta*. *Trans. Am. Fish. Soc.* 115:492–495.

Henderson, M. A., and A. J. Cass. 1991. Effect of smolt size on smolt-to-adult survival for Chilko Lake sockeye salmon (*Oncorhynchus nerka*). *Can. J. Fish. Aquat. Sci.* 48:988–994.

Henderson, M. A., and T. G. Northcote. 1985. Visual prey detection and foraging in sympatric cutthroat trout (*Salmo clarki clarki*) and Dolly Varden (*Salvelinus malma*). *Can. J. Fish. Aquat. Sci.* 42:785–790.

———. 1988. Retinal structures of sympatric and allopatric populations of cutthroat trout (*Salmo clarki clarki*) and Dolly Varden char (*Salvelinus malma*) in relation to their spatial distribution. *Can. J. Fish. Aquat. Sci.* 45:1321–1326.

Hendry, A. P. 2001. Adaptive divergence and the evolution of reproductive isolation in the wild: an empirical demonstration using introduced sockeye salmon. *Genetica* 112–113:515–534.

Hendry, A. P., and O. K. Berg. 1999. Secondary sexual characters, energy use, senescence, and the cost of reproduction in sockeye salmon. *Can. J. Zool.* 77:1663–1675.

Hendry, A. P., O. K. Berg, and T. P. Quinn. 1999. Breeding date, life history, and energy allocation in a population of sockeye salmon (*Oncorhynchus nerka*). *Oikos* 85:499–514.

Hendry, A. P., V. Castric, M. T. Kinnison, and T. P. Quinn. 2004. The evolution of philopatry and dispersal: homing versus straying in salmonids. Pages 52–91 in A. P. Hendry and S. Stearns, eds., *Evolution illuminated: Salmon and their relatives*. Oxford University Press, Oxford.

Hendry, A. P., J. K. Wenburg, P. Bentzen, E. C. Volk, and T. P. Quinn. 2000. Rapid evolution of reproductive isolation in the wild: Evidence from introduced salmon. *Science* 290:516–518.

Hicks, B. J., and J. D. Hall. 2003. Rock type and channel gradient structure salmonid populations in the Oregon Coast Range. *Trans. Am. Fish. Soc.* 132:468–482.

Higgs, D. A., J. S. Macdonald, C. D. Levings, and B. S. Dosanjh. 1995. Nutrition and feeding habits in relation to life history stage. Pages 159–315 in C. Groot, L. Margolis, and W. C. Clarke, eds., *Physiological ecology of Pacific salmon*. University of British Columbia Press, Vancouver.

Hilborn, R., T. P. Quinn, D. E. Schindler, and D. E. Rogers. 2003. Biocomplexity and fisheries sustainability. *Proc. Nat. Acad. Sci.* (U.S.) 100:6564–6568.

Hilderbrand, G. V., S. D. Farley, C. T. Robbins, T. A. Hanley, K. Titus, and C. Servheen. 1996. Use of stable isotopes to determine diets of living and extinct bears. *Can. J. Zool.* 74:2080–2088.

Hilderbrand, G. V., T. A. Hanley, C. T. Robbins, and C. C. Schwartz. 1999. Role of brown bears (*Ursus arctos*) in the flow of marine nitrogen into a terrestrial ecosystem. *Oecologia*. 121:546–550.

Hilderbrand, G. V., S. G. Jenkins, C. C. Schwartz, T. A. Hanley, and C. T. Robbins. 1999. Effect of seasonal differences in dietary meat intake on changes in body mass and composition in wild and captive brown bears. *Can. J. Zool.* 77:1623–1630.

Hilderbrand, G. V., C. C. Schwartz, C. T. Robbins, and T. A. Hanley. 2000. Effect of hibernation and reproductive status on body mass and condition of coastal brown bears. *J. Wildl. Mgmt.* 64:178–183.

Hilderbrand, G. V., C. C. Schwartz, C. T. Robbins, M. E. Jacoby, T. A. Hanley, S. M. Arthur, and C. Servheen. 1999. The importance of meat, particularly salmon, to body size, population productivity, and conservation of North American brown bears. *Can. J. Zool.* 77:132–138.

Hinch, S. G., R. E. Diewert, T. J. Lissmore, A. M. J. Prince, M. C. Healey, and M. A. Henderson. 1996. Use of electromyogram telemetry to assess difficult passage areas for river-migrating adult sockeye salmon. *Trans. Am. Fish. Soc.* 125:253–260.

Hinch, S. G., and P. S. Rand. 1998. Swim speeds and energy use of upriver-migrating sockeye salmon (*Oncorhynchus nerka*): Role of local environment and fish characteristics. *Can. J. Fish. Aquat. Sci.* 55:1821–1831.

Hinch, S. G., E. M. Standen, M. C. Healey, and A. P. Farrell. 2002. Swimming patterns and behaviour of upriver-migrating adult pink (*Oncorhynchus gorbuscha*) and sockeye (*O. nerka*) salmon as assessed by EMG telemetry in the Fraser River British Columbia, Canada. *Hydrobiologia* 483:147–160.

Hindar, K., B. Jonsson, J. H. Andrew, and T. G. Northcote. 1988. Resource utilization of sympatric and experimentally allopatric cutthroat trout and Dolly Varden charr. *Oecologia* 74:481–491.

Hiramatsu, K., and Y. Ishida. 1989. Random movement and orientation in pink salmon (*Oncorhynchus gorbuscha*) migrations. *Can. J. Fish. Aquat. Sci.* 46:1062–1066.

Hoar, W. S. 1958. The evolution of migratory behavior among juvenile salmon of the genus *Oncorhynchus. J. Fish. Res. Board Can.* 15:391–428.

———. 1976. Smolt transformation: evolution, behavior, and physiology. *J. Fish. Res. Board Can.* 33:1233–1252.

Hodgson, S. 2000. Marine and freshwater climatic influences on the migratory timing of adult sockeye salmon. M.S. thesis, University of Washington, Seattle.

Hodgson, S., and T. P. Quinn. 2002. The timing of adult sockeye salmon migration into fresh water: adaptations by populations to prevailing thermal regimes. *Can. J. Zool.* 80: 542–555.

Holtby, L. B., B. C. Andersen, and R. K. Kadowaki. 1990. Importance of smolt size and early ocean growth to interannual variability in marine survival of coho salmon (*Oncorhynchus kisutch*). *Can. J. Fish. Aquat. Sci.* 47:2181–2194.

Holtby, L. B., and M. C. Healey. 1986. Selection for adult size in female coho salmon (*Oncorhynchus kisutch*). *Can. J. Fish. Aquat. Sci.* 43:1946–1959.

———1990. Sex-specific life history tactics and risk-taking in coho salmon. *Ecology* 71:678–690.

Honda, H. 1982. On the female pheromones and courtship behaviour in the salmonids, *Oncorhynchus masou* and *O. rhodurus. Bull. Jap. Soc. Sci. Fish.* 48:47–49.

Hughes, N. F. 2004. The wave-drag hypothesis: An explanation for size-biased lateral segregation during the upstream migration of salmonids. *Can. J. Fish. Aquat. Sci.* 61:103–109.

Hume, J. M. B., and E. A. Parkinson. 1988. Effects of size at and time of release on the survival and growth of steelhead fry in stocked streams. *N. Am. J. Fish. Mgmt.* 8:50–57.

Humpesch, U. H. 1985. Inter- and intra-specific variation in hatching success and embryonic development of five species of salmonids and *Thymallus thymallus. Arch. Hydrobiol.* 104:129–144.

Hunter, J. G. 1959. Survival and production of pink and chum salmon in a coastal stream. *J. Fish. Res. Board Can.* 16:835–885.

Huntsman, A. G. 1942. Return of a marked salmon from a distant place. *Science* 95:381–382.

Hutchinson, I. 1988. Estuarine marsh dynamics in the Puget Trough—implications for habitat management. Pages 455–462 in *Proceedings of the first annual meeting on Puget Sound research.* Puget Sound Water Quality Authority, Seattle, Washington.

Hutchison, M. J., and M. Iwata. 1997. A comparative analysis of aggression in migratory and non-migratory salmonids. *Env. Biol. Fish.* 50:209–215.

Hyatt, K. D., and B. E. Riddell. 2000. The importance of "stock" conservation definitions to the concept of sustainable fisheries. Pages 51–62 in E. E. Knudsen, C. R. Steward, D. D. MacDonald, J. E. Williams, and D. W. Reiser, eds., *Sustainable fisheries management: Pacific salmon.* Lewis Publishers, Boca Raton, FL.

Hyatt, K. D., and J. G. Stockner. 1985. Responses of sockeye salmon (*Oncorhynchus nerka*) to fertilization of British Columbia coastal lakes. *Can. J. Fish. Aquat. Sci.* 42:320–331.

Idler, D. R., and W. A. Clemens. 1959. The energy expenditures of Fraser River sockeye salmon during the spawning migration to Stuart and Chilko lakes. *Int. Pac. Salmon Fish. Comm. Prog. Rept.* 6. New Westminster, BC. 80 p.

Ingram, B. L., and P. K. Weber. 1999. Salmon origin in California's Sacramento–San Joaquin river system as determined by otolith strontium isotopic composition. *Geology* 27:851–854.

Irvine, J. R., and B. R. Ward. 1989. Patterns of timing and size of wild coho salmon (*Oncorhynchus kisutch*) smolts migrating from the Keogh River watershed on northern Vancouver Island. *Can. J. Fish. Aquat. Sci.* 46:1086–1094.

Ivankov, V. N., Y. A. Mitrofanov, and V. P. Bushuyev. 1975. An instance of the pink salmon (*Oncorhynchus gorbuscha*) reaching maturity at an age of less than 1 year. *J. Ichthyol.* 15:497–499.

Ivankov, V. N., S. N. Padetskiy, and V. S. Chikina. 1977. On the postspawning neotenic males of the masu, *Oncorhynchus masu. J.: Ichthyol.* 15:673–678.

Iwamoto, R. N., B. A. Alexander, and W. K. Hershberger. 1984. Genotypic and environmental effects on the incidence of sexual precocity in coho salmon (*Oncorhynchus kisutch*). *Aquaculture* 43: 105–121.

Iwata, M., and S. Komatsu. 1984. Importance of estuarine residence for adaptation of chum salmon (*Oncorhynchus keta*) fry to seawater. *Can. J. Fish. Aquat. Sci.* 41:744–749.

Jaenicke, H. W., and A. G. Celewycz. 1994. Marine distribution and size of juvenile Pacific salmon in southeast Alaska and northern British Columbia. *Fish. Bull.* 92:79–90.

Jaenicke, H. W., A. G. Celewycz, J. E. Baily, and J. A. Orsi. 1984. Paired open beach seines to study estuarine migrations of juvenile salmon. *Mar. Fish. Rev.* 46(3):62–67.

Jamon, M. 1990. A reassessment of the random hypothesis in the ocean migration of Pacific salmon. *J. Theoret. Biol.* 143:197–213.

Jefferts, K. B., P. K. Bergman, and H. F. Fiscus. 1963. A coded wire identification system for macro-organisms. *Nature* 198:460–462.

Johnsen, P. B. 1982. A behavioral control model for homestream selection in migratory salmonids. Pages 266–273 in E. L. Brannon and E. O. Salo, eds., *Salmon and trout migratory behavior symposium.* University of Washington School of Fisheries, Seattle.

Johnsen, R. C. 1964. Direction of movement of salmon in the North Pacific Ocean and Bering Sea as indicated by surface gillnet catches, 1959–1960. *Int. N. Pac. Fish. Comm. Bull.* 14:33–48.

Johnson, L. 1980. The arctic charr, *Salvelinus alpinus.* Pages 15–98 in E. K. Balon, ed., *Charrs: Salmonid fishes of the genus* Salvelinus. Dr. W. Junk, Publishers, The Hague.

Johnson, O. W., W. S. Grant, R. G. Kope, K. Neely, F. W. Waknitz, and R. S. Waples. 1997. Status review of chum salmon from Washington, Oregon, and California. National Marine Fisheries Service, NOAA Tech. Memo. NMFS-NWFSC-32. Seattle. 280 p.

Johnson, O. W., M. H. Ruckelshaus, W. S. Grant, F. W. Waknitz, A. M. Garrett, G. J. Bryant, K. Neely, and J. J. Hard. 1999. Status review of coastal cutthroat from Washington, Oregon, and California. National Marine Fisheries Service, NOAA Tech. Memo. NMFS-NWFSC-37. Seattle. 292 p.

Johnson, S. C., R. B. Blaylock, J. Elphick, and K. D. Hyatt. 1996. Disease induced by the sea louse (*Lepeophtheirus salmonis*) (Copepoda: Caligidae) in wild sockeye salmon (*Oncorhynchus nerka*) stocks of Alberni Inlet, British Columbia. *Can. J. Fish. Aquat. Sci.* 53:2888–2897.

Johnson, S. C., G. A. Chapman., and D. G. Stevens. 1989. Relationships between temperature units and sensitivity to handling for coho salmon and rainbow trout embryos. *Prog. Fish-Cult.* 51:61–68.

Johnson, S. L. 1984. The effects of the 1983 El Niño on Oregon's coho and chinook salmon. Oregon Dept. Fish Wildl., Fish Division, Info. Rept. 84-8. Portland. 40 p.

Johnson, S. W., J. F. Thedinga, and K. V. Koski. 1992. Life history of juvenile ocean-type chinook salmon (*Oncorhynchus tshawytscha*) in the Situk River, Alaska. *Can. J. Fish. Aquat. Sci.* 49:2621–2629.

Johnston, G. 2002. *Arctic charr aquaculture.* Fishing News Books, Oxford.

Johnston, J. M. 1982. Life histories of anadromous cutthroat with emphasis on migratory behavior. Pages 123–127 in E. L. Brannon and E. O. Salo, eds., *Salmon and trout migratory behavior symposium.* University of Washington School of Fisheries, Seattle.

Johnston, S. V., and J. S. Hopelain. 1990. The application of dual-beam target tracking and Doppler-shifted echo processing to assess upstream salmonid migration in the Klamath River, California. *Rapp. P.- v. Réun. Cons. int. Explor. Mer.* 189:210–222.

Jones, D. E. 1972. A study of steelhead-cutthroat trout in Alaska. Pages 91–108 in Alaska Dept. Fish Game, Sport Fish Division, Rept. G-11-1. Juneau.

———. 1973. Steelhead and sea-run cutthroat life history in southeast Alaska. Alaska Dept. Fish Game, Division of Sport Fish, Rept. AFS-42-1. Juneau. 18 p.

Jones, J. A., and G. E. Grant. 1996. Peak flow responses to clear-cutting and roads in small and large basins, western Cascades, Oregon. *Water Resour. Res.* 32:959–974.

Jonsson, B., and N. Jonsson. 1993. Partial migration: Niche shift versus sexual maturation in fishes. *Rev. Fish Biol. Fish.* 3:348–365.

———. 2001. Polymorphism and speciation in Arctic charr. *J. Fish Biol.* 58:605–638.

Jonsson, N., B. Jonsson, J. Skurdal, and L. P. Hansen. 1994. Differential response to water current in offspring of inlet- and outlet-spawning brown trout *Salmo trutta. J. Fish Biol.* 45:356–359.

Juday, C., W. H. Rich, G. I. Kemmerer, and A. Mann. 1932. Limnological studies of Karluk Lake, Alaska 1926–1930. *Bull. U.S. Bureau Fish.* 47:407–436.

Kaczynski, V. W., R. J. Fellar, J. Clayton, and R. J. Gerke. 1973. Trophic analysis of juvenile pink and chum salmon (*Oncorhynchus gorbuscha* and *O. keta*) in Puget Sound. *J. Fish. Res. Board Can.* 30:1003–1008.

Kadri, S., J. E. Thorpe, and N. B. Metcalfe. 1997. Anorexia in one-sea-winter Atlantic salmon (*Salmo salar*) during summer, associated with sexual maturation. *Aquaculture* 151:405–409.

Kaeriyama, M. 1998. Dynamics of chum salmon, *Oncorhynchus keta*, populations released from Hokkaido, Japan. *N. Pac. Anad. Fish Comm. Bull.* 1:90–102.

———. 1999. Hatchery programmes and stock management of salmonid populations in Japan. Pages 153–167 in B. R. Howell, E. Moksness, and T. Svasand, eds., *Stock enhancement and sea ranching.* Blackwell Science, London.

Kahler, T. H., P. Roni, and T. P. Quinn. 2001. Summer movement and growth of juvenile anadromous salmonids in small western Washington streams. *Can. J. Fish. Aquat. Sci.* 58:1947–1956.

Karpenko, V. I. 2003. Review of Russian marine investigations of juvenile Pacific salmon. *N. Pac. Anad. Fish Comm. Bull.* 3:69–88.

Kato, F. 1991 Life histories of masu and amago salmon (*Oncorhynchus masou* and *Oncorhynchus rhodurus*). Pages 448–520 in C. Groot and L. Margolis, eds., *Pacific salmon life histories*. University of British Columbia Press, Vancouver.

Keeley, E. R. 2001. Demographic responses to food and space competition by juvenile steelhead trout. *Ecology* 82:1247–1259.

Keeley, E. R., and J. W. A. Grant. 1995. Allometric and environmental correlates of territory size in juvenile Atlantic salmon. *Can. J. Fish. Aquat. Sci.* 52:186–196.

———. 2001. Prey size of salmonid fishes in streams, lakes, and oceans. *Can. J. Fish. Aquat. Sci.* 58: 1122–1132.

Keenleyside, M. H. A., and H. M. C. Dupuis. 1988. Comparison of digging behavior by female pink salmon (*Oncorhynchus gorbuscha*) before and after spawning. *Copeia* 1988:1092–1095.

Kerns, O. E. J., and J. R. Donaldson. 1968. Behavior and distribution of spawning sockeye salmon on island beaches in Iliamna Lake, Alaska. *J. Fish. Res. Board Can.* 24:485–494.

Kesner, W. D., and R. A. Barnhart. 1972. Characteristics of the fall–run steelhead trout (*Salmo gairdneri gairdneri*) of the Klamath River system with emphasis on the half-pounder. *Cal. Fish Game* 58: 204–220.

Killick, S. R. 1955. The chronological order of Fraser River sockeye salmon during migration, spawning and death. *Int. Pac. Salmon Fish. Comm. Bull.* 7. New Westminster, BC. 95 p.

Kinnison, M. T., M. J. Unwin, A. P. Hendry, and T. P. Quinn. 2001. Migratory costs and the evolution of egg size and number allocation in new and indigenous salmon populations. *Evolution* 55:1656–1667.

Kinnison, M. T., M. J. Unwin and T. P. Quinn. 2003. Migratory costs and contemporary evolution of reproductive allocation in male chinook salmon. *J. Evol. Biol.* 16:1257–1269.

Kjelson, M. A., P. F. Raquel, and F. W. Fisher. 1982. Life history of fall-run juvenile chinook salmon, *Oncorhynchus tshawytscha*, in the Sacramento–San Joaquin estuary, California. Pages 393–411 in V. S. Kennedy, ed., *Estuarine comparisons*. Academic Press, New York.

Kline, T. C. J., J. J. Goering, O. A. Mathisen, and P. Poe. 1990. Recycling of elements transported upstream by runs of Pacific salmon: I. δ^{15} N and δ^{13} C evidence in Sashin Creek, southeastern Alaska. *Can. J. Fish. Aquat. Sci.* 47:136–144.

Kline, T. C. J., J. J. Goering, O. A. Mathisen, P. Poe, P. L. Parker, and R. S. Scalan. 1993. Recycling of elements transported upstream by runs of Pacific salmon: II. δ^{15} N and δ^{13} C evidence in the Kvichak River watershed, Bristol Bay, southern Alaska. *Can. J. Fish. Aquat. Sci.* 50:2350–2365.

Knapp, R. A., and V. T. Vredenburg. 1996. Spawning by California golden trout: characteristics of spawning fish, seasonal and daily timing, redd characteristics, and microhabitat preferences. *Trans. Am. Fish. Soc.* 125:519–531.

Knudsen, E. E. 2000. Managing Pacific salmon escapements: the gaps between theory and reality. Pages 237–272 in E. E. Knudsen, C. R. Steward, D. D. MacDonald, J. E. Williams, and D. W. Reiser, eds., *Sustainable fisheries management: Pacific salmon.* Lewis Publishers, Boca Raton, FL.

Knudsen, E. E., C. R. Steward, D. D. MacDonald, J. E. Williams and D. W. Reiser, eds. 2000. *Sustainable fisheries management: Pacific salmon.* Lewis Publishers, Boca Raton, FL.

Koenings, J. P., and R. D. Burkett. 1987. Population characteristics of sockeye salmon (*Oncorhynchus nerka*) smolts relative to temperature regimes, euphotic volume, fry density, and forage base within Alaskan lakes. *Can. Spec. Publ. Fish. Aquat. Sci.* 96:216–234.

Koenings, J. P., H. J. Geiger, and J. J. Hasbrouck. 1993. Smolt-to-adult survival patterns of sockeye salmon (*Oncorhynchus nerka*): Effect of smolt length and geographic latitude when entering the sea. *Can. J. Fish. Aquat. Sci.* 50:600–611.

Kondolf, G. M., M. J. Sale, and M. G. Wolman. 1994. Modification of fluvial gravel size by spawning salmonids. *Water Resour. Res.* 29:2265–2274.

Kondolf, G. M., and M. G. Wolman. 1993. The sizes of salmonid spawning gravels. *Water Resour. Res.* 29:2275–2285.

Koo, T. S. Y. 1962a. Age designation in salmon. Pages 37–48 in T. S. Y. Koo, ed., *Studies of Alaska red salmon*. University of Washington Press, Seattle.

———. 1962b. Age and growth studies of red salmon scales by graphical means. Pages 53–121 in T. S. Y. Koo, ed., *Studies of Alaska red salmon*. University of Washington Press, Seattle.

Koslow, J. A., A. J. Hobday, and G. W. Boehlert. 2002. Climate variability and marine survival of coho salmon (*Oncorhynchus kisutch*) in the Oregon production area. *Fish. Oceanogr.* 11:65–77.

Kurenkov, S. I. 1977. Two reproductively isolated groups of kokanee salmon, *Oncorhynchus nerka kennerlyi*, from Lake Kronotskiy. *J. Ichthyol.* 17:526–534.

Kwain, W. 1987. Biology of pink salmon in the North American Great Lakes. Pages 57–65 in Am. Fish. Soc. Symp. 1. Bethesda, MD.

Labelle, M. 1992. Straying patterns of coho salmon (*Oncorhynchus kisutch*) stocks from southeast Vancouver Island, British Columbia. *Can. J. Fish. Aquat. Sci.* 49:1843–1855.

Lackey, R. T. 2001. Defending reality. *Fisheries* 26(6):26–27.

LaLanne, J. J. 1971. Marine growth of chum salmon. *Int. N. Pac. Fish. Comm. Bull.* 27:71–91.

Landingham, J. H., M. V. Sturdevant, and R. D. Brodeur. 1998. Feeding habits of juvenile Pacific salmon in marine waters of southeastern Alaska and northern British Columbia. *Fish. Bull.* 96:285–302.

Larkin, P. A. 1975. Some major problems for further study on Pacific salmon. *Int. N. Pac. Fish. Comm. Bull.* 32:3–9.

Latterell, J. J. 2001. Distribution, constraints and population genetics of native trout in unlogged and clear-cut headwater streams. M.S. thesis, University of Washington, Seattle.

Latterell, J. J., K. D. Fausch, C. Gowan, and S. C. Riley. 1998. Relationship of trout recruitment to snowmelt runoff flows and adult trout abundance in six Colorado mountain streams. *Rivers* 6:240–250.

Lear, W. H. 1980. The pink salmon transplant experiment in Newfoundland. Pages 213–243 in J. E. Thorpe, ed., *Salmon ranching*. Academic Press, London.

LeBrasseur, R. J. 1966. Stomach contents of salmon and steelhead trout in the northeastern Pacific Ocean. *J. Fish. Res. Board Can.* 23:85–100.

———. 1969. Growth of juvenile chum salmon (*Oncorhynchus keta*) under different feeding regimes. *J. Fish. Res. Board Can.* 26:1631–1645.

Leggett, W. C. 1977. The ecology of fish migrations. *Ann. Rev. Ecol. Syst.* 8:285–308.

———. 1984. Fish migrations in coastal and estuarine environments: A call for new approaches to the study of an old problem. Pages 159–178 in J. D. McCleave, G. P. Arnold, J. J. Dodson, and W. H. Neill, eds., *Mechanisms of migration in fishes*. Plenum Press, New York.

Leman, V. N. 1993. Spawning sites of chum salmon, *Oncorhynchus keta*: microhydrological regime and viability of progeny in redds (Kamchatka River basin). *J. Ichthyol.* 33:104–117.

Leonetti, F. E. 1997. Estimation of surface and intragravel water flow at sockeye salmon spawning beaches in Iliamna Lake, Alaska. *N. Am. J. Fish. Mgmt.* 17:194–201.

Lestelle, L. C., M. L. Rowse, and C. Weller. 1993. Evaluation of natural stock improvement measures for Hood Canal coho salmon. Point No Point Treaty Council, Tech. Rept. TR 93-1. Kingston, WA.

Levings, C. D. 1994. Feeding behaviour of juvenile salmon and significance of habitat during estuary and early sea phase. Nordic *J. Freshw. Res.* 69:7–16.

Levings, C. D., C. D. McAllister, and B. D. Chang. 1986. Differential use of the Campbell River estuary, British Columbia, by wild and hatchery-reared juvenile chinook salmon (*Oncorhynchus tshawytscha*). *Can. J. Fish. Aquat. Sci.* 43:1386–1397.

Levy, D. A. 1987. Review of the ecological significance of diel vertical migrations by juvenile sockeye salmon (*Oncorhynchus nerka*). *Can. Spec. Publ. Fish. Aquat. Sci.* 96:44–52.

———. 1990. Sensory mechanisms and selective advantage for diel vertical migration in juvenile sockeye salmon, *Oncorhynchus nerka*. *Can. J. Fish. Aquat. Sci.* 47:1796–1802.

Levy, D. A., and A. D. Cadenhead. 1995. Selective tidal stream transport of adult sockeye salmon (*Oncorhynchus nerka*) in the Fraser River estuary. *Can. J. Fish. Aquat. Sci.* 52:1–12.

Levy, D. A., and T. G. Northcote. 1982. Juvenile salmon residency in a marsh area of the Fraser estuary. *Can. J. Fish. Aquat. Sci.* 39:270–276.

Lichatowich, J. A. 1999. Salmon without rivers: A history of the Pacific salmon crisis. Island Press, Washington, D.C.

Lichatowich, J. A., and S. Cramer. 1979. Parameter selection and sample sizes in studies of anadromous salmonids. Oregon Dept. Fish Wildl., Fish Division, Info. Rept. 80-1. Portland. 25 p.

Liley, N. R., P. Tamkee, R. Tsai, and D. J. Hoysak. 2002. Fertilization dynamics in rainbow trout (*Oncorhynchus mykiss*): Effect of male age, social experience, and sperm concentration and motility on in vitro fertilization. *Can. J. Fish. Aquat. Sci.* 59:144–152.

Lindsey, C. C., and J. D. McPhail 1986. Zoogeography of fishes of the Yukon and Mackenzie basins. Pages 639–674 in C. H. Hocutt and E. O. Wiley, eds., *The zoogeography of North American freshwater fishes*. John Wiley and Sons, New York.

Linley, T. J. 1993. Patterns of life history variation among sockeye salmon (*Oncorhynchus nerka*) in the Fraser River, British Columbia. Ph.D. dissertation, University of Washington, Seattle.

Lisle, T. E., and J. Lewis. 1992. Effects of sediment transport on survival of salmonid embryos in a natural stream: a simulation approach. *Can. J. Fish. Aquat. Sci.* 49:2337–2344.

Lister, D. B., and H. S. Genoe. 1970. Stream habitat utilization by cohabiting underyearlings of chinook (*Oncorhynchus tshawytscha*) and coho (*O. kisutch*) salmon in the Big Qualicum River, British Columbia. *J. Fish. Res. Board Can.* 27:1215–1224.

Loch, J. J., and D. R. Miller. 1988. Distribution and diet of sea-run cutthroat trout captured in and adjacent to the Columbia River plume, May–July 1980. *Northwest Sci.* 62:41–48.

Lohmann, K. J., S. D. Cain, S. A. Dodge, and C. M. F. Lohmann. 2001. Regional magnetic fields as navigational markers for sea turtles. *Science* 294:364–366.

Lough, M. J. 1983. Radio telemetry studies of summer steelhead trout in the Cranberry, Kispiox, Kitwanga and Zymoetz rivers and Toboggan Creek, 1980. British Columbia Fish Wildl. Branch, Skeena Fish. Rept. 80-04. Smithers.

MacCrimmon, H. R. 1971. World distribution of rainbow trout (*Salmo gairdneri*). *J. Fish. Res. Board Can.* 28:663–704.

MacCrimmon, H. R., and J. S. Campbell. 1969. World distribution of brook trout, *Salvelinus fontinalis*. *J. Fish. Res. Board Can.* 26:1699–1725.

MacCrimmon, H. R., and B. L. Gots. 1979. World distribution of Atlantic salmon, *Salmo salar*. *J. Fish. Res. Board Can.* 36:422–457.

MacCrimmon, H. R., and T. C. Marshall. 1968. World distribution of brown trout, *Salmo trutta*. *J. Fish. Res. Board Can.* 25:2527–2548.

Macdonald, J. S., ed. 2000. Mortality during the migration of Fraser River sockeye salmon (*Oncorhynchus nerka*): A study of the effect of ocean and river environmental conditions in 1997. *Can. Tech. Rept. Fish. Aquat. Sci.* 2315. Burnaby, BC. 120 p.

Macdonald, J. S., I. K. Birtwell, and G. M. Kruzynski. 1987. Food and habitat utilization by juvenile salmonids in the Campbell River estuary. *Can. J. Fish. Aquat. Sci.* 44:1233–1246.

Macdonald, J. S., M. G. G. Foreman, T. Farrell, I. V. Williams, J. Grout, A. Cass, J. C. Woodey, H. Enzenhofer, W. C. Clarke, R. Houtman, E. M. Donaldson, and D. Barnes. 2000. The influence of extreme water temperatures on migrating Fraser River sockeye salmon during the 1998 spawning season. *Can. Tech. Rept. Fish. Aquat. Sci.* 2326. Burnaby, BC. 131 p.

Macdonald, J. S., C. D. Levings, C. D. McAllister, U. H. M. Fagerlund, and J. R. McBride. 1988. A field experiment to test the importance of estuaries for chinook salmon (*Oncorhynchus tshawytscha*) survival: short term results. *Can. J. Fish. Aquat. Sci.* 45:1366–1377.

MacFarlane, R. B., and E. C. Norton. 2002. Physiological ecology of juvenile chinook salmon (*Oncorhynchus tshawytscha*) at the southern end of their distribution, the San Francisco Estuary and Gulf of the Farallones, California. *Fish. Bull.* 100:244–257.

Machidori, S., and F. Kato. 1984. Spawning populations and marine life of masu salmon (*Oncorhynchus masou*). *Int. N. Pac. Fish. Comm. Bull.* 43. Vancouver, BC. 138 p.

Madison, D. M., R. M. Horrall, A. B. Stasko, and A. D. Hasler. 1972. Migratory movements of adult sockeye salmon (*Oncorhynchus nerka*) in coastal British Columbia as revealed by ultrasonic tracking. *J. Fish. Res. Board Can.* 29:1025–1033.

Magnusson, A., and R. Hilborn. 2003. Estuarine influence on survival rates of coho (*Oncorhynchus kitsutch*) and chinook salmon (*Oncorhynchus tshawytscha*) from hatcheries on the U.S. Pacific coast. *Estuaries* 26:1094–1103.

Major, R.L., J. Ito, S. Ito, and H. Godfrey. 1978. Distribution and origin of chinook salmon (*Oncorhynchus tshawytscha*) in offshore waters of the North Pacific Ocean. *Int. N. Pac. Fish. Comm. Bull.* 38. Vancouver, BC. 54 p.

Mantua, N. J., S. R. Hare, Y. Zhang, J. M. Wallace, and R. C. Francis. 1997. A Pacific interdecadal climate oscillation with impacts on salmon production. *Bull. Am. Meteor. Soc.* 78:1069–1079.

Manzer, J. I. 1968. Food of Pacific salmon and steelhead trout in the Northeast Pacific Ocean. *J. Fish. Res. Board Can.* 25:1085–1089.

Margolis, L. 1963. Parasites as indicators of the geographical origin of sockeye salmon, *Oncorhynchus nerka* (Walbaum), occurring in the North Pacific Ocean and adjacent seas. *Int. N. Pac. Fish. Comm. Bull.* 11:101–156.

———. 1998. Are naturally-occurring parasite "tags" stable? An appraisal from four case histories involving Pacific salmonids. *N. Pac. Anad. Fish Comm. Bull.* 1:205–212.

Marshall, D. E., and E. W. Britton. 1990. Carrying capacity of coho salmon streams. *Can. MS. Rept. Fish. Aquat. Sci.* 2058. Vancouver, BC. 32 p.

Martel, G. 1996. Growth rate and influence of predation risk on territoriality in juvenile coho salmon (*Oncorhynchus kisutch*). *Can. J. Fish. Aquat. Sci.* 53:660–669.

Martin, D. J. 1985. Production of cutthroat trout (*Salmo clarki*) in relation to riparian vegetation in Bear Creek, Washington. Ph.D. dissertation, University of Washington, Seattle.

Maser, C., and J. R. Sedell. 1994. *From the forest to the sea: The ecology of wood in streams, rivers, estuaries, and oceans.* St. Lucie Press, Delray Beach, FL.

Mason, J. C. 1974. Behavioral ecology of chum salmon fry (*Oncorhynchus keta*) in a small estuary. *J. Fish. Res. Board Can.* 31:83–92.

———. 1976a. Some features of coho salmon, *Oncorhynchus kisutch*, fry emerging from simulated redds and concurrent changes in photobehavior. *Fish. Bull.* 74:167–175.

———. 1976b. Response of underyearling coho salmon to supplemental feeding in a natural stream. *J. Wildl. Mgmt.* 40:775–788.

Mathisen, O. A. 1962. The effect of altered sex ratios on the spawning of sockeye salmon. Pages 137–248 in T. S. Y. Koo, ed., *Studies of Alaska red salmon*. University of Washington Press, Seattle.

Matthews, K. R., and N. H. Berg. 1997. Rainbow trout responses to water temperature and dissolved oxygen stress in two southern California stream pools. *J. Fish Biol.* 50:50–67.

Mayama, H. 1989. Reciprocal transplantation experiment of masu salmon (*Oncorhynchus masou*) population. 1. Comparison of biological characteristics between two masu populations, the Shari River on the Okhotsk Sea coast and the Shiribetsu River on the Japan Sea coast, Hokkaido. *Sci. Repts. Hokkaido Salmon Hatchery* 43:75–97.

Mayama, H., and Y. Ishida. 2003. Japanese studies on the early ocean life of juvenile salmon. *N. Pac. Anad. Fish Comm. Bull.* 3:41–67.

Mayama, H., T. Nomura, and K. Ohkuma. 1989. Reciprocal transplantation experiment of masu salmon (*Oncorhynchus masou*) population. 2. Comparison of seaward migrations and adult returns of local stock and transplanted stock of masu salmon. *Sci. Repts. Hokkaido Salmon Hatchery* 43:99–113.

Maynard, D. J., T. A. Flagg, and C. V. W. Mahnken. 1995. A review of seminatural culture strategies for enhancing the postrelease survival of anadromous salmonids. Pages 307–314 in Am. Fish. Soc. Symp. 15. Bethesda, MD.

Mazumder, A., and J. A. Edmundson. 2002. Impact of fertilization and stocking on trophic interactions and growth of juvenile sockeye salmon (*Oncorhynchus nerka*). *Can. J. Fish. Aquat. Sci.* 59:1361–1373.

McCabe, G. T., R. L. Emmett, W. D. J. Muir, and T. H. Blahm. 1986. Utilization of the Columbia River estuary by subyearling chinook salmon. *Northwest Sci.* 60:113–124.

McCabe, G. T., W. D. J. Muir, R. L. Emmett, and J. T. Durkin. 1983. Interrelationships between juvenile salmonids and nonsalmonid fish in the Columbia River estuary. *Fish. Bull.* 81:815–826.

McDonald, J. G. 1960. The behavior of Pacific salmon fry during their downstream migration to freshwater and saltwater nursery areas. *J. Fish. Res. Board Can.* 17:655–676.

———. 1969. Distribution, growth, and survival of sockeye fry (*Oncorhynchus nerka*) produced in natural and artificial stream environments. *J. Fish. Res. Board Can.* 26:229–267.

———. 1981. The stock concept and its application to British Columbia salmon fisheries. *Can. J. Fish. Aquat. Sci.* 38:1657–1664.

McDowall, R. M. 1988. *Diadromy in fishes*. Timber Press, Portland, Oregon.

———. 1990. *New Zealand freshwater fishes*. Heinemann Reed, Auckland.

———. 1997. The evolution of diadromy in fishes (revisited) and its place in phylogenetic analysis. *Rev. Fish Biol. Fish.* 7:443–462.

———. 2001. Anadromy and homing: Two life-history traits with adaptive synergies in salmonid fishes? *Fish and Fisheries* 2:78–85.

———. 2002. The origin of the salmonid fishes: marine, freshwater . . . or neither? *Rev. Fish Biol. Fish.* 11:171–179.

———. 2003. Impacts of introduced salmonids on native galaxiids in New Zealand upland streams: a new look at an old problem. *Trans. Am. Fish. Soc.* 132:229–238.

McGurk, M. D. 1996. Allometry of marine mortality of Pacific salmon. *Fish. Bull.* 94:77–88.

———. 1999. Size dependence of natural mortality rate of sockeye salmon and kokanee in freshwater. *N. Am. J. Fish. Mgmt.* 19:376–396.

McInerney, J. E. 1964. Salinity preference: an orientation mechanism in salmon migration. *J. Fish. Res. Board Can.* 21:995–1018.

McIsaac, D. O. 1990. Factors affecting the abundance of 1977–1979 brood wild fall chinook salmon (*Oncorhynchus tshawytscha*) in the Lewis River, Washington. Ph.D. dissertation, University of Washington, Seattle.

McIsaac, D. O., and T. P. Quinn. 1988. Evidence for a hereditary component in homing behavior of chinook salmon (*Oncorhynchus tshawytscha*). *Can. J. Fish. Aquat. Sci.* 45:2201–2205.

McKeown, B. A. 1984. *Fish migration*. Timber Press, Portland, Oregon.

McKinstry, C. A. 1993. Forecasting migratory timing and abundance of pink salmon (*Oncorhynchus gorbuscha*) runs using sex-ratio information. M.S. thesis, University of Washington, Seattle.

McLean, J. E., P. Bentzen, and T. P. Quinn. 2003. Differential reproductive success of sympatric, naturally-spawning hatchery and wild steelhead trout (*Oncorhynchus mykiss*) through the adult stage. *Can. J. Fish. Aquat. Sci.* 60:433–440.

McMahon, T. E., and L. B. Holtby. 1992. Behaviour, habitat use, and movements of coho salmon (*Oncorhynchus kisutch*) smolts during seaward migration. *Can. J. Fish. Aquat. Sci.* 49:1478–1485.

McPhail, J. D. 1996. The origin and speciation of *Oncorhynchus* revisited. Pages 29–38 in D. J. Stouder, P. A. Bisson, and R. J. Naiman, eds., *Pacific salmon and their ecosystems: Status and future options*. Chapman and Hall, New York.

McPhail, J. D., and C. C. Lindsey. 1970. Freshwater fishes of northwestern Canada and Alaska. *Fish. Res. Board Can. Bull.* 173. Ottawa. 381 p.

———. 1986. Zoogeography of the freshwater fishes of Cascadia (the Columbia system and rivers north to the Stikine). Pages 615–637 in C. H. Hocutt and E. O. Wiley, eds., *The zoogeography of North American freshwater fishes*. John Wiley and Sons, New York.

Mead, R. W., and W. L. Woodall. 1968. Comparison of sockeye salmon fry produced by hatcheries, artificial channels and natural spawning areas. *Int. Pac. Salmon Fish. Comm. Prog. Rept.* 20. New Westminster, BC. 41 p.

Meehan, E. P., E. E. Seminet-Reneau, and T. P. Quinn. Forthcoming. Bear predation on Pacific salmon facilitates colonization of carcasses by fly maggots. *Amer. Midl. Nat.*

Meehan, W. R., ed. 1991. *Influences of forest and rangeland management on salmonid fishes and their habitats*. Am. Fish. Soc. Spec. Publ. 19. Bethesda, MD. 751 p.

Meisner, J. D., J. S. Rosenfeld, and H. A. Regier. 1988. The role of groundwater in the impact of climate warming on stream salmonines. *Fisheries* 13(3):2–8.

Meka, J. M., E. E. Knudsen, D. C. Douglas, and R. P. Benter. 2003. Variable migratory patterns of different adult rainbow trout life history types in a southwest Alaska watershed. *Trans. Am. Fish. Soc.* 132:717–732.

Meyer, K. A., D. J. Schill, F. S. Elle, and J. A. Lamansky. 2003. Reproductive demographics and factors that influence length at sexual maturity of Yellowstone cutthroat trout in Idaho. *Trans. Am. Fish. Soc.* 132:183–195.

Miller, D. R., J. G. Williams, and C. W. Sims. 1983. Distribution, abundance and growth of juvenile salmonids off the coast of Oregon and Washington, summer 1980. *Fish. Res.* 2:1–17.

Miller, J. A., and C. A. Simenstad. 1997. A comparative assessment of a natural and created estuarine slough as rearing habitat for juvenile chinook and coho salmon. *Estuaries* 20:792–806.

Milligan, P. A., W. O. Rublee, D. D. Cornett, and R. A. C. Johnston. 1985. The distribution and abundance of chinook salmon (*Oncorhynchus tshawytscha*) in the upper Yukon River basin as determined by a radio-tagging and spaghetti tagging program: 1982–1983. *Can. Tech. Rept. Fish. Aquat. Sci.* 1352. New Westminster, BC. 161 p.

Milner, A. M., and R. G. Bailey. 1989. Salmonid colonization of new streams in Glacier Bay National Park, Alaska. *Aquacult. Fish. Mgmt.* 20:179–192.

Milner, A. M., E. E. Knudsen, C. Soiseth, A. L. Robertson, D. Schell, I. T. Phillips, and K. Magnusson. 2000. Colonization and development of stream communities across a 200–year gradient in Glacier Bay National Park, Alaska, U.S.A. *Can. J. Fish. Aquat. Sci.* 57:2319–2335.

Milner, J. W. 1876. The propagation and distribution of the shad. Operations in the distribution of shad in 1874. Pages 323–326 in *Report of the commissioner for 1873–4 and 1874–5*. U.S. Commission of Fish and Fisheries, Washington, DC.

Minckley, W. L., D. A. Hendrickson, and C. E. Bond. 1986. Geography of western North American freshwater fishes: Description and relationships to intracontinental tectonism. Pages 519–613 in C. H. Hocutt and E. O. Wiley, eds., *The zoogeography of North American freshwater fishes*. John Wiley and Sons, New York.

Miyakoshi, Y., M. Nagata, and S. Kitada. 2001. Effect of smolt size on postrelease survival of hatchery–reared masu salmon *Oncorhynchus masou*. *Fish. Sci.* 67:134–137.

Mobrand, L. E., J. A. Lichatowich, L. C. Lestelle, and T. S. Vogel. 1997. An approach to describing ecosystem performance "through the eyes of salmon." *Can. J. Fish. Aquat. Sci.* 54:2964–2973.

Molyneaux, D. B., and L. DuBois. 1998. *Salmon age, sex and length catalog for the Kuskokwim area, 1996–1997 progress report*. Alaska Dept. Fish Game, Reg. Info. Rept. 3A-98-15. Anchorage. 351 p.

Montgomery, D. R., E. M. Beamer, G. Pess, and T. P. Quinn. 1999. Channel type and salmon spawning distribution. *Can. J. Fish. Aquat. Sci.* 56:377–387.

Montgomery, D. R., and J. M. Buffington. 1997. Channel-reach morphology in mountain drainage basins. *Geol. Soc. Am. Bull.* 109:596–611.

Montgomery, D. R., J. M. Buffington, N. P. Peterson, D. Schuett-Hames, and T. P. Quinn. 1996. Streambed scour, egg burial depths, and the influence of salmonid spawning on bed surface mobility and embryo survival. *Can. J. Fish. Aquat. Sci.* 53:1061–1070.

Montgomery, D. R., J. M. Buffington, R. D. Smith, K. M. Schmidt, and G. Pess. 1995. Pool spacing in forest channels. *Water Resour. Res.* 31:1097–1105.

Moore, K. M. S., and S. V. Gregory. 1988. Response of young-of-the-year trout to manipulation of habitat structure in a small stream.–*Trans. Am. Fish. Soc.* 117:162–170.

Morán, P., A. M. Pendás, E. Beall, and E. García-Vázquez. 1996. Genetic assessment of the reproductive success of Atlantic salmon precocious parr by means of VNTR loci. *Heredity* 77:655–660.

Morbey, Y. 2000. Protandry in Pacific salmon. *Can. J. Fish. Aquat. Sci.* 57:1252–1257.

Morley, R. B., H. T. Bilton, A. S. Coburn, D. Brouwer, J. van Tyne, and W. C. Clarke. 1988. The influence of time and size at release of juvenile coho salmon (*Oncorhynchus kisutch*) on returns at maturity: results of studies on three brood years at Quinsam Hatchery, B.C. *Can. Tech. Rept. Fish. Aquat. Sci.* 1620. Nanaimo, BC. 120 p.

Mortensen, D., A. Wertheimer, S. Taylor, and J. Landingham. 2000. The relation between early marine growth of pink salmon, *Oncorhynchus gorbuscha*, and marine water temperature, secondary production, and survival to adulthood. *Fish. Bull.* 98:319–335.

Moscrip, A. L., and D. R. Montgomery. 1997. Urbanization, flood frequency, and salmon abundance in Puget lowland streams. *J. Am. Water Resour. Assoc.* 33:1289–1297.

Moser, M. L., A. F. Olson, and T. P. Quinn. 1991. Riverine and estuarine migratory behavior of coho salmon (*Oncorhynchus kisutch*) smolts. *Can. J. Fish. Aquat. Sci.* 48:1670–1678.

Moulton, F. R., ed. 1939. *The migration and conservation of salmon.* Am. Assoc. Advanc. Sci. Publ. 8. Lancaster, PA. 106 p.

Moyle, P. B., and B. Herbold. 1987. Life-history patterns and community structure in stream fishes of western North America: Comparisons with eastern North America and Europe. Pages 25–32 in W. J. Matthews and D. C. Heins, eds., *Community and evolutionary ecology of North American stream fishes.* University of Oklahoma Press, Norman.

Mueter, F. J., R. M. Peterman, and B. J. Pyper. 2002. Opposite effects of ocean temperature on survival rates of 120 stocks of Pacific salmon (*Oncorhynchus* spp.) in northern and southern areas. *Can. J. Fish. Aquat. Sci.* 59:456–463.

Muir, W. D., S. G. Smith, J. G. Williams, E. E. Hockersmith, and J. R. Skakski. 2001. Survival estimates for migrant yearling chinook salmon and steelhead tagged with passive integrated transponders in the lower Snake and lower Columbia rivers, 1993–1998. *N. Am. J. Fish. Mgmt.* 21:269–282.

Mullan, J. W., A. Rockhold, and C. R. Chrisman. 1992. Life histories and precocity of chinook salmon in the mid–Columbia River. *Prog. Fish-Cult.* 54:25–28.

Murphy, J. M., and W. R. Heard. 2001. *Chinook salmon data storage tag studies in southeast Alaska, 2001.* N. Pac. Anad. Fish Comm. Doc. 555. Juneau, AK. 21 p.

Murray, C. B., and J. D. McPhail. 1988. Effect of temperature on the development of five species of Pacific salmon (*Oncorhynchus*) embryos and alevins. *Can. J. Zool.* 66:266–273.

Murray, C. B., and M. L. Rosenau. 1989. Rearing of juvenile chinook salmon in nonnatal tributaries of the lower Fraser River, British Columbia. *Trans. Am. Fish. Soc.* 118:284–289.

Myers, J. M., R. G. Kope, G. J. Bryant, D. Teel, L. J. Lierheimer, T. C. Wainwright, W. S. Grant, F. W. Waknitz, K. Neely, S. T. Lindley, and R. S. Waples. 1998. *Status review of chinook salmon from Washington, Idaho, Oregon and California.* National Marine Fisheries Service, NOAA Tech. Memo. NMFS-NWFSC-35. Seattle. 443 p.

Myers, K. W., K. Y. Aydin, R. V. Walker, S. Fowler, and M. L. Dahlberg. 1996. *Known ocean ranges of stocks of Pacific salmon and steelhead as shown by tagging experiments, 1956–1995.* University of Washington School of Fisheries, Fish. Res. Inst., FRI-UW-9614. Seattle. 229 p.

Myers, K. W., and H. F. Horton. 1982. Temporal use of an Oregon estuary by hatchery and wild juvenile salmon. Pages 377–392 in V.S. Kennedy, ed., *Estuarine comparisons.* Academic Press, New York.

Naiman, R. J., R. E. Bilby, D. E. Schindler, and J. M. Helfield. 2002. Pacific salmon, nutrients, and the dynamics of freshwater and riparian ecosystems. *Ecosystems* 5:399–417.

Naiman, R. J., D. G. Lonzarich, T. J. Beechie, and S. C. Ralph. 1992. General principles of classification and the assessment of conservation potential of rivers. Pages 93–123 in P. J. Boon, P. Calow, and G. E. Petts, eds., *River conservation and management.* John Wiley and Sons, New York.

Nakano, S., and M. Murakami. 2001. Reciprocal subsidies: dynamic interdependence between terrestrial and aquatic food webs. *Proc. Nat. Acad. Sci.* 98:166–170.

Narver, D. W. 1970. Diel vertical movement and feeding of underyearling sockeye salmon and the limnetic zooplankton in Babine Lake, British Columbia. *J. Fish. Res. Board Can.* 27:281–316.

Neave, F. 1943. Diurnal fluctuations in the upstream migration of coho and spring salmon. *J. Fish. Res. Board Can.* 6:158–163.

———. 1955. Notes on the seaward migration of pink and chum fry. *J. Fish. Res. Board Can.* 12:369–374.

———. 1958. The origin and speciation of *Oncorhynchus. Trans. Royal Soc. Can.* 52(3):25–39.

———. 1964. Ocean migrations of Pacific salmon. *J. Fish. Res. Board Can.* 21:1227–1244.

Neave, F., T. Yonemori, and R.G. Bakkala. 1976. Distribution and origin of chum salmon in offshore waters of the North Pacific Ocean. *Int. N. Pac. Fish. Comm. Bull.* 35. Vancouver, BC. 79 p.

Nehlsen, W., J. E. Williams, and J. A. Lichatowich. 1991. Pacific salmon at the crossroads: stocks at risk from California, Oregon, Idaho and Washington. *Fisheries* 16(2):4–21.

Nelson, J. S. 1984. *Fishes of the world.* John Wiley and Sons, New York.

Nelson, P. R. 1959. Effects of fertilizing Bare Lake, Alaska, on growth and production of red salmon (*O. nerka*). *Fish. Bull.* 60:59–86.

Nero, R. W., and M. E. Huster. 1996. Low-frequency acoustic imaging of Pacific salmon on the high seas. *Can. J. Fish. Aquat. Sci.* 53:2513–1523.

Nevitt, G. A., A. H. Dittman, T. P. Quinn, and W. J. Moody, Jr. 1994. Evidence for a peripheral olfactory memory in imprinted salmon. *Proc. Nat. Acad. Sci.* 91:4288–4292.

Nichol, L. M., and D. M. Shackleton. 1996. Seasonal movements and foraging behaviour of northern resident killer whales (*Orcinus orca*) in relation to the inshore distribution of salmon (*Oncorhynchus* spp.) in British Columbia. *Can. J. Zool.* 74:983–991.

Nicholas, J. W., and D. G. Hankin. 1989. *Chinook salmon populations in Oregon's coastal river basins.* Oregon Dept. Fish Wildl., Portland. 359 p.

Nickelson, T. E. 1986. Influences of upwelling, ocean temperature, and smolt abundance on marine survival of coho salmon (*Oncorhynchus kisutch*) in the Oregon production area. *Can. J. Fish. Aquat. Sci.* 43:527–535.

Nickelson, T. E., J. W. Nicholas, A. M. McGie, R. B. Lindsay, D. L. Bottom, R. J. Kaiser, and S. E. Jacobs. 1992. *Status of anadromous salmonids in Oregon coastal basins.* Oregon Dept. Fish Wildl., Portland. 83 p.

Nielsen, J. L. 1992. Microhabitat-specific foraging behavior, diet, and growth of juvenile coho salmon. *Trans. Am. Fish. Soc.* 121:617–634.

———, ed. 1995. *Evolution and the aquatic ecosystem.* Am. Fish. Soc. Symp. 17. Bethesda, MD. 433 p.

Nilsson, N. A. and T. G. Northcote. 1981. Rainbow trout (*Salmo gairdneri*) and cutthroat trout (*S. clarki*) interactions in coastal British Columbia lakes. *Can. J. Fish. Aquat. Sci.* 38:1228–1246.

Nomura, T., S. Urawa, and Y. Ueno. 2000. Variations in muscle lipid content of high-seas chum and pink salmon in winter. *N. Pac. Anad. Fish Comm. Bull.* 2:347–352.

Nordeng, H. 1971. Is the local orientation of anadromous fishes determined by pheromones? *Nature* 223:411–413.

———. 1977. A pheromone hypothesis for homeward migration in anadromous salmonids. *Oikos* 28:155–159.

———. 1989. Salmonid migration: Hypotheses and principles. Pages 1–8 in E. Brannon and B. Jonsson, eds., *Proceedings of the salmonid migration and distribution symposium.* University of Washington School of Fisheries, Seattle.

Norris, J. G., S.-Y. Hyun, and J. J. Anderson. 2000. Ocean distribution of Columbia River upriver bright fall chinook salmon stocks. *N. Pac. Anad. Fish Comm. Bull.* 2:221–232.

Northcote, T. G. 1978. Migratory strategies and production in freshwater fishes. Pages 326–359 in S. D. Gerking, ed., *Ecology of freshwater fish populations.* John Wiley and Sons, New York.

———. 1992. Migration and residency in stream salmonids—some ecological considerations and evolutionary consequences. *Nordic J. Freshw. Res.* 67:5–17.

Nowak, G. M., and T. P. Quinn. 2002. Diel and seasonal patterns of horizontal and vertical movements of telemetered cutthroat trout (*Oncorhynchus clarki*) in Lake Washington, Washington. *Trans. Am. Fish. Soc.* 131:452–462.

Nowak, G. M., R. A. Tabor, E. J. Warner, K. L. Fresh and T. P. Quinn. 2004. Ontogenetic shifts in habitat and diet of cutthroat trout in Lake Washington, Washington. *N. Am. J. Fish. Mgmt* 24:624–635.

NRC (National Research Council). 1996. *Upstream: Salmon and society in the Pacific Northwest.* National Academy Press, Washington, D.C.

Nunan, C. P., and D. L. G. Noakes. 1987. Effects of light on movement of rainbow trout embryos within, and on their emergence from, artificial redds. Pages 151–156 in Am. Fish. Soc. Symp. 2. Bethesda, MD.

NWPPC (Northwest Power Planning Council). 1986. *Compilation of information on salmon and steelhead losses in the Columbia River basin.* Northwest Power Planning Council, Portland.

Oakley, T. H., and R. B. Phillips. 1999. Phylogeny of Salmonidae fishes based on growth hormone introns: Atlantic (*Salmo*) and Pacific (*Oncorhynchus*) salmon are not sister taxa. *Molec. Phylogen. Evol.* 11:381–393.

Ogura, M., and Y. Ishida. 1992. Swimming behavior of coho salmon, *Oncorhynchus kisutch*, in the open sea as determined by ultrasonic telemetry. *Can. J. Fish. Aquat. Sci.* 49:453–457.

———. 1995. Homing behavior and vertical movements of four species of Pacific salmon (*Oncorhynchus* spp.) in the central Bering Sea. *Can. J. Fish. Aquat. Sci.* 52:532–540.

O'Keefe, T. C., and R. T. Edwards. 2002. Evidence for hyporheic transfer and removal of marine-derived nutrients in a sockeye stream in southwest Alaska. Pages 99–107 in Am. Fish. Soc. Symp. 33. Bethesda, MD.

Olesiuk, P. F. 1993. Annual prey consumption by harbor seals (*Phoca vitulina*) in the Strait of Georgia, British Columbia. *Fish. Bull.* 91:491–515.

Olla, B. L., M. W. Davis, and C. H. Ryer. 1998. Understanding how the hatchery environment represses or promotes the development of behavioral survival skills. *Bull. Mar. Sci.* 62:531–550.

Olson, A. F., and T. P. Quinn. 1993. Vertical and horizontal movements of adult chinook salmon *Oncorhynchus tshawytscha* in the Columbia River estuary. *Fish. Bull.* 91:171–178.

O'Neill, S. M., and K. D. Hyatt. 1987. An experimental study of competition for food between sockeye salmon (*Oncorhynchus nerka*) and the threespine sticklebacks (*Gasterosteus aculeatus*) in a British Columbia coastal lake. *Can. Spec. Publ. Fish. Aquat. Sci.* 96:143–160.

Orsi, J. A., and H. W. Jaenicke. 1996. Marine distribution and origin of prerecruit chinook salmon, *Oncorhynchus tshawytscha*, in southeastern Alaska. *Fish. Bull.* 94:482–497.

Orsi, J. A., and A. C. Wertheimer. 1995. Marine vertical distribution of juvenile chinook and coho salmon in southeastern Alaska. *Trans. Am. Fish. Soc.* 124:159–169.

Parker, R. R. 1965. Estimation of sea mortality rates for the 1961 brood-year pink salmon of the Bella Coola area, British Columbia. *J. Fish. Res. Board Can.* 22:1523–1554.

———. 1968. Marine mortality schedules of pink salmon of the Bella Coola River, Central British Columbia. *J. Fish. Res. Board Can.* 25:757–794.

Parker, R. R., and P. A. Larkin. 1959. A concept of growth in fishes. *J. Fish. Res. Board Can.* 16:721–745.

Parkyn, D. C., J. D. Austin, and C. W. Hawryshyn. 2003. Acquisition of polarized-light orientation in salmonids under laboratory conditions. *Anim. Behav.* 65:895–904.

Pascual, M. A., and T. P. Quinn. 1991. Evaluation of alternative models of the coastal migration of adult Fraser River sockeye salmon (*Oncorhynchus nerka*). *Can. J. Fish. Aquat. Sci.* 48:799–810.

———. 1994. Geographical patterns of straying of fall chinook salmon (*Oncorhynchus tshawytscha*) from Columbia River (U.S.A.) hatcheries. *Aquacult. Fish. Mgmt.* 25 (Suppl. 2):17–30.

Patten, B. G. 1975. Comparative vulnerability of fry of Pacific salmon and steelhead trout to predation by torrent sculpin in stream aquaria. *Fish. Bull.* 73:931–934.

———. 1977. Body size and learned avoidance as factors affecting predation on coho salmon, *Oncorhynchus kisutch*, fry by torrent sculpin, *Cottus rhotheus*. *Fish. Bull.* 75:457–459.

Pearcy, W. G. 1991. Biology of the Transition Zone. Pages 39–55 in J. A. Wetherall, ed., *Biology, oceanography, and fisheries of the North Pacific Transition Zone and Subarctic Frontal Zone*. National Marine Fisheries Service, NOAA Tech. Rept. 105. Honolulu. 111 p.

———. 1992. *Ocean ecology of North Pacific salmonids*. Washington Sea Grant Program, University of Washington, Seattle.

Pearcy, W. G., R. D. Brodeur, and J. P. Fisher. 1990. Distribution and biology of juvenile cutthroat trout *Oncorhynchus clarki clarki* and steelhead *O. mykiss* in coastal waters off Oregon and Washington. *Fish. Bull.* 88:697–711.

Pearcy, W. G., R. D. Brodeur, J. M. Shenker, W. W. Smoker, and Y. Endo. 1988. Food habits of Pacific salmon and steelhead trout, midwater trawl catches and oceanographic conditions in the Gulf of Alaska, 1980–1985. *Bull. Oceanogr. Res. Inst.* 26:29–78.

Pearcy, W. G., and J. P. Fisher. 1988. Migrations of coho salmon, *Oncorhynchus kisutch*, during their first summer in the ocean. *Fish. Bull.* 86:173–195.

———. 1990. *Distribution and abundance of juvenile salmonids off Oregon and Washington, 1981–1985*. National Marine Fisheries Service, NOAA Tech. Rept. 93. Corvallis, OR. 83 p.

Pearcy, W. G., T. Nishiyama, T. Fujii, and K. Masuda. 1984. Diel variations in the feeding habits of Pacific salmon caught in gill nets during a 24-hour period in the Gulf of Alaska. *Fish. Bull.* 82:391–399.

Pearcy, W. G., C. D. Wilson, A. W. Chung, and J. W. Chapman. 1989. Residence times, distribution, and production of juvenile chum salmon, *Oncorhynchus keta*, in Netarts Bay, Oregon. *Fish. Bull.* 87: 553–568.

Pearsons, T. N., and A. L. Fritts. 1999. Maximum size of chinook salmon consumed by juvenile coho salmon. *N. Am. J. Fish. Mgmt.* 19:165–170.

Peterman, R. M. 1987. Review of the components of recruitment of Pacific salmon. Pages 417–429 in Am. Fish. Soc. Symp. 1. Bethesda, MD.

Peterman, R. M., S. G. Marinone, K. A. Thomson, I. D. Jardine, R. N. Crittenden, P. H. LeBlond, and C. J. Walters. 1994. Simulation of juvenile sockeye salmon (*Oncorhynchus nerka*) migrations in the Strait of Georgia, British Columbia. *Fish. Oceanogr.* 3:221–235.

Petersen, J. H., and J. F. Kitchell. 2001. Climate regimes and water temperature changes in the Columbia River: bioenergetic implications for predators of juvenile salmon. *Can. J. Fish. Aquat. Sci.* 58: 1831–1841.

Peterson, N. P. 1982a. Immigration of juvenile coho salmon (*Oncorhynchus kisutch*) into riverine ponds. *Can. J. Fish. Aquat. Sci.* 39:1308–1310.

———. 1982b. Population characteristics of juvenile coho salmon (*Oncorhynchus kisutch*) overwintering in riverine ponds. *Can. J. Fish. Aquat. Sci.* 39:1303–1307.

Peterson, N. P., and T. P. Quinn. 1996. Spatial and temporal variation in dissolved oxygen in natural egg pockets of chum salmon, *Oncorhynchus keta* (Walbaum), in Kennedy Creek, Washington. *J. Fish Biol.* 48:131–143.

Peven, C. M. 1987. Downstream migration timing of two stocks of sockeye salmon on the mid–Columbia River. *Northwest Sci.* 61:186–190.

Peven, C. M., R. R. Whitney, and K. R. Williams. 1994. Age and length of steelhead smolts from the mid–Columbia River basin, Washington. *N. Am. J. Fish. Mgmt.* 14:77–86.

Phillips, R. B., K. A. Pleyte, and M. R. Brown. 1992. Salmonid phylogeny inferred from ribosomal DNA restriction maps. *Can. J. Fish. Aquat. Sci.* 49:234–2353.

Pielou, E. C. 1991. *After the Ice Age: The return of life to glaciated North America.* University of Chicago Press, Chicago.

Pinkas, L., M. S. Oliphant, and I. L. K. Iverson. 1971. *Food habits of albacore, bluefin tuna, and bonito in California waters.* California Dept. Fish Game, Fish Bull. 152. Sacramento. 105 p.

Pyper, B. J., and R. M. Peterman. 1999. Relationship among adult body length, abundance, and ocean temperature for British Columbia and Alaska sockeye salmon (*Oncorhynchus nerka*), 1967–1997. *Can. J. Fish. Aquat. Sci.* 56:1716–1720.

Pyper, B. J., R. M. Peterman, M. F. Lapointe, and C. J. Walters. 1999. Patterns of covariation in length and age at maturity of British Columbia and Alaska sockeye salmon (*Oncorhynchus nerka*) stocks. *Can. J. Fish. Aquat. Sci.* 56:1046–1057.

Quinn, T. P. 1980. Evidence for celestial and magnetic compass orientation in lake migrating sockeye salmon fry. *J. Comp. Physiol.* 137:243–248.

———. 1982. A model for salmon navigation on the high seas. Pages 229–237 in E. L. Brannon and E.O. Salo, eds., *Salmon and trout migratory behavior symposium.* University of Washington School of Fisheries, Seattle.

———. 1984. Homing and straying in Pacific salmon. Pages 357–362 in J. D. McCleave, G. P. Arnold, J. J. Dodson, and W. H. Neill, eds., *Mechanisms of migration in fishes.* Plenum Press, New York.

———. 1988. Estimated swimming speeds of migrating adult sockeye salmon. *Can. J. Zool.* 66: 2160–2163.

———. 1990a. Current controversies in the study of salmon homing. *Ethol. Ecol. Evol.* 2:49–63.

———. 1990b. Migratory behavior of Pacific salmon in estuaries: Recent results with ultrasonic telemetry. Pages 13–25 in C. A. Simenstad, ed., *Effects of dredging on anadromous Pacific Coast fishes.* Washington Sea Grant Program, University of Washington, Seattle.

———. 1991. Models of Pacific salmon orientation and navigation on the open ocean. *J. Theoret. Biol.* 150:539–545.

———. 1999. Variation in Pacific salmon reproductive behaviour associated with species, sex and levels of competition. *Behaviour* 136:179–204.

Quinn, T. P., and D. J. Adams. 1996. Environmental changes affecting the migratory timing of American shad and sockeye salmon. *Ecology* 77:1151–1162.

Quinn, T. P., and E. L. Brannon. 1982. The use of celestial and magnetic cues by orienting sockeye salmon smolts. *J. Comp. Physiol.* 147:547–552.

Quinn, T. P., and C. A. Busack. 1985. Chemosensory recognition of siblings in juvenile coho salmon (*Oncorhynchus kisutch*). *Anim. Behav.* 33:51–56.

Quinn, T. P., and C. J. Foote. 1994. The effects of body size and sexual dimorphism on the reproductive behaviour of sockeye salmon (*Oncorhynchus nerka*). *Anim. Behav.* 48:751–761.

Quinn, T. P., and K. Fresh. 1984. Homing and straying in chinook salmon (*Oncorhynchus tshawytscha*) from Cowlitz River Hatchery, Washington. *Can. J. Fish. Aquat. Sci.* 41:1078–1082.

Quinn, T. P., S. M. Gende, G. T. Ruggerone, and D. E. Rogers. 2003. Density dependent predation by brown bears (*Ursus arctos*) on sockeye salmon (*Oncorhynchus nerka*). *Can. J. Fish. Aquat. Sci.* 60: 553–562.

Quinn, T. P., A. P. Hendry, and G. B. Buck. 2001. Balancing natural and sexual selection in sockeye salmon: interactions between body size, reproductive opportunity and vulnerability to predation by bears. *Evol. Ecol. Res.* 3:917–937.

Quinn, T. P., A. P. Hendry, and L. A. Wetzel. 1995. The influence of life history trade–offs and the size of incubation gravels on egg size variation in sockeye salmon (*Oncorhynchus nerka*). *Oikos* 74: 425–438.

Quinn, T. P., M. T. Kinnison, and M. J. Unwin. 2001. Evolution of chinook salmon (*Oncorhynchus tshawytscha*) populations in New Zealand: Pattern, rate, and process. *Genetica* 112/113:493–513.

Quinn, T. P., R. S. Nemeth, and D. O. McIsaac. 1991. Patterns of homing and straying by fall chinook salmon in the lower Columbia River. *Trans. Am. Fish. Soc.* 120:150–156.

Quinn, T. P., and N. P. Peterson. 1996. The influence of habitat complexity and fish size on over-winter survival and growth of individually-marked juvenile coho salmon (*Oncorhynchus kisutch*) in Big Beef Creek, Washington. *Can. J. Fish. Aquat. Sci.* 53:1555–1564.

Quinn, T. P., and B. A. terHart. 1987. Movements of adult sockeye salmon (*Oncorhynchus nerka*) in British Columbia coastal waters in relation to temperature and salinity stratification: Ultrasonic telemetry results. *Can. Spec. Publ. Fish. Aquat. Sci.* 96:61–77.

Quinn, T. P., B. A. terHart, and C. Groot. 1989. Migratory orientation and vertical movements of homing adult sockeye salmon, *Oncorhynchus nerka*, in coastal waters. *Anim. Behav.* 37:587–599.

Quinn, T. P., M. J. Unwin, and M. T. Kinnison. 2000. Evolution of temporal isolation in the wild: genetic divergence in timing of migration and breeding by introduced chinook salmon populations. *Evolution* 54:1372–1385.

Quinn, T. P., E. C. Volk, and A. P. Hendry. 1999. Natural otolith microstructure patterns reveal precise homing to natal incubation sites by sockeye salmon (*Oncorhynchus nerka*). *Can. J. Zool.* 77:766–775.

Quinn, T. P., L. A. Vøllestad, J. Peterson, and V. Gallucci. 2004. Influences of fresh water and marine growth on the egg size–egg number tradeoff in coho and chinook salmon. *Trans. Am. Fish. Soc.* 133:55–65.

Quinn, T. P., L. A. Wetzel, S. Bishop, K. Overberg, and D. E. Rogers. 2001. Influence of breeding habitat on bear predation, and age at maturity and sexual dimorphism of sockeye salmon populations. *Can. J. Zool.* 79:1782–1793.

Quinn, T. P., C. C. Wood, L. Margolis, B. E. Riddell, and K. D. Hyatt. 1987. Homing in wild sockeye salmon (*Oncorhynchus nerka*) populations as inferred from differences in parasite prevalence and allozyme allele frequencies. *Can. J. Fish. Aquat. Sci.* 44:1963–1971.

Raleigh, R. F. 1967. Genetic control in the lakeward migrations of sockeye salmon (*Oncorhynchus nerka*) fry. *J. Fish. Res. Board Can.* 24:2613–2622.

Randall, R. G., M. C. Healey, and J. B. Dempson. 1987. Variability in length of freshwater residence of salmon, trout, and char. Pages 27–41 in Am. Fish. Soc. Symp. 1. Bethesda, MD.

Raymond, H. L. 1988. Effects of hydroelectric development and fisheries enhancement on spring and summer chinook salmon and steelhead in the Columbia River basin. *N. Am. J. Fish. Mgmt.* 8:1–24.

Reese, C. D., and B. C. Harvey. 2002. Temperature-dependent interactions between juvenile steelhead and Sacramento pikeminnow in laboratory streams. *Trans. Am. Fish. Soc.* 131:599–606.

Reeves, G. H., F. H. Everest, and J. D. Hall. 1987. Interactions between the redside shiner (*Richardsonius balteatus*) and the steelhead trout (*Salmo gairdneri*) in Western Oregon: the influence of water temperature. *Can. J. Fish. Aquat. Sci.* 44:1603–1613.

Reeves, G. H., F. H. Everest, and J. R. Sedell. 1993. Diversity of juvenile anadromous salmonid assemblages in coastal Oregon basins with different levels of timber harvest. *Trans. Am. Fish. Soc.* 122:309–317.

Reimchen, T. E. 2000. Some ecological and evolutionary aspects of bear–salmon interactions in coastal British Columbia. *Can. J. Zool.* 78:448–458.

Reimchen, T. E., D. Mathewson, M. D. Hocking, J. Moran, and D. Harris. 2003. Isotopic evidence for enrichment of salmon-derived nutrients in vegetation, soil, and insects in riparian zones in coastal British Columbia. Pages 59–69 in Am. Fish. Soc. Symp. 34. Bethesda, MD.

Reimers, P. E. 1971. The length of residence of juvenile fall chinook salmon in Sixes River, Oregon. Ph.D. dissertation, Oregon State University, Corvallis. 99 p.

Reisenbichler, R. R. 1988. Relation between distance transferred from natal stream and recovery rate for hatchery coho salmon. *N. Am. J. Fish. Mgmt.* 8:172–174.

Reisenbichler, R. R., and S. P. Rubin. 1999. Genetic changes from artificial propagation of Pacific salmon affect the productivity and viability of supplemented populations. *ICES J. Mar. Sci.* 56:459–466.

Rennie, C. D., and R. G. Millar. 2000. Spatial variability of stream bed scour and fill: a comparison of scour depth in chum salmon (*Oncorhynchus keta*) redds and adjacent bed. *Can. J. Fish. Aquat. Sci.* 57:928–938.

Rhodes, J. S., and T. P. Quinn. 1998. Factors affecting the outcome of territorial contests between hatchery and naturally reared coho salmon parr in the laboratory. *J. Fish Biol.* 53:1220–1230.
Rich, W. H. 1939. Local populations and migration in relation to the conservation of Pacific salmon in the western states and Alaska. Pages 45–50 in Am. Assoc. Advanc. Sci. Publ. 8. Lancaster, PA.
Ricker, W. E. 1938. "Residual" and kokanee salmon in Cultus Lake. *J. Fish. Res. Board Can.* 4:192–218.
———. 1950. Cycle dominance among the Fraser sockeye. *Ecology* 31:6–26.
———. 1954. Stock and recruitment. *J. Fish. Res. Board Can.* 11:559–623.
———. 1972. Hereditary and environmental factors affecting certain salmonid populations. Pages 19–160 in R. C. Simon and P. A. Larkin, eds., *The stock concept in Pacific salmon*. H. R. MacMillan Lectures in Fisheries, University of British Columbia, Vancouver.
———. 1976. Review of the rate of growth and mortality of Pacific salmon in salt water, and noncatch mortality caused by fishing. *J. Fish. Res. Board Can.* 33:1483–1525.
———. 1980. Causes of the decrease in age and size of chinook salmon (*Oncorhynchus tshawytscha*). *Can. Tech. Rept. Fish. Aquat. Sci.* 944. Nanaimo, BC. 25 p.
———. 1981. Changes in the average size and average of Pacific salmon. *Can. J. Fish. Aquat. Sci.* 38: 1636–1656.
———. 1987. Effects of the fishery and of obstacles to migration on the abundance of Fraser River sockeye salmon (*Oncorhynchus nerka*). *Can. Tech. Rep. Fish. Aquat. Sci.* 1522. Nanaimo, BC. 75 p.
———. 1995. Trends in the average size of Pacific salmon in Canadian catches. *Can. Spec. Publ. Fish. Aquat. Sci.* 121:593–602.
Riddell, B. E. 1993. Spatial organization of Pacific salmon: what to conserve? Pages 23–41 in J. G. Cloud and G. H. Thorgaard, eds., *Genetic conservation of salmonid fishes*. Plenum Press, New York.
Rieman, B. E., and D. L. Myers. 1992. Influence of fish density and relative productivity on growth of kokanee in ten oligotrophic lakes and reservoirs in Idaho. *Trans. Am. Fish. Soc.* 121:178–191.
Rieman, B. E., D. L. Myers, and R. L. Nielsen. 1994. Use of otolith microchemistry to discriminate *Oncorhynchus nerka* of resident and anadromous origin. *Can. J. Fish. Aquat. Sci.* 51:68–77.
Rigby, P., J. McConnaughey, and H. Savikko. 1991. Alaska commercial salmon catches, 1878–1991. Alaska Dept. Fish Game, Reg. Info. Rept. 5J91-16. Juneau. 88 p.
Robards, M. D., and T. P. Quinn. 2002. The migratory timing of adult summer-run steelhead trout (*Oncorhynchus mykiss*) in the Columbia River: Six decades of environmental change. *Trans. Am. Fish. Soc.* 131:523–536.
Robertson, O. H. 1957. Survival of precociously mature king salmon male parr (*Oncorhynchus tshawytscha* Juv.) after spawning. *Cal. Fish Game* 43:119–130.
Rogers, D. E. 1973. Abundance and size of juvenile sockeye salmon, *Oncorhynchus nerka*, and associated species in Lake Aleknagik, Alaska, in relation to their environment. *Fish. Bull.* 71:1061–1075.
———. 1980. Density-dependent growth of Bristol Bay sockeye salmon. Pages 267–283 in W. J. McNeil and D. C. Himsworth, eds., *Salmonid ecosystems of the North Pacific*. Oregon State University Press, Corvallis.
———. 1987. The regulation of age at maturity in Wood River sockeye salmon (*Oncorhynchus nerka*). *Can. Spec. Publ. Fish. Aquat. Sci.* 96:78–89.
———. 1988. Bristol Bay smolt migrations: timing and size composition and the effects on distribution and survival at sea. Pages 87–101 in W. J. McNeil, ed., *Salmon production, management, and allocation*. Oregon State University Press, Corvallis.
———. 2001. *Estimates of annual salmon runs from the North Pacific, 1951–2001*. University of Washington School of Aquatic and Fishery Sciences, SAFS-UW-0115. Seattle. 11 p.
Rogers, D. E., and G. T. Ruggerone. 1993. Factors affecting marine growth of Bristol Bay sockeye salmon. *Fish. Res.* 18:89–103.
Roni, P. 1992. Life history and spawning habitat in four stocks of large-bodied chinook salmon (*Oncorhynchus tshawytscha*). M.S. thesis, University of Washington, Seattle.
———. 2002. Habitat use by fishes and Pacific giant salamanders in small western Oregon and Washington streams. *Trans. Am. Fish. Soc.* 131:743–761.
Roni, P., and A. Fayram. 2000. Estimating winter salmonid abundance in small western Washington streams: a comparison of three techniques. *N. Am. J. Fish. Mgmt.* 20:683–692.
Roni, P., and T. P. Quinn. 2001. Density and size of juvenile salmonids in response to placement of large woody debris in western Oregon and Washington streams. *Can. J. Fish. Aquat. Sci.* 58: 282–292.
Roos, J. F. 1991. *Restoring Fraser River salmon*. Pacific Salmon Commission, Vancouver, British Columbia.
Roper, B. B., and D. L. Scarnecchia. 1999. Emigration of age-0 chinook salmon (*Oncorhynchus tshawytscha*) smolts from the upper South Umpqua River basin, Oregon, U.S.A. *Can. J. Fish. Aquat. Sci.* 56:939–946.

Rosenfeld, J. S., and S. Boss. 2001. Fitness consequences of habitat use for juvenile cutthroat trout: energetic costs and benefits in pools and riffles. *Can. J. Fish. Aquat. Sci.* 58:585–593.

Rounsefell, G. A. 1958. Anadromy in North American Salmonidae. *Fish. Bull.* 58:171–185.

Rowe, D. K., and B. L. Chisnall. 1995. Effects of oxygen, temperature and light gradients on the vertical distribution of rainbow trout, *Oncorhynchus mykiss*, in two North Island, New Zealand, lakes differing in trophic status. *N. Z. J. Mar. Freshw. Res.* 29:421–434.

Royce, W. F., L. S. Smith, and A. C. Hartt. 1968. Models of oceanic migrations of Pacific salmon and comments on guidance mechanisms. *Fish. Bull.* 66:441–462.

Ruggerone, G. T. 1986. Consumption of migrating juvenile salmonids by gulls foraging below a Columbia River dam. *Trans. Am. Fish. Soc.* 115:736–742.

———. 1989. Coho salmon predation on juvenile sockeye salmon in the Chignik Lakes, Alaska. Ph.D. dissertation, University of Washington, Seattle.

Ruggerone, G. T., R. Hanson, and D. E. Rogers. 2000. Selective predation by brown bears (*Ursus arctos*) foraging on spawning sockeye salmon (*Oncorhynchus nerka*). *Can. J. Zool.* 78:974–981.

Ruggerone, G.T., T. P. Quinn, I. A. McGregor, and T. S. Wilkinson. 1990. Horizontal and vertical movements of adult steelhead trout, *Oncorhynchus mykiss*, in Dean and Fisher channels, British Columbia. *Can. J. Fish. Aquat. Sci.* 47:1963–1969.

Ruggerone, G. T., and D. E. Rogers. 1984. Arctic char predation on sockeye salmon smolts at Little Togiak River, Alaska. *Fish. Bull.* 82:401–410.

———. 1992. Predation on sockeye salmon fry by juvenile coho salmon in the Chignik Lakes, Alaska: Implications for salmon management. *N. Am. J. Fish. Mgmt.* 12:89–102.

———. 2003. Multi-year effects of high densities of sockeye salmon spawners on juvenile salmon growth and survival: A case study from the *Exxon Valdez* oil spill. *Fish. Res.* 63:379–392.

Ruggerone, G. T., M. Zimmermann, K. W. Myers, J. L. Nielsen, and D. E. Rogers. 2003. Competition between Asian pink salmon (*Oncorhynchus gorbuscha*) and Alaskan sockeye salmon (*O. nerka*) in the North Pacific Ocean. *Fish. Oceanogr.* 12:209–219.

Ryan, B. A., J. W. Ferguson, R. D. Ledgerwood, and E. P. Nunnallee. 2001. Detection of passive integrated transponder tags from juvenile salmonids on piscivorous bird colonies in the Columbia River Basin. *N. Am. J. Fish. Mgmt.* 21:417–421.

Ryan, B. A., S. G. Smith, J. Butzerin, and J. W. Ferguson. 2003. Relative vulnerability to avian predation of PIT-tagged juvenile salmonids in the Columbia River estuary, 1998–2000. *Trans. Am. Fish. Soc.* 132:275–288.

Sabo, J. L., and G. B. Pauley. 1997. Competition between stream-dwelling cutthroat trout (*Oncorhynchus clarki*) and coho salmon (*Oncorhynchus kisutch*): Effects of relative size and population origin. *Can. J. Fish. Aquat. Sci.* 54:2609–2617.

Saila, S. B., and R. A. Shappy. 1963. Random movement and orientation in salmon migration. *Journal du Conseil International pour l'Exploration de la Mer* 28:153–166.

Salo, E. O. 1991. Life history of chum salmon (*Oncorhynchus keta*). Pages 231–309 in C. Groot and L. Margolis, eds., *Pacific salmon life histories.* University of British Columbia Press, Vancouver.

Salo, E. O., and T. W. Cundy, eds. 1987. *Streamside management: Forestry and fishery interactions.* University of Washington College of Forest Resources, Seattle.

Sandercock, F. K. 1991. Life history of coho salmon (*Oncorhynchus kisutch*). Pages 395–445 in C. Groot and L. Margolis, eds., *Pacific salmon life histories.* University of British Columbia Press, Vancouver.

Satou, M., A. Shiraishi, T. Matsushima, and N. Okumoto. 1991. Vibrational communication during spawning behavior in hime salmon (landlocked red salmon, *Oncorhynchus nerka*). *J. Comp. Physiol.* 168:417–428.

Scheel, D., and K. R. Hough. 1997. Salmon fry predation by seabirds near an Alaskan hatchery. *Mar. Ecol. Prog. Ser.* 150:35–48.

Scheuerell, M. D., and D. E. Schindler. 2003. Diel vertical migration by juvenile sockeye salmon: empirical evidence for the anti-predation window. *Ecology* 84:1713–1720.

Schindler, D. E., D. E. Rogers, M. D. Scheuerell, and C. A. Abrey. Forthcoming. Effects of changing climate on zooplankton and growth of juvenile sockeye salmon in southwestern Alaska. *Ecology.*

Schmidt, D. C., S. R. Carlson, and G. B. Kyle. 1998. Influence of carcass-derived nutrients on sockeye salmon productivity of Karluk Lake, Alaska: Importance in the assessment of an escapement goal. *N. Am. J. Fish. Mgmt.* 18:743–763.

Scholz, A. T., J. C. Cooper, R. M. Horrall, and A. D. Hasler. 1978. Homing of morpholine-imprinted brown trout, *Salmo trutta*. *Fish. Bull.* 76:293–295.

Scholz, A. T., C. K. Gosse, J. C. Cooper, R. M. Horrall, A. D. Hasler, R. I. Daly, and R. J. Poff. 1978. Homing of rainbow trout transplanted in Lake Michigan: A comparison of three procedures used for imprinting and stocking. *Trans. Am. Fish. Soc.* 107:439–443.

Scholz, A.T., R. M. Horrall, J. C. Cooper, and A. D. Hasler. 1976. Imprinting to chemical cues: the basis for homestream selection in salmon. *Science* 192:1247–1249.

Schroder, S. L. 1981. The role of sexual selection in determining the overall mating patterns and mate choice in chum salmon. Ph.D. dissertation, University of Washington, Seattle.

Schuett-Hames, D. E., N. P. Peterson, R. Conrad, and T. P. Quinn. 2000. Patterns of gravel scour and fill after spawning by chum salmon in a western Washington stream. *N. Am. J. Fish. Mgmt.* 20: 610–617.

Scott, W. B., and E. J. Crossman. 1973. Freshwater Fishes of Canada. *Fish. Res. Board Can. Bull.* 184. Ottawa. 966 p.

Scrivener, J. C., T. G. Brown, and B. C. Anderson. 1994. Juvenile salmon *(Oncorhynchus tshawytscha)* utilization of Hawks Creek, a small and nonnatal tributary of the upper Fraser River. *Can. J. Fish. Aquat. Sci.* 51:1139–1146.

Scrivener, J. C., and M. J. Brownlee. 1989. Effects of forest harvesting on spawning gravel and incubation survival of chum (*Oncorhynchus keta*) and coho salmon (*O. kisutch*) in Carnation Creek, British Columbia. *Can. J. Fish. Aquat. Sci.* 46:681–696.

Scudder, G. G. E. 1989. The adaptive significance of marginal populations: A general perspective. *Can. Spec. Publ. Fish. Aquat. Sci.* 105:180–185.

Seamons, T. R., P. Bentzen, and T. P. Quinn. 2004. The effects of adult length and arrival date on individual reproductive success in wild steelhead trout, *Oncorhynchus mykiss*. *Can. J. Fish. Aquat. Sci.* 61:193–204.

Seidelman, D. L., P. B. Cunningham, and R. B. Russell. 1973. *Life history studies of rainbow trout in the Kvichak drainage of Bristol Bay.* Alaska Dept. Fish Game, Sport Fish Division, Vol. 14, Study G-11. Juneau. 50 p.

Seiler, D. 1995. *Estimation of Cedar River sockeye salmon fry production*. Washington Dept. Fish Wildl. Annual Rept., Olympia.

Seiler, D., and L. Kishimoto. 1997. *1997 Cedar River sockeye salmon fry production evaluation.* Washington Dept. Fish Wildl. Annual Rept., Olympia.

Seiler, D., G. Volkhardt, L. Fleischer, and L. Kishimoto. 2002. *2000 Cedar River sockeye salmon fry production evaluation*. Washington Dept. Fish Wildl. Annual Rept. Olympia.

Seiler, D., G. Volkhardt, S. Neuhauser, P. Hanratty, and L. Kishimoto. 2002. *2002 wild coho forecasts for Puget Sound and Washington coastal systems.* Washington Dept. Fish Wildl., Olympia.

Shapovalov, L., and A. C. Taft. 1954. *The life histories of the steelhead rainbow trout* (Salmo gairdneri gairdneri) *and silver salmon* (Oncorhynchus kisutch) *with special reference to Waddell Creek, California, and recommendations regarding their management.* California Dept. Fish Game, Fish Bull. 98. Sacramento. 375 p.

Sharma, R., and R. Hilborn. 2001. Empirical relationships between watershed characteristics and coho salmon (*Oncorhynchus kisutch*) smolt abundance in 14 western Washington streams. *Can. J. Fish. Aquat. Sci.* 58:1453–1463.

Shearer, W. M. 1990. The Atlantic salmon (*Salmo salar* L.) of the North Esk with particular reference to the relationship between river and sea age and time of return to home waters. *Fish. Res.* 10:93–123.

Shedlock, A. M., J. D. Parker, D. A. Crispin, T. W. Pietsch, and G. C. Burmer. 1992. Evolution of the salmonid mitochondrial control region. *Molec. Phylogen. Evol.* 1:179–192.

Sheridan, W. 1962. Relation of stream temperatures to timing of pink salmon escapements in Southeast Alaska.Pages 87–102 in N. J. Wilimovsky, ed., *Symposium on Pink salmon.* University of British Columbia, Vancouver.

Shreffler, D. K., C. A. Simenstad, and R. M. Thom. 1990. Temporary residence by juvenile salmon in a restored estuarine wetland. *Can. J. Fish. Aquat. Sci.* 47:2079–2084.

Shumway, D. L., C. E. Warren, and P. Doudoroff. 1964. Influence of oxygen concentration and water movement on the growth of steelhead trout and coho salmon embryos. *Trans. Am. Fish. Soc.* 93:342–356.

Sibert, J. R. 1979. Detritus and juvenile salmon production in the Nanaimo estuary: II. Meiofauna available as food to juvenile chum salmon (*Oncorhynchus keta*). *J. Fish. Res. Board Can.* 36:497–503.

Sibert, J. R., T. J. Brown, M. C. Healey, and B. A. Kask. 1977. Detritus-based food webs: Exploitation by juvenile chum salmon (*Oncorhynchus keta*). *Science* 196:649–650.

Siitonen, L., and G. A. E. Gall. 1989. Response to selection for early spawn date in rainbow trout, *Salmo gairdneri*. *Aquaculture* 78:153–161.

Simenstad, C. A., K. L. Fresh, and E. O. Salo. 1982. The role of Puget Sound estuaries in the life history of Pacific salmon: An unappreciated function. Pages 343–364 in V. S. Kennedy, ed., *Estuarine comparisons.* Academic Press, New York.

Simenstad, C. A., W. J. Kinney. S. S. Parker, E. O. Salo, J. R. Cordell, and H. Buechner. 1980. *Prey community structures and trophic ecology of outmigrating juvenile chum and pink salmon in Hood Canal, Washington: A synthesis of three years' studies, 1977–1979.* Final report to the Washington Department of Fisheries. University of Washington School of Fisheries, Fish. Res. Inst., FRI-UW-8026. Seattle.

Simenstad, C. A., L. F. Small, and C. D. McIntire. 1990. Consumption processes and food web structure in the Columbia River Estuary. *Prog. Oceanogr.* 25:271–297.

Simms, S. E., and P. A. Larkin. 1977. Simulation of dispersal of sockeye salmon *(Oncorhynchus nerka)* underyearlings in Babine Lake. *J. Fish. Res. Board Can.* 34:1379–1388.

Sims, C. W., and F. J. Ossiander. 1981. *Migrations of juvenile chinook salmon and steelhead trout in the Snake River from 1973 to 1979, a research summary.* Report to the U.S. Army Corps of Engineers by the National Marine Fisheries Service, Seattle. 31 p.

Skalski, J. R. 1998. Estimating season-wide survival rates of outmigrating salmon smolt in the Snake River, Washington. *Can. J. Fish. Aquat. Sci.* 55:761–769.

Slaney, T. L., K. D. Hyatt, T. G. Northcote, and R. J. Fielden. 1996. Status of anadromous salmon and trout in British Columbia and Yukon. *Fisheries* 21(10):20–35.

Slater, D. W. 1963. *Winter-run chinook salmon in the Sacramento River, California with notes on water temperature requirements at spawning.* U.S. Fish Wildl. Serv., Spec. Sci. Rept.—Fisheries No. 461. Washington, D.C. 9 p.

Slatick, E., D. L. Park, and W. J. Ebel. 1975. Further studies regarding effects of transportation on survival and homing of Snake River chinook salmon and steelhead trout. *Fish. Bull.* 73:925–931.

Smith, G. R., and R. F. Stearley. 1989. The classification and scientific names of rainbow and cutthroat trouts. *Fisheries* 14(1):4–10.

Smith, G. W., A. D. Hawkins, G. G. Urquhart, and W. M. Shearer. 1981. Orientation and energetic efficiency in the offshore movements of returning Atlantic salmon *Salmon salar* L. Scot. Fish. Res. Rept. 21. Aberdeen, Scotland. 22 p.

Smith, R. W., and J. S. Griffith. 1994. Survival of rainbow trout during their first winter in the Henrys Fork of the Snake River, Idaho. *Trans. Am. Fish. Soc.* 123:747–756.

Smith, S. G., W. D. Muir, J. G. Williams, and J. R. Skalski. 2002. Factors associated with travel time and survival of migrant yearling chinook salmon and steelhead in the lower Snake River. *N. Am. J. Fish. Mgmt.* 22:385–505.

Smoker, W. A. 1955. Effects of streamflow on silver salmon production in western Washington. Ph.D. dissertation, University of Washington. Seattle.

Smoker, W. W., A. J. Gharrett, and M. S. Stekoll. 1998. Genetic variation of return date in a population of pink salmon: a consequence of fluctuating environment and dispersive selection? *Alaska Fish. Res. Bull.* 5:46–54.

Solazzi, M. F., T. E. Nickelson, and S. L. Johnson. 1991. Survival, contribution, and return of hatchery coho salmon (*Oncorhynchus kisutch*) released from freshwater, estuarine, and marine environments. *Can. J. Fish. Aquat. Sci.* 48:248–253.

Solazzi, M. F., T. E. Nickelson, S. L. Johnson, and J. D. Rodgers. 2000. Effects of increasing winter rearing habitat on abundance of salmonids in two coastal Oregon streams. *Can. J. Fish. Aquat. Sci.* 57:906–914.

Sowden, T. K., and G. Power. 1985. Prediction of rainbow trout embryo survival in relation to groundwater seepage and particle size of spawning substrates. *Trans. Am. Fish. Soc.* 114:804–812.

Spidle, A.P., T. P. Quinn, and P. Bentzen. 1998. Sex-biased marine survival and growth in a population of coho salmon (*Oncorhynchus kisutch*). *J. Fish Biol.* 52:907–915.

Stasko, A. B., R. M. Horrall, and A. D. Hasler. 1976. Coastal movements of adult Fraser River sockeye salmon *(Oncorhynchus nerka)* observed by ultrasonic tracking. *Trans. Am. Fish. Soc.* 105:64–71.

Stearley, R. F., and G. R. Smith. 1993. Phylogeny of the Pacific trouts and salmons (*Oncorhynchus*) and genera of the family Salmonidae. *Trans. Am. Fish. Soc.* 122:1–33.

Steinhart, G. B., and W. A. Wurtsbaugh. 1999. Under-ice vertical migrations of *Oncorhynchus nerka* and their zooplankton prey. *Can. J. Fish. Aquat. Sci.* 56 (Suppl. 1):152–161.

St-Hillaire, S., M. Boichuk, D. Barnes, M. Higgins, R. Devlin, R. Withler, J. Khattra, S. Jones, and D. Kieser. 2002. Epizootiology of *Parvicapsula minibicornis* in Fraser River sockeye salmon, *Oncorhynchus nerka* (Walbaum). *J. Fish Dis.* 25:107–120.

Steen, R. P., and T. P. Quinn. 1999. Egg burial depth by sockeye salmon (*Oncorhynchus nerka*): implications for survival of embryos and natural selection on female body size. *Can. J. Zool.* 77:836–841.

Stober, Q. J., and A. H. Hamalainen. 1980. *Cedar River sockeye salmon production.* Project completion report to the Washington Department of Fisheries. University of Washington School of Fisheries, Fish. Res. Inst., FRI-UW-8016. Seattle. 59 p.

Stockner, J. G. 1987. Lake fertilization: the enrichment cycle and lake sockeye salmon (*Oncorhynchus nerka*) production. *Can. Spec. Publ. Fish. Aquat. Sci.* 96:198–215.

Stockner, J. G., and E. A. MacIsaac. 1996. British Columbia lake enrichment programme: Two decades of habitat enhancement for sockeye salmon. *Reg. Rivers: Res. Mgmt.* 12:547–561.

Stouder, D. J., P. A. Bisson, and R. J. Naiman, eds. 1996. *Pacific salmon and their ecosystems: Status and future options.* Chapman and Hall, New York.

Straty, R. R. 1975. *Migratory routes of adult sockeye salmon,* Oncorhynchus nerka, *in the eastern Bering Sea and Bristol Bay.* National Marine Fisheries Service, NOAA Tech. Rept. NMFS-SSRF-690. Seattle. 32 p.

Straty, R. R., and H. W. Jaenicke. 1980. Estuarine influence of salinity, temperature, and food on the behavior, growth, and dynamics of Bristol Bay sockeye salmon. Pages 247–265 in W. J. McNeil and D. C. Himsworth, eds., *Salmonid ecosystems of the North Pacific.* Oregon State University Press, Corvallis.

Sumner, F. H. 1962. Migration and growth of the coastal cutthroat trout in Tillamook County, Oregon. *Trans. Am. Fish. Soc.* 91:77–83.

Swales, S., and C. D. Levings. 1989. Role of off-channel ponds in the life cycle of coho salmon (*Oncorhynchus kisutch*) and other juvenile salmonids in Coldwater River, British Columbia. *Can. J. Fish. Aquat. Sci.* 46:232–242.

Tadokoro, K., Y. Ishida, N. D. Davis, S. Ueyanagi, and T. Sugimoto. 1996. Change in chum salmon (*Oncorhynchus keta*) stomach contents associated with fluctuation of pink salmon (*O. gorbuscha*) abundance in the central subarctic Pacific and Bering Sea. *Fish. Oceanogr.* 5:89–99.

Takagi, K., K. V. Aro, A. C. Hartt, and M. B. Dell. 1981. Distribution and origin of pink salmon (*Oncorhynchus gorbuscha*) in offshore waters of the North Pacific Ocean. *Int. N. Pac. Fish. Comm. Bull.* 40. Vancouver, BC. 195 p.

Tallman, R. F. 1986. Genetic differentiation among seasonally distinct spawning populations of chum salmon, *Oncorhynchus keta. Aquaculture* 57:211–217.

Tallman, R. F., and M. C. Healey. 1991. Phenotypic differentiation in seasonal ecotypes of chum salmon, *Oncorhynchus keta. Can. J. Fish. Aquat. Sci.* 48:661–671.

———. 1994. Homing, straying, and gene flow among seasonally separated populations of chum salmon *(Oncorhynchus keta). Can. J. Fish. Aquat. Sci.* 51:577–588.

Tallman, R. F., F. Saurette, and T. Thera. 1996. Migration and life history variation in Arctic charr, *Salvelinus alpinus. Écoscience* 3:33–41.

Tanaka, H., Y. Takagi, and Y. Naito. 2000. Behavioural thermoregulation of chum salmon during homing migration in coastal waters. *J. Exp. Biol.* 203:1825–1833.

Taylor, E. B. 1989. Precocial male maturation in laboratory-reared populations of chinook salmon, *Oncorhynchus tshawytscha. Can. J. Zool.* 67:1665–1669.

———. 1990a. Environmental correlates of life-history variation in juvenile chinook salmon, *Oncorhynchus tshawytscha* (Walbaum). *J. Fish Biol.* 37:1–17.

———. 1990b. Phenotypic correlates of life-history variation in juvenile chinook salmon, *Oncorhynchus tshawytscha. J. Anim. Ecol.* 59:455–468.

———. 1991. A review of local adaptation in Salmonidae, with particular reference to Pacific and Atlantic salmon. *Aquaculture* 98:185–207.

Taylor, E. B., C. J. Foote, and C. C. Wood. 1996. Molecular genetic evidence for parallel life-history evolution within a Pacific salmon (sockeye salmon and kokanee, *Oncorhynchus nerka*). *Evolution* 50:401–416.

Taylor, S. G. 1980. Marine survival of pink salmon fry from early and late spawners. *Trans. Am. Fish. Soc.* 109:79–82.

Thedinga, J. F., A. C. Wertheimer, R. A. Heintz, J. M. Maselko, and S. D. Rice. 2000. Effects of stock, coded-wire tagging, and transplant on straying of pink salmon (*Oncorhynchus gorbuscha*) in southeastern Alaska. *Can. J. Fish. Aquat. Sci.* 57:2076–2085.

Thomas, W. K., R. E. Withler, and A. T. Beckenbach. 1986. Mitochondrial DNA analysis of Pacific salmonid evolution. *Can. J. Zool.* 64:1058–1064.

Thompson, W. F. 1945. Effect of the obstruction at Hell's Gate on the sockeye salmon of the Fraser River. *Int. Pac. Salmon Fish. Comm. Bull.* 1. New Westminster, BC. 175 p.

———. 1962. The research program of the Fisheries Research Institute in Bristol Bay, 1945–1958. Pages 1–36 in T. S. Y. Koo, ed., *Studies of Alaska red salmon.* University of Washington Press, Seattle.

Thomson, K. A., W. J. Ingraham, Jr., M. C. Healey, P. H. LeBlond, C. Groot, and C. G. Healey. 1992. The influence of ocean currents on latitude of landfall and migration speed of sockeye salmon returning to the Fraser River. *Fish. Oceanogr.* 1:163–179.

Thomson, K. A., W. J. Ingraham, Jr., M. C. Healey, P. H. LeBlond, C. Groot, and C. G. Healey. 1994. Computer simulations of the influence of ocean currents on Fraser River sockeye salmon (*Oncorhynchus nerka*) return times. *Can. J. Fish. Aquat. Sci.* 51:441–449.

Thorne, R. E. 1983. Application of hydroacoustic assessment techniques to three lakes with contrasting fish distributions. *FAO Fish. Rept.* 300:269–277.

Thorpe, J. E. 1987. Smolting versus residency: developmental conflict in salmonids. Pages 244–252 in Am. Fish. Soc. Symp. 1. Bethesda, MD.

———. 1994. Salmonid fishes and the estuarine environment. *Estuaries* 17:76–93.

Tipping, J. M. 1991. Heritability of age at maturity in steelhead. *N. Am. J. Fish. Mgmt.* 11:105–108.

Tipping, J. M., and H. L. Blankenship. 1993. Effect of condition factor at release on smolt-to-adult survival of hatchery sea-run cutthroat trout. *Prog. Fish-Cult.* 55:184–186.

Traxler, G. S., and J. B. Rankin. 1989. An infectious hematopoietic necrosis epizootic in sockeye salmon *Oncorhynchus nerka* in Weaver Creek spawning channel, Fraser River system, B.C., Canada. *Dis. Aquat. Org.* 6:221–226.

Trotter, P. C. 1989. Coastal cutthroat trout: a life history compendium. *Trans. Am. Fish. Soc.* 118:463–473.

Truscott, B., D. R. Idler, Y. P. So, and J. M. Walsh. 1986. Maturational steroids and gonadotropin in upstream migratory sockeye salmon. *Gen. Comp. Endocrinol.* 62:99–110.

Tschaplinski, P. J., and G. F. Hartman. 1983. Winter distribution of juvenile coho salmon (*Oncorhynchus kisutch*) before and after logging in Carnation Creek, British Columbia, and some implications for overwinter survival. *Can. J. Fish. Aquat. Sci.* 40:452–461.

Tsiger, V. V., V. I. Skirin, N. I. Krupyanko, K. A. Kashlin, and A. Y. Semenchenko. 1994. Life history form of male masu salmon (*Oncorhynchus masou*) in South Primoré, Russia. *Can. J. Fish. Aquat. Sci.* 51:197–208.

Ueda, H., O. Hiroi, K. Yamauchi, A. Hara, H. Kagawa, S. Adachi, and Y. Nagahama. 1991. Changes in serum steroid levels and *in vitro* steroid hormone production in the ovaries of female chum salmon *Oncorhynchus keta* during spawning migration. *Nippon Suisan Gakkaishi* 57:881–884.

Ueda, K., T. J. Hara, and A. Gorbman. 1967. Electroencephalographic studies on olfactory discrimination in adult spawning salmon. *Comp. Biochem. Physiol.* 21:133–143.

Ueno, Y. 1992. Deepwater migrations of chum salmon (*Oncorhynchus keta*) along the Pacific coast of northern Japan. *Can. J. Fish. Aquat. Sci.* 49:2307–2312.

U.S. Commission of Fish and Fisheries. 1876. *Report of the commissioner for 1873–4 and 1874–5.* U.S. Commission of Fish and Fisheries, Washington, DC.

U.S. Forest Service. 1989. *Biology of the bull trout,* Salvelinus confluentus. Willamette National Forest, Eugene, OR.

Unwin, M. J., M. T. Kinnison, N. C. Boustead, and T. P. Quinn. 2003. Genetic control over survival in Pacific salmon (*Oncorhynchus* spp.): Experimental evidence between and within populations of New Zealand chinook salmon (*O. tshawytscha*). *Can. J. Fish. Aquat. Sci.* 60:1–11.

Unwin, M. J., M. T. Kinnison, and T. P. Quinn. 1999. Exceptions to semelparity: postmaturation survival, morphology and energetics of male chinook salmon (*Oncorhynchus tshawytscha*). *Can. J. Fish. Aquat. Sci.* 56:1172–1181.

Unwin, M. J., and D. H. Lucas. 1993. Scale characteristics of wild and hatchery chinook salmon *(Oncorhynchus tshawytscha)* in the Rakaia River, New Zealand, and their use in stock identification. *Can. J. Fish. Aquat. Sci.* 50:2475–2484.

Unwin, M. J., and T. P. Quinn. 1993. Homing and straying patterns of chinook salmon (*Oncorhynchus tshawytscha*) from a New Zealand hatchery: spatial distribution of strays and effects of release date. *Can. J. Fish. Aquat. Sci.* 50:1168–1175.

Unwin, M. J., T. P. Quinn, M. T. Kinnison, and N. C. Boustead. 2000. Divergence in juvenile growth and life history in two recently colonized and partially isolated chinook salmon populations. *J. Fish Biol.* 57:943–960.

Urawa, S., K. Nagasawa, L. Margolis, and A. Moles. 1998. Stock identification of chinook salmon (*Oncorhynchus tshawytscha*) in the North Pacific Ocean and Bering Sea by parasite tags. *N. Pac. Anad. Fish. Comm. Bull.* 1:199–204.

Utter, F. M. 1991. Biochemical genetics and fishery management: an historical perspective. *J. Fish Biol.* 39 (Suppl. A):1–20.

Utter, F. M., and F. W. Allendorf. 1994. Phylogenetic relationships among species of *Oncorhynchus*: a consensus view. *Cons. Biol.* 8:864–867

Utter, F. M., F. W. Allendorf, and H. O. Hodgins. 1973. Genetic variability and relationships in Pacific salmon and related trout based on protein variations. *Syst. Zool.* 22:257–270.

Utter, F. M., J. E. Seeb, and L. W. Seeb. 1993. Complementary uses of ecological and biochemical genetic data in identifying and conserving salmon populations. *Fish. Res.* 18:59–76.

van den Berghe, E. P., and M. R. Gross. 1989. Natural selection resulting from female breeding competition in a Pacific salmon (coho: *Oncorhynchus kisutch*). *Evolution* 43:125–140.

Vannote, R. L., G. W. Minshall, K. W. Cummins, J. R. Sedell, and C. E. Cushing. 1980. The river continuum concept. *Can. J. Fish. Aquat. Sci.* 37:130–137.

Velsen, F. P. J. 1987. Temperature and incubation in Pacific salmon and rainbow trout: compilation of data on median hatching time, mortality and embryonic staging. *Can. Data Rept. Fish. Aquat. Sci.* 626. Nanaimo, BC. 58 p.

Verhoeven, L. A., and E. B. Davidoff. 1962. Marine tagging of Fraser River sockeye salmon. *Int. Pac. Salmon Fish. Comm. Bull.* 13. New Westminster, BC. 132 p.

Vernon, E. H. 1958. An examination of factors affecting the abundance of pink salmon in the Fraser River. *Int. Pac. Salmon Fish. Comm. Progr. Rept.* 5. New Westminster, BC. 49 p.

Volk, E. C., S. L. Schroder, and J. J. Grimm. 1994. Use of a bar code symbology to produce multiple thermally induced otolith marks. *Trans. Am. Fish. Soc.* 123:811–816.

Vøllestad, L. A., J. Peterson, and T. P. Quinn. 2004. Effects of fresh water and marine growth rates on early maturity in male coho and chinook salmon. *Trans. Am. Fish. Soc* 133:495–503.

Vreeland, R. R., R. J. Wahle, and A. H. Arp. 1975. Homing behavior and contribution to Columbia River fisheries of marked coho salmon released at two locations. *Fish. Bull.* 73:717–725.

Wagner, H. H. 1969. Effect of stocking location of juvenile steelhead trout, *Salmo gairdnerii*, on adult catch. *Trans. Am. Fish. Soc.* 98:27–34.

Wahle, R. J., E. Chaney, and R. E. Pearson. 1981. Areal distribution of marked Columbia River basin spring chinook salmon recovered in fisheries and at parent hatcheries. *Mar. Fish. Rev.* 43(12):1–9.

Walker, M. M., C. E. Diebel, C. V. Haugh, P. M. Pankhurst, J. C. Montgomery, and C. R. Green. 1997. Structure and function of the vertebrate magnetic sense. *Nature* 390:371–376.

Walker, R. V., K. W. Myers, N. D. Davis, K. Y. Aydin, K. D. Friedland, H. R. Carlson, G. W. Boehlert, S. Urawa, Y. Ueno, and G. Anma. 2000. Diurnal variation in thermal environment experienced by salmonids in the North Pacific as indicated by data storage tags. *Fish. Oceanogr.* 9:171–186.

Waltemyer, D. L. 1994. *Age, sex and size composition of chinook, sockeye, coho, and chum salmon returning to upper Cook Inlet, Alaska in 1991*. Alaska Dept. Fish Game, Tech. Fish. Rept. 94-11. Juneau. 95 p.

Walters, C. J., R. Hilborn, R. M. Peterman, and M. J. Staley. 1978. Model for examining early ocean limitation of Pacific salmon production. *J. Fish. Res. Board Can.* 35:1303–1315.

Waples, R. S. 1991. Pacific salmon, *Oncorhynchus* spp., and the definition of "species" under the Endangered Species Act. *Mar. Fish. Rev.* 53(3):11–22.

———. 1995. Evolutionarily significant units and the conservation of biological diversity under the Endangered Species Act. Pages 8–27 in Am. Fish. Soc. Symp. 17. Bethesda, MD.

Waples, R. S., R. G. Gustafson, L. A. Weitkamp, J. M. Myers, O. W. Johnson, P. J. Busby, J. J. Hard, G. J. Bryant, F. W. Waknitz, K. Neely, D. Teel, W. S. Grant, G. A. Winans, S. Phelps, A. Marshall, and B. M. Baker. 2001. Characterizing diversity in salmon from the Pacific Northwest. *J. Fish Biol.* 59:1–41.

Waples, R. S., D. J. Teel, J. M. Myers, and A. R. Marshall. 2004. Life history divergence in chinook salmon: Historic contingency and parallel evolution. *Evolution* 58: 386–403.

Ward, B. R., P. A. Slaney, A. R. Facchin, and R. W. Land. 1989. Size-biased survival in steelhead trout (*Oncorhynchus mykiss*): Back-calculated lengths from adults' scales compared to migrating smolts at the Keogh River, British Columbia. *Can. J. Fish. Aquat. Sci.* 46:1853–1858.

Ward, R. D., M. Woodwark, and D. O. F. Skibinski. 1994. A comparison of genetic diversity levels in marine, freshwater, and anadromous fishes. *J. Fish Biol.* 44:213–232.

WDF (Washington Department of Fisheries), Washington Department of Wildlife, and Western Washington Treaty Indian Tribes. 1993. *1992 Washington State salmon and steelhead stock inventory*. Olympia. 211 p.

Wedemeyer, G. A., R. L. Saunders, and W. C. Clarke. 1980. Environmental factors affecting smoltification and early marine survival of anadromous salmonids. *Mar. Fish. Rev.* 42:1–14.

Weitkamp, L. A., and K. Neely. 2002. Coho salmon (*Oncorhynchus kisutch*) ocean migration patterns: insight from marine coded-wire tag recoveries. *Can. J. Fish. Aquat. Sci.* 59:1100–1115.

Weitkamp, L. A., T. C. Wainwright, G. J. Bryant, G. B. Milner, D. J. Teel, R. G. Kope, and R. S. Waples. 1995. *Status review of coho salmon from Washington, Oregon, and California*. National Marine Fisheries Service, NOAA Tech. Memo. NMFS-NWFSC-24. Seattle. 258 p.

Welch, D. W., and T. R. Parsons. 1993. δ^{13}C - δ^{15}N values as indicators of trophic position and competitive overlap for Pacific salmon (*Oncorhynchus* spp.). *Fish. Oceanogr.* 2:11–23.

Wells, B. K., B. E. Rieman, J. L. Clayton, D. L. Horan, and C. M. Jones. 2003. Relationship between water, otolith, and scale chemistries of westslope cutthroat trout from the Coeur d'Alene River, Idaho: the potential application of hard-part chemistry to describe movements in freshwater. *Trans. Am. Fish. Soc.* 132:409–424.

West, C. J., and P. A. Larkin. 1987. Evidence for size-selective mortality of juvenile sockeye salmon (*Oncorhynchus nerka*) in Babine Lake, British Columbia. *Can. J. Fish. Aquat. Sci.* 44:712–721.

Westerberg, H. 1982. Ultrasonic tracking of Atlantic salmon (*Salmo salar* L.): II. Swimming depth and temperature stratification. *Inst. Freshw. Res. Drottningholm* 60:102–120.

Whitman, R. P. 1987. An analysis of smoltification indices in fall chinook salmon (*Oncorhynchus tshawytscha*). M.S. thesis, University of Washington, Seattle.

Willette, T. M. 2001. Foraging behaviour of juvenile pink salmon (*Oncorhynchus gorbuscha*) and size-dependent predation risk. *Fish. Oceanogr.* 10 (Suppl. 1):110–131.

Willette, T. M., R. T. Cooney, V. Patrick, D. M. Mason, G. L. Thomas, and D. Scheel. 2001. Ecological processes influencing mortality of juvenile pink salmon (*Oncorhynchus gorbuscha*) in Prince William Sound, Alaska. *Fish. Oceanogr.* 10 (Suppl. 1):14–41.

Williams, I. V. 1987. Attempts to re-establish sockeye salmon (*Oncorhynchus nerka*) populations in the Upper Adams River, British Columbia, 1949–1984. *Can. Spec. Publ. Fish. Aquat. Sci.* 96:235–242.

Williams, I. V., and D. F. Amend. 1976. A natural epizootic of infectious hematopoietic necrosis in fry of sockeye salmon (*Oncorhynchus nerka*) at Chilko Lake, British Columbia. *J. Fish. Res. Board Can.* 33:1564–1567.

Willson, M. F., S. M. Gende, and B. H. Marston. 1998. Fishes and the forest—expanding perspectives on fish-wildlife interactions. *BioScience* 48:455–462.

Willson, M. F., and K. C. Halupka. 1995. Anadromous fish as keystone species in vertebrate communities. *Cons. Biol.* 9:489–497.

Winans, G. A., and R. S. Nishioka. 1987. A multivariate description of change in body shape of coho salmon (*Oncorhynchus kisutch*) during smoltification. *Aquaculture* 66:235–245.

Winemiller, K. O., and K. A. Rose. 1992. Patterns of life-history diversification in North American fishes: implications for population regulation. *Can. J. Fish. Aquat. Sci.* 49:2196–2218.

Wipfli, M. S. 1997. Terrestrial invertebrates as salmonid prey and nitrogen sources in streams: contrasting old-growth and young-growth riparian forests in southeastern Alaska, U.S.A. *Can. J. Fish. Aquat. Sci.* 54:1259–1269.

Wipfli, M. S., J. Hudson, and J. Caouette. 1998. Influence of salmon carcasses on stream productivity: response of biofilm and benthic macroinvertebrates in southeastern Alaska, U.S.A. *Can. J. Fish. Aquat. Sci.* 55:1503–1511.

Wisby, W. J., and A. D. Hasler. 1954. Effect of olfactory occlusion on migrating silver salmon (*O. kisutch*). *J. Fish. Res. Board Can.* 11:472–478.

Wissmar, R. C., and C. A. Simenstad. 1988. Energetic constraints of juvenile chum salmon (*Oncorhynchus keta*) migrating in estuaries. *Can. J. Fish. Aquat. Sci.* 45:1555–1560.

Withler, F. C. 1982. Transplanting Pacific salmon. *Can. Tech. Rept. Fish. Aquat. Sci.* 1079. Nanaimo, BC. 27 p.

Wood, C. C. 1987a. Predation of juvenile Pacific salmon by the common merganser (*Mergus merganser*) on eastern Vancouver Island. I: Predation during the seaward migration. *Can. J. Fish. Aquat. Sci.* 44:941–949.

———. 1987b. Predation of juvenile Pacific salmon by the common merganser (*Mergus merganser*) on eastern Vancouver Island. II: Predation of stream-resident juvenile salmon by merganser broods. *Can. J. Fish. Aquat. Sci.* 44:950–959.

———. 1995. Life history variation and population structure in sockeye salmon. Pages 195–216 in Am. Fish. Soc. Symp. 17. Bethesda, MD.

Wood, C. C., and C. J. Foote. 1996. Evidence for sympatric genetic divergence of anadromous and nonanadromous morphs of sockeye salmon (*Oncorhynchus nerka*). *Evolution* 50:1265–1279.

Woodey, J. C. 1972. Distribution, feeding, and growth of juvenile sockeye salmon in Lake Washington. Ph.D. dissertation, University of Washington, Seattle.

Wootton, R. J. 1984. Introduction: strategies and tactics in fish reproduction. Pages 1–12 in G. W. Potts and R. J. Wootton, eds., *Fish reproduction: Strategies and tactics*. Academic Press, London.

Wydoski, R. S., and R. R. Whitney 2003. *Inland fishes of washington*. 2nd ed. University of Washington Press, Seattle, and American Fisheries Society, Bethesda, MD.

Yamamoto, S., K. Morita, and A. Goto. 1999. Marine growth and survival of white-spotted charr, *Salvelinus leucomaenis*, in relation to smolt size. *Ichthyol. Res.* 46:85–92.

Yamamoto, T., K. Edo, and H. Ueda. 2000. Lacustrine forms of mature male masu salmon, *Oncorhynchus masou* Brevoort, in Lake Toya, Hokkaido, Japan. *Ichthyol. Res.* 47:407–410.

Yamamoto, T., H. Ueda, and S. Higashi 1998. Correlation among dominance status, metabolic rate and otolith size in masu salmon. *J. Fish Biol.* 52:281–290.

Yoshiyama, R. M., F. W. Fisher, and P. B. Moyle. 1998. Historical abundance and decline of chinook salmon in the Central Valley region of California. *N. Am. J. Fish. Mgmt.* 18:487–521.

Young, K. A. 1999. Environmental correlates of male life history variation among coho salmon populations from two Oregon coastal basins. *Trans. Am. Fish. Soc.* 128:1–16.

Zaugg, W. S., E. F. Prentice, and F. W. Waknitz. 1985. Importance of river migration to the development of seawater tolerance in Columbia River anadromous salmonids. *Aquaculture* 51:33–47.

Zaugg, W. S., and C. V. W. Mahnken. 1991. The importance of smolt development to successful marine ranching of Pacific salmon. Pages 89–97 in R. S. Svrjeek, ed., *Marine ranching: Proceedings of the 17th U.S.-Japan meeting on aquaculture.* National Marine Fisheries Service, NOAA Tech. Rept. NMFS 102. Seattle.

Zhou, S. 2002. Size-dependent recovery of chinook salmon in carcass surveys. *Trans. Am. Fish. Soc.* 131:1194–1202.

Zimmerman, C. E., and H. Mosegaard. 1992. Initial feeding in migratory brown trout (*Salmo trutta* L.) alevins. *J. Fish Biol.* 40:647–650.

Zimmerman, C. E., and G. H. Reeves. 2000. Population structure of sympatric anadromous and nonanadromous *Oncorhynchus mykiss*: Evidence from spawning surveys and otolith microchemistry. *Can. J. Fish. Aquat. Sci.* 57:2152–2162.

Index

Italicized page numbers indicate figures and tables.

Index